The Comparative Biology of Aging

Norman S. Wolf
Editor

The Comparative Biology of Aging

Springer

Editor
Norman S. Wolf
Seattle WA 98195-7470
C-423 Health Sciences Bldg.
USA
normwolf@u.washington.edu

Cover photo credits:

Human: 45 year old Werner's Syndrome patient, International Registry of Werner Syndrome, courtesy of Dru Leistritz.Rhesus monkey: 29 year old male (M. mulatta), NIA Study on Primate Aging, courtesy of Dr. Julie Mattison, primate aging colony, NIA. Dog: "Tiger", 13 year old West Highland White terrier, courtesy of Dan and Pat McCutcheon and Jeremy and Michele Wolf. Mouse, BalbC strain: Courtesy of Dr. Norm Wolf laboratory, Department of Pathology, University of Washington. Fruit fly (D. melanogaster): Courtesy of Dr. Rolf Bodmer, Development and Aging, Burnham Institute, La Jolla, Ca. Round worm (C. elegans): Courtesy of Drs. Matt Kaeberlein and Brian Kennedy, Department of Pathology, University of Washington.

ISBN 978-90-481-3464-9 e-ISBN 978-90-481-3465-6
DOI 10.1007/978-90-481-3465-6
Springer Dordrecht Heidelberg London New York

Library of Congress Control Number: 2009942696

Printed on acid-free paper

Springer is part of Springer Science+Business Media (www.springer.com)

Contents

Contributors

Rozalyn Anderson Wisconsin National Primate Research Center, University of Wisconsin-Madison, Madison, WI, USA, rmanderson5@wisc.edu

Steven N. Austad Department of Cellular & Structural Biology, Barshop Institute for Longevity and Aging Research, University of Texas Health Science Center at San Antonio, San Antonio, TX, USA, austad@uthscsa.edu

Andrzej Bartke Geriatric Research, Department of Internal Medicine, Southern Illinois University School of Medicine, Springfield, IL 62794, USA; Department of Physiology, Southern Illinois University School of Medicine, Springfield, IL 62794, USA, abartke@siumed.edu

Rolf Bodmer Burnham Institute, Development and aging program, La Jolla, CA, USA

Rochelle Buffenstein Department of Physiology, Cellular and Structural Biology, Barshop Institute for Longevity and Aging Studies, University of Texas Health Science Center at San Antonio, San Antonio, TX, USA

Brent Calder Department of Genetics, Albert Einstein College of Medicine, Bronx, NY 10461, USA

Luka Cicin-Sain Vaccine and Gene Therapy Institute and the Oregon National Primate Research Center, Oregon Health and Science University, Beaverton, OR 97006, USA

Ricki Colman Wisconsin National Primate Research Center, University of Wisconsin-Madison, Madison, WI, USA; Department of Medicine, University of Wisconsin-Madison, Madison, WI, USA

Ana Maria Cuervo Department of Developmental and Molecular Biology, Marion Bessin Liver Research Center, Institute for Aging Research, Albert Einstein College of Medicine, Bronx, NY, USA, amcuervo@aecom.yu.edu

Dao-Fu Dai Department of Pathology, University of Washington, Seattle, WA, USA

Martijn Dollé National Institute of Public Health and the Environment, Bilthoven, The Netherlands

Ana Maria Garcia University of Texas at San Antonio, San Antonio, TX, USA

N.M.V. Gomes Department of Cell Biology, The University of Texas Southwestern Medical Center at Dallas, Dallas, TX 75390-9039, USA

Yuji Ikeno Barshop Institute and University of Texas Science Center at San Antonio, San Antonio, TX, USA

Matt Kaeberlein Department of Pathology, University of Washington, Seattle, WA, USA, kaeber@u.washington.edu

Mark P. Mattson Laboratory of Neurosciences, National Institute on Aging Intramural Research Program, Baltimore, MD, USA, mattsonm@grc.nia.nih.gov

David J. Marcinek Department of Radiology, University of Washington, Seattle, WA 98195, USA, dmarc@u.washington.edu

Julie Mattison Laboratory of Experimental Gerontology, NIA, Baltimore, MD, USA

Janko Nikolich-Zugich Department of Immunobiology and the Arizona Center on Aging, University of Arizona College of Medicine, Tucson, AZ 85718, USA; Vaccine and Gene Therapy Institute and the Oregon National Primate Research Center, Oregon Health and Science University, Beaverton, OR 97006, USA, nikolich@email.arizona.edu

Samantha J. Orenstein Department of Developmental and Molecular Biology, Marion Bessin Liver Research Center, Institute for Aging Research, Albert Einstein College of Medicine, Bronx, NY, USA

Beverly Paigen JAX Laboratories, Bar Harbor, ME, USA

Peter S. Rabinovitch Department of Pathology, University of Washington, Seattle, WA, USA, PeterR@medicine.washington.edu

Lara S. Shamieh Department of Pathology, University of Washington, Seattle, WA, USA

J.W. Shay Department of Cell Biology, The University of Texas Southwestern Medical Center at Dallas, Dallas, TX 75390-9039, USA

Yun Shi Department of Physiology, Barshop Institute for Longevity and Aging Studies, University of Texas Health Science Center at San Antonio, San Antonio, TX, USA

Daniel L. Smith Department of Biochemistry and Molecular Genetics, School of Medicine, University of Virginia Health System, Charlottesville, VA 22908, USA

Jeffrey S. Smith Department of Biochemistry and Molecular Genetics, School of Medicine, University of Virginia Health System, Charlottesville, VA 22908, USA, jss5y@virginia.edu

Adam Spong Geriatric Research, Department of Internal Medicine, Southern Illinois University School of Medicine, Springfield, IL 62794, USA

Holly Van Remmen Department of Physiology, Cellular and Structural Biology, Barshop Institute for Longevity and Aging Studies, University of Texas Health Science Center at San Antonio; Audie Murphy Division, South Texas Veterans Health Care System, San Antonio, TX, USA, vanremmen@uthscsa.edu

Jan Vijg Department of Genetics, Albert Einstein College of Medicine, Bronx, NY 10461, USA, jvijg@aecom.yu.edu

Jason J. Villarin Department of Radiology, University of Washington, Seattle, WA 98195, USA

Jonathan Wanagat Division of Gerontology and Geriatric Medicine, University of Washington, Seattle, WA 98195, USA

Richard Weindruch Wisconsin National Primate Research Center, University of Wisconsin-Madison, Madison, WI, USA; Department of Medicine, University of Wisconsin-Madison, Madison, WI, USA; GRECC VA Hospital Madison, University of Wisconsin-Madison, Madison, WI, USA

Robert J. Wessells Department of Internal Medicine, University of Michigan, Ann Arbor, MI, USA

Norman S. Wolf Department of Pathology, University of Washington, Seattle, WA 91895, USA, normwolf@u.washington.edu

Catherine A. Wolkow Laboratory of Neurosciences, National Institute on Aging Intramural Research Program, Baltimore, MD, USA, wolkowca@grc.nia.nih.gov

W.E. Wright Department of Cell Biology, The University of Texas Southwestern Medical Center at Dallas, Dallas, TX 75390-9039, USA, woodring.Wright@UTsouthwestern.edu

Rong Yuan JAX Laboratories, Bar Harbor, ME, USA

Sige Zou Laboratory of Experimental Gerontology, National Institute on Aging Intramural Research Program, Baltimore, MD, USA

Introduction: Lifespans and Pathologies Present at Death in Laboratory Animals

Norman S. Wolf[1] and Steven Austad[2]
[1]*Department of Pathology, University of Washington, Seattle, WA*
[2]*Barshop Institute and University of Texas Science Center at San Antonio, San Antonio, TX*

Abstract This initial chapter introduces those that follow with a summary of the life spans and end of life pathologies of the several species that are included in later chapters. It is not presented as a complete coverage, but rather as background for what follows, as gathered from the literature and personal information. Included here is that information on yeast as Saccharomyces cervisiae, round worms as Caenorhabditis. elegans, the fruit fly, Drosophila melanogaster, mice, rats, dogs, primates, and some bats and birds. The following chapters will compare findings in their specific area of coverage for those species for which such data is available.

This brief introduction provides generally accepted species wild type life spans and the end of life pathologies, along with some special attributes that lend a species to a comparative biology approach.

Keywords Aging · Lifespan · End of life conditions · Humans · Dogs · Mice · Rates · Fish · Amphibians · Cause of death

Laboratory Strains of Yeast (*Saccharomyces cervisiae*) Wild Type Only

The lifespan of *S. cervisiae* varies with both strain and media used (especially glucose content). The lifespan of yeast may be determined by two different means: the number of divisions accomplished by the mother cell (replicative lifespan), or by the survival of individual cells in a non-dividing, quiescence-like state (chronological lifespan). Replicative life span is determined by physical separation of daughter cells from mother cells via micromanipulation, while tallying the number of daughter cells produced by each mother [1]. Chronological life span is typically determined

---

Data Contributors: Julie Mattison, Laboratory of Experimental Gerontology, NIA; Mathew Kaeberlein, Department of Pathology, University of Washington, Seattle, WA; Beverly Paigen and Rong Yuan, JAX Laboratories, Bar Harbor, ME; Yuji Ikeno, Barshop Institute and University of Texas Science Center at San Antonio, TX; Rochelle Buffenstein, Department of Physiology and Barshop Institute, University of Texas Science Center at San Antonio TX

N.S. Wolf (ed.), *The Comparative Biology of Aging*,
DOI 10.1007/978-90-481-3465-6_1,

by culturing cells into stationary phase and then monitoring the ability of cells to reestablish vegetative growth upon return to nutrient rich conditions [2]. The ability to independently model aging of both dividing and non-dividing cell types is a useful property of the yeast model system.

Longevity is known to vary widely among different laboratory strains. Median replicative lifespan for wild type laboratory strains is commonly 18–26 days, while median chronological lifespan varies from one to several weeks, depending strain background on genotype and culture conditions [3]. The most extensively utilized yeast strains are the parental strains of the yeast ORF deletion collection, which are closely related to the *S. cerevisiae* type strain S288C [4]. This collection has been used for genome-wide screens for single-gene deletions that increase either chronological life span or replicative life [5, 6].

In addition to genetic background effects, environmental parameters also influence aging in yeast. Media composition, particularly glucose levels, affect both replicative and chronological life span. In both aging paradigms, reducing the glucose level from 2 to 0.5% or lower significantly increases life span, and has been referred to as a yeast model of caloric restriction [7–11]. Amino acid abundance, aeration, and temperature are other environmental parameters that can also influence aging in yeast.

The replicative life span of yeast is determined, at least in part, by the accumulation of extrachromosomal ribosomal DNA circles in the mother cell nucleus [12, 13]. Extrachromosomal ribosomal DNA circles are formed by homologous recombination between tandem rDNA repeats, are asymmetrically segregated to mother cells, and are self-replicating; however, the mechanism by which they induce toxicity remains unknown. In addition to extrachromosomal rDNA circles, other unknown factors also contribute to replicative senescence, potentially including age-associated genomic instability, mitochondrial retrograde signaling, and accumulation of oxidatively damaged proteins in the mother cell [14–18]. Replicative senescence is associated with a loss of transcriptional silencing near telomeres, sterility, increased mother cell size, and a primarily G1 cell cycle arrest.

Chronological aging is thought to be largely determined by resistance to oxidative and other types of stress [19]. Mitochondrial function also appears to be particularly important for chronological longevity, as respiratory deficient cells are short-lived, while pre-adaptation to respiratory growth is sufficient to increase life span [20]. Although the proximal molecular cause of death is not known, chronological senescence of yeast cells is associated with an activation of an apoptotic like cell death pathway [21]. Recently, Burtner et al. have shown that the accumulation of acetic acid in the media serves as a primary molecular cause of chronological senescence [22].

Laboratory Round Worm (*Caenorhabditis elegans*)

The lifespan of the roundworm *C. elegans* is generally defined as the length of survival from the final larval molt (L4) until death (adult lifespan), where death is determined by an inability to move in response to touch. *C. elegans* develop through four larval stages following hatching and prior to adulthood. Adult *C. elegans* are reproductive for about the first week of adulthood followed by approximately two weeks of post-reproductive adulthood prior to death. Life span is most commonly measured in the laboratory by maintaining the worms on the surface of a nutrient-agar medium (Nematode Growth Medium, NGM) with *E. coli* OP50 as the bacterial food source (REF). Alternative culture conditions have been described in liquid media; however, these are not widely used for longevity studies.

Longevity of the commonly used wild type *C. elegans* hermaphrodite (N2) varies from 16 to 23 days under standard laboratory conditions (20°C, NGM agar, *E. coli* OP50 food source). Life span can be increased by maintaining animals at lower ambient temperatures and shortened by raising the ambient temperature. Use of a killed bacterial food source, rather than live *E. coli*, increases lifespan by 2–4 days, and growth of adult animals in the absence of bacteria (axenic growth or bacterial deprivation) increases median life span to 32–38 days [3, 23, 24]. Under both standard laboratory conditions and bacterial deprivation conditions, wild-derived *C. elegans* hermaphrodites exhibit longevity comparable to N2 animals [25].

More than 250 genes have been reported to modulate aging in *C. elegans*, a majority of which were identified from genome-wide RNAi screens for increased life span [26–29]. Many of these genes can be broadly classified into one of four groups based on the proteins they encode: (1) genes involved in insulin/IGF-1-like signaling, (2) genes involved in mitochondrial function, (3) genetic mimics of dietary restriction, and (4) genes that promote mRNA translation. The precise genetic and molecular relationships between these epistasis groups remain unclear; however, current opinion is that they represent four, at least partially, distinct functional classes of longevity genes. The lifespan of the roundworm, *C. elegans* is influenced by factors that affect the length of the larval or resting phase known as the dauer, and by those that affect the adult phase of life span. The most important pathway in the adult stage is that of IGF-1 homologue and its receptor. Curran and Ruvkun [30] found 63 genes among the 2,700 screened that affected a lengthening of lifespan when inactivated post-developmentally, i.e., that were considered to act to shorten lifespan in the wild type worm. Among these the most potent fell into the class affecting protein synthesis. Classes of genes that affected signaling and/or expression, included Daf-2, Daf-16, and mitochondrial function also affected lifespan. The IGF1 signaling pathway via the FOXO transcription factor Daf-16 plays a role in the survival of *C. elegans*. Wild type lifespan of the *C. elegans* hermaphrodite varies from 12 to 17 days [8], but doubling or tripling of life span can be accomplished by manipulation of the IGF-1 homologue pathway during adulthood [30]. However, other mechanisms can be used to extend lifespan, not all of which use this pathway [31]. The neuronal control of *C. elegans* lifespan via the IGF-1 homologue pathway is well described by Finch and Ruvkun [32].

End of life conditions for *C. elegans* are not fully known, but the most commonly reported phenotype associated with old age is a gradual decrease in movement, ultimately resulting in paralysis of all but the head and tail regions. Analysis of tissue specific aging in *C. elegans* has led to the conclusion that neuronal cells largely retain function, even in very old animals, while muscle cells show a gradual decline in function beginning in many animals near the transition to the post-reproductive stage of adulthood [33]. Associated with this general decline is muscle function is a decrease in pharyngeal pumping, resulting in reduced food consumption [34, 35], and an accumulation of lipofuscin throughout the [36]. If a live food source is used, bacterial colonization of the gut can also contribute to senescence; however, the relevance of this to normal aging is unclear, as animals fed a killed bacterial food source show a similar progression of age-associated phenotypes with life span extended by only a few days [37, 38].

Laboratory Fruit Fly (*Drosophila melanogaster*)

The lifespan of the wild type and untreated fruitfly varies from 20 to 40 days, depending upon laboratory conditions, such as temperature and reproductive status [39]. General conditions affecting survival include temperature, food type and intake, and degree of activity. The IGF1 signaling pathway via the transcription factor dFOXO strongly affects lifespan in the fly. This is related to its fat body expression. Reference [40] and this channel are involved in cardiac function. While apoptosis of muscle and fat cells occurs during lifespan in this mostly post-mitotic animal and contributes to weakness, this may not be the essential end of life event. Normal neuronal signaling is thought to be essential to the condition of other organs and autophagy of neurons may contribute to death of the organism [41]. The effect of this on life span is not presently known. Senescent changes occur in heart function, sleep patterns, and metabolism. The Bodmer laboratory has reported that cardiac dysfunction and fibrillation that is related to potassium channel defects develops in old wild type *D. melanogaster* [42, 43]. The senescence of this model at the organ level and related to the TOR and insulin signaling pathways are of particular interest [44].

Laboratory Mouse (*Mus musculus*)

The mouse, as used in the laboratory, consists of a large number of strains developed over the past century. While various types of carcinomas are most often the cause of death in laboratory mice, the type of cancer that is most commonly found varies among the mouse strains, as noted below. In addition, degenerative conditions, especially those of the heart and kidney are commonly found, as well as cognitive deficiencies with aging (while the latter are seldom the cause of death under laboratory conditions). The reader is referred to The Aging Mouse published by AFIP and additional data on physiological and biochemical values for 15 mouse strains published by the Jackson Laboratories that is available at the Jackson Laboratories web site. There are well over 100 strains in use. For this reason, only four strains in common use will be discussed and the reader is referred to the Jackson Laboratories web site for more details on these and other strains. The life spans of most strains in common laboratory use can be found at the JAX laboratories web site under the Mouse Phenome Data Base as MPD-project, see also [45].

The C57BL/6 strain. In the *C57BL/6 strain* under non-SPF conditions the mean lifespan is from 27 to 30 months of age, while the maximum lifespan (last 10% of the population remaining) is commonly 37–40 months. The most common end of life pathology in the C57BL/6 strain is lymphosarcoma., although sometimes referred to as simply lymphoma. This condition is most evident as an extremely enlarged spleen, although peripheral nodes are usually involved also. This strain and all other mouse strains develop age-related cataracts that are primarily situated in the lens cortex, while extension into the lens nucleus is not unusual. The neurological and muscular efficiencies and capacities are reduced with aging, and development of an enlarged heart in extreme old age is common.

Bllackwell et al. [46] have published both survival curves and cause of death tables for ad lib fed (AL) and calorically restricted (CR) C57BL/6 mice. In this study the mean lifespan for AL mice was 27.5 months for males and 26.9 months for females. This is slightly less than the approximately 870 days (about 28 months) given in the JAX lab summary of strains. Maximum lifespan, reported as the age a which 10% of the cohort remained alive, was 34.8 months for males and 34.1 months for females. Values for mean lifespan for the CR animals exceeded those for AL in the males by about 4 months and 6.6 months in the females, for maximum lifespan the differences were 6.2 for males and 6.1 in females. These authors also gave detailed distributions for lesions present at time of death. As might be expected in this strain, the reported cause of death in AL mice was lymphoma in 60% of males and 57% of females, with the remaining degenerative and proliferative lesions distributed as minor percentages among hemangiosarcoma, nephropathy, liver neoplasms, inflammation, pituitary adenoma, heart thrombus, and unknown factors. Further and more detailed data for the C57BL/6 strain are available from this group [47].

The DBA strain. This strain is relatively short lived, when compared to the C57BL/6. Open angle glaucoma occurs due to blockage of the trabecular meshwork and the canal of Schlemm by melanin containing particles that originate in the iris, ciliary body, and possibly in the retina [48]. Apparently related to this are the development of cataracts at a somewhat earlier age (mid-life span), and senechia formation (adherence between the lens and iris or iris and ciliary body). Thinning of the retina and the loss of optic nerve substance is correlated with the degree of glaucomatous intra-ocular pressure [49, 50]. Recently, the accumulation of amyloid in the retinas of old DBA mice, as compared to controls was reported [50]. The DBA strain has a mean life span under SPF conditions, reported as 629 days (20.6 months) for males and 719 days (23.6 months) for females [51]. Spontaneous calcific heart lesions progress with age and are present by 1 year [51] and 90% of individuals are reported as being affected.

The C3H strain. This strain is somewhat shorter lived than the C57BL/6 strain and malignant hepatomas are a common finding in animals dying after middle age. However, the mean life span is about 830 days (27.2 months).

The C57BL/6xC3H hybrid. Sheldon et al. [52] reported in detail the life span and tumor incidence in 1,064 SPF B6C3F1mice of both sexes, both ad libitum and 40% food restricted. Caloric restriction resulted in reduced tumor incidence and a 36% increase in median life span in both sexes, and increased survival among the last 10% survivors by 21.5% in males and 32.5% in females. End of life pathologies found in ad libitum fed mice in the longitudinal survival cohorts were commonly lymphomas and histiocytic sarcomas, as might be expected in a partial C57BL/6 background, while the C3H background probably encouraged the development of liver and lung tumors that were among equally common in the male ad libitum fed mice. Male mice did not have pituitary or thyroid tumors but females did. A much smaller number of tumors were classified as vascular, musculoskeletal, skin and subcutaneous, kidney, and Harderian gland.

The AKR strain is quite short lived with deaths due to thymic origin lymphoma that is characteristic of this strain. Mean life span is about 200–250 days (6.6 months) with most animals dead by 300 days (8.2 months).

The 129 strain is often used to initiate genetic alterations and these are commonly transferred to the C57BL/6 strain by hybridization followed by inbreeding. The 129 strain itself is relatively long lived with a mean life span of around 850 days (27.9 months) (see JAX lab life span curves for exact data).

Gartner [53] gives the following lifespan data for the following strains, with natural deaths reported as 727 ± 215 days for C57BL/6, 638 ± 260 days for BALB/c, 630 ± 187 days for CBA, 560 ± 230 days for DBA/2, and 317 ± 62 days for AKR where early cause of deaths is due to thymic origin lymphoma. It was notable that medium-sized animals were found to have a greater life expectancy than small or large ones. DeHaan and coworkers [54] in a paper comparing bone marrow precursor cell number and cycling status with related life spans give the "maximum" life spans of the strains used by them in days as follows: C3H/He, 500; CBA, 512; DBA, 710; BALB/c, 745; C57BL/6, 789. However, the mean survival times (50% mortality curves) as published in 1999 by Tuturro et al. [47] for groups of 56 mice each are more robust: measurements from published figures indicate approximately 800 days for male DBA/2 and 725 for females; 820 days for male C57BL/6-NIH and 750 for females; 970 for male B6D2F1 and 900 for females; 1,000 for male B6C3F1 and 900 for females. In all comparisons for strain and sex mice placed on a vitamin supplemented but calorically restricted diet (60% of ad lib intake from 14 to 17 weeks of age induction period onward) survived significantly longer, as measured for either mean or maximum life spans. Maximum body weight was reached in males at approximately 18 months of age in all but the DBA/2 mice, where it was closer to 12 months. In males of all strains it began to gradually fall within the next 3–4 months. In females maximum body weight, also in ad libitum fed mice, was much lower, and reached at 10–12 months in DBA/2 females; 18 months in C57BL/6 and B6C3F1; and 25 months in B6D2F1.

Lipman [55] has remarked on the importance of recording and considering the importance of all lesions seen in mice and rats. In comparing lesion burden in a group of B6C3F1 mice seen at necropsy she notes the steady increase in lesions seen at months 6 (2.1), 12 (4.5), 18 (8.3), 24 (13.0), and 30 (15.6), while a calorically restricted cohort develop only 1/3 as many lesions at 6 months and only 1/2 as many at the later times. Values for lifespan and for end of life lesions (as opposed to incidence of disease at various times of sacrifice) for the various strains are surprisingly rare in the literature. Tables 1 and 2 on life span and body weights of the various mouse strains were made available courtesy of Drs. Beverly Paigen and Rong Yuan, Jackson Laboratories, Bar Harbor, Maine.

In addition, survival curves (see JAX web site) were drawn using the Kaplan Meier method with the inclusion of censored mice. From these curves, median lifespans were calculated with 95% confidence intervals (CI) and the ages of 25 and 75% death. CAST male is excluded from this study, because of the aggressive fighting issue. Median lifespan and age of 75% death of WSB female are not available due to the insufficient number of dead mice.

Body weights for the strains were also calculated by the JAX group and the following table was prepared by the authors credited above.

Lymphoma is a common cause of death in MOLF mice. MOLF. This strain was set up later than other strains.

The following end of life necropsy data for ad libitum and calorically restricted mice was furnished by Dr. Yuji Ikeno and applies only to the C57BL/6 strain (Table 3). The predominance of lymphomas is notable and is a common end of life finding in several strains of mice and rats. However, other forms of cancer may be the apparent cause of death in several strains. Among these are hepatomas in C3H mice, lung tumors in A strain mice, and the kidney degenerative disease in old Fisher and Sprague Dawley rats that is subject to quantity and quality of protein content in the diet.

Laboratory Rat (*Ratus norvegicus*)

The most commonly used pure stains of rats are the Norway, the Fischer 344, the Lewis, the Long-Evans hooded and hybrids of these strains, especially the Norway-Fisher F1. A commonly used non-inbred strain is the Sprague Dawley. The (Brown Norway x Fisher 344) F1 is the most used and is presently stocked by the National Institutes on Aging rodent colonies. Mean survival times for the Fischer 344 rat are 31 months for males and 29 months for females (Boolean search Jax data base). For the Brown Norway the equivalent times are 32 and 32 months, and for the F1 cross with Fisher 344 they are 34 and 29 months, respectively [47]. Maximum lifespans are more difficult to obtain. However, if the maximum is set at 10% survival in the cohort it is not unusual for this to reach 36–38 months in the BN and F344BN F1 strains, while less in the F344 (personal observation, N. Wolf).

End of life lesions and diseases: The strains of rats held and supplied by the NIA aging colonies are the Brown Norway, the Fisher 344, and the hybrid cross between these strains. Age related lesions and their comparative distributions for these three strains have been published [56]. The end of life pathologies found in rats, like

Table 1 Median lifespan with 95% confidence interval and the ages of 25 and 75% death (days)

Group	Female				Male			
	Median lifespan (CI)		25% Death	75% Death	Median lifespan (CI)		25% Death	75% Death
129S1	791	704,855	651	920	882	798,919	778	984
A	619	534,673	480	739	594	555,652	514	700
AKR	258	242,266	219	308	323	288,330	267	385
B10	902	804,973	705	1,056	771	733,818	727	852
B6	922	838,999	772	1,020	901	888,929	817	973
BALB	798	736,819	707	882	664	512,721	440	763
BLK	860	734,888	608	917	827	811,881	770	932
BRCD	877	842,905	792	933	865	834,933	777	943
BTBR	611	595,638	536	665	583	530,694	456	743
BUB	538	390,628	341	726	477	354,768	307	873
C3H	652	581,729	512	795	693	595,729	456	820
C57L	721	707,729	679	749	740	723,751	644	775
CAST	399	199,589	190	747	NA			
CBA	637	532,706	476	783	679	540,750	508	808
D2	680	635,773	442	823	701	525,746	407	799
FVB	701	612,829	518	883	598	553,693	476	755
KK	607	593,621	564	656	565	531,649	469	686
LP	815	750,912	700	973	826	798,855	723	893
MRL	555	514,575	466	626	645	599,665	555	690
NOD.B10	640	556,667	260	748	696	550,780	427	878
NON	764	695,833	631	858	863	806,905	793	919
NZO	576	476,680	418	700	415	364,484	286	598
NZW	732	691,782	600	825	769	685,915	586	992
P	661	559,728	513	811	619	478, NA	526	645
PL	471	400,512	360	563	469	408,512	372	563
PWD	854	697,921	596	967	794	639,834	535	902

Table 1 (continued)

Group	Female				Male			
	Median lifespan (CI)		25% Death	75% Death	Median lifespan (CI)		25% Death	75% Death
RIIIS	820	777,863	691	891	905	826,926	765	940
SJL	501	473,529	379	624	212	133,351	85	514
SM	729	712,763	642	820	775	670,798	523	825
SWR	630	571,694	499	791	718	651,812	498	904
WSB	NA		328	NA	1,005	8,711,091	662	1,091

Table 2 Body weight of inbred strains at 6-, 12- and 18-month

	Female									Male								
	6 month			12 month			18 month			6 month			12 month			18 month		
Strain	Mean	SD	*n*	Mean	SD	*n*	Mean	SD	*n*	Mean	SD	*n*	Mean	SD	*n*	Mean	SD	*n*
129S1	23.4	1.4	8	25.3	2.1	8	26.8	2.7	7	30.8	2.7	8	31.4	3.3	8	29.3	4.4	14
A	26.8	1.6	8	28.1	3.7	8	26.0	2.2	7	27.8	3.2	8	30.3	1.6	6	27.6	3.8	7
AKR	37.0	4.1	8	Short-lived strain						42.8	6.2	7	Short-lived strain					
B10	26.0	2.1	16	23.1	1.2	7	23.2	3.5	7	27.5	3.5	8	29.6	4.3	8	27.7	4.0	7
B6	25.6	2.6	8	25.0	2.1	8	27.8	4.0	7	32.1	1.6	8	31.5	1.7	8	31.7	1.5	8
BALB	24.5	2.2	8	26.2	2.4	8	27.3	2.9	8	31.8	4.0	8	28.8	2.4	7	26.8	4.4	8
BLK	22.4	1.3	8	22.9	1.5	8	22.7	1.6	7	27.4	2.0	8	29.1	2.9	8	29.3	2.7	8
BRCD	26.5	2.2	5	28.8	5.9	8	28.6	2.6	8	37.3	5.0	4	36.0	4.8	8	36.1	4.7	8
BTBR	35.2	5.9	8	43.6	6.4	8	34.7	5.7	7	41.5	4.5	8	39.5	5.0	7	34.9	2.5	6
BUB	25.4	2.8	8	27.4	2.5	3	25.9	2.0	11	36.6	1.3	4	32.2	2.5	5	31.6	1.5	6
C3	26.1	3.3	8	25.6	4.9	8	23.7	3.2	8	33.5	3.0	8	30.8	1.7	8	28.6	1.6	8
C57L	26.6	3.4	8	30.6	4.3	8	28.6	2.3	7	37.1	4.5	8	38.8	3.5	8	33.4	2.9	6
CAST	15.0	0.7	7	14.6	0.7	9	16.1	2.3	3	Not available								
CBA	26.7	3.2	8	27.6	3.5	8	36.4	6.4	9	42.4	2.5	8	39.9	5.4	6	30.1	2.7	6
D2	26.2	3.8	8	27.6	4.6	7	24.9	2.7	5	32.5	4.3	8	28.5	4.5	8	28.4	4.1	6
FVB	26.7	2.6	8	28.3	2.6	8	29.1	3.4	7	29.8	2.4	8	29.7	3.3	7	29.1	5.9	4
KK	32.3	6.6	8	29.9	7.4	7	29.4	11.3	7	32.9	7.1	4	34.4	1.9	7	29.0	4.8	7
LP	20.7	2.0	8	22.4	1.9	8	22.9	1.6	7	27.4	2.1	8	27.4	1.5	8	28.0	1.7	8
MOLF	12.8	1.6	6	Not available						13.7	0.6	6	Not available					
MRL	51.3	5.2	8	41.0	8.5	7	24.8	3.8	6	48.8	2.5	8	42.4	4.5	8	29.5	6.1	8
NOD.B10	28.5	1.5	8	39.7	4.0	8	33.0	1.7	6	34.0	1.6	8	40.7	3.3	7	33.8	2.7	9
NON	34.0	5.1	8	33.3	6.4	7	30.7	5.9	6	40.7	8.8	8	41.5	10.0	8	32.4	8.2	7
NZO	67.4	5.4	8	70.2	6.0	6	62.8	7.6	7	66.8	2.9	7	60.3	22.2	4	62.2	15.3	7
NZW	36.6	4.9	8	39.6	5.0	8	41.4	4.7	7	34.8	2.6	8	35.6	3.3	7	40.2	3.6	6
P	23.5	1.5	8	24.7	1.1	7	24.0	2.8	7	32.7	3.6	8	32.1	4.8	3	26.9	1.5	4

Table 2 (continued)

Strain	Female									Male								
	6 month			12 month			18 month			6 month			12 month			18 month		
	Mean	SD	*n*	Mean	SD	*n*	Mean	SD	*n*	Mean	SD	*n*	Mean	SD	*n*	Mean	SD	*n*
PL	22.7	2.3	7	22.1	2.0	7	21.1	1.6	6	26.7	2.2	8	26.6	2.5	5	19.9	3.6	4
PWD	19.2	3.7	9	21.9	4.6	9	21.9	2.5	7	22.7	3.9	7	25.2	3.1	12	23.5	2.9	7
RIIIS	22.9	2.1	8	22.5	2.6	8	22.5	4.0	8	30.8	3.7	8	31.3	3.3	8	30.2	3.0	8
SJL	22.3	2.3	8	23.5	3.4	7	21.8	3.0	6	25.4	1.9	7	25.1	3.2	8	22.7	3.2	5
SM	19.0	1.8	8	25.9	5.3	8	27.3	2.3	6	32.1	1.9	8	31.5	5.8	8	28.7	4.2	9
SWR	22.5	1.8	8	24.7	3.0	7	23.2	2.0	5	28.0	2.0	8	29.2	2.3	8	27.5	1.6	7
WSB	14.3	1.2	7	16.6	1.9	9	16.5	2.6	8	18.9	0.9	7	22.3	2.9	9	20.7	2.0	9

Table 3 Causes of deaths determined in C57BL/6 mice (Y. Ukeno)

	AL	CR
Neoplasm	21	11
Lymphoma	15	8
Hemangioma	4	1
Adenocarcinoma (lung)	0	2
Others	2	0
Non-neoplasm	3	10
Thrombus	1	0
Glomerulosclerosis	2	0
Prolapse of rectum	0	7
Acidophilic macrophage	0	3
Neoplasm and non-neoplasm	1	0
Lymphoma & Glomerulosclerosis	1	0
Undetermined	5	9
Total	30	30

mice, are generally carcinomas. Most common in the obesity prone Sprague Dawley strain are mammary cancer, pituitary adenomas, adrenal tumors, brain tumors and notably lymphocytic malignancies [57–59]. Non-neoplastic lesions consisted of cardiomyopathies, nethropathies [60].

Useful references to life span and age-related lesions in rats of several commonly used strains can be found at the following references [61, 62]: for Wistar rats (high levels of kidney sclerosis), which Lipman et al. reported on pathology and end of life pathology of the 3 strains supplied by the NIA, i.e. Brown Norway, Fisher, and BNF1 cross [63]. A presentation of aging-associated pathologies and changes in the individual organ systems of rats may be found in 2 volumes issued by the International Life Science Press [64–66].

The order of normal life spans from longest to shortest is Brown Norway x Fisher cross (BNF1), Brown Norway, Fischer 344, Lewis, and Sprague Dawley. The Sprague Dawley strain (outbred) life span is affected by its tendency to obesity, ad libitum fed animals, often exceeding 500 g by middle age. Mean lifespans for ad libitum fed Brown Norway males are 129 and 133 weeks in male and females respectively; for Fisher 344 they are 103 and 116, respectively; for the hybrid of these strains they are 145 and 137, respectively []. The most common end of life pathology in the Fisher strain was glomerulonephritis, while this lesion was milder in the BN rats, where end of life pathology that could be life ending was often pituitary adenoma, renal pelvis pathology, or cardiac degeneration (leukemia incidence was not reported) [55, 63]. The end of life pathologies in the hybrid BNF1 rats was generally found to be the same conditions as the in parent strains, but less severe, compatible with the longer lifespan. End of life pathology for the Sprague Dawley strain was also often affected by chronic progressive nephrosis eventuating in death [67]. This strain is also susceptible to neoplasms, some life ending. Nakasawa [68] found their mean survival times to be 89–105 weeks of age. Their total tumor incidences were 70–76.7% and 87–95.8% in males and females, respectively. The common neoplasms were pituitary adenoma and adrenal pheochromocytoma in both sexes, testicular Leydig cell tumor in males and mammary gland tumors, thyroidal C-cell adenoma and uterine stromal polyp in females. Lymphocytic leukemia is a common cause of death in the F344 strain, especially in male animals.

Baum [69] followed the pathologies and survivals of Lewis rats and A total of 629 LEW/Han rats (305 females and 324 males) from a specified pathogen-free breeding colony were kept from weaning up to their natural death under defined environmental conditions. A complete histological examination was performed on all organs and macroscopically altered tissues of all animals that died during the first three years of the study. These were 296 female (98%) and 213 male (66%) rats. The mean lifespan of the females (27.7 ± 5.1 months) was significantly shorter than that of the males (32.5 ± 6.6 months). In both sexes, the lifespan was mainly determined by the occurrence of neoplasms. Of the large spectrum of 52 histologically different tumour types, the highest incidences were observed for adenomas of the pituitary gland and adenomas/adenocarcinomas of the adrenal cortex in both sexes, mammary gland tumours and endometrial carcinomas in females, and C-cell adenomas/adenocarcinomas of the thyroid gland and tumors of the haemopoietic system in males. Of these, the high incidences of tumors of the haemopoietic system in males (27.7%) and of endometrial carcinomas in females (45.2%) should be considered as characteristic features of the strain. Tuturro et al. 1999 [47] reported mean life spans (approximated from figures) for males as follows: Brown Norway 940 days; Fisher 344 at 800 days; BNF1 1,000 days. Female mean life spans were: Brown Norway 940 days; Fisher 344 at 815 days; BNF1 910 days. Mice on caloric restriction lived significantly longer in all genotypes and sexes. Body weights in the Brown Norway rats peaked at 25 months of age, in Fisher 344 at 20 months, in the BNF1 at 28 months. The weight curve for all of the 3 genotype females remained flat from early adulthood until advanced old age.

Lipman [55] compared the mean life spans in weeks for male and female ad libitum fed Brown Norway rats (129 and 133), Fisher 344 (103 and 116), and BNF hybrids (145 and 137), respectively. Again, the hybrid animal lived longer than either parent, rather than taking a midpoint position.

Wistar strain rats have been used in anitoxidant studies and Quiles et al. [70] reported 20 months for mean life span and 24 for maximum on PUFA diet fed rats. Life span according to Altun et al. [71] is comparable to that of the Sprague Dawley at a mean of 29–30 months.

The end of life necropsy data below are provided by Dr. Yuji Ikeno but do not necessarily indicate final death cause.

Naked Mole-Rat (*Heterocephalus glaber*)

The naked mole-rat, weighing on average 35 g, has a mean lifespan of approximately 25 years, with a maximum lifespan of >30 years. Naked mole-rats in captivity show attenuated age-related declines in behavior, as well as physiological and biochemical function, maintaining good health for at least 83% of their maximum lifespan. Older animals tend to be less active, however sleep patterns of both young and old tend to be more random and more closely resemble that of larger, longer living organisms than small mammals. Age-related changes are markedly and consistently reduced in all the variables examined over a 24 year age range; these assessments at the organism level include sustained vascular youthfulness, lean mass, fertility, gut function, cartilage and bone quality, hormone profiles, glucose handling and metabolism [72]. Furthermore, naked mole-rats show remarkable resistance to spontaneous neoplasia. To date no tumors have been observed in more than 1,000 necropsies. Unlike skin fibroblasts from mice and humans, naked mole-rat fibroblasts when infected with retroviruses containing two oncogenes known to induce tumors, do not form tumors when implanted into immunocompromised mice (presently unpublished data Rochelle Buffenstein and Peter Hornsby) confirming that naked mole-rat cells are resistant to cancer induction. This may be due to maintenance of better genomic surveillance and/or genomic stability than do other shorter-living species, and suggests that these rodents maintain better genomic surveillance and genomic stability. Importantly, age-related detrimental cardiovascular changes, such as arterial relaxation diminishment and superoxide and H2O2 production, were unaffected by age in this animal, and apoptosis rates only moderately so. At the same time, cGPX activity taken alone was 70 times lower than that in the than that in mice. In the heart, antioxidant enzymes and mitochondrial mass were not altered with age even in 26 year old naked mole rats [73]. Evidence exists that this animal can live a long life in spite of the accrual of oxidative damage,

Table 4

F344	($N=60$)	Sprague-Dawley	($N=24$)
Neoplastic	33 (55%)	*Neoplastic*	11 (45.8%)
Leukemia	17	Leukemia	3
Pituitary adenoma	6	Pituitary adenoma	2
Subcutaneous tumor	4	Subcutaneous tumor	2
Others	6	Islet tumor	3
		Others	1
Non-neoplasm	8		
Chronic nephropathy	2	*Non-neoplasm*	10 (41.7%)
Thrombus	4	Nephropathy	7 (29.2%)
Suppurative inflammation	2	Polyarteritis	3
		Others	3
Neoplasm and non-neoplasm	10	Undetermined	3
		Total	24
Leukemia and thrombus	5		
Pituitary adenoma and chronic nephropathy	5		
Others	9		
Undetermined	9		
Total	60		

such as high carbonyl levels in several tissues, low GSH and GSH/GSSG levels and 10-fold higher levels of lipid peroxidation than that in mice. Thus, NMR provides a model for testing the effect of body size, cancer incidence, the presence of oxidative damage and energy production on senescence and lifespan [74]. A series of papers [72, 74–76] note pronounced resistance of this animal to oxidative driven tissue and organ degenerative changes, in spite of the accumulation of oxidative molecules in its tissues. Thus, the NMR has 10-fold higher levels of in vivo lipid oxidation, as well as accumulated oxidative adducts to DNA and proteins, when compared to physiologically-age matched mice [74]. Naked mole-rat cells are extremely resistant to a wide variety of genotoxins, heat, oxidative stressors and heavy metals [77]. Fibroblasts from naked mole-rats are resistant to multiple forms of cell injury, but sensitive to peroxide, UV light, and ER stress [77]. The mechanisms facilitating this resilience are to date unknown and the end of life conditions that may contribute death are also not fully known. The further determination of NMR status in oxidant and antioxidant conditions, its presence of or resistance to DNA lesions in key tissues, the molecular basis of its peculiar metabolic status, as well as related conditions and metabolic pathways that suffer senescent changes in most animal models and in humans, and differ in this model, may provide much information on the conditions that affect organ and organismal aging events and life span.

Domestic Dog (*Canis familiaris*)

The expected normal mean life span of *Canis familiaris* is quite variable due to breed heredity, and perhaps is the most so among the various non-mutant mammals used or referred to in aging research. The term "non-mutant" is, perhaps considered in a special way here, since the history of the dog breeds is one of encouragement or suppression of gene expression by selective breeding for specific traits. Much of the size and skeletal differences appear to be a result of the general and localized effect of IGF-1 in the growing animal, and many of the conformational differences between dog breeds are largely under the influence of the IGF-1 gene and its regional and proportional influence [78]. IGF-2 activity is largely important in the fetus and the juvenile [79]. The canine life span varies from a mean of 14 years and a maximum of 18 years in several small breeds, especially the terriers, and a mean of 7–9 years in the very large breeds, such as the Irish Wolfhound, Great Dane, St. Bernard, Mastiff, Bull Mastiff, and Bernese Mountain dog, with a maximum of 12 years. Mid-size and middle weight breeds' mean lifespans fall in between these two extremes [80], (manuscript submitted, S. Urfer, K. Greer, and N. Wolf, and data from various breed clubs and the AKC). There are, of course variations on this theme among the several breeds in each general category. However, the general correlation between body size/weight and mean lifespan makes this species valuable for research on GH and IGF1 hormonal effect studies, and on which diseases affecting the several organs and metabolic pathways appear to be linked to the hormonal and the related organ growth rate pathways. Rapid growth and adult size in large breeds affects predispositions due to large internal organs (volvulus), skeletal development (osteosarcoma), and overall body weight (osteoarthritis). The lifespan data and pathologies dominant for individual breeds may also be obtained or calculated from data the referred to above and, also, at the Vet Med Data Base web site.

Breed club and American Kennel Club records indicate life span differences that overall coincide inversely with body size and weight. The very large breeds, and especially those, such as the Great Dane and Irish Wolfhound, that experience rapid growth of body size and organ size correlated with high serum levels of growth hormone (GH) and IGF-1 are subject to life-ending conditions that may be related to their rapid growth and size, such as osteosarcoma, cardiomyopathies, bloat or volvulus [65]. Taken as a group, the small and medium size breeds of dogs suffer from the same general array of those conditions as do humans, with cause of death largely distributed among neoplasms and cardiac conditions (but with atherosclerosis rare), making them useful models for those life-ending conditions. While a small subset of the small breeds predominantly display particular conditions, such as a high incidence of cardiomyopathies, this seems to be due to poor breeder selection. The end of life conditions for small and mid-size breeds are more broadly dispersed among the degenerative and the neoplastic diseases. The presence of beta amyloid plaques in the frontal cortex and other regions of the brain, along with the presence of soluable and insoluable forms of A beta 40 and 42 has been noted [81, 82]. A description of the types of end of life diseases for small versus large breeds is listed in Deeb and Wolf [65]. Non-lethal, but late life degenerative conditions, such as hearing loss, cataract, and senile dementia that is accompanied by amyloid deposits in specific brain regions are common accompaniments of old age in the canine, making this species an extremely useful, but under-utilized model for these aging conditions in the human. Recently, several studies of aging have provided such useful information [82–84]. In particular, the use of caloric restriction (CR) in this species has provided useful insight into the effect of CR on several aging conditions and on lifespan in dogs, but requires further exploration.

Rhesus Monkey (*Macaca mulata*)

Several species of monkeys under the listings of rhesus, and it is noted that Macaca nemestrima is in common use in biological research. The most careful studies of aging have been carried out with and without caloric restriction at the National Institutes on Aging and at the University of Wisconsin School of Medicine. These studies have not reached finality. However, several publications have been forthcoming from both sites. A summation of aging changes as affected by caloric restriction and as observed in both the NIA and the University of Wisconsin studies are included in the chapter on CR in this publication.

Aging changes are recorded in the NIA colony of M. mulata in which some animals have now reached 26 years of age (rough mean life span expected as 25–28 years and maximum life span of about 30–33 years).

At the time of this book preparation the status of ad libitum fed monkeys at the National Institutes on Aging colony and their severe end of life conditions revealed at necropsy are noted below. The authors are grateful to Dr. Julie Mattison and her colleagues at NIA for the data.

At the time of this writing, out of 14 deaths in female Macaca monkeys in the ad libitum fed group in the aging and caloric restriction study at the NIA, seven were listed with endometriosis at end of life. While this does not mean that this condition was certainly the cause of death, it was the outstanding pathology upon necropsy and listed as a probable cause. There were also two cases of abdominal lesions that may have been related to endometriosis. Three cases of abdominal adenocarcinoma of intestinal origin were among the other terminal events, also one of bloat with endometriosis present, and 1of hepatic failure due to amyloidosis. The findings to-date suggest that endometriosis is a major problem in ad libitum fed female Macaques. While most of the monkeys in this study that had died at this time were from 20 to 35 years of age, endometriosis was not confined to the oldest animals and was present as a terminal event in a 12 year old monkey.

Among nine male monkeys dead at this time 4 died with adenocarcinoma, 1 each of hepatocelluar carcinoima, pancreatic carcinoma, and two of intestinal origin carcinoma. The remaining five animals died of either cardiovascular disease [2], or with amyloidosis, hemorrhagic colitis, or secondary to advanced diabetes. Remarks on the calorically restricted monkeys will appear later in the chapter on caloric restriction.

The age of death and end of life pathologies in the Southwest Foundation for Biomedical Research chimpanzee colony, consisting of 87 males and 116 females, Hubbard, Lee and Eichberg [85] reported as follows: The primary causes of death since 1982 were heart disease [6], trauma [5], and respiratory disease [4]. The heart lesions included myocarditis, necrosis, fibrosis, and mineralization. The respiratory disease deaths were due to Streptococcus pneumoniae and Klebsiella pneumoniae. The traumatic deaths were primarily in young chimpanzees and were caused by adults. There were two cases of placenta previa and one of abruptio placentae. Clinical conditions not leading to death included respiratory disease, parasitism, alopecia, diarrhea, maternal rejection, and trauma. The most significant commonly isolated bacteria were Staphylococcus, Streptococcus, Haemophilus, Klebsiella, Citrobacter, and Campylobacter. The most common intestinal parasites were Balantidium, Entamoeba, Chilomastix, Iodamoeba, Giardia, Trichuris, Enterobius, and Strongyloides.

Baboon (*Papio hamadryas*)

Baboons have been used in numerous aging-related studies, but not in formal, prospective lifespan studies like rhesus and squirrel monkeys. However, a large colony of baboons has existed since the 1960s at the Southwest Foundation for Biomedical Research in San Antonio, Texas and the demography of this colony (from several thousand records of each sex) as well as two wild populations of baboons has been analyzed [86, 87]. Demographic information is only available on females in the wild populations, as males disperse from their natal group and move outside local study areas. Infant mortality is quite high in both captive and field populations, but if females reach age five years of age (i.e. adulthood), then life expectancy is 21 years in captivity compared with 20 years in one wild population (Gombe, Tanzania) not subject to predation and 12 years in a high predation wild population (Amboseli, Kenya). Record female longevity is 33 years in the captive colony, 27 years in both wild populations. Interestingly, the rate of increase in mortality rate with age is similar between wild and captive populations. The fact that captive life expectancy in a wild population is not that different from a captive population suggests that captive husbandry is still rudimentary in its development. Even the difference in record longevity (6 years) between captive and wild populations may be an artifact of the 4–8-fold larger numbers of complete life spans from the captive colony [86]. Male life expectancy similarly contingent on survival to age 5 is 17 years in the captive colony, some 4 years less than that of females. Female baboons undergo menopause at about age 26 in captivity with an extended perimenopausal period beginning at about age 19 [88]. Wild populations undergo menopause at about the same age [89].

Baboons die of a variety of diseases including chronic colitis (primarily young animals) [90] and a range of neoplasias (primarily old animals) [91]. The most common neoplasias are those of the hematopoetic and urigenital organs. The incidence of neoplasia in baboons is considerably lower than that of humans, however this may reflect captive husbandry and usage that does not yet allow sufficient numbers of baboons to reach the advanced ages at which tumors are most likely to occur. As is common in laboratory confined rhesus monkeys, endometriosis is common in baboon females, with all of the complications seen in human females. A review by Dick et al. [92]. reported that this condition was present in 40% of mostly middle aged female baboons. A similar study [93] found a maximum of 17% incidence among 100 females examined. The Dick study also found widespread vertebral disc degeneration in animals as early as 14 years of age. Spinal bone density also declines with age, but this is not present in the forearm [94]. As expected, hormonal responses are delayed with aging as in seen in other mammalian species [95].

Human (*Homo sapiens*)

Human lifespan studies are generally retrospective, often considering earlier diet, life style, and other factors. An exception to this is the recent set of intervention studies on diet restriction in human volunteers. While these studies have revealed favorable changes in body weight, insulin and blood glucose levels, blood pressure and blood lipids, they have generally found that test subjects to experience a sense of coldness and, in some cases, lessened energy. Any effect on lifespan cannot yet be determined [95].

The mean human lifespan varies on the basis of place of residence and includes diet and exposure to infectious diseases. In the Western industrialized countries mean life span tables generally list the female mean at 77 years and the male mean at 75 years for the United States, with somewhat longer lifespans for Japan, France, Scandinavia, and Canada. In general, in Western industrialized countries the lifespan is becoming longer. However, the mean lifespan is increasing with little or no increase in the maximum lifespan. The overall longest male and female lifespans are reliably recorded for the island of Okinawa, Japan, and thereby among a largely fish-eating population. This introduction will not deal in detail with end of life diseases in humans. It is generally conceded that cardiovascular conditions are the most common cause of death in western humans, followed closely by various types of cancer, with mammary cancer the leading cause in females and lung cancer in males, then in females, followed in both sexes by colon carcinoma. As the western population ages, Alzheimer's disease and its complications has become a major factor. Animal models are increasingly designed to mimic the human age-associated conditions, and neurological defect and disease models are high on this list. While humans cannot be considered laboratory animals, testing of drugs designed to combat age-related conditions is increasing. This includes drugs and regimens aimed at extending lifespan. Among the latter are self-imposed caloric restriction, as well as prospective and retrospective studies of pharmacologic agents such as catechins, polyphenols, and antioxidant vitamins, as well as several fruits and fruit extracts.

Little Brown Bat (*Myotis lucifugus*)

Bats have long been known to be exceptionally long-lived [96]. In fact, corrected for body size, bats are the longest-lived order of mammals even though most bat longevity records come from wild populations whereas most other mammal longevity records come from zoos. The longest-lived bat known to date is a 41 year old male Brandt's bat (*Myotis brandti*) from Siberia. Adult Brandt's bats weigh about 7 g [97]. Although their exceptional longevity was previously attributed to their habit of hibernating for a substantial fraction of their lives, this turns out to be a rather minor effect. Even nonhibernating bats are long-lived [96]. Virtually all information on bat longevity has been acquired incidentally as part of studies of other aspects of their biology, hence it is possible that most species records are considerably underestimates of their true longevity. Of the roughly 1,000 species of bats, the most likely candidate to become commonly used as a research model of extended health and longevity is the little brown bat of North America. This 7–8 g species is found throughout the United States and its physiology has been more thoroughly studied than that of any other bat [98]. One male little brown bat was recaptured in 1995, 34 years after his original banding as part of a marked cohort of several thousand animals in upstate New York [99]. In the genus *Myotis*, males appear to survive longer than females in the field, although whether this has to do with fundamental physiological differences or simply the greater vulnerability to predators and greater energy demands of pregnant females is unknown [97]. Causes of death for bats in either nature or captivity are largely unknown. The little brown bat has been shown to produce fewer reactive oxygen species in kidney, heart, and brain tissue than similar-sized, shorter-lived mammals, a finding consistent with a role for reduced oxidative stress as a contributor to their exceptional longevity [100].

Whole genome sequencing of the little brown bat is currently underway and is expected to be finished by late 2008. The whole genome sequence may well provide molecular clues as to mechanisms of its exceptional longevity and will certainly provide molecular tools to facilitate its further investigation.

Budgerigar (*Melopsittacus undulatus*)

Birds as a group live about three times as long as similar-sized mammals [101]. They achieve this longevity despite having traits that many mechanistic hypotheses of aging would suggest should lead to shorter lives. For instance, bird metabolic rates are nearly twice as high as that of mammals, their body temperatures are typically 3–5°C higher than mammals, and their blood glucose concentration is 2–5 times higher than mammals. A calculation of "lifetime oxygen burden", that is oxygen consumed per gram body mass per lifetime – a crude measure of potential reactive oxygen species exposure – reveals that bird tissues consume as much as four times as much oxygen over a lifetime as even long-lived mammals such as humans [102]. Birds apparently manage this oxygen burden by minimizing tissue level oxygen radical production per amount of oxygen consumed by as yet unknown mechanisms [103]. In addition, long-lived bird tissues appear to contain membrane lipids that are more resistant to oxidative damage than such lipids in shorter-lived mammals [104].

A suitable bird model for the mechanistic study of long-life and extended health is the budgerigar, or as it is more commonly known the pet store "parakeet" (*Melopsittacus undulatus*). Budgerigars are the most commonly kept cage bird in the world, and as such they are inexpensive and materials for their care and maintenance and fundamental husbandry practices are well developed [105]. Budgerigars are about the same size (30–50 g) as laboratory mice and live up to 20 years. Their metabolic rate is about 1.5 times as high as laboratory mice. Except for chickens, the physiology and pathophysiology of budgerigars is better known than that of any other bird species [106]. As with other cage birds, considerable information is available on end-of-life pathology in the Veterinary Medical Database (http://www.vmdb.org/vmdb.html), a compilation of health data from veterinary teaching hospitals in the United States. It is noted that Budgerigars commonly die of a wide variety of tumors [107].

Quail (*Coturnix* spp.)

Bobwhite (*Colinus virginianus*) and Japanese quail (*Coturnix japonica*) have more often been used in feeding and toxicity research. However this animal is becoming more in use for lifespan studies because of its lifespan that is relatively short for a bird model, and for studies on atherosclerosis. The mean lifespan of the Japanese quail is 3–4 years, with a maximum of 7 years [108]. That of the bobwhite quail is similar, with Ottinger [109] reporting a lifespan for females as 2.5–3 years and for males 4–5 years. This provides a contrast with the long-lived parrot species (50–60 years) and sea birds such as the storm petrel (20 or more years) [110].

A number of hormonal studies related to aging and caloric restriction have been carried out on the bobwhite.

Zebra Fish (*Danio rerio*)

The zebra fish has become a common research animal in aging studies. In non-aeriated 20 gallon content tank aquarium conditions the mean lifespan is given as 42 months and maximum 66 months for Nothobranchius rachovii. However, in small tanks with an aeriation system the lifespan is much shorter, and may be as short as 2 months [111]. Thus, environmental conditions and whether inbred or outbred strains are used makes a great difference in survival. Furthermore, there are several genuses and species of annual fish that are in use.

Amphibians

While this is a very large class, the commonly used members in research are the Bull frog (*Xenopus laevus*) 16 years average lifespan, the leopard frog (*Rana pipiens*) approximately 6 years. In general, the length of tadpole development depends upon environmental conditions. Amphibians are used in regeneration studies, as they are capable of extensive regeneration of tissues, such as spinal cord even in the adult [112]. They are susceptible to several lethal virus infections, especially in captivity or limited space in the wild.

References

1. Mortimer RK and Johnston JR (1959). Life span of individual yeast cells. *Nature* 183: 1751–1752.
2. Fabrizio P and Longo VD (2003). The chronological life span of *Saccharomyces cerevisiae*. *Aging Cell* 2: 73–81.
3. Kaeberlein M (2006). Longevity and aging in the budding yeast. In *Handbook of Models for Human Aging*. (PM Conn, ed). Boston: Elsevier Press, pp. 109–120.
4. Kaeberlein M et al. (2005). Genes determining yeast replicative life span in a long-lived genetic background. *Mech Ageing Dev* 126(4): 491–504.
5. Kaeberlein M et al. (2005). Regulation of yeast replicative life span by TOR and Sch9 in response to nutrients. *Science* 310(5751): 1193–1196.
6. Powers RW 3rd et al. (2006). Extension of chronological life span in yeast by decreased TOR pathway signaling. *Genes Dev* 20(2): 174–184.
7. Kaeberlein M et al. (2004). Sir2-independent life span extension by calorie restriction in yeast. *PLoS Biol* 2(9): E296.
8. Gems D and Riddle DL (2000). Defining wild-type life span in *Caenorhabditis elegans*. *J Gerontol A Biol Sci Med Sci* May; 55(5): B215–B219.
9. Murakami CJ et al. (2008). A method for high-throughput quantitative analysis of yeast chronological life span. *J Gerontol A Biol Sci Med Sci* 63(2): 113–121.

10. Smith DL Jr. et al. (2007). Calorie restriction extends the chronological lifespan of *Saccharomyces cerevisiae* independently of the Sirtuins. *Aging Cell* 6(5): 649–652.
11. Lin SJ, Defossez PA, and Guarente L (2000). Requirement of NAD and SIR2 for life-span extension by calorie restriction in *Saccharomyces cerevisiae*. *Science* 289(5487): 2126–2128.
12. Defossez PA et al. (1999). Elimination of replication block protein Fob1 extends the life span of yeast mother cells. *Mol Cell* 3(4): 447–455.
13. Sinclair DA and Guarente L (1997). Extrachromosomal rDNA circles–a cause of aging in yeast. *Cell* 91(7): 1033–1042.
14. Aguilaniu H et al. (2003). Asymmetric inheritance of oxidatively damaged proteins during cytokinesis. *Science* 299(5613): 1751–1753.
15. Kaeberlein M, McVey M, and Guarente L (1999). he SIR2/3/4 complex and SIR2 alone promote longevity in *Saccharomyces cerevisiae* by two different mechanisms. *Genes Dev* 13(9): 2570–2580.
16. Kirchman PA et al. (1999). Interorganelle signaling is a determinant of longevity in *Saccharomyces cerevisiae*. *Genetics* 152(1): 179–190.
17. McMurray MA and Gottschling DE (2003). An age-induced switch to a hyper-recombinational state. *Science* 301(5641): 1908–1911.
18. Kaeberlein TL et al. (2006). Lifespan extension in *Caenorhabditis elegans* by complete removal of food. *Aging Cell* 5(6): 505–514.
19. Fabrizio P et al. (2001). Regulation of longevity and stress resistance by Sch9 in yeast. *Science* 292(5515): 288–290.
20. Piper PW, Harris NL, and MacLean M (2006). Preadaptation to efficient respiratory maintenance is essential both for maximal longevity and the retention of replicative potential in chronologically ageing yeast. *Mech Ageing Dev* 127(9): 733–740.
21. Herker E et al. (2004). Chronological aging leads to apoptosis in yeast. *J Cell Biol* 164(4): 501–507.
22. Burtner CR, Murakami CJ, Kennedy BK, and Kaeberlein M (2009). A molecular mechanism of chronological aging in yeast. *Cell Cycle* Apr 15; 8(8): 1256–1270.
23. Vanfleteren JR and Braeckman BP (1999). Mechanisms of life span determination in *Caenorhabditis elegans*. *Neurobiol Aging* 20(5): 487–502.
24. Lee GD et al. (2006). Dietary deprivation extends lifespan in *Caenorhabditis elegans*. *Aging Cell* 5(6): 515–524.
25. Sutphin GL and Kaeberlein M (2008). Dietary restriction by bacterial deprivation increases life span in wild-derived nematodes. *Exp Gerontol* 43(3): 130–135.
26. Dillin A et al. (2002). Rates of behavior and aging specified by mitochondrial function during development. *Science* 298(5602): 2398–2401.
27. Hamilton B et al. (2005). A systematic RNAi screen for longevity genes in *C. elegans*. *Gemes Dev* 19(13): 1544–1555.
28. Hansen M et al. (2005). New genes tied to endocrine, metabolic, and dietary regulation of lifespan from a *Caenorhabditis elegans* genomic RNAi screen. *PLoS Genet* 1(1): 119–128.
29. Lee SS et al. (2003). A systematic RNAi screen identifies a critical role for mitochondria in *C. elegans* longevity. *Nature Genet* 33(1): 40–48.
30. Curran SP and Ruvkun G (2007). Lifespan regulation by evolutionarily conserved genes essential for viability. *PLoS Genet* Apr 6; 3(4): e56.
31. Kenyon C (2005). The plasticity of aging: Insights from long-lived mutants. *Cell* Feb 25; 120(4): 449–460.
32. Finch CE and Ruvkun G (2001). The genetics of aging. *Annu Rev Genomics Hum Genet* 2: 435–462.
33. Herndon LA et al. (2002). Stochastic and genetic factors influence tissue-specific decline in ageing *C. elegans*. *Nature* 419(6909): 808–814.
34. Huang C, Xiong C, and Kornfeld K (2004). Measurements of age-related changes of physiological processes that predict lifespan of *Caenorhabditis elegans*. *Proc Natl Acad Sci U S A* 101(21): 8084–8089.

35. Kenyon C et al. (1993). A *C. elegans* mutant that lives twice as long as wild type. *Nature* 366(6465): 461–464.
36. Klass MR (1977). Aging in the nematode *Caenorhabditis elegans*: Major biological and environmental factors influencing life span. *Mech Ageing Dev* 6(6): 413–429.
37. Garigan D et al. (2002). Genetic analysis of tissue aging in *Caenorhabditis elegans*: A role for heat-shock factor and bacterial proliferation. *Genetics* 161(3): 1101–1112.
38. Garsin DA et al. (2003). Long-lived *C. elegans* daf-2 mutants are resistant to bacterial pathogens. *Science* 300(5627): 1921.
39. Bauer JH, Goupil S, Garber GB, and Helfand SL (2004). An accelerated assay for the identification of lifespan-extending interventions in *Drosophila melanogaster*. *Proc Natl Acad Sci U S A* Aug 31; 101(35): 12980–12985.
40. Giannakou ME and Partridge L (2004). The interaction between FOXO and SIRT1: Tipping the balance towards survival. *Trends Cell Biol* Aug; 14(8): 408–412.
41. Takacs-Vellai K, Bayci A, and Vellai T (2006). Autophagy in neuronal cell loss: A road to death. *Bioessays* Nov; 28(11): 1126–1131.
42. Ocorr KA, Crawley T, Gibson G, and Bodmer R (2007). Genetic variation for cardiac dysfunction in Drosophila. *PLoS One* 2(7): e601.
43. Wessells RJ and Bodmer R (2007). Cardiac aging. *Semin Cell Dev Biol* Feb; 18(1): 111–116.
44. Shaw P, Ocorr K, Bodmer R, and Oldham S (2008). Drosophila aging 2006/2007. *Exp Gerontol* Jan; 43(1): 5–10.
45. Gregorova S, Divina P, Storchova R, Trachtulec Z, Fotopulosova V, Svenson KL et al. (2008). Mouse consomic strains: Exploiting genetic divergence between Mus m. musculus and Mus m. domesticus subspecies. *Genome Res* Mar; 18(3): 509–515.
46. Blackwell BN, Bucci TJ, Hart RW, and Turturro A (1995). Longevity, body weight, and neoplasia in ad libitum-fed and diet-restricted C57BL6 mice fed NIH-31 open formula diet. *Toxicol Pathol* Sep–Oct; 23(5): 570–582.
47. Turturro A, Witt WW, Lewis S, Hass BS, Lipman RD, and Hart RW (1999). Growth curves and survival characteristics of the animals used in the Biomarkers of Aging Program. *J Gerontol A Biol Sci Med Sci* Nov; 54(11): B492–B501.
48. John SW, Smith RS, Savinova OV, Hawes NL, Chang B, Turnbull D et al. (1998). Essential iris atrophy, pigment dispersion, and glaucoma in DBA/2 J mice. *Invest Ophthalmol Vis Sci* May; 39(6): 951–962.
49. Inman DM, Sappington RM, Horner PJ, and Calkins DJ (2006). Quantitative correlation of optic nerve pathology with ocular pressure and corneal thickness in the DBA/2 mouse model of glaucoma. *Invest Ophthalmol Vis Sci* Mar; 47(3): 986–996.
50. Bhattacharya SK, Peachey NS, and Crabb JW (2005). Cochlin and glaucoma: A mini-review. *Vis Neurosci* Sep–Oct; 22(5): 605–613.
51. Jackson Research Laboratories J (2007). Mouse genome informatics. *Mouse Genome Informatics* 2007: 1–100.
52. Sheldon WG, Bucci TJ, Hart RW, and Turturro A (1995). Age-related neoplasia in a lifetime study of ad libitum-fed and food-restricted B6C3F1 mice. *Toxicol Pathol* Jul–Aug; 23(4): 458–476.
53. Gartner K (1992). Life expectancy, its relation to sexual activity and body weight in male inbred mice. *J Exp Anim Sci* Dec; 35(3): 125–135.
54. de Haan G, Nijhof W, and Van Zant G (1997). Mouse strain-dependent changes in frequency and proliferation of hematopoietic stem cells during aging: Correlation between lifespan and cycling activity. *Blood* Mar 1; 89(5): 1543–1550.
55. Lipman RD (1997). Pathobiology of aging rodents: Inbred and hybrid models. *Exp Gerontol* Jan–Apr; 32(1–2): 215–228.
56. Lipman RD, Dallal GE, and Bronson RT (1999). Effects of genotype and diet on age-related lesions in ad libitum fed and calorie-restricted F344, BN, and BNF3F1 rats. *J Gerontol A Biol Sci Med Sci* Nov; 54(11): B478–B491.

57. Belpoggi F, Soffritti M, Tibaldi E, Falcioni L, Bua L, and Trabucco F (2006). Results of long-term carcinogenicity bioassays on Coca-Cola administered to Sprague-Dawley rats. *Ann N Y Acad Sci* Sep; 1076: 736–752.
58. Son WC (2004). Factors contributory to death of young Sprague-Dawley rats in carcinogenicity studies. *Toxicol Lett* Nov 2; 153(2): 213–219.
59. Duffy PH, Lewis SM, Mayhugh MA, Trotter RW, Hass BS, Latendresse JR et al. (2008). Neoplastic pathology in male Sprague-Dawley rats fed AIN-93 M diet ad libitum or at restricted intakes. *Nutr Res* Jan; 28(1): 36–42.
60. Duffy PH, Lewis SM, Mayhugh MA, Trotter RW, Hass BS, Thorn BT et al. (2008). Nonneoplastic pathology in male Sprague-Dawley rats fed the American Institute of Nutrition-93 M purified diet at ad libitum and dietary-restricted intakes. *Nutr Res* Mar; 28(3): 179–189.
61. Rao GN, Haseman JK, Grumbein S, Crawford DD, and Eustis SL (1990). Growth, body weight, survival, and tumor trends in F344/N rats during an eleven-year period. *Toxicol Pathol* 18(1 Pt 1): 61–70.
62. Roth GS, Brennecke LH, French AW, Williams NG, Waggie KS, Spurgeon HA et al. (1993). Pathological characterization of male Wistar rats from the gerontology research center. *J Gerontol* Nov; 48(6): B213–B230.
63. Lipman RD, Chrisp CE, Hazzard DG, and Bronson RT (1996). Pathologic characterization of brown Norway, brown Norway x Fischer 344, and Fischer 344 x brown Norway rats with relation to age. *J Gerontol A Biol Sci Med Sci* Jan; 51(1): B54–B59.
64. Mohr U, Dungworth DL, and Capen CC (Eds.) (1992). *Pathobiology of the Aging Rat*, Vol. 1. Washington, DC: International Life Sciences Institute.
65. Deeb BJ and Wolf NS (1994). Studying longevity and morbidity in giant and small breeds of dogs. *Vet Med* 89(Supplement): 702–713.
66. Mohr U, Dungworth DL, and Capen CC (Eds.) (1994). *Pathology of the Aging Rat*, Vol. 2. Washington, DC: Internal Life Sciences Institute.
67. Gray JE, van Zwieten MJ, and Hollander CF (1982). Early light microscopic changes in chronic progressive nephrosis in several strains of aging laboratory rats. *J Gerontol* Mar; 37(2): 142–150.
68. Nakazawa M, Tawaratani T, Uchimoto H, Kawaminami A, Ueda M, Ueda A et al. (2001). Spontaneous neoplastic lesions in aged Sprague-Dawley rats. *Exp Anim* Apr; 50(2): 99–103.
69. Baum A, Pohlmeyer G, Rapp KG, and Deerberg F (1995). Lewis rats of the inbred strain LEW/Han: Life expectancy, spectrum and incidence of spontaneous neoplasms. *Exp Toxicol Pathol* Jan; 47(1): 11–18.
70. Quiles JL, Ochoa JJ, Huertas JR, and Mataix J (2004). Coenzyme Q supplementation protects from age-related DNA double-strand breaks and increases lifespan in rats fed on a PUFA-rich diet. *Exp Gerontol* Feb; 39(2): 189–194.
71. Altun M, Bergman E, Edstrom E, Johnson H, and Ulfhake B (2007). Behavioral impairments of the aging rat. *Physiol Behav* Dec 5; 92(5): 911–923.
72. Buffenstein R (2008). Negligible senescence in the longest living rodent, the naked mole-rat: Insights from a successfully aging species. *J Comp Physiol [B]* Jan 8; 178(4): 439–445.
73. Csiszar A, Labinskyy N, Orosz Z, Xiangmin Z, Buffenstein R, and Ungvari Z (2007). Vascular aging in the longest-living rodent, the naked mole rat. *Am J Physiol Heart Circ Physiol* Aug; 293(2): H919–H927.
74. Andziak B and Buffenstein R (2006). Disparate patterns of age-related changes in lipid peroxidation in long-lived naked mole-rats and shorter-lived mice. *Aging Cell* Dec; 5(6): 525–532.
75. Lambert AJ, Boysen HM, Buckingham JA, Yang T, Podlutsky A, Austad SN et al. (2007). Low rates of hydrogen peroxide production by isolated heart mitochondria associate with long maximum lifespan in vertebrate homeotherms. *Aging Cell* Oct; 6(5): 607–618.
76. Buffenstein R (2005). The naked mole-rat: A new long-living model for human aging research. *J Gerontol A Biol Sci Med Sci* Nov; 60(11): 1369–1377.

77. Salmon AB, Sadighi Akha AA, Buffenstein R, and Miller RA (2008). Fibroblasts from naked mole-rats are resistant to multiple forms of cell injury, but sensitive to peroxide, ultraviolet light, and endoplasmic reticulum stress. *J Gerontol A Biol Sci Med Sci* Mar; 63(3): 232–241.
78. Sutter NB, Bustamante CD, Chase K, Gray MM, Zhao K, Zhu L et al. (2007). A single IGF1 allele is a major determinant of small size in dogs. *Science* Apr 6; 316(5821): 112–115.
79. Favier RP, Mol JA, Kooistra HS, and Rijnberk A (2001). Large body size in the dog is associated with transient GH excess at a young age. *J Endocrinol* Aug; 170(2): 479–484.
80. Greer KA, Canterberry SC, and Murphy KE (2007). Statistical analysis regarding the effects of height and weight on life span of the domestic dog. *Res Vet Sci* Apr; 82(2): 208–214.
81. Head E, Rofina J, and Zicker S (2008). Oxidative stress, aging, and central nervous system disease in the canine model of human brain aging. *Vet Clin North Am Small Anim Pract* Jan; 38(1): 167–178.
82. Pugliese M, Geloso MC, Carrasco JL, Mascort J, Michetti F, and Mahy N (2006). Canine cognitive deficit correlates with diffuse plaque maturation and S100beta (-) astrocytosis but not with insulin cerebrospinal fluid level. *Acta Neuropathol* Jun; 111(6): 519–528.
83. Rofina J, van Andel I, van Ederen AM, Papaioannou N, Yamaguchi H, and Gruys E (2003). Canine counterpart of senile dementia of the Alzheimer type: Amyloid plaques near capillaries but lack of spatial relationship with activated microglia and macrophages. *Amyloid* Jun; 10(2): 86–96.
84. Milgram NW, Siwak-Tapp CT, Araujo J, and Head E (2006). Neuroprotective effects of cognitive enrichment. *Ageing Res Rev* Aug; 5(3): 354–369.
85. Hubbard GB, Lee DR, and Eichberg JW (2005). Diseases and pathology of chimpanzees at the Southwest Foundation for Biomedical Research. *Am J Primatol* June; 24(3–4): 273–282.
86. Bronikowski AM, Alberts SC, Altmann J, Packer C, Carey KD, and Tatar M (2002). The aging baboon: Comparative demography in a non-human primate. *Proc Natl Acad Sci USA* 99: 9591–9595.
87. Martin LJ, Mahaney MC, Bronikowski AM, Dee CK, Dyke B, and Comuzzie AG (2002). Lifespan in captive baboons is heritable. *Mech Ageing Dev* 123: 1461–1467.
88. Martin LJ, Carey KD, and Comuzzie AG (2003). Variation in menstrual cycle length and cessation of menstruation in captive raised baboons. *Mech Ageing Dev* 2003(124): 865–871.
89. Packer C, Tatar M, and Collins A (1998). Reproductive cessation in female mammals. *Nature* 392: 807–811.
90. Rubio CA and Hubbard GB (2001). Chronic colitis in baboons: Similarities with chronic colitis in humans. *In Vivo* 15: 109–116.
91. Cianciolo RE, Butler SD, Eggers JS, Dick EJ, Jr, Leland MM, and de la Garza M (2007). Spontaneous neoplasia in the baboon (*Papio* spp.). *J Med Primatol* 36: 61–79.
92. Dick EJ Jr., Hubbard GB, Martin LJ, and Leland MM (2003). Record review of baboons with histologically confirmed endometriosis in a large established colony. *J Med Primatol* Feb; 32(1): 39–47.
93. D'Hooghe TM, Bambra CS, De Jonge I, Lauweryns JM, and Koninckx PR (1996). The prevalence of spontaneous endometriosis in the baboon (Papio anubis, Papio cynocephalus) increases with the duration of captivity. *Acta Obstet Gynecol Scand* Feb; 75(2): 98–101.
94. Havill LM, Mahaney MC, Czerwinski SA, Carey KD, Rice K, and Rogers J (2003). Bone mineral density reference standards in adult baboons (*Papio hamadryas*) by sex and age. *Bone* Dec; 33(6): 877–888.
95. Goncharova ND and Lapin BA (2004). Age-related endocrine dysfunction in nonhuman primates. *Ann N Y Acad Sci* Jun; 1019: 321–325.
96. Wilkinson GS and South JM (2002). Life history, ecology and longevity in bats. *Aging Cell* 1: 124–131.
97. Podlutsky AJ, Khritankov AM, Ovodov ND, and Austad SN (2005). A new field record for bat longevity. *J Gerontol A Biol Sci Med Sci* 60: 1366–1368.
98. Fenton M and Barclay R (1980). *Myotis lucifugus*. *Mammalian Species* 142: 1–8.

99. Davis WH and Hitchcock HB (1995). A new longevity record for the bat *Myotis lucifugus*. *Bat Research News* 36: 6.
100. Brunet-Rossinni AK (2004). Reduced free-radical production and extreme longevity in the little brown bat (*Myotis lucifugus*) versus two non-flying mammals. *Mech Ageing Dev* 125: 11–20.
101. Holmes DJ, Fluckiger R, and Austad SN (2001). Comparative biology of aging in birds: An update. *Exp Gerontol* Apr; 36(4–6): 869–883.
102. Holmes DJ and Ottinger MA (2003). Birds as long-lived animal models for the study of aging. *Exp Gerontol* Nov–Dec; 38(11–12): 1365–1375.
103. Pamplona R, Portero-Otin M, Riba D, Ledo F, Gredilla R, Herrero A et al. (1999). Heart fatty acid unsaturation and lipid peroxidation, and aging rate, are lower in the canary and the parakeet than in the mouse. *Aging (Milano)* 11: 44–49.
104. Pamplona R and Barja G (2007). Highly resistant macromolecular components and low rate of generation of endogenous damage: Two key traits of longevity. *Ageing Res Rev* 6: 189–210.
105. Kavanaugh JL (1987). *Lovebirds, Cockatiels, Budgerigars: Behavior and Evolution*. Los Angeles: Science Software Systems, Inc.
106. Ritchie BW, Harrison GJ, and Harrison LR (1994). *Avian Medicine: Principles and Applications*. Lake Worth, FL: Wingers Publishing.
107. Holmes DJ and Austad SN (1995). Birds as animal models for the comparative biology of aging: A prospectus. *J Gerontol A Biol Sci Med Sci* 50: B59–B66.
108. Nagata J, Oku H, Toda T, and Chinen I (1996). Effect of dietary choresterol on the activities of key enzymes of choresterol metabolism in hyperlipidemia and atherosclerosis-prone Japanese quail. *J Nutr Sci Vitaminol* 42: 287–300.
109. Ottinger MA (2001). Quail and other short-lived birds. *Exp Gerontol* 36: 859–868.
110. Haussmann MF, Winkler DW, and Vleck CM (2005). Longer telomeres associated with higher survival in birds. *Biol Lett* Jun 22; 1(2): 212–214.
111. Herrera MaJ P (2004). Annual fish as a genetic model for aging. *J Gerontol, Biol Sci* 59A: 101–107.
112. Chernoff EA, Sato K, Corn A, and Karcavich RE (2002). Spinal cord regeneration: Intrinsic properties and emerging mechanisms. *Semin Cell Dev Biol* Oct; 13(5): 361–368.

Animal Size, Metabolic Rate, and Survival, Among and Within Species

Steven N. Austad

Abstract The positive interspecific relation between mammalian body size and longevity was described more than a century ago and remains one of the most robust patterns known in the biology of aging. Hypotheses about the role of metabolic rate or relative brain size in explaining this pattern have not been supported by detailed analyses. This pattern may be due to an inverse relation between mitochondrial oxygen radical production and body size, although evidence for this hypothesis is sparse. On a less mechanistic level, evolutionary senescence theory provides a compelling rationale that species regardless of size that are less prone to environmental hazards evolve longevity assurance mechanisms leading to longer life- and health span. Considerable evidence suggests that the opposite pattern – smaller size associated with longer life – obtains within species, although detailed information is available for only a few species. Within the species – mice, dogs, and horses – in which this relationship is well-established, the deleterious effects of growth hormone acting either autonomously or through its effect on IGF-I signaling provide a possible explanatory mechanism. Evidence from humans does not appear to conform to this pattern, perhaps because of the dominance of cardiovascular disease as a human cause-of-death.

Keywords Body size · Comparative biology · Energetics · Oxygen radicals · Growth hormone · IGF-I · Mammal longevity · Metabolic rate · Rate of living

Even a casual observer will recognize that among mammals, size, aging, and longevity are intimately related. Mice live accelerated lives compared to dogs. That is, mice reach maturity, reproduce, grow frail and feeble, and die more quickly than do dogs. Similarly, dogs live accelerated lives compared to horses. But a closer look at mammalian longevity clouds this seemingly simple relationship. So, for instance,

S.N. Austad (✉)
Department of Cellular & Structural Biology, Barshop Institute for Longevity and Aging Research, University of Texas Health Science Center at San Antonio, San Antonio, TX, USA
e-mail: austad@uthscsa.edu

N.S. Wolf (ed.), *The Comparative Biology of Aging*,
DOI 10.1007/978-90-481-3465-6_2, © Springer Science+Business Media B.V. 2010

horses live accelerated lives compared with much smaller humans. Mice live accelerated lives compared with much smaller bats. To further complicate the picture, sometimes size is associated with acceleration of some parts of the life history and deceleration in others. For instance, small dogs typically reach sexual maturity more quickly than large dogs, but they remain healthy and live considerably longer. This chapter will discuss the history of ideas about the mechanisms underlying the relationship between body size, health, and longevity and try to bring some coherence to what is known about this relationship at present.

First, it is convenient to dichotomize the discussion into size-health-longevity patterns *between* species – the *inter*specific pattern – such as horses versus dogs versus mice and patterns *within* species – the *intra*specific pattern – such as small dogs versus large dogs, horses versus ponies, or one mouse genotype versus another. Although it is theoretically possible that the between-species and within-species relationship between body size, health, and longevity would be the same, there is no particular reason to expect it to be. Within-species variation in body size represents fairly specific variation in hormone action during development overlaid on the same basic body plan, ecology, physiology, and underlying genetics. By contrast, differences between species can be due to a wide range of genetic factors and dramatically different body plans, developmental strategies, physiologies and ecologies.

Body Size and Interspecific Variation in Mammalian Longevity

Although it was probably casually observed earlier, the first systematic statistical investigation of the relation between body size and longevity among mammals was performed by George Sacher, who analyzed 63 taxonomically diverse species and found that variation in body weight accounted for 60% of the variation in maximum longevity on a log-log plot [1]. "Maximum longevity," in this case, is defined as the maximum longevity record of any individual within a species. The rationale for using this longevity metric was that it was likely to be a more stable measure of how long individuals could live under optimum conditions than mean or median longevity, which can vary dramatically depending on the quality of care and husbandry they receive, something that is poorly understood for many mammal species even today.

A similar pattern between size and longevity is found *within* taxonomic groups of mammals (e.g. primates, ungulates, carnivores, rodents), although the slope and elevation of the relationship may be slightly different [2]. Bird species display the same positive relationship between size and longevity, although they live substantially longer for any given body size [3, 4]. Although Sacher's sample of species was relatively small, and many of the longevity records short by current standards, the same basic pattern has been found in many subsequent, larger and more current analyses [5, 3, 6–8] (Fig. 1).

An obvious question is what mechanism or process might be responsible for the robust positive longevity-body size relationship? Note that this relationship is not hypothesized to be a causal one, meaning that the functional relationship is no doubt

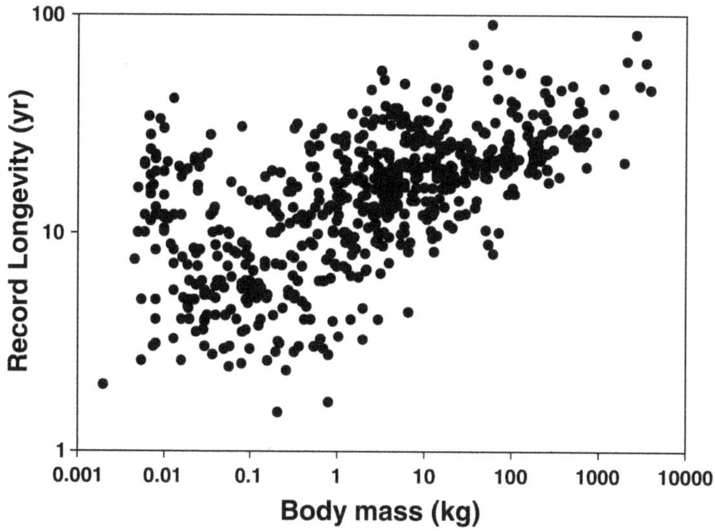

Fig. 1 Body size and longevity records for 605 species of mammals

due to one or more third variables that are themselves linked to both body size and longevity. It is also worth noting that many analyses of this pattern assume that the body size-longevity relationship is linear throughout its range. The one study that used a statistical approach (LOWESS regression) to assess whether the relationship was truly linear found that the body size-longevity relationship disappeared among species less than 1 kg unless both bats and marsupials were removed from the analysis [6].

The first and most influential hypothesis about the mechanistic link between body size and longevity ascribed that relationship to variation in metabolic rate. This notion – that the use of energy to support life's processes was inherently destructive – was an ancient, intuitively satisfying one, hinted at even by Aristotle [8]. It became considerably more plausible when oxygen free radicals generated by normal metabolism provided a plausible cellular mechanism for metabolism-as-damaging-agent [9]. A folk version of this idea links metabolic rate to heart rate and holds that all mammals have about the same number of heartbeats per lifetime. Small species with their rapid heart beats expend their lifetime allotment quickly, larger species with slower heart rates expend it slowly. Empirical support for the idea linking metabolism and longevity derived from observations that in poikilothermic species, ambient temperature varied inversely with longevity [10, 11] and that even a small sample of six mammal species performed a century ago by physiologist Max Rubner indicated that small species had both higher mass-specific metabolic rates and shorter lives compared with larger species [12, 13].

Pearl [11] called the idea that metabolism dictates longevity the "rate of living theory" (ROLT). There are strong and weak forms of the ROLT. The strong form suggests that all species of a group, for instance mammals or birds or primates, will be limited by similar lifetime energy expenditure per gram of metabolically active body mass [12]. In modern terms, we might re-formulate this idea as "lifetime energy expenditure per cell limits the lifespan of animals." For mammals, this idea was clearly wrong even in Rubner's day as humans were known to expend several times more energy per gram or cell over the course of a lifetime than cows, horses, dogs, cats, or Guinea pigs. But human exceptionalism was assumed in the early twentieth century, so the theory in its strong form lived on.

Pearl [11] stated his version of the ROLT less specifically than Rubner, noting that "*in general*, the duration of life varies inversely as the rate of energy expenditure during its continuance." Sacher [1] made the mechanistic link of the hypothesis more specific by phrasing the ROLT as "the lifespan of a [mammal] species varies inversely as its ***basal*** metabolic rate (emphasis added)." He noted that the exponent of the *increase* in longevity with body size among species was close to the negative exponent of the *decrease* in mass-specific basal metabolic rate with size, suggesting that mammals have a fixed energetic potential per cell per lifetime. Small species expend this potential quickly and die early, large species expend it more slowly and die later. Note that this idea relied on measures of basal metabolic rate (BMR), defined loosely as the metabolic rate of an animal at rest, but more rigorously as the minimal metabolic rate of a nonreproductive, fasted, inactive animal at its thermoneutral environmental temperature [14] – a rate, it should be noted, at which animals rarely live. Subsequently, several other writers claimed that energy expended per unit body mass per lifetime is relatively constant across large subgroups of mammals [15, 16].

A weaker form of the same idea holds that there will be some general relation between metabolic rate and longevity, such that mammal species that live longer than expected for their size will have lower than normal metabolic rates, and conversely [8].

As we shall see shortly, the determination of life- and health-span by relative metabolic rate either in its strong or weak form fails on many fronts, however first we might ask what the relevance of BMR might be for animals that do move, eat, reproduce, and experience a variety of temperatures in the real world. Clearly animals seldom experience BMR. In fact, among small mammals, resting metabolism contributes only about 40% of total energy expenditure [8]. Yet resting and active energy expenditure involve the same cellular machinery, so the rationale for considering BMR rather than total energy expenditure is obscure. Most species longevity records derive from captive animals in zoos or research facilities. Although animals in such facilities are considerably less active than they would be in nature, they usually still do move, eat, reproduce and live at temperatures outside their thermoneutral zone. So it is difficult to imagine that resting metabolic rate or BMR even in captive animals represents a meaningful metabolic measure unless it is an indicator of some underlying metabolic tuning that suggests generally higher or lower rates of a broad range of cellular activities. An alternative approach would be to evaluate total energy expenditure of animals and observe how this links to longevity. Total energy expenditure can be easily measured in either captive or free-living animals using doubly-labeled water and has now been done so for dozens of species [8]. Thus, total energy expenditure of captive animals might be a better approximation of the actual values that might be causally related to life span.

As noted above, the ROLT fails on many fronts. Lifetime resting (or basal) metabolic expenditure varies by at least 30-fold across mammal species, an observation that is difficult to reconcile with the strong form of the ROLT. Further, if one looks at the major evolutionary subgroups of mammals, the 250 or so marsupial species display a lower (70–80%) BMR compared with similar-sized eutherian mammals (all 5,000+ other mammals except for the three monotreme species). The ROLT predicts then that marsupials should live longer than eutherians, but in fact they live shorter [6]. If one expands the analysis to consider birds, problems with the ROLT are even greater. Birds maintain about twice the BMR of similar-sized mammals, hence the ROLT predicts they should be shorter lived, but in fact they live about three times as long as mammals [17].

A more sophisticated analysis of mammal interspecific data would stress perhaps the weaker form of the ROLT and take into account the potential confounding effects of variation due to body size independent of metabolic rate, might employ measures of total energy expenditure rather than BMR, and might account statistically for evolutionary relationships among species. Historically analyses of interspecific patterns tended to treat species as independent data points when of course species are related to one another evolutionarily and may share traits because they have had a recent common ancestor rather than for other reasons. Without accounting for evolutionary relationships statistical degrees of freedom and consequent statistical power will be artificially inflated. One can remove phylogenetic effects by using the method of phylogenetically-independent contrasts which incorporates evolutionary history into the analysis [18].

Because body size itself is correlated with a host of physiological variables from heart rate to body composition to the number and size of various blood cells [3] and one might be interested in mechanisms of aging separate from body size itself, it becomes useful to statistically remove via residual analysis the effects of body size from the analysis and focus on whether species that live longer or shorter than expected for their size expend more or less energy than expected for their size.

John Speakman has performed analyses controlling for body size and phylogeny on both mammals and birds [8, 4]. When employing these statistical controls, the association between resting metabolic rate and longevity in both birds and mammals disappears [8]. If he uses total (rather than resting) metabolic rate (as measured by doubly-labeled water) then lifetime energy expenditure per gram body mass is not remotely constant. Small mammals expend considerably more total energy per gram of body mass over the course of their lives than do large mammals [8]. For birds, the results are similar [4]. The same sort of sophisticated analyses using resting metabolic rate finds no relationship between metabolic rate and longevity [19, 8]. Thus energy expenditure per se appears not to play a dominant role in the determination of longevity.

It is still possible that some aspect of energy expenditure – something like the rate of free radical generation – could be causally involved in longevity determination. There have been few rigorous comparative studies of this issue. The most thorough study to date, examined hydrogen peroxide production by isolated heart mitochondria from 10 species of mammals and 2 bird species [20]. The study found a weak, but statistically significant, negative association between the rate of hydrogen peroxide production and species longevity. Evaluation of the general relationship between body size and radical production is not possible in this study as the species studied were deliberately selected to have exceptionally long or short lives for their particular body size.

Another possibility is that some component of body size, rather than total body size, is a key to underlying mechanism(s) of aging and longevity. A long-time suspect in this regard is brain size, the reasonable assumption being that because the brain is involved in neuroendocrine regulation of many physiological processes, larger brains may provide more precise homeostatic regulation. Perhaps intelligence actively facilitates longevity in some way.

In Sacher's original analysis, he noted that whereas body size alone could account for 60% of the variation in longevity among species, brain size alone accounted for 79% of the variation. If he combined body weight and brain weight, nearly 85% of the variation could now be statistically accounted for [1]. He focused on brain size specifically because he had noticed that in his sample of 63 species, primates invariably fell above the regression line for all mammals and primates are notable for their large brains relative to body weight. He concluded that body weight, probably via its effect on metabolic rate, and relative brain weight (i.e. brain weight relative to that expected for a given body size), probably via its regulatory function, represented two independent determinants of longevity.

The idea that longevity was determined to some extent by relative brain size had enormous appeal, possibly not least because humans had such large brains and long lives. Many papers followed using other data sets [21–23, 24]. A warning note that life might not be so simple was sounded by Economos [25], who noted that if one looked across mammalian orders from primates to carnivores to ungulates to rodents, relative brain size was not consistently related to longevity [26]and that weight of other organs such as liver and adrenal gland did almost as well or better than brain weight in predicting longevity [25]. Perhaps the strength of these correlations had more to do with the range of their values (the range of body weights among mammals is 10–50 times greater than the range of organ weights) or the fact that body weights for some species might be subject to the vicissitudes of captive obesity, hence brain weight might be a better indicator of real body size than body weight. A much larger analysis than had been performed previously (587 species) indicated that except among primates, relative brain weight does not correlate significantly with longevity (except in the ungulates where the correlation is negative) [27]. Moreover, the weight of heart, liver, kidney and spleen accounted for as much or more of the variation in species longevity than did brain weight. Primates seem to be a special case. Relative brain weight in primates is much greater than in any other mammalian order and brain weight correlated more significantly with primate longevity than did other organs, although the correlation was considerably weaker than reported in other studies [27]. Something about primate brains may contribute to their longevity, although what this might be remains elusive.

A less mechanistic approach to understanding the body size-longevity relationship derives from evolutionary senescence theory [6]. That theory, developed by Peter Medawar, George Williams, and W.D. Hamilton more than half a century ago, is based on the declining power of natural selection to affect the fate of allelic variants with deleterious effects on health late in life [28]. A corollary of that theory is that other things being equal, species subject to high levels of extrinsic mortality (mortality due to environmental hazards unrelated to age, such as famine, flood, pestilence and predators) should evolve faster aging and higher intrinsic mortality compared with species subject to lower levels of extrinsic mortality [29]. Empirical evidence supporting the basic tenets of evolutionary senescence theory are manifold [30, 31].

How is evolutionary senescence theory relevant to the body size-longevity relationship? One could easily imagine that extrinsic hazards vary consistently with species size for a diversity of reasons. First, an indirect effect of decreasing mass-specific metabolic rate with increasing body size is that larger animals are capable of withstanding longer periods of food and water deprivation than small animals. Thus large animals are buffered from famine and drought when compared with small animals. Second, large animals have a smaller surface-to-volume ratio than small animals, meaning that other things being equal they will also be more buffered from climatic temperature extremes. Third, larger animals occupy larger home ranges than smaller animals, hence live at lower population densities [32]. Low population densities expose animals to fewer infectious diseases from other members of their species [33]. Finally, larger animals by virtue of their size and strength will be vulnerable to a reduced range of predator species. All-in-all, then there are many reasons to expect that larger animals will be exposed to fewer environmental hazards, which according to evolutionary senescence theory suggests they should evolve slower aging than smaller species [6], which is of course consistent with the general body size-longevity relationship.

One way to explore the validity of the idea that the body size-longevity relationship is at least partially due to the evolutionary consequences of reduced vulnerability to environmental hazards is to consider the longevity of species that for reasons unrelated to body size might be expected to be exposed to reduced hazards. Most obviously, we know that birds are longer-lived than equivalent size mammals and that weak-flying or nonflying birds are shorter-lived than strong fliers [34]. Flight is a particularly fast and energetically efficient method for moving long distances [35] and an effective means of escape from terrestrial predators. Therefore, flight is an exquisite mechanism to escape a deteriorating environment or flush of predators. It is no surprise then that flight in mammals might also be expected to be coupled with the evolution of long life, as indeed it is [36]. Bats are by far the longest-lived mammals for their body size, even though longevity records for most mammal species come from zoo populations whereas the majority of bat records come from the wild.

In addition to powered flight, we might expect mammal species capable of aerial sailing such as flying squirrels and sugar gliders, especially when combined with a nocturnal arboreal life style, to share some of the resistance to extrinsic hazards with flying mammals. Indeed, aerially sailing species are considerably longer-lived than expected for their body size, although not as long-lived as bats [6]. Similarly many mammals protected by defensive spines or quills, or covered in protective armor, or live in colonies underground are also exceptionally long-lived relative to a "standard" mammal. A striking example of this phenomenon is the naked mole-rat, the longest-lived known rodent (notably, a porcupine is the second longest-lived) which lives underground in large colonies on the African equator [37].

Although the mechanism(s) underlying the interspecific relationship between body size and longevity remains elusive, a large body of evidence suggests that the evolutionary reason seems to be that body size may be one rough surrogate variable for vulnerability to extrinsic dangers.

Body Size and Intraspecific Variation in Mammalian Longevity

Although the evidence is not as conclusive as for the interspecies relationship between body size and longevity, a handful of well-studied species suggests that within a species there may well be a general pattern such that smaller individuals tend to live longer than larger individuals [38].

It is important that we define precisely what is meant by larger and smaller individuals within a species. There are two issues of interest. First, does the source of the size difference matter? That is, individuals of a given species can differ in size because of nutritional or other environmental factors, because of genetic factors, or because of some combination of both. Small birth size or neonatal size due to maternal undernutrition has been linked to a range of late life diseases [39], but nutritional restriction in early life has also been linked to improved health and longer life [40]. So this issue is obviously pretty complicated, may differ among species, and may subtly depend on the nature, degree, and timing of undernutrition. Also, some size differences represent natural variation within a species, but more of what we know about size and longevity comes from strains or breeds that have been purposefully bred by humans to be different in size. This is true, for instance, of the three species I will focus on most intensely in this section – dogs, horses, and mice. Second, what measure of body size is most appropriate? Between-species studies typically use body mass as a size indicator because it is one metric that is available for many species, but for intraspecific analysis there are other options, such as height or some other measure of linear dimension. Using mass by itself confounds differences in linear size with relative leanness or obesity, which themselves have effects on health independent of stature. As this chapter is not focused on dietary or life style issues, virtually all of what follows considers adult size due primarily to genetics and longevity.

More information is available with respect to genetic body size, healthspan, and longevity in dogs than in any other mammal species [38]. Dogs have been selectively bred by humans into a diversity of sizes and shapes for a variety of purposes for millennia and over the past 150 years more than 400 breeds have been created [41]. As a consequence, there is nearly a 100-fold difference in adult body mass between the largest and smallest dog breeds – the largest such range within any mammal species. A host of degenerative conditions of dogs bear a striking similarity to those in humans [42].

Multiple large studies have confirmed that small dog breeds live as much as 50% longer than typical large breeds [38, 43]. This longevity difference appears to represent not merely changes in survival, but actual differences in aging rate, as multiple degenerative conditions appear earlier in large as compared with small breeds. Indeed, the age at which clinical veterinarians consider dogs to require "geriatric" care depends on the animal's size [44]. Clinical veterinarians consider giant dog breeds to require "geriatric care" by ages 6–9, whereas smaller breeds require such care at 9–13 years of age [44].

In humans, plasma insulin-like growth factor (IGF-1) and its secretagogue growth hormone (GH) are major contributors to growth rate and this appears to be true in dogs as well, as large breeds have higher circulating levels of IGF-1 than small breeds. Furthermore, a recent genomic analysis found a strong association between dog breed size and IGF-1 genetic haplotypes, suggesting that these genetic variants differentially activated the GH/IGF-1 axis. Surprisingly, a single IGF-1 haplotype (of 14 identified) was found in all small dog breeds [45]. The mechanism by which small body size is linked to longer life and preserved health in dogs remains unknown, although evidence from mice (see below) suggests that GH/IGF-1 signaling may be a major contributor.

It is worth noting that although the effects of GH signaling on both adult size and longevity is often assumed to operate via its stimulatory effect on circulating IGF-1 secreted from the liver, these two hormones have independent as well as overlapping functions. Research in mice suggest in fact that the impact of GH reduction on life span is substantially greater than IGF-1 alone (see below).

Body size also seems to be inversely related to longevity in horses. Although data on this topic are sparse, one of the things that "everybody knows" in the world of horse breeders, riders, and veterinarians is that ponies, which are by definition small (less than 58 in. at the withers) horses, live longer, healthier lives than regular size or large horses. The hormonal mechanisms underlying the smaller body size of ponies is not entirely elucidated, although again GH/IGF-1 signaling is a strong suspect. Some breeds of ponies, like some small dogs, display lower levels of circulating IGF-1, but others do not [46]. It may well be that the breeds without reduced IGF-1 have defects at the IGF- or GH-receptor level or even some more downstream components in that signaling pathway.

What we know about body size and longevity within domestic horse breeds can be easily summarized. Life insurance records on more than 100,000 Swedish horses ranging up to age 20 (horses are seldom insured beyond this age, even though some live considerably longer) revealed that only 28% of nonracing female riding horses insured before age 1 survived to 20 years of age, 47% of pony mares did so [47]. Sufficient data were not available to compare male horses and ponies across this age range. For females insured later in life (before age four), 26% of nonracing thoroughbreds and other riding horses survived to age 20 compared with 55% of ponies.

Table 1 Longevity in horses versus ponies

Age category	Non-ponies (%)	Ponies (%)	References
20–29 years	87	13	Brosnahan and Paradis [49]
≥ 30 years	52	48	
15–19 years	97	3	Williams [48]
≥ 30 years	67	33	

Brosnahan and Paradis [49] present percent of 467 geriatric horses (regular size horses and ponies) visiting teaching hospital veterinarians for health problems. The Williams [48] study represents age-at-death for 817 animals. Non-ponies are regular horses, donkeys, mules, or horse-pony hybrids

Two further studies also suggest that the folk wisdom of slower pony aging is valid (Table 1). Of more than 800 necropsies performed on equids 15 years old or older at the Kentucky Livestock Disease Diagnostic Center, ponies represented only 3% of animals that died between 15 and 19 years of age, whereas they represented one-third of animals that died at greater than 30 years of age [48]. The oldest animal necropsied in this sample was a 45 year old pony. Another study simply documented the ages at which older horses versus ponies were brought to a large teaching hospital for veterinary clinical evaluation. Whereas ponies formed only 13% of animals evaluated at ages 20–29 years, they formed nearly half of animals seen at ages thirty years or older [49].

A surprising phenomenon which may or may not be linked with aging is that young ponies are also known to have substantially faster wound healing than horses. Specifically, a standardized surgical wound had healed in all five experimental ponies by 7–9 weeks whereas it was 12 weeks before the first (of 5) horse's wound had healed [50, 51]. Ponies differ from horses metabolically too, in that they develop an exaggerated hyperlipidemia and become insulin resistant when fasted (M.R. Paradis, personal communication, 7/08). There are no reports of faster wound healing in smaller dogs compared with larger ones.

The one species in which the inverse relation between size and longevity appears to be fairly well understood is the house mouse [38]. In recent years, a substantial body of research on single gene mouse mutants has established that reduced GH signaling, and to a lesser extent IGF-1, signaling, increases life- and healthspan. This is most cleanly shown by genetically inactivating the GH receptor, which has been found to decrease body size, of course, and also increase longevity by 20–50% in both sexes in multiple genetic backgrounds in multiple survival experiments within labs and in multiple labs [52, 53]. Complete inactivation of IGF-1 signaling alone by contrast is lethal. However less dramatic reduction of IGF-1 signaling increases life span in female mice but not males [54]. A report of a more dramatic increase in life span associated with genetically reduced IGF-1 signaling has been reported but the longevity of both control and mutant animals in that report was considerably shorter than expected in mouse colonies maintained under high husbandry standards, making interpretation of the results difficult [55].

Mouse mutants with multiple endocrine deficiencies due to developmental inhibition of full pituitary function such as the Ames and Snell dwarfs which lack the cell types that produce GH, prolactin, and thyroid stimulating hormone are small as well as dramatically long-lived – up to 60% longer-lived – compared to normal size littermates, but because these mutants affect only pituitary-derived hormone secretion but not local tissue production, it is difficult to ascribe their life-extending effects to reduced activity of any specific hormone. For instance, although Ames dwarf mice lack circulating GH or IGF-1, the concentration of these hormones in the hippocampus is equal to or higher than in normal-sized controls [56].

Among a heterogeneous stock of mice derived from inbred laboratory strains, body weight of individuals at 2–5 months of age was strongly predictive of subsequent longevity, although this pattern did not reach statistical significance if one measured maximum body weight [57]. Unfortunately, linear size was not reported in that study. By contrast, researchers observed no correlation between body mass and longevity in ad lib-fed, genetically uniform strains of mice and rats, but found a significant positive relation relationship between variables in calorically-restricted mice [58, 59]. Similarly in wild-derived mice, there was no correlation between body size and longevity in an ad lib-fed population, but when wild-derived mice were calorically-restricted, a significant positive relationship was observed [60].

As suggested by the above, even genetically small mice are not invariably longer-lived. An array of mouse mutants show stunted growth and short-lives as would be expected if a mutation simply interfered with normal development. Even natural variation in body size is not necessarily correlated with life span. For instance, wild mice display much smaller body size than laboratory mice even when both are maintained on life-long ad lib feeding in the laboratory. Yet some of these wild-derived mouse stocks live longer than laboratory controls, and some do not [61]. On the other hand, laboratory selection for increased or decreased juvenile growth rate in an outbred mouse background produced stocks with an array of adult body sizes, among which there was a significant inverse correlation with longevity [62]. In these cases, it is not known whether the variation in body size is a function only of GH/IGF-1 signaling or something more complicated. Without question, the most comprehensive investigation of the general trend for body size and longevity to be inversely correlated has been that of Rollo [63], who reviewed about 400 studies each of rats and mice, their body weight, and maximum longevity, performed during the twentieth century. Given the variation in diet, genotype, and husbandry details among labs and studies over this time period, it is rather remarkable that he found a highly significant negative relationship between body weight and longevity, which explained 9–25% of the variation in longevity among these studies [63].

Genetically small body size appears often, but not invariably, to associate with longer life in dogs, horses, and mice. What about humans? As usual, human studies provide difficulties due to lack of experimental controls. Also, differences in adult size may be due early environmental factors, the timing and degree of those factors, and their interaction with genetic regulators of body size.

It is important to try to disentangle body weight from measures of linear dimension such as height. Extreme obesity is well-known to increase mortality, but so does excessive leanness even when corrected for smoking status and pre-existing conditions [64]. As obesity is not the focus of this chapter, I will focus on the relationship between longevity and height or stature.

Economists and historians often use height as an indicator of the nutritional health and disease exposure of human populations and as such an indicator in dozens of studies, greater height correlates with longer life [65–67]. Samaras, a strident advocate for the opposite view – that shorter humans live longer than taller ones – reaches his conclusion by comparing heights of different sexes or different countries or ethnic groups within a country with one another [68]. Due to variation in hormonal milieu, diet, lifestyle, and multiple other factors in his analyses, it is difficult to evaluate these claims in the face of a mountain of opposing epidemiological evidence.

There is a less consistent trend with height and specific causes of death than with overall mortality. Although generally height and cardiovascular disease mortality are inversely correlated [67], the opposite may be true for cancer mortality [38] although studies are by no means unanimous on these patterns. Assuming these patterns are real, it would be intriguing to note whether they are due to differences in environmental or genetic factors. The best study of this issue to date evaluated the relationship between height and coronary heart disease mortality in 35,000 Scandinavian twin pairs and determined that the environmental factors, not genetic factors, explained the lower mortality of taller people [67].

Compared with evidence from mice, dogs, and horses, data on the relationship between height, health, and longevity from humans remains confusing. The most convincing human pattern is that height is inversely correlated with mortality – the opposite of what is observed in the small selection of other mammal species for which we have substantial evidence.

Conclusions

The mechanistic basis for the positive interspecies relationship between body size and longevity in mammals remains mysterious, although oxygen radical production may play a role. By contrast, the negative within-species relationship between body size and longevity, now defined for a small sample of mammal species, appears to be due to signaling through the growth hormone pathway. Current evidence suggests that this pattern does not hold in humans.

References

1. Sacher GA (1959) Relation of life span to brain weight and body weight in mammals. In (GEW Wolstenholme and M O'Connor (Eds.)). *CIBA Foundation Colloquia on Ageing*. London: Churchill, pp. 115–133.
2. Pagel MD and Harvey PH (1989). Taxonomic differences in the scaling of brain on body weight among mammals. *Science* 244: 1589–1593.

3. Calder WAI (1986) *Size, Function, and Life History*, New ed. Mineola, NY: Dover.
4. Furness LJ and Speakman JR (2008). Energetics and longevity in birds. *Age* 30: pp. 75–87.
5. Prothero J and Jürgens KD (1987). Scaling of Maximal Lifespan in Mammals: A Review. In (AD Woodhead and KH Thompson (Eds.)). *Evolution of longevity in animals*. New York: Plenum Press.
6. Austad SN and Fischer KE (1991). Mammalian aging, metabolism, and ecology: Evidence from the bats and marsupials. *J Gerontol* 46: B47–B53.
7. de Magalhaes JP, Costa J, and Church GM (2007b). An analysis of the relationship between metabolism, developmental schedules, and longevity using phylogenetic independent contrasts. *J Gerontol A Biol Sci Med Sci* 62: 149–160.
8. Speakman JR (2005). Body size, energy metabolism and lifespan. *J Exp Biol* 208: 1717–1730.
9. Harman D (1956). Aging: A theory based on free radical and radiation chemistry. *J Gerontol* 11: 298–300.
10. Sohal RS (1986) The rate of living theory: A contemporary interpretation. In (K-GCollatz and RS Sohal (Eds.)). *Insect Aging*. Berlin: Springer-Verlag, pp. 23–44.
11. Pearl R (1928). *The Rate of Living*. New York: A.A. Knopf.
12. Rubner M (1908). *Das Problem der Lebensdauer und seine Beziehungen zum Wachstum und Ernahrung*. Munich: Oldenbourg.
13. Kleiber M (1956). Energy metabolism. *Annu Rev Physiol* 18: 35–52.
14. McNab BK (2008). An analysis of the factors that influence the level and scaling of mammalian BMR. *Comp Biochem Physiol A Mol Integr Physiol* 151: 5–28.
15. Cutler RG (1979b) Evolutionary biology of senescence. In (JA Behnke, CE Finch, and GB Moment (Eds.)). *The biology of aging*. New York: Plenum Press, pp. 1–388.
16. Stahl WR (1962). Similarity and dimensional methods in biology. *Science* 137: 205–212.
17. Holmes DJ and Ottinger MA (2003). Birds as long-lived animal models for the study of aging. *Exp Gerontol* 38: 1365–1375.
18. Garland T Jr., Harvey PH, and Ives AR (1992). Procedures for the analysis of comparative data using phylogenetically independent contrasts. *Syst Biol* 41: 18–32.
19. de Magalhaes JP, Costa J, and Church GM (2007a). An analysis of the relationship between metabolism, developmental schedules, and longevity using phylogenetic independent contrasts. *J Gerontol A Biol Sci Med Sci* 62: 149–160.
20. Lambert AJ, Boysen HM, Buckingham JA, Yang T, Podlutsky A, Austad SN et al. (2007). Low rates of hydrogen peroxide production by isolated heart mitochondria associate with long maximum lifespan in vertebrate homeotherms. *Aging Cell* 6: 607–618.
21. Cutler RG (1979a). Evolution of human longevity: A critical overview. *Mech Ageing Dev* 9: 337–354.
22. Mallouk RS (1975). Letter: Longevity in vertebrates is proportional to relative brain weight. *Fed Proc* 34: 2102–2103.
23. Hofman MA (1983). Energy metabolism, brain size and longevity in mammals. *Q Rev Biol* 58: 495–512.
24. Cutler RG (1975). Evolution of human longevity and the genetic complexity governing aging rate. *Proc Natl Acad Sci USA* 72: 4664–4668.
25. Economos AC (1980a). Brain-life span conjecture: A reevaluation of the evidence. *Gerontology* 26: 82–89.
26. Economos AC (1980b). Taxonomic differences in the mammalian life span-body weight relationship and the problem of brain weight. *Gerontology* 26: 90–98.
27. Austad SN and Fischer KE (1992). Primate longevity: Its place in the mammalian scheme. *Am J Primatol* 28: 251–261 (Ref Type: Generic).
28. Rose MR (1991). *Evolutionary Biology of Aging*. Oxford: Oxford University Press.
29. Charlesworth B (2000). Fisher, Medawar, Hamilton and the evolution of aging. *Genetics* 156: 927–931.
30. Kirkwood TB and Austad SN (2000). Why do we age?. *Nature* 408: 233–238.

31. Austad SN and Kirkwood TBL (2008) Evolutionary theory in aging research. In (LPGuarente, L Partridge, and DC Wallace (Eds.)). *Molecular Biology of Aging*. Cold Spring Harbor: Cold Spring Harbor Laboratory Press, pp. 95–111.
32. McNab BK and Brown JH (2002). *The Physiological Ecology of Vertebrates: A View from Energetics*. Ithaca, NY: Cornell University Press.
33. Anderson RM (1995). Evolutionary pressures in the spread and persistence of infectious agents in vertebrate populations. *Parasitology* 111(Suppl): S15–S31.
34. Holmes DJ and Austad SN (1995). Birds as animal models for the comparative biology of aging: A prospectus. *J Gerontol A Biol Sci Med Sci* 50: B59–B66.
35. Schmidt-Nielsen K (1990) *Animal Physiology: Adaptation and Environment*, 4th ed. Cambridge, UK: Cambridge University Press.
36. Wilkinson GS and South JM (2002). Life history, ecology and longevity in bats. *Aging Cell* 1: 124–131.
37. Buffenstein R (2008). Negligible senescence in the longest living rodent, the naked mole-rat: Insights from a successfully aging species. *J Comp Physiol [B]* 178: 439–445.
38. Miller RA and Austad SN (2006). Growth and aging: Why do big dogs die young?. In (EJ Masoro and SN Austad (Eds.)). *Handbook of the Biology of Aging*, 6th ed San Diego: Academic Press, pp. 512–533.
39. Kajantie E, Barker DJ, Osmond C, Forsen T, and Eriksson JG (2008). Growth before 2 years of age and serum lipids 60 years later: The Helsinki Birth Cohort study. *Int J Epidemiol* 37: 280–289.
40. Ozanne SE and Hales CN (2004). Lifespan: Catch-up growth and obesity in male mice. *Nature* 427: 411–412.
41. Ellegren H (2005). Genomics: The dog has its day. *Nature* 438: 745–746.
42. Waters DJ and Wildasin K (2006). Cancer clues from pet dogs. *Sci Am* 295: 94–101.
43. Greer KA, Canterberry SC, and Murphy KE (2007). Statistical analysis regarding the effects of height and weight on life span of the domestic dog. *Res Vet Sci* 82: 208–214.
44. Hoskins JD and McCurnin DM (1997). Geriatric care in the late 1990s. *Vet Clin North Am Small Anim Pract* 27: 1273–1284.
45. Sutter NB, Bustamante CD, Chase K, Gray MM, Zhao K, Zhu L et al. (2007). A single IGF1 allele is a major determinant of small size in dogs. *Science* 316: 112–115.
46. Malinowski K, Kristensen RA, Hafs HD, and Scanes CG (1996). Age and breed differences in thyroid hormones, insulin-like growth factor (IGF)-I and IGF binding proteins in female horses. *J Anim Sci* 74: 1936–1942.
47. Egenvall A, Penell JC, Bonnett BN, Olson P, and Pringle J (2006). Mortality of Swedish horses with complete life insurance between 1997 and 2000: Variations with sex, age, breed and diagnosis. *Vet Rec* 158: 397–406.
48. Williams N (2000). Disease conditions in geriatric horses. *Equine Pract* 22: 32.
49. Brosnahan MM and Paradis MR (2003). Demographic and clinical characteristics of geriatric horses: 467 cases (1989–1999). *J Am Vet Med Assoc* 223: 93–98.
50. Wilmink JM, van Weeren PR, Stolk PW, Van Mil FN, and Barneveld A (1999). Differences in second-intention wound healing between horses and ponies: Histological aspects. *Equine Vet J* 31: 61–67.
51. Wilmink JM and van Weeren PR (2005). Second-intention repair in the horse and pony and management of exuberant granulation tissue. *Vet Clin North Am Equine Pract* 21: 15–32.
52. Bartke A (2005). Minireview: Role of the growth hormone/insulin-like growth factor system in mammalian aging. *Endocrinology* 146: 3718–3723.
53. Coschigano KT, Holland AN, Riders ME, List EO, Flyvbjerg A, and Kopchick JJ (2003). Deletion, but not antagonism, of the mouse growth hormone receptor results in severely decreased body weights, insulin, and insulin-like growth factor I levels and increased life span. *Endocrinology* 144: 3799–3810.
54. Holzenberger M, Dupont J, Ducos B, Leneuve P, Geloen A, Even PC et al. (2003). IGF-1 receptor regulates lifespan and resistance to oxidative stress in mice. *Nature* 421: 182–187.

55. Conover CA and Bale LK (2007). Loss of pregnancy-associated plasma protein A extends lifespan in mice. *Aging Cell* 6: 727–729.
56. Sun LY, Al-Regaiey K, Masternak MM, Wang J, and Bartke A (2005). Local expression of GH and IGF-1 in the hippocampus of GH-deficient long-lived mice. *Neurobiol Aging* 26: 929–937.
57. Miller RA, Harper JM, Galecki A, and Burke DT (2002b). Big mice die young: Early life body weight predicts longevity in genetically heterogeneous mice. *Aging Cell* 1: 22–29.
58. Bertrand HA, Lynd FT, Masoro EJ, and Yu BP (1980). Changes in adipose mass and cellularity through the adult life of rats fed ad libitum or a life-prolonging restricted diet. *J Gerontol* 35: 827–835.
59. Weindruch R, Walford RL, Fligiel S, and Guthrie D (1986). The retardation of aging in mice by dietary restriction: Longevity, cancer, immunity and lifetime energy intake. *J Nutr* 116: 641–654.
60. Harper JM, Leathers CW, and Austad SN (2006). Does caloric restriction extend life in wild mice?. *Aging Cell* 5: 441–449.
61. Miller RA, Harper JM, Dysko RC, Durkee SJ, and Austad SN (2002a). Longer life spans and delayed maturation in wild-derived mice. *Exp Biol Med (Maywood.)* 227: 500–508.
62. Miller RA, Chrisp C, and Atchley W (2000). Differential longevity in mouse stocks selected for early life growth trajectory. *J Gerontol A Biol Sci Med Sci* 55: B455–B461.
63. Rollo CD (2002). Growth negatively impacts the life span of mammals. *Evol Dev* 4: 55–61.
64. Flegal KM, Graubard BI, Williamson DF, and Gail MH (2005). Excess deaths associated with underweight, overweight, and obesity. *J Am Med Assoc* 293: 1861–1867.
65. Finch CE and Crimmins EM (2004). Inflammatory exposure and historical changes in human life-spans. *Science* 305: 1736–1739.
66. Fogel RW (2003). Secular trends in physiological capital: Implications for equity in health care. *Perspect Biol Med* 46: S24–S38.
67. Silventoinen K, Zdravkovic S, Skytthe A, McCarron P, Herskind AM, Koskenvuo M et al. (2006). Association between height and coronary heart disease mortality: A prospective study of 35,000 twin pairs. *Am J Epidemiol* 163: 615–621.
68. Samaras TT, Elrick H, and Storms LH (2003). Is height related to longevity?. *Life Sci* 72: 1781–1802.

Hormonal Influences on Aging and Lifespan

Adam Spong and Andrzej Bartke

Abstract This chapter summarizes the present understanding of the role of IGF-1, insulin, and homologous signaling in the control of aging in a worm, *C. elegans*, a fly, *Drosophila melanogaster*, in the yeast *Saccharomyces cerevisiae*, and in the mouse, and identifies some of the known or suspected mechanisms linking the actions of these hormones to longevity. The activities of these hormones and their receptors, and their various effects on metabolism and growth are covered. A discussion of the applicability of these findings among all species investigated, including the human, is also included.

Keywords Insulin/IGF-1 signaling (IIS) · Growth hormone (GH) · Insulin receptor substrates (IRS) 1 and 2 · Calorie restriction (CR) · *C. elegans* · *Drosophila* · *Saccharomyces cerevisiae* · Mice

Introduction

While aging occurs in all living organisms, it is only within the last 20 years that the existence of common genetic and endocrine mechanisms of aging has been discovered within widely divergent species. Pioneering studies of Johnson in a microscopic worm, *Caenorhabditis elegans* [1], and Jazwinski and Guarente in yeast [2, 3] demonstrated that mutation of a single gene can cause a major increase in longevity. This opened the door to the possibility of identifying genetic mechanisms and cellular signaling pathways responsible for the control of aging. A list of genes and gene mutations that influence longevity in yeast, worms, insects, and mammals increased rapidly. In 1997, Kimura et al., working in Ruvkun's laboratory, reported that *daf-2*, one of the key genes which affect the longevity of *C. elegans*, exhibits homology to

A. Bartke (✉)
Geriatric Research, Department of Internal Medicine, Southern Illinois University School of Medicine, Springfield, IL 62794, USA
e-mail: abartke@siumed.edu

N.S. Wolf (ed.), *The Comparative Biology of Aging*,
DOI 10.1007/978-90-481-3465-6_3,

genes coding for two important hormone receptors in mammals, insulin receptor and insulin-like growth factor-1 (IGF-1) receptor [4]. This exciting discovery led to the now widely accepted proposal that closely related cellular signaling mechanisms, often referred to as the insulin/IGF-like signaling (IIS) pathways, play a fundamental role in the control of aging in organisms as diverse as a worm consisting of fewer than 1,000 somatic cells and a mouse. The IIS pathway is an evolutionarily ancient system involved in broadly similar functions across species, including growth, development, stress resistance, metabolic homeostasis, and reproduction. Demonstration of the evolutionarily conserved and perhaps universal role of this pathway in aging inspired a great amount of research and rekindled interest in the complex interplay between growth, metabolism, aging and longevity. In this chapter, we will briefly summarize the present understanding of the role of IGF-1, insulin, and homologous signaling in the control of aging in a worm, *C. elegans*, a fly, *Drosophila melanogaster*, in the yeast *Saccharomyces cerevisiae*, and in the mouse, and identify some of the known or suspected mechanisms linking the actions of these hormones to longevity. We will also discuss the applicability of these findings to other species, including the human.

C. elegans

Caenorhabditis elegans (C. elegans) is a small roundworm, typically 1–1.5 mm long, that is found in soil, feeds on bacteria, and reproduces either by self-fertilization of hermaphrodites or by the mating of hermaphrodites with males, which arise rarely by meiotic non-disjunction of the sex chromosome [5]. It is a very simple organism, consisting of only 959 somatic cells in adult hermaphrodites, and its extensive use in developmental biology led to the identification of the lineage of each cell in its body. Larval development in *C. elegans* consists of four stages and an alternative larval stage known as "dauer," which is stress-resistant, hypometabolic, and long-lived, allowing for prolonged survival in unfavorable environmental conditions (reviewed in [6, 7]).

Studies in *C. elegans* laid a foundation for the present understanding of the genetic control of aging in multicellular organisms. The existence of mutations affecting dauer formation or adult lifespan led to the identification of age-1, daf-2, daf-16, and a large number of other genes that can influence longevity (reviewed in [5]). Mutations in many of these genes are associated with significant, often impressive increases of lifespan. Research on the effects of individual mutations and their epistatic relationships demonstrated that the normal products of these genes constitute elements of several signaling pathways within the cell. Prominent among them is the pathway involving multiple endogenous ligands interacting with the DAF-2 (dauer formation 2) receptor, first discovered for its role in the developmental control of dauer diapause [8].

The identification of AGE-1 as a PI3K (phosphatidylinositol-3 kinase) ortholog in 1996 by Morris et al. in the Ruvkun laboratory [9], and the demonstration by the same group in 1997 that *daf-2* exhibits extensive homology to genes coding for mammalian insulin and insulin-like growth factor 1 (IGF-1) receptors [4], led to the realization that the *daf-2/age-1/daf-16* signaling pathway is homologous to pathways transmitting insulin and IGF-1 signals in mammalian cells. The independent cloning of *daf-16* by the Ruvkun and Kenyon labs [10, 11] later showed DAF-16 to be homologous to members of the mammalian FOXO family of forkhead transcription factors, which have important roles in energy metabolism, cell cycle arrest, apoptosis and stress resistance, and which are major targets of the insulin and IGF-1 pathways (reviewed in [12]).

The basic components of the IIS pathway are highly conserved from *C. elegans* to *Drosophila* to mammals. Upon activation through binding of an insulin-like ligand, the *C. elegans* receptor tyrosine kinase DAF-2 (insulin-like receptor) causes the recruitment of AGE-1 (PI3K) to the cell membrane. AGE-1 phosphorylates its phosphatidylinositide (PI) substrates to produce either PI 3,4-bisphosphate or PI 3,4,5-trisphosphate [13]. These membrane phospholipids then act as docking sites to recruit to the membrane the serine/threonine kinases PDK-1 (PI3K-dependent kinase), SGK-1 (serum- and glucocorticoid-inducible kinase 1, homologous to SGK in humans), AKT-1, and AKT-2 (AKT in humans, also known as protein kinase B, PKB). PDK-1 functions to activate, by phosphorylation, both SGK-1 and the AKT proteins [14]. These in turn are thought to phosphorylate the critical IIS target, DAF-16 (FOXO), thereby causing its cytoplasmic, as opposed to nuclear, localization. By this pathway, activated IIS acts to turn off the longevity-promoting transcriptional activity of the DAF-16/FOXO transcription factor.

The receptor tyrosine kinases, including the insulin-like receptors, are unique to animals and first appeared in the sponges [15]. The evolution of these receptors, by allowing more complex signaling networks, may have played a significant role in the subsequent rapid evolution of multicellular animals. Other components of the IIS pathway are even more ancient – PI3K genes are found from yeast to slime molds to plants to humans [16], and forkhead genes are found in yeast (but not in plants). The number of forkhead genes in a species rises with increasing anatomical complexity, from 4 in yeast to over 35 in mice and humans [17].

A key feature of the IIS pathway in *C. elegans* is that multiple endogenous ligands – as many as 39 – signal through a single receptor, DAF-2 [18]. This resembles the pathway in *Drosophila*, in which there are seven insulin-like peptides and one receptor, the insulin-like receptor, InR [19]. In both species, these peptides include both agonists and antagonists of their unique receptor. In contrast to these invertebrate species, the corresponding signaling in mammals involves only three ligands, which function exclusively as agonists, and four distinct receptors plus their heterodimers. While the intracellular components of the IIS pathway are encoded by single genes in *C. elegans* and *Drosophila* (an exception is the duplication of the AKT protein in *C. elegans* [20]), the orthologous proteins in mammals are represented by multiple isoforms. Though the multiplicity of receptors and cellular proteins in the mammalian IIS pathway likely reflects the evolutionary development of more complex metabolic pathways [21], the functional role of the diversity of ligands in *C. elegans* and *Drosophila* is not well understood.

The isolation in 1988 of the first long-lived *C. elegans* IIS mutant by Friedman and Johnson, out of a series of strains previously created by Klass, marked the first known instance of a mutation in a single gene extending the lifespan of an animal [22, 1]. This gene, named *age-1*, was later found to encode the p110 catalytic subunit of PI3K [9]. Although initially reported to extend maximum lifespan by 110%, a null mutation in the age-1 gene has recently been found to result in a nearly 10-fold extension of *C. elegans*' maximum lifespan – the current record for life extension in any animal [23]. These exceptionally long-lived *age-1* mutants are infertile, but are normally active at ages more than eight times the median lifespan of the wild-type controls; in addition, they are more resistant to oxidative and electrophilic (but not thermal) stress.

The *C. elegans* IIS receptor, DAF-2, is named for the fact that certain mutant daf-2 alleles result in a partially penetrant, constitutive developmental arrest in the alternative third larval stage known as dauer. The dauer phenotype is characterized by stress resistance, the absence of feeding associated with closure of the oral orifice and cessation of pharyngeal pumping, a specialized cuticle and a distinct, radially constricted shape, and by a markedly increased lifespan (3–6 months) [6, 7, 24]. In normal animals, dauer development is induced by adverse environmental stimuli, specifically overcrowding, food shortage, and heat; if favorable conditions are restored, wild-type *C. elegans* dauer will develop into reproductively capable adults with normal adult lifespan. In 1993, Kenyon et al., discovered that mutations in daf-2 resulted in dramatic life-extension in adult animals that did not undergo dauer, and that this lifespan increase required the transcription factor daf-16 [25]. This finding demonstrated that the effects of the *daf-2/daf-16* pathway on aging and longevity could be uncoupled from its role in dauer development and growth arrest, and bolstered the view that genetic pathways responding to environmental cues could play a role in the plasticity of aging.

In vitro work has suggested that the DAF-2 effectors SGK-1, AKT-1 and AKT-2 may form a multimeric protein complex which acts as a unit to regulate the localization of DAF-16 [14]. In the same study, Hertweck et al. found that either SGK-1(RNAi) or a loss-of-function mutation in PDK-1 resulted in an increase in *C. elegans* mean and maximum lifespan. In contrast, this study found that while inactivation of both AKT-1 and AKT -2 together led to a modest lifespan increase, inactivating either AKT gene individually produced no life extension, suggesting that SGK-1 may be a critical component in the effect of IIS on longevity [14]. However, a subsequent, large scale RNAi screen for *C. elegans* longevity genes found that AKT-1(RNAi) produced robust life extension [26]. This discrepancy might be related to the fact that the original report utilized an AKT-1 deletion mutant rather than RNAi. Further evidence for AKT-independent effects of DAF-2 signaling on lifespan are provided by the finding that disruption of the AKT-consensus phosphorylation sites on DAF-16 results in nuclear localization of DAF-16 but has little effect on lifespan [27, 6, 7].

In addition to its dependence on DAF-16 activity, the lifespan extension resulting from mutation in *daf-2* requires AAK-2, which is an alpha subunit of the *C. elegans* AMP-activated protein kinase (AMPK) [28]. AMPK functions as an energy sensor by responding to a high AMP:ATP ratio to activate mechanisms involved in the response to low energy availability. AAK-2 was found to be required or partially required for a number of life-extending interventions in addition to daf-2 mutation, including overexpression of the sirtuin sir-2.1, mutation in clk-1, and a high-temperature pulse (which lowers energy levels) [28, 29]. Overexpression of AAK-2 modestly increased *C. elegans* lifespan, by 13% [28]. These results suggest that *aak-2* may function in parallel with *daf-16* to partially mediate the life-extending effects of *daf-2* mutation.

A major focus of current research on the role of the IIS pathway in aging concerns the effects on longevity of insulin-like signaling in specific tissues and specific developmental stages. A large body of evidence indicates that IIS in specific, key tissues can function cell non-autonomously to regulate the aging process of the whole animal. In particular, studies in roundworms, flies and mammals have implicated neuronal tissue and adipose tissue as especially critical tissue types in the endocrine control of aging.

The first indication of a tissue-specific role for the IIS pathway in aging came from a study by Apfeld and Kenyon in 1998 in which mosaic *C. elegans* carrying a mutation in *daf-2* in only a subset of cells were generated [30]. The study found that *daf-2* inactivation in partial sets of tissues only was sufficient for life extension. Subsequently, Wolkow et al. reported that restoring *daf-2* or *age-1* function to neuronal tissue (but not muscle or intestine) in whole-body mutants for *daf-2* or *age-1*, respectively, returned lifespan to that of wild-type animals [31], suggesting that IIS in neuronal tissue is critically important to the regulation of aging. Consistent with a role for an IIS-mediated neuroendocrine signal, mutation in either of two proteins involved in Ca^{2+}-regulated secretion primarily in neurons, a homolog of syntaxin and a CAPS (Ca^{2+}-dependent activator protein for secretion) protein, both extend *C. elegans* lifespan in a DAF-16-dependent manner [32]. In contrast to the findings of Wolkow et al. [31], Libina et al. found that restoring DAF-16 function specifically to the neurons in short-lived *daf-2;daf-16* double mutants resulted in only a modest (5–20%) life extension, compared to a 50–60% extension from restored DAF-16 activity in the intestine, which in *C. elegans* provides a storage function analogous to that of adipose tissue [33].

The importance of signals from the reproductive system in *C. elegans* aging was indicated by the finding that ablation of the germline gonad extends lifespan by 60%, provided the somatic component of the gonad remains [34]. This effect on lifespan is dependent on DAF-16, but the finding that germline ablation further extends the lifespan of long-lived *daf-2* mutants in a synergistic fashion suggests that IIS may act partially in parallel to a gonad-derived signal. Another tissue-specific intervention that can extend *C. elegans* lifespan is the disruption of chemosensory neurons or of sensory signal transduction, and this intervention also partially requires DAF-16 [35]. Lin et al. found that both ablation of the germline gonad and sensory neuron disruption resulted in nuclear localization of DAF-16, although the nuclear localization patterns are different for the two interventions [27].

In addition to the tissue-specific effects of IIS, the developmental timing of DAF-2 effects on lifespan has been investigated through the use of *daf-2* or *daf-16* RNAi initiated at various times in the *C. elegans* life cycle [36]. DAF-2 signaling during larval development was found to regulate fecundity and dauer formation with no effect on adult lifespan. In contrast, reduced IIS initiated in adulthood, including at advanced ages, extended lifespan, indicating that the effects of IIS on lifespan can be uncoupled from its effects on reproduction and development.

The identification of candidate cellular, endocrine and metabolic mechanisms linking various *C. elegans* IIS-related mutations with longevity and defining the role of different organs in this process is somewhat hampered by difficulties in conducting biochemical or metabolic studies in these microscopic organisms, and by limited understanding of nematode physiology. Nevertheless, available information strongly suggests that the neuronal system, the cells homologous to adipocytes, and the gonads are importantly involved in the control of longevity.

Drosophila

Although the existence of life-extending mutations in the fruit fly, *Drosophila melanogaster*, has been known since the earlier studies of Seymour Benzer [37], the evidence for involvement of the IIS pathway in the control of aging in this species is relatively recent. In 2001, Tatar et al. reported extended longevity of flies with a mutation in the *Drosophila* insulin-like receptor (InR) gene [38], and Clancy et al. reported in the same year that flies mutant for the *chico* gene, an ortholog of the mammalian insulin receptor substrate (IRS) genes, live longer than normal flies [39].

Mutation in the InR gene resulted in an 85% increase in the lifespan of females only, compared to a reduction in males of late-life age-specific mortality with no change in maximum lifespan [38]. This sex-specific effect on lifespan is a general finding for the *Drosophila* IIS pathway; most life-extending interventions in IIS benefit female flies to a much greater extent than males. Although a number of InR alleles were analyzed, Tatar et al. found that only a specific heteroallelic combination resulted in life extension. This may be similar to the situation with *C. elegans* DAF-2, for which there are a wide variety of alleles having distinct pleiotropic effects on a number of physiological processes, including differential effects on dauer formation and lifespan [40]. Interestingly, long-lived InR mutants, as well as homozygous chico mutants [39], have a dwarf phenotype with increased content of body fat and reduced or absent fertility, and thus share three prominent phenotypic characteristics with long-lived growth hormone (GH) resistant and GH-deficient mutant mice. As is the case with mice, however, studies of other life-extending interventions in IIS have shown that these phenotypes can be uncoupled from the beneficial effects on lifespan (reviewed in [18]).

The *chico* mutation causes increased lifespan in both heterozygous and homozygous females (up to a 48% increase in median lifespan) [39]. Homozygous male mutants are not long-lived. Although initially reported to have only a minor increase in lifespan, heterozygous males have been found by Tu et al. to live up to 50%

longer than wild type [41]. While homozygous chico mutants are less than half the size of normal animals, heterozygotes of both sexes are long-lived but have normal body size and near-normal fecundity [39]. Clancy et al. also examined the effects of mutation in PKB, the *Drosophila* ortholog of mammalian Akt/PKB, and found that homozygous PKB mutants had a dwarf phenotype but were in fact short-lived relative to controls. This has been suggested to be due to a possible requirement for precise manipulation of IIS in the control of lifespan [42], a possibility that may be supported by the finding in *C. elegans* that *akt-1*(RNAi) results in life extension, whereas a deletion mutation in *akt-1* does not [14, 26].

The studies of Tatar et al. [38] and Clancy et al. [39] were quickly followed by a detailed characterization of the involvement of the IIS pathway in the control of aging and longevity in *Drosophila* (reviewed in [42]). The wealth of information about *Drosophila* genetics, the relative ease of genetic manipulations in this species, and its short lifespan allowed impressive progress in the study of genetic control of aging in this organism. In the leading laboratories of Partridge and Tatar, the emphasis of these studies now includes the identification of the role of IIS genes in specific organs and specific developmental stages in the regulation of aging.

The tissue-specific role of dFOXO, the *Drosophila* ortholog of the mammalian FOXO proteins and *C. elegans* DAF-16, has been investigated using a method which allows inducible expression in specific tissues of the adult fly. Hwangbo et al. [43] found that dFOXO expression specific to the adult cerebral fat body was sufficient to extend median lifespan in both males and females by up to 56%. The *Drosophila* fat bodies serve the functional roles of both liver and adipose tissue, and are divided into the cerebral and peripheral fat bodies. Interestingly, overexpression of dFOXO in the cerebral fat body was found to also result in increased nuclear localization of dFOXO in the peripheral fat body [43], suggesting that diminished IIS in the cerebral fat body can regulate, by an unknown endocrine mechanism, the activity of IIS in the peripheral fat body. In the same study, overexpression of dPTEN specific to the adult cerebral fat body extended lifespan by 20%. dPTEN is homologous to mammalian PTEN (phosphatase and tensin homolog deleted on chromosome ten), which antagonizes IIS by dephosphorylating the phosphatidylinositol products of PI3K. Giannakou et al. [44] have investigated the effects on mortality rate of increasing or decreasing dFOXO levels in the fat body at various stages of adulthood. Switching the status of fat body-specific IIS in early adulthood was found to result in complete conversion to the mortality rate of animals chronically exposed to that level of IIS. However, the extent of these mortality rate changes progressively declined when alterations in dFOXO were made at increasing ages.

In addition to the fat body, neuroendocrine tissues of the brain have been implicated in *Drosophila* aging. Ablation of a cell type which produces three of the seven *Drosophila* insulin-like peptides (DILPs), specifically dilp2, dilp3, and dilp5, resulted in extension of mean and maximum lifespan [45]. These flies had elevated circulating glucose and increased carbohydrate and lipid stores, along with increased oxidative stress resistance. In contrast to the phenotypic characteristics of these insulin-producing, neuron-ablated flies, another tissue-specific life-extending intervention – ablation of the germ cells – results in flies with upregulated neural

DILPs and hypoglycemia [46]. This discrepancy may be due to a compensatory upregulation of DILP production in the germ cell-ablated flies in response to a reduction in downstream elements of the IIS pathway, as indicated by an observed upregulation of dFOXO target genes. In conjunction with the observation of DAF-16-dependent life extension by germ-line ablation in *C. elegans*, these findings suggest an evolutionarily conserved role of the germ-line reproductive tissue in modulating IIS and aging.

Insulin-like peptides have been characterized in many insect species besides *Drosophila* (reviewed in [47]), and while their functions are not well understood, some evidence suggests a possible conserved role in aging. Injection of vertebrate insulin into the butterfly *Pieris brassicae* induced adult development, the termination of diapause, and a shortened lifespan [48], resembling the effects of IIS on development in *C. elegans* and *Drosophila*. Early work identified honey bee royal jelly as the first substance in insects found to display insulin bioactivity [49]. Royal jelly, a secreted substance fed to developing honey bee queens, is known to be a key determinant in specifying a queen's developmental fate. Adult honey bee (*Apis mellifera*) queens, which can live up to 3 years, compared to the 3–6 month lifespan of the worker caste, have been shown to express lower levels of an insulin-like peptide, AmILP-1, and its putative receptors in the head compared to workers [50]. In contrast, queen larvae expressed high levels of AmILP-1 and low AmILP-2 levels compared to worker larvae, in the period of development in which nutritional and hormonal input can affect caste specification [51]. Thus, IIS may play an important role in the regulation of reproductive status and longevity in the honey bee.

Yeast

Although the budding yeast *Saccharomyces cerevisiae* lacks a true insulin-like signaling pathway, yeast aging is significantly influenced by glucose- and nutrient-responsive mechanisms. These pathways may represent evolutionary precursors to the IIS systems in higher organisms. Yeast replicative lifespan (the number of daughter cells produced by a mother cell through mitotic division) and chronological lifespan (the survival time of nondividing cells in stationary phase) can be extended simply by limiting the availability of glucose in the media, a finding consistent with the general efficacy of calorie restriction across many diverse species. Yeast grown on media containing 0.5% glucose had a higher mean and maximum replicative lifespan than yeast grown in 2% glucose; moreover, limiting glucose utilization by inactivating the hexokinase gene *hxk2*, a putative genetic model of caloric restriction, also extended the maximum lifespan of yeast on 2% glucose [52]. While the complex metabolism of glucose involves numerous biochemical pathways, the discovery of several key genes has helped to identify specific pathways involved in mediating the effect of glucose on lifespan.

Yeast mutants lacking the glucose-responsive serine/threonine kinase Sch9, a homolog of the mammalian IIS protein AKT, have an extension of both chronological and replicative lifespan and exhibit an increased resistance to oxidative and thermal stress [53, 54]. High levels of glucose, in addition to activating the Sch9 pathway, also stimulate a pathway in yeast consisting of the monomeric G protein Ras2, Cyrl (adenylate cyclase), and PKA (protein kinase A). *RAS2* is a yeast ortholog of the mammalian proto-oncogene *Ras*, a key target of the IIS pathway in mammals. A null mutation in *RAS2* results in increased chronological lifespan and greater stress resistance [55, 56]. Underexpression of the homologous gene *RAS1* leads to increased replicative lifespan, but overexpression of *RAS2* in fact leads to increased replicative lifespan [57], in contrast to the observed effect of *RAS2* mutation on longevity by the measure of chronological lifespan. *CYR1*, a downstream target of *RAS2*, encodes for adenylate cyclase, and functions to stimulate PKA activity through the second messenger cAMP. Mutation in *CYR1* can extend yeast chronological lifespan up to threefold, and a variety of mutations resulting in decreased PKA activity increase replicative longevity [58, 52]. The *RAS2/PKA* pathway is activated, in response to high glucose levels, by the G protein-coupled receptor GPR1, in conjunction with the heterotrimeric G protein GPA2. Mutation in either *GPR1* or *GPA2* extends yeast replicative lifespan [52]. Downstream of PKA stress resistance found in *RAS2/PKA* or *SHC9* mutants, via a number of transcriptional targets including SOD2 (superoxide dismutase 2) and heat shock proteins [56]. Taken together, the glucose-sensing mechanisms in yeast consisting of *RAS2/PKA* and *SHC9* appear to constitute a pathway that is largely parallel in structure, and often in genetic homology, to the IIS pathways of higher organisms, suggesting a high degree of conservation in the role of these fundamental nutrient-responsive pathways in aging. Figure 1 depicts the essentially conserved nature of this system, from yeast to humans.

Mammals

In contrast to the situation in invertebrates, mammals produce only three ligands from the IGF/insulin "family," (insulin, IGF-1, and IGF-2) but these ligands signal through separate rather than a common receptor. Moreover, the function of these ligands is clearly divergent, particularly during the postnatal period. Thus, IGF-1 primarily controls growth via effects on cell proliferation and apoptotic death, while insulin affects mainly carbohydrate and lipid metabolism. Moreover, the levels of IGF-1 in several organs and in peripheral circulation are controlled by stimulatory input of growth hormone (GH), a secretory product of the anterior pituitary that has no known counterparts in worms or insects, and which exerts anti-insulinemic actions. The picture is further complicated by the still-not-fully-explained paradox that while elimination of GH signaling in mice delays aging and increases lifespan, the levels of GH normally decline with age. This decline is believed to contribute to changes in body composition and various functional deficits that develop during aging. Because of these complexities, the effects of GH, IGF-1, and insulin on aging in mammals will be discussed separately.

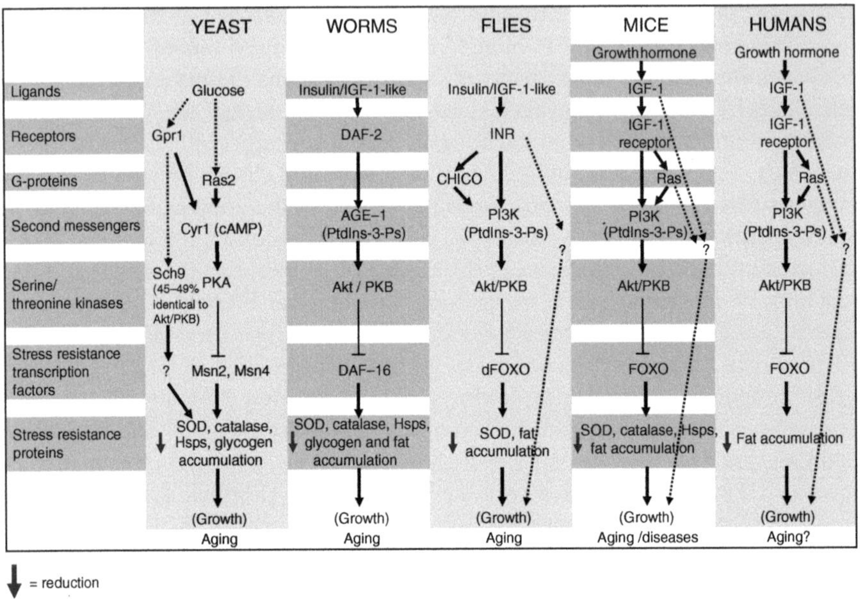

Fig. 1 Homology of signaling pathways involved in the control of longevity in different species. Adapted from Longo and Finch [140]. Reprinted with permission from AAAS

Growth Hormone

The role of GH in the normal (physiological) control of mammalian aging was first suggested on the basis of the observation that Ames dwarf mice with hereditary deficiency of GH, prolactin (PRL) and thyrotropin (TSH) live much longer than their normal siblings [59], while transgenic mice overexpressing GH are short-lived and exhibit various symptoms of accelerated aging [60, 61]; reviewed in [141]. The role of GH in the control of aging was unequivocally demonstrated by documentation of increased lifespan in Laron dwarf mice, which have targeted disruption of the GH receptor gene and consequent GH resistance [62, 63], and in "little" mice with isolated GH deficiency [64]. The evidence linking GH signaling and longevity is particularly strong for the GHRKO mice, in which significant increases in the median, average and, most importantly, maximal lifespan in both females and males have been recorded in several studies conducted in two different labs, using GH receptor knockouts on three different genetic backgrounds and using several different diet formulations [63, 65–67]. Key characteristics of long-lived hypopituitary Ames dwarf and Snell dwarf (Pit1dw) mice, as well as Laron dwarf (Ghr/ghbp—/—; GHRKO) and "little" mice are summarized in Table 1. While no attempts to reverse the long-lived phenotype of these mutants by life-long GH substitution therapy or genetic rescue have been reported, treatment of juvenile Snell dwarf mice with GH for 11 weeks did not alter their lifespan [68]. These findings suggest that actions of GH during the pre-weaning period and/or during adulthood

Table 1 Phenotypic characteristics of mice with mutations affecting longevity and IGF-1/insulin signaling

	Ames and Snell dwarfs	Little	Laron (GHR–/–)
Key characteristic	GH, PRL & TSH deficiency	GH deficiency	GH resistance
Percentage of increase in lifespan	40–60%	25%	38–55%
Body weight	↓↓	↓	↓↓
Relative brain weight	↑	?	↑
Adiposity (% body fat)	↑	↑	↑↑
Plasma GH	↓↓	↓↓	↑↑
Plasma IGF-1	↓↓	↓↓	↓↓
Plasma insulin	↓	↓	↓↓
Plasma glucose	↓	↓	↓ or –
Insulin sensitivity	↑	↑	↑↑

? = Unknown (not reported).
GH = growth hormone, GHR = growth hormone receptor, PRL = prolactin, TSH = thyroid stimulating hormone.

must be important for the determination of lifespan. Indirect support for the importance of GH, and more broadly, somatic growth in the control of longevity in mice is provided by the very consistent and extensively documented negative relationship between adult body weight and lifespan in comparisons of different strains, stocks, lines [69, 70 reviewed in 71] or individual animals within a genetically heterogeneous population [72]. Surprisingly, lifespan was not affected in transgenic mice expressing a GH antagonist [63]. This has been tentatively ascribed to obesity counteracting the effects of reduced GH signaling on insulin sensitivity in these animals.

It is important to emphasize that increased longevity in hypopituitary Snell and Ames dwarf mice and in GH-resistant GHRKO mice is accompanied by various indices of delayed aging. These include delays in the aging of the immune system and in age-related alterations of collagen [64], reduced osteoarthritis [73], delay in the development of fatal neoplastic disease [74], and maintenance of various aspects of cognitive function into advanced chronological age [75–77]. Collectively, these data indicate that increase in the lifespan of these mutants is accompanied by a significant extension of "healthspan," a period of life free of age-related disease and physical functional deterioration.

There is little information on the role of GH in the control of aging and longevity in mammals other than the mouse. Growth hormone-deficient dwarf rats have normal lifespan, but treatment with GH limited to 11 weeks after weaning, designed to produce an animal model of adult GH deficiency, increased their longevity [78]. Transgenic rats heterozygous for expression of missense GH live longer than normal controls, while homozygous carriers of the same gene live shorter, apparently due to early development of leukemia [79]. Although these findings indicate that a role of GH in the determination of lifespan is less pronounced in rats than in mice, adult body weight is negatively correlated with longevity in both species [71].

Negative correlation of size and lifespan applies also to other mammalian species and is particularly strong and well-documented in domestic dogs. Comparisons between different breeds and between cross-bred individuals show striking differences in the life expectancy of small and large dogs, with small dogs living longer [80, 81]. Notably, allelic variation in IGF-1 has been found to be a major determinant of body size in dogs [82], and body weight and plasma IGF-1 levels are significantly correlated across breeds [83].

In the human, the relationship between height and longevity is complicated by effects of diet, health care, socio-economical status and perhaps also ethnicity, and any generalizations are considered controversial. However, Samaras and his colleagues provided numerous examples of short individuals within a population having a significant longevity advantage [84, 85] and it is well documented that shorter individuals are at a reduced risk of developing several common types of cancer [86–88].

It is challenging to explain why the relationship of body size to lifespan is opposite in individuals from the same species when compared to relationships observed in different species. With few exceptions (e.g., bats and some subterranean rodents), large mammals such as whales, elephants, horses or cattle live considerably longer than small mammals such as most rodents, insectivores or small carnivores. The complex relationships between body size, life histories, environment and longevity are discussed in another chapter in this book [89].

There is little data on the longevity of humans with dwarfism of various etiologies. Individuals with GH resistance (Laron dwarfism) or hypopituitarism due to mutations of the Prop1 gene (the same gene which is mutated in Ames dwarf mice) can reach very advanced age [90, 91], but it is unclear how their average or maximal life expectancy compare to life expectancy of normal individuals from the same population. Individuals with isolated GH deficiency in a cohort studied by Besson et al. [92] had significantly reduced life expectancy due primarily to cardiovascular disease. However, GH-deficient individuals in a cohort studied by Menezes Oliveira et al. [93] were protected from atherosclerosis even though they had serum lipid and body composition changes that normally predict increased risk for this disease. Resembling the findings in mice, GH-resistant individuals were recently reported to be protected from cancer [94]. Reduced hepatic expression and circulating levels of IGF-1 provide the most likely explanation for the reduced incidence of tumors and delayed occurrence of fatal neoplastic disease in these animals [74, 68] as well as for their resistance to experimentally induced tumors [95, 96].

Mammalian target of rapamycin (mTOR) signaling, which is importantly involved in the regulation of translation, growth, and responses to nutritional signals, is reduced in long-lived Ames dwarf mice [97]. Homologous signaling was linked to extended longevity in worms and flies [98, 99]. Reduced GH, IGF-1 and mTOR actions undoubtedly account for reduced growth rate and diminutive adult body size of long-lived GH-deficient and GH-resistant mutants. Association of reduced body size with extended longevity in mice and in other species was mentioned earlier and is discussed in more detail elsewhere in this book [89]. In terms of mechanisms, it most likely represents a marker of some underlying process or processes pertinent to determination of lifespan, rather than a primary determinant or a *bona fide* mechanism of delayed aging.

We suspect that enhanced insulin sensitivity with a concomitant reduction in circulating insulin levels represents important mechanisms of extended longevity in GH-deficient and GH-resistant mice, as will be discussed in more detail later in this chapter. Enhanced insulin sensitivity can be traced to the lack of well-documented and extensively studied anti-insulinemic effects of GH in these animals and increased levels of adiponectin [100, 101], likely reflecting a lack of GH action in the adipocytes [102], as well as reduced insulin levels. Circulating insulin levels and capacity of the pancreatic islets to secrete insulin in response to glucose or food ingestion are reduced in GH-deficient and GH-resistant mice, possibly due to the absence or attenuation of GH and IGF-1 signals that normally promote islet development [103–106].

Reduced generation of reactive oxygen species [107], combined with increased activity of antioxidant enzymes [108] in Ames dwarf mice likely contributes to their longevity and extended healthspan and may explain their increased resistance to administration of paraquat [109–111].

A causative role of reduced GH signaling in enhanced stress resistance of these mutants is suggested by the results of our recent studies involving GH replacement therapy. Dermal fibroblasts derived from Ames dwarf, Snell dwarf and GHRKO mice are more resistant to multiple forms of cytotoxic stress than fibroblasts derived from normal siblings of these mutants [112, 113]. In contrast, fibroblasts derived from Ames dwarfs that had been injected with GH failed to exhibit enhanced resistance to most of the tested stressful stimuli (Masternak, Miller and Bartke, unpublished).

Additional support for a role of GH in the control of longevity is provided by the consequences of abnormally elevated levels of GH. Giant transgenic mice over-expressing GH are short-lived and, in addition to GH-related pathological changes, exhibit numerous symptoms of accelerated aging [61, 141]. Many characteristics of these short-lived giant mice are opposite to those of long-lived GH-deficient or GH-resistant dwarfs. This includes hyperinsulinemia, insulin resistance [114], reduced adiponectin levels [115] and increased sensitivity to paraquat (Panici and Bartke, unpublished).

In the human, excessive GH release in individuals affected by gigantism or acromegaly leads to reduced life expectancy due to increased incidence of cardio-vascular disease, diabetes and cancer [116, 117]. It is unclear whether, in addition to an altered risk of age-related disease, the process of biological aging is accelerated or advanced in these syndromes.

Insulin-Like Growth Factor-1

Reports of increased longevity of GHRKO, Ames dwarf, Snell dwarf and "little" mice in which hepatic IGF-1 expression and circulating IGF-1 levels are dramatically reduced (usually to near or below detectability limits) strongly suggested a role of IGF-1 in the control of mammalian longevity. Direct evidence for a role of IGF-1 in this regard was subsequently provided by demonstration of extended longevity in female mice heterozygous for the disruption of the IGF-1 receptor [118]. Partial IGF-1 resistance in these animals had little effect on growth and no effect on the examined parameters of fertility but led to increased resistance to paraquat toxicity and an increase in lifespan [118]. In further support for a role of IGF-1 in the control of aging, female mice expressing a hypomorphic mutation of the IGF-1 gene, which have reduced IGF-1 levels, are also long-lived [119]. Surprisingly, lifespan was not affected in transgenic mice expressing a GH antagonist in spite of reduced levels of IGF-1 [63]. It remains to be determined whether this may have been related to a relatively modest reduction of IGF-1 levels or to extreme obesity of GH antagonist transgenic mice.

It is unclear why reduction of GH signaling in GHRKO, Ames dwarf and Snell dwarf mice, and the resulting suppression of peripheral IGF-1 levels, leads to extended longevity in both sexes while primary defects in IGF-1 signaling increase lifespan only in females. Reduced GH signaling impacts biosynthesis of IGF-1 in the liver, with a smaller or negligible impact on local IGF-1 expression in other organs, while a mutation of the IGF-1 gene or a deletion of the IGF-1 receptor gene presumably affect every cell and organ in the body. However, it is not obvious how these differences could relate to differential responses of male mice to reduced IGF-1 vs. reduced GH signaling. Perhaps the impact of GH on adipose tissue, body composition, and insulin resistance is more relevant to the regulation of longevity in females than in males.

Very intriguing recent results suggest that some aspects of IGF-1 signaling are important for the control of aging and longevity in both sexes. Conover and Bale [120] reported that lifespan was significantly increased in both male and female mice by deletion of pregnancy-associated plasma protein A (PAPP-A). This plasma protein is a protease that cleaves IGFBP4, one of the IGF-1 binding proteins, and is believed to regulate tissue availability of IGF-1 (reviewed in [121]). Presumably in PAPP-A knockout mice, IGFBP levels are increased, thus increasing the proportion of IGF-1 bound to IGFBP, and the amount of free, bioactive IGF-1 at the tissue level is reduced.

Potential mechanisms linking reduced IGF-1 signaling with extended longevity overlap those suspected of mediating the "anti-aging" effects of reduced GH action (Fig. 1) with an important exception of enhanced insulin sensitivity. In contrast to the anti-insulinemic actions of GH, IGF-1 mimics some of the effects of insulin and enhances responsiveness to insulin in some of its target organs. The well-documented effects of IGF-1 on proliferation and survival of cancer cells suggest that reduced IGF-1 signaling may offer significant protection from neoplastic disease, a leading cause of death in most mouse strains and an important influence on longevity in other species, including the human.

Extended longevity of animals with severely reduced somatotropic (GH-IGF-1) signaling brings up a broader issue of genetic control of aging and lifespan. At first glance, it is paradoxical that any benefits could be derived from an absence of hormones which have well-documented, major roles in normal growth and maturation, and in the control of metabolism and body composition. A very plausible explanation of this paradox is provided by the concept of antagonistic pleiotropy, which posits that genes with detrimental effects during adult and particularly post-reproductive life could have been selected "for" rather than "against" if they confer benefits on early reproductive fitness. Physiological actions of genes related to somatotropic signaling seem to conform to this concept, since GH and IGF-1 promote early growth, sexual maturation and fertility (including large litter size in mice) early in life but increase cancer risk and exert various "pro-aging" effects (importantly including induction of insulin resistance by GH) later in life. Under natural conditions, early puberty, large litter size, and large body size (especially in males) would have likely been strongly selected for as characteristics promoting reproductive fitness. In contrast, negative effects of these genes on lifespan would have been subjected to little if any selective pressure, because they are manifested primarily during the post-reproductive period and at ages that are unlikely to be attained under conditions of predation and various environmental challenges.

Insulin

There are many reasons to suspect that insulin has an important role in the control of mammalian aging. Hyperinsulinemia and insulin resistance are key elements of the metabolic syndrome and have been associated with increased risk of age-associated disease, including cardiovascular problems, type 2 diabetes, and cancer. As discussed earlier in the chapter, improved insulin sensitivity, combined with reduced insulin secretion, is suspected to contribute to the increased longevity of GH-deficient and GH-resistant mutant mice.

However, very few studies have specifically addressed the role of insulin signaling in the control of mammalian longevity. Deletion of insulin or insulin receptors in all organs in mammals is not compatible with survival, and the consequences of partial or organ-specific reduction of insulin signaling on aging and longevity remain to be explored, except for very intriguing studies conducted by Blüher et al. [122]. These authors reported increased lifespan of FIRKO mice, which have a targeted disruption of the insulin receptor gene in the adipocytes. The authors suggested that this increase may have been due to the reduced adiposity of these animals [122], but this interpretation was challenged [123]. The suspected insulin resistance of macrophages in FIRKO mice [124], along with altered levels of adipocyte-derived factors that control insulin signaling in other organs and that exert pro- or anti-inflammatory effects, may have contributed to the extended longevity of these animals.

Regardless of the mechanisms involved, the results obtained in FIRKO mice provide direct evidence for the importance of insulin signaling in the control of mammalian longevity. These results also suggest that the impact of altered insulin action on lifespan can be dissociated from alterations in growth, adult body size, and other characteristics of hypoinsulinemic GH-deficient and GH-resistant mutants discussed earlier in this chapter. Particularly striking is the contrast between the increased adiposity of GHRKO mice [125] and reduced adiposity of FIRKO animals. Subsequent studies by the same group revealed major alterations in age-related profiles of the expression of nuclear-encoded mitochondrial genes in FIRKO, as compared to normal mice [126]. The authors suggested that maintenance of mitochondrial activity in adipose tissue during aging and increased oxygen consumption may lead to the increased longevity of these animals. They also reported that FIRKO mice remain insulin-sensitive as they age [126].

Interestingly, one study found that the longest living rodent (at ~28 years), the naked mole rat, has an extreme degree of insulin sensitivity and a level of insulin which is undetectable by ELISA, combined with glucose intolerance [127].

Insulin Receptor Substrates (IRS) 1 and 2

As discussed earlier in this chapter, the role of genes downstream from the IIS receptors in the control of longevity was characterized in considerable detail in *C. elegans* and *Drosophila*. In particular, the role of the homologs of mammalian IRSs, protein kinase B (Akt), and the FOXO family of forkhead transcription factors in the control of aging in these organisms has been conclusively established. In contrast, until recently there was little direct evidence for the involvement of genes in the IRS-PI3K-Akt-FOXO signaling pathway in the control of mammalian aging, apart from the documented or suspected role of Akt2 and FOXO in mediating the beneficial effects of calorie restriction [128].

Against this background, two recent studies of the effects of deleting IRS1 or IRS2 on longevity in mice are of particular interest. Selman and his colleagues [129] reported that female mice homozygous for the deletion of IRS1 live longer than normal animals. $IRS^{-/-}$ animals are small and, in contrast to long-lived GH-related mutants, do not exhibit enhanced insulin sensitivity except for being protected from age-related insulin resistance that developed in control mice between the ages of 450 and 700 days. In the same study, mice heterozygous for either IRS1 or IRS2 deletion had normal lifespans.

In contrast to these observations, Taguchi et al. [130] reported significantly increased lifespan in $IRS2^{+/-}$ mice. These animals had essentially normal body weight and food intake, and increased insulin sensitivity as indicated by improved glucose and insulin tolerance, as well as reduced levels of both insulin and glucose [130].

One possible reason for the different effects of heterozygous IRS2 deletion on longevity in different laboratories is the use of a "high energy" (9% fat) diet in the study of Taguchi et al. [130] and a standard (5% fat) diet in the study of Selman et al. [129]. Increased fat content of the diet promotes insulin resistance, and this may alter the consequences of experimentally produced alterations in insulin and/or IGF signaling.

Taguchi et al. also reported that selective homozygous or heterozygous deletion of IRS2 in the brain increases lifespan of mice. Interestingly, brain (b) $IRS2^{-/-}$ and $^{+/-}$ mice were insulin resistant. The authors suggested that these animals' brains may have been shielded from the deleterious actions of insulin and obesity.

Additional studies will be required to elucidate the mechanisms linking deletion of IRSs with longevity, to define the role of IRS1 and IRS2 expression in different organs in the control of aging, to relate these findings specifically to IGF-1 signaling or insulin signaling – or both – and to reconcile differences between results obtained in different laboratories. However, the evidence available to date allows several important, if somewhat tentative conclusions. These results indicate that IRSs, and thus presumably the IRS-PI3K:Akt-FOXO signaling pathway, is involved in the control of longevity in mammals, as it is in invertebrates. Moreover, disruption of these signaling events only in the central nervous system can increase lifespan, again echoing some findings in *C. elegans* and *Drosophila*.

Interactions of Nutrients and Nutritional Status with IIS Signaling

In general, nutrient availability and feeding promote growth and reproduction while nutrient shortage activates various survival mechanisms. The role of IIS in mediating the effects of nutritional status on aging and longevity is of considerable interest, and elucidating these relationships may hold promise for the development of effective anti-aging interventions.

Feeding activates IIS pathways by stimulating the biosynthesis and release of endogenous ligands of IIS receptors. In both *C. elegans* and *Drosophila*, sensory perception of food-derived signals activates release of insulin-like peptides from specialized neurons which produce them. In mammals, post-prandial increase in circulating nutrient levels stimulates insulin release, and circulating IGF-1 levels are positively related to nutritional status, although food deprivation produces divergent responses on GH release in different species, including suppression in rodents and stimulation in primates.

Chronic reduction in food intake or calorie restriction (CR, also called dietary restriction, DR), reliably slows aging and increases lifespan in virtually every species examined to date, as long as malnutrition or starvation are avoided. However, the role of IIS in mediating the effects of CR appears to be quite different in different taxonomic groups.

In mammals, CR reduces circulating insulin and IGF-1 levels. Reduced insulin level is among the most consistent effects of CR in organisms ranging from mice to men and, together with enhanced insulin sensitivity, is believed to contribute to or account for many beneficial actions of CR. Reduction in IGF-1 levels is related to CR-induced changes in GH release and in hepatic sensitivity to GH stimulation. The effects of CR on GH release are complex and species-specific. In the rat, similarly to other species, CR suppresses GH release but also opposes the age-related decline in GH pulsatility [131]. In the recent studies of CR in overweight humans, secretory episodes of GH monitored by frequent blood sampling were enhanced rather than suppressed by CR [132].

Studies of the effects of CR in long-lived hypopituitary and GH-resistant mutant mice support the role of altered insulin signaling in mediating the effects of CR on aging. In GH-deficient Ames dwarf mice, CR produced further improvements in both insulin sensitivity and longevity, thus resembling its effects in normal animals [109, 110]. In contrast, in the GH resistant Ghr—/— mice, an identical regimen of CR failed to enhance their extreme insulin sensitivity, and had no effects on the longevity in males and only minor effects on the longevity of females [67].

Insulin, IGF-1 and Human Aging

Although the major role of IIS signaling in the control of aging is conserved from worms to mammals, the role of insulin and IGF-1 signaling in the control of human aging is poorly understood. Severe deficiencies of insulin secretion in type 1 and type 2 diabetes are serious disease conditions and are associated with reduced life expectancy. In contrast, enhanced insulin sensitivity, modest reduction of insulin levels and prevention of dietary-induced or aging-associated hyperinsulinemia are important lifestyle intervention goals aimed at preventing age-associated disease. The effects of IGF-1 on aging are difficult to separate from the role of GH, the key determinant of IGF-1 levels and peripheral circulation. Potentially relevant to the control of aging, GH exerts anti-insulinemic effects and induces insulin resistance, while IGF-1 mimics some of the effects of insulin and enhances insulin signaling.

The relationship of GH to human aging and particularly its suggested utility as an anti-aging therapy are hotly debated and highly controversial [133, 134]. Reduced IGF-1 levels due to congenital GH deficiency or resistance have been associated with both enhancement [92] and protection from aging-associated disease [93, 135, 94]. Studies in centenarians suggest that genetic polymorphisms leading to reduced IIS signaling [136, 137, 100], increased adiponectin levels, GH resistance (Atzmon et al., unpublished), or IGF-1 resistance [138] play a role in the attainment of extreme longevity.

Conclusion

Insulin and insulin-like signaling have emerged as fundamental regulatory systems connecting development, metabolism, and aging. The dramatic degree of life extension achievable by alteration of IIS, and the highly conserved role of the pathway in aging across species, have contributed greatly to the current understanding of aging as an intelligible process amenable to modification. This remarkable plasticity of aging in response to changes in a single genetic pathway has far-reaching implications for the relationship between genes and the aging process, which is nevertheless not thought to be genetically programmed. Most importantly, the proliferation of IIS interventions in several animals has resulted in extended life and decreased susceptibility to pathology, offering a clear cause for optimism at the prospects for pharmacological or other measures available to improve human health and increase lifespan. This is particularly the case given the numerous examples in several species in which IIS alteration can act during adulthood to affect lifespan and can be uncoupled from deleterious reproductive and other side effects.

While the IIS pathway is the best-understood case of genetic effects on lifespan, enormous challenges remain in identifying the direct causal role of IIS in aging. Recent work on the tissue specificity of IIS has implicated adipose, neural, and germ-line tissues as being of critical importance in multiple species. Characterizing the effects of these tissues on longevity will require elucidation of the specific endocrine signals originating from them, as well as the interactions between them. Particularly challenging will be the identification of the causally important downstream effectors of FOXO signaling and IIS generally (reviewed in [18]). This task is complicated by the finding, based on cross-species gene expression analysis, that specific IIS target genes are widely divergent across species, while significant conservation of gene categories exists at the process level [139]. Also of key importance are the complex interactions of IIS with other pathways known to be involved in aging, such as the TOR or sirtuin pathways (reviewed in [124, 8, 42]) and with caloric restriction.

Despite these complexities, progress in understanding the role of IIS in aging, and in our capacity to harness that knowledge to achieve substantial lifespan extension in complex animals, has been remarkably rapid. The biotechnological tools now available to biogerontology and the increasingly evident demonstration of conservation of IIS involvement in aging hold the promise of accelerating that progress toward a practical clinical intervention in aging.

References

1. Friedman DB and Johnson TE (1988). A mutation in the age-1 gene in *Caenorhabditis elegans* lengthens life and reduces hermaphrodite fertility. *Genetics* Jan 1; 118(1): 75–86.
2. Jazwinski SM (1996). Longevity, genes, and aging. *Science* 273(5271): 54–59.
3. Guarente L (2000). Sir2 links chromatin silencing, metabolism, and aging. *Genes Dev* May 1; 14(9): 1021–1026.

4. Kimura KD, Tissenbaum HA, Liu Y, and Ruvkun G (1997). daf-2, an Insulin Receptor-like gene that regulates longevity and diapause in *Caenorhabditis elegans*. *Science* Aug 15; 277(5328): 942–946.
5. Braeckman BP and Vanfleteren JR (2007). Genetic control of longevity in *C. elegans*. *Exp Gerontol* 42(1–2): 90–98.
6. Hu PJ, Dauer (August 08, 2007), WormBook, ed. The *C. elegans* Research Community, WormBook, doi/10.1895/wormbook.1.144.1, http://www.wormbook.org.
7. Hu PJ Dauer. WormBook [online review]. 2007;8: 1–19. Available from: http://www.wormbook.org/chapters/www_dauer/dauer.html.
8. Baumeister R, Schaffitzel E, and Hertweck M (2006). Endocrine signaling in *Caenorhabditis elegans* controls stress response and longevity. *J Endocrinol* Aug 1; 190(2): 191–202.
9. Morris JZ, Tissenbaum HA, and Ruvkun GA (1996). phosphatidylinositol-3-OH kinase family member regulating longevity and diapause in *Caenorhabditis elegans*. *Nature* 382(6591): 536–539.
10. Ogg S, Paradis S, Gottlieb S, Patterson GI, Lee L, Tissenbaum HA et al. (1997). The Fork head transcription factor DAF-16 transduces insulin-like metabolic and longevity signals in *C. elegans*. *Nature* 389(6654): 994–999.
11. Lin K, Dorman JB, Rodan A, and Kenyon C (1997). daf-16: An HNF-3/forkhead family member that can function to double the life-span of *Caenorhabditis elegans*. *Science* Nov 14; 278(5341): 1319–1322.
12. Salih DAM and Brunet A (2008). FoxO transcription factors in the maintenance of cellular homeostasis during aging. *Curr Opin Cell Biol* 20(2): 126–136.
13. Gami MS and Wolkow CA (2006). Studies of *Caenorhabditis elegans* DAF-2/insulin signaling reveal targets for pharmacological manipulation of lifespan. *Aging Cell* 5(1): 31–37.
14. Hertweck M, Göbel C, and Baumeister RC (2004). elegans SGK-1 is the critical component in the Akt/PKB kinase complex to control stress response and life span. *Dev Cell* 6(4): 577–588.
15. Skorokhod A, Gamulin V, Gundacker D, Kavsan V, Muller IM, and Muller WEG (1999). Origin of insulin receptor-like tyrosine kinases in marine sponges. *Biol Bull* Oct 1; 197(2): 198–206.
16. Leevers SJ, Weinkov D, MacDougall LK, Hafen E, and Waterfield MD (1996). *Drosophila* phosphoinositide 2-kinase Dp110 promotes cell growth. *EMBO J* Dec 2; 15(23): 6584–6594.
17. Wijchers PJEC, Burbach JPH, and Smidt MP (2006). In control of biology: Of mice, men and foxes. *Biochem J* Jul 15; 397(2): 233–246.
18. Piper MD, Selman C, McElwee JJ, and Partridge L (2008). Separating cause from effect: How does insulin/IGF signalling control lifespan in worms, flies and mice? *J Intern Med* 263(2): 179–191.
19. Brogiolo W, Stocker H, Ikeya T, Rintelen F, Fernandez R, and Hafen E (2001). An evolutionarily conserved function of the *Drosophila* insulin receptor and insulin-like peptides in growth control. *Curr Biol* 11(4): 213–221.
20. Paradis S and Ruvkun G (1998). *Caenorhabditis elegans* Akt/PKB transduces insulin receptor-like signals from AGE-1 PI3 kinase to the DAF-16 transcription factor. *Genes Dev* Aug 15; 12(16): 2488–2498.
21. Rincon M, Muzumdar R, Atzmon G, and Barzilai N (2004). The paradox of the insulin/IGF-1 signaling pathway in longevity. *Mech Ageing Dev* 125(6): 397–403.
22. Klass MR (1983). A method for the isolation of longevity mutants in the nematode *Caenorhabditis elegans* and initial results. *Mech Ageing Dev* Jul–Aug; 22(3–4): 279–286.
23. Ayyadevara S, Alla R, Thaden JJ, and Shmookler RRJ (2008). Remarkable longevity and stress resistance of nematode PI3K-null mutants. *Aging Cell* 7(1): 13–22.
24. Rottiers V and Antebi A (2006). Control of *Caenorhabditis elegans* life history by nuclear receptor signal transduction. *Exp Gerontol* Oct; 41(10): 904–909.

25. Kenyon C, Chang J, Gensch E, Rudner A, and Tabtiang RAC (1993). Elegans mutant that lives twice as long as wild type. *Nature* 366(6454): 461–464.
26. Hamilton B, Dong Y, Shindo M, Liu W, Odell I, Ruvkun G et al. (2005). A systematic RNAi screen for longevity genes in *C. elegans*. *Genes Dev* Jul 1; 19(13): 1544–1555.
27. Lin K, Hsin H, Libina N, and Kenyon C (2001). Regulation of the *Caenorhabditis elegans* longevity protein DAF-16 by insulin/IFG-1 and germline signaling. *Nat Genet* 28(2): 139–145.
28. Apfeld J, O'Connor G, McDonagh T, DiStefano PS, and Curtis R (2004). The AMP-activated protein kinase AAK-2 links energy levels and insulin-like signals to lifespan in *C. elegans*. *Genes Dev* Dec 15; 18(24): 3004–3009.
29. Curtis R, O'Connor G, and DiStefano PS (2006). Aging networks in *Caenorhabditis elegans*: AMP-activated protein kinase (*aak-2*) links multiple aging and metabolism pathways. *Aging Cell* 5(2): 119–126.
30. Apfeld J and Kenyon C (1998). Cell nonautonomy of *C. elegans* daf-2 function in the regulation of diapause and life span. *Cell* 95(2): 199–210.
31. Wolkow CA, Kimura KD, Lee M-S, and Ruvkun G (2000). Regulation of *C. elegans* lifespan by insulinlike signaling in the nervous system. *Science* Oct 6; 290(5489): 147–150.
32. Ailion M, Inoue T, Weaver CI, Holdcraft RW, and Thomas JH (1999). Neurosecretory control of aging in *Caenorhabditis elegans*. *Proc Natl Acad Sci USA* Jun 22; 96(13): 7394–7397.
33. Libina N, Berman JR, and Kenyon C (2003). Tissue-specific activities of *C. elegans* DAF-16 in the regulation of lifespan. *Cell* 115(4): 489–502.
34. Hsin H and Kenyon C (1999). Signals from the reproductive system regulate the lifespan of *C. elegans*. *Nature* 399(6734): 362–366.
35. Apfeld J and Kenyon C (1999). Regulation of lifespan by sensory perception in *Caenorhabditis elegans*. *Nature* Dec 16; 402(6763): 804–809.
36. Dillin A, Crawford DK, and Kenyon C (2002). Timing requirements for insulin/IGF-1 signaling in *C. elegans*. *Science* Oct 25; 298(5594): 830–834.
37. Lin Y-J, Seroude L, and Benzer S (1998). Extended life-span and stress resistance in the *Drosophila* mutant Methuselah. *Science* Oct 30; 282(5390): 943–946.
38. Tatar M, Kopelman A, Epstein D, Tu MP, Yin CM, and Garofalo RS (2001). A mutant *Drosophila* insulin receptor homolog that extends life-span and impairs neuroendocrine function. *Science* Apr 6; 292(5514): 107–110.
39. Clancy D, Gems D, Harshman LG, Oldham S, Hafen E, Leevers SJ et al. (2001). Extension of lifespan by loss of chico, a *Drosophila* insulin receptor substrate protein. *Science* 292: 104–106.
40. Gems D, Sutton AJ, Sundermeyer ML, Albert PS, King KV, Edgley ML et al. (1998). Two pleiotropic classes of daf-2 mutation affect larval arrest, adult behavior, reproduction and longevity in *Caenorhabditis elegans*. *Genetics* Sep 1; 150(1): 129–155.
41. Tu M-P, Epstein D, and Tatar M (2002). The demography of slow aging in male and female *Drosophila* mutant for the insulin-receptor substrate homologue *chico*. *Aging Cell* 1(1): 75–80.
42. Giannakou ME and Partridge L (2007). Role of insulin-like signalling in *Drosophila* lifespan. *Trends Biochem Sci* 32(4): 180–188.
43. Hwangbo DS, Gersham B, Tu M-P, Palmer M, and Tatar M (2004). *Drosophila* dFOXO controls lifespan and regulates insulin signalling in brain and fat body. *Nature* 429(6991): 562–566.
44. Giannakou ME, Goss M, Jacobson J, Vinti G, Leevers SJ, and Partridge L (2007). Dynamics of the action of *dFOXO* on adult mortality in *Drosophila*. *Aging Cell* 6(4): 429–438.
45. Broughton SJ, Piper MDW, Ikeya T, Bass TM, Jacobson J, Driege Y et al. (2005). Longer lifespan, altered metabolism, and stress resistance in *Drosophila* from ablation of cells making insulin-like ligands. *Proc Natl Acad Sci U S A* Feb 22; 102(8): 3105–3110.

46. Flatt T, Min K-J, D'Alterio C, Villa-Cuesta E, Cumbers J, Lehmann R et al. (2008). *Drosophila* germ-line modulation of insulin signaling and lifespan. *Proc Natl Acad Sci U S A* Apr 29; 105(17): 6368–6373.
47. Wu Q and Brown MR (2006). Signaling and function of insulin-like peptides in insects. *Ann Rev Entomol* 51(1): 1–24.
48. Arpagaus M (1987). Vertebrate insulin induces diapause termination in Pieris brassicae pupae. *Roux's Arch Dev Biol* 196: 527–530.
49. Dixit PK and Patel NG (1964). Insulin-like activity in larval foods of the honeybee. *Nature* 202: 189–190.
50. Corona M, Velarde RA, Remolina S, Moran-Lauter A, Wang Y, Hughes KA et al. (2007). Vitellogenin, juvenile hormone, insulin signaling, and queen honey bee longevity. *Proc Natl Acad Sci U S A* Apr 24; 104(17): 7128–7133.
51. Wheeler DE, Buck N, and Evans JD (2006). Expression of insulin pathway genes during the period of caste determination in the honey bee, *Apis mellifera. Insect Mol Biol* 15(5): 597–602.
52. Lin SJ, Defossez PA, and Guarente L (2000). Requirement of NAD and SIR2 for life-span extension by calorie restriction in *Saccharomyces cerevisiae. Science* Sep 22; 289(5487): 2126–2128.
53. Fabrizio P, Pozza F, Pletcher SD, Gendron CM, and Longo VD (2001). Regulation of longevity and stress resistance by Sch9 in yeast. *Science* Apr 13; 292(5515): 288–290.
54. Kaeberlein M, Kirkland KT, Fields S, and Kennedy BK (2005). Genes determining yeast replicative life span in a long-lived genetic background. *Mech Ageing Dev* 126(4): 491–504, Apr.Epub 2005Jan 7.
55. Longo VD (1999). Mutations in signal transduction proteins increase stress resistance and longevity in yeast, nematodes, fruit flies, and mammalian neuronal cells. *Neurobiol Aging* Sep–Oct; 20(5): 479–486.
56. Fabrizio P, Liou LL, Moy VN, Diaspro A, Valentine JS, Gralla EB, and Longo VD (2003). SOD2 functions downstream of Sch9 to extend longevity in yeast. *Genetics* Jan; 163(1): 35–46.
57. Sun J, Kale SP, Childress AM, Pinswasdi C, and Jazwinski SM (1994). Divergent roles of RAS1 and RAS2 in yeast longevity. *J Biol Chem* Jul 15; 269(28): 18638–18645.
58. Thevelein JM and de Winde JH (1999). Novel sensing mechanisms and targets for the cAMP-protein kinase A pathway in the yeast *Saccharomyces cerevisiae. Mol Microbiol* Sep; 33(5): 904–918.
59. Brown-Borg HM, Borg KE, Meliska CJ, and Bartke A (1996). Dwarf mice and the ageing process. *Nature* 384: 33.
60. Quaife CJ, Mathews LS, Pinkert CA, Hammer RE, Brinster RL, and Palmiter RD (1989). Histopathology associated with elevated levels of growth hormone and insulin-like growth factor I in transgenic mice. *Endocrinology* 124: 40–48.
61. Wolf E, Kahnt E, Ehrlein J, Hermanns W, Brem G, and Wanke R (1993). Effects of long-term elevated serum levels of growth hormone on life expectancy of mice: Lessons from transgenic animal models. *Mech Ageing Dev* 68: 71–87.
62. Coschigano KT, Clemmons D, Bellush LL, and Kopchick JJ (2000). Assessment of growth parameters and life span of GHR/BP gene-disrupted mice. *Endocrinology* 141: 2608–2623.
63. Coschigano KT, Holland AN, Riders ME, List EO, Flyvbjerg A, and Kopchick JJ (2003). Deletion, but not antagonism, of the mouse growth hormone receptor results in severely decreased body weights, insulin, and insulin-like growth factor I levels and increased life span. *Endocrinology* 144: 3799–3810.
64. Flurkey K, Papaconstantinou J, Miller RA, and Harrison DE (2001). Lifespan extension and delayed immune and collagen aging in mutant mice with defects in growth hormone production. *Proc Natl Acad Sci U S A* 98: 6736–6741.
65. Bartke A, Chandrashekar V, Bailey B, Zaczek D, and Turyn D (2002). Consequences of growth hormone (GH) overexpression and GH resistance. *Neuropeptides* 36: 201–208.

66. Bartke A, Peluso MR, Moretz N, Wright C, Bonkowski M, Winters TA et al. (2004). Effects of soy-derived diets on plasma and liver lipids, glucose tolerance, and longevity in normal, long-lived and short-lived mice. *Horm Metab Res* 36: 550–558.
67. Bonkowski MS, Rocha JS, Masternak MM, Al-Regaiey KA, and Bartke A (2006). Targeted disruption of growth hormone receptor interferes with the beneficial actions of calorie restriction. *Proc Natl Acad Sci U S A* 103: 7901–7905.
68. Vergara M, Smith-Wheelock M, Harper JM, Sigler R, and Miller RA (2004). Hormone-treated Snell dwarf mice regain fertility but remain long lived and disease resistant. *J Gerontol A Biol Sci Med Sci* 59: 1244–1250.
69. Roberts RC (1961). The lifetime growth and reproduction of selected strains of mice. *Heredity* 16: 369–381.
70. Eklund J and Bradford CE (1977). Longevity and lifetime body weight in mice selected for rapid growth. *Nature* 265: 48–49.
71. Rollo CD (2002). Growth negatively impacts the life span of mammals. *Evol Dev* 4: 55–61.
72. Miller RA, Harper JM, Galecki A, and Burke DT (2002). Big mice die young: Early life body weight predicts longevity in genetically heterogeneous mice. *Aging Cell* 1: 22–29.
73. Silberberg R (1972). Articular aging and osteoarthrosis in dwarf mice. *Pathol Microbiol (Basel)* 38: 417–430.
74. Ikeno Y, Bronson RT, Hubbard GB, Lee S, and Bartke A (2003). Delayed occurrence of fatal neoplastic diseases in Ames dwarf mice: Correlation to extended longevity. *J Gerontol A Biol Sci Med Sci* 58A: 291–296.
75. Kinney BA, Coschigano KT, Kopchick JJ, and Bartke A (2001a). Evidence that age-induced decline in memory retention is delayed in growth hormone resistant GH-R-KO (Laron) mice. *Physiol Behav* 72: 653–660.
76. Kinney BA, Meliska CJ, Steger RW, and Bartke A (2001b). Evidence that Ames dwarf mice age differently from their normal siblings in behavioral and learning and memory parameters. *Horm Behav* 39: 277–284.
77. Kinney-Forshee B, Kinney N, Steger R, and Bartke A (2004). Could a deficiency in growth hormone signaling be beneficial to the aging brain? *Physiol Behav* 80: 589–594.
78. Sonntag WE, Carter CS, Ikeno Y, Ekenstedt K, Carlson CS, Loeser RF et al. (2005). Adult-onset growth hormone and insulin-like growth factor I deficiency reduces neoplastic disease, modifies age-related pathology, and increases life span. *Endocrinology* 146: 2920–2932.
79. Shimokawa I, Higami Y, Utsuyama M, Tuchiya T, Komatsu T, Chiba T, and Yamaza H (2002). Lifespan extension by reduction of the growth hormone-insulin-like growth factor-1 axis: Relation to caloric restriction. *Am J Pathol* 160: 2259–2265.
80. Patronek GJ, Waters DJ, and Glickman LT (1997). Comparative longevity of pet dogs and humans: Implications for gerontology research. *J Gerontol A Biol Sci Med Sci* 52A: B171–B178.
81. Greer KA, Canterberry SC, and Murphy KE (2007). Statistical analysis regarding the effects of height and weight on life span of the domestic dog. *Res Vet Sci* 82: 208–214.
82. Sutter NB, Bustamante CD, Chase K, Gray MM, Zhao K, Zhu L et al. (2007). A single IGF1 allele is a major determinant of small size in dogs. *Science* Apr 6; 316(5821): 112–115.
83. Eigenmann JE, Amador A, and Patterson DF (1988). Insulin-like growth factor I levels in proportionate dogs, chondrodystrophic dogs and in giant dogs. *Acta Endocrinol (Copenh)* May; 118(1): 105–108.
84. Samaras TT (2007). *Human Body Size and the Laws of Scaling: Physiological, Performance, Growth, Longevity and Ecological Ramifications*, 1st ed. New York: Nova Science Publishers, Inc.
85. Samaras TT (2009). Should we be concerned over increasing body height and weight? *Exp Gerontol* Jan–Feb; 44(1–2): 83–92.
86. Tretli S (1989). Height and weight in relation to breast cancer morbidity and mortality. A prospective study of 570,000 women in Norway. *Int J Cancer* 44: 23–30.

87. Gunnell D, Okasha M, Davey Smith G, Oliver SE, Sandhu J, and Holly JMP (2001). Height, leg length, and cancer risk: A systematic review. *Epidemiol Rev* 23: 313–342.
88. Littman AJ, White E, and Kristal AR (2007). Anthropometrics and prostate cancer risk. *Am J Epidemiol* 165: 1271–1279.
89. S. Austad (2010) Animal size, metabolic rate, and survival, among and within species. In *Comparative Biology of Aging* N.S. Wolf (Ed.) Springer: Dordrecht.
90. Krzisnik C, Kolacio Z, Battelino T, Brown M, Parks JS, and Laron Z (1999). The "Little People" of the island of Krk – revisited. Etiology of hypopituitarism revealed. *J Endo Gen* 1: 9–19.
91. Laron Z (2005). Do deficiencies in growth hormone and insulin-like growth factor-1 (IGF-1) shorten or prolong longevity? *Mech Ageing Dev* 126: 305–307.
92. Besson A, Salemi S, Gallati S, Jenal A, Horn R, Mullis PS, and Mullis PE (2003). Reduced longevity in untreated patients with isolated growth hormone deficiency. *J Clin Endocrinol Metab* 88: 3664–3667.
93. Menezes Oliveira JL, Marques-Santos C, Barreto-Filho JA, Ximenes Filho R, de Oliveira Britto AV, Oliveira Souza AH et al. (2006). Lack of evidence of premature atherosclerosis in untreated severe isolated growth hormone (GH) deficiency due to a GH-releasing hormone receptor mutation. *J Clin Endocrinol Metab* 91: 2093–2099.
94. Shevah O and Laron Z (2007). Patients with congenital deficiency of IGF-I seem protected from the development of malignancies: A preliminary report. *Growth Horm IGF Res* 17: 54–57.
95. Bielschowsky F and Bielschowsky M (1961). Carcinogenesis in the pituitary of dwarf mouse. The response to dimethylbenzanthracene applied to the skin. *Br J Cancer* 15: 257–262.
96. Wang Z, Prins GS, Coschigano KT, Kopchick JJ, Green JE, Ray VH et al. (2005). Disruption of growth hormone signaling retards early stages of prostate carcinogenesis in the C3(1)/T antigen mouse. *Endocrinology* 146: 5188–5196.
97. Sharp ZD and Bartke A (2005). Evidence for down-regulation of phosphoinositide 3-kinase/Akt/mammalian target of rapamycin (PI3K/Akt/mTOR)-dependent translation regulatory signaling pathways in Ames dwarf mice. *J Gerontol A Biol Sci Med Sci* 60: 293–300.
98. Kapahi P, Zid BM, Harper T, Koslover D, Sapin V, and Benzer S (2004). Regulation of lifespan in *Drosophila* by modulation of genes in the TOR signaling pathway. *Curr Biol* May 25; 14(10): 885–890.
99. Vellai T, Takacs-Vellai K, Zhang Y, Kovacs AL, Orosz L, and Muller F (2003). Genetics: Influence of TOR kinase on lifespan in *C. elegans*. *Nature* 426(6967): 620.
100. Al-Regaiey KA, Masternak MM, Bonkowski M, Sun L, and Bartke A (2005). Long-lived growth hormone receptor knockout mice: Interaction of reduced insulin-like growth factor 1/insulin signaling and caloric restriction. *Endocrinology* 146: 851–860.
101. Wang Z, Masternak MM, Al-Regaiey KA, and Bartke A (2006). The regulation of fat metabolism in Ames dwarf mice, In 59th Annual Scientific Meeting, Dallas, TX: The Gerontological Society of America.
102. Nilsson L, Binart N, Bohlooly YM, Bramnert M, Egecioglu E, Kindblom J et al. (2005). Prolactin and growth hormone regulate adiponectin secretion and receptor expression in adipose tissue. *Biochem Biophys Res Commun* 331: 1120–1126.
103. Borg KE, Brown-Borg HM, and Bartke A (1995). Assessment of the primary adrenal cortical and pancreatic hormone basal levels in relation to plasma glucose and age in the unstressed Ames dwarf mouse. *Proc Soc Exp Biol Med* 210: 126–133.
104. Parsons JA, Bartke A, and Sorenson RL (1995). Number and size of islets of Langerhans in pregnant, human growth hormone-expressing transgenic, and pituitary dwarf mice: Effect of lactogenic hormones. *Endocrinology* 136: 2013–2021.

105. Zhou Y, Xu BC, Maheshwari HG, He L, Reed M, Lozykowski M et al. (1997). A mammalian model for Laron syndrome produced by targeted disruption of the mouse growth hormone receptor/ binding protein gene (the Laron mouse. *Proc Nat Acad Sci USA* 94: 13215–13220.
106. Liu J-L, Coschigano KT, Robertson K, Lipsett M, Guo Y, Kopchick JJ, Kumar U, and Liu YL (2004). Disruption of growth hormone receptor gene causes diminished pancreatic islet size and increased insulin sensitivity in mice. *Am J Physiol Endocrinol Metab* 287: E405–E413.
107. Brown-Borg H, Johnson W, Rakoczy S, and Romanick M (2001). Mitochondrial oxidant generation and oxidative damage in Ames dwarf and GH transgenic mice. *J Amer Aging Assoc* 24: 85–96.
108. Brown-Borg HM, Rakoczy SG, Sharma S, and Bartke A (2009). Long-living growth hormone receptor knockout mice: Potential mechanisms of altered stress resistance. *Exp Gerontol* 44(1–2): 10–19.
109. Bartke A, Brown-Borg H, Mattison J, Kinney B, Hauck S, and Wright C (2001). Prolonged longevity of hypopituitary dwarf mice. *Exp Gerontol* 36: 21–28.
110. Bartke A, Wright JC, Mattison JA, Ingram DK, Miller RA, and Roth GS (2001). Longevity: Extending the lifespan of long-lived mice. *Nature* 414: 412.
111. Bokov AF (2008). The role of somatotropic and estrogen signaling in longevity and resistance to oxidative stress. San Antonio, TX: University of Texas Health Science Center at San Antonio.
112. Murakami S, Salmon A, and Miller R (2003). Multiplex stress resistance in cells from long-lived dwarf mice. *FASEB* 17: 1565–1566.
113. Salmon AB, Murakami S, Bartke A, Kopchick J, Yasumura K, and Miller RA (2005). Fibroblast cell lines from young adult mice of long-lived mutant strains are resistant to multiple forms of stress. *Am J Physiol Endocrinol Metab* 289: E23–E29.
114. Dominici FP, Cifone D, Bartke A, and Turyn D (1999). Loss of sensitivity to insulin at early events of the insulin signaling pathway in the liver of growth hormone-transgenic mice. *J Endocrinol* Jun 1; 161(3): 383–392.
115. Wang Z, Masternak MM, Al-Regaiey KA, and Bartke A (2007). Adipocytokines and the regulation of lipid metabolism in growth hormone transgenic and calorie-restricted mice. *Endocrinology* 148: 2845–2853.
116. Jadresic A, Banks LM, Child DF, Diamant L, Doyle FH, Fraser TR, and Joplin GF (1982). The acromegaly syndrome. *Quart J Med* 202: 189–204.
117. Orme SM, McNally RJQ, Cartwright RA, and Belchetz PE (1998). Mortality and cancer incidence in acromegaly: A retrospective cohort study. *J Clin Endocrinol Metab* 83: 2730–2734.
118. Holzenberger M, Dupont J, Ducos B, Leneuve P, Geloen A, Evens P, Cervera P, and LeBouc YIGF- (2003). 1 receptor regulates lifespan and resistance to oxidative stress in mice. *Nature* 421: 182–187.
119. Sell C and Lorenzini S (2007). Aging in IGF-1 hypomorphic mice, In The American Aging Association 36th Annual Meeting, San Antonio, TX: American Aging Association.
120. Conover CA and Bale LK (2007). Loss of pregnancy-associated plasma protein A extends lifespan in mice. *Aging Cell* 6: 727–729.
121. Boldt HB and Conover CA (2007). Pregnancy-associated plasma protein-A (PAPP-A): A local regulator of IGF bioavailability through cleavage of IGFBPs. *Growth Horm IGF Res* 17(1): 10–18.
122. Blüher M, Kahn B, and Kahn CR (2003). Extended longevity in mice lacking the insulin receptor in adipose tissue. *Science* 299: 572–574.
123. Masoro EJ (2003). A forum for commentaries on recent publications. FIRKO mouse report: Important new model—but questionable interpretation. *J Gerontol A Biol Sci Med Sci* 58: B.
124. Russell SJ and Kahn CR (2007). Endocrine regulation of ageing. *Nat Rev Mol Cell Biol* 8(9): 681–691.

125. Berryman DE, List EO, Coschigano KT, Behar K, Kim JK, and Kopchick JJ (2004). Comparing adiposity profiles in three mouse models with altered GH signaling. *Growth Horm IGF Res* 14: 309–318.
126. Katic M, Kennedy AR, Leykin I, Norris A, McGettrick A, Gesta S, Russell SJ, Bluher M, Maratos-Flier E, and Kahn CR (2007). Mitochondrial gene expression and increased oxidative metabolism: Role in increased lifespan of fat-specific insulin receptor knock-out mice. *Aging Cell* 6: 827–839.
127. Buffenstein R, Kang J, and Biney A (2007). Glucose tolerance and insulin sensitivity in an extremely long-living rodent, the naked mole-rat. *FASEB J* Apr 1; 21(6): A1423-c-.
128. McCurdy CE and Cartee GD (2005). Akt2 is essential for the full effect of calorie restriction on insulin-stimulated glucose uptake in skeletal muscle. *Diabetes* 54: 1349–1356.
129. Selman C, Lingard S, Choudhury AI, Batterham RL, Claret M, Clements M et al. (2008). Evidence for lifespan extension and delayed age-related biomarkers in insulin receptor substrate 1 null mice. *FASEB J* 22(3): 807–818, fj.07-9261com.
130. Taguchi A, Wartschow LM, and White MF (2007). Brain IRS2 signaling coordinates life span and nutrient homeostasis. *Science* Jul 20; 317(5836): 369–372.
131. Sonntag WE, Xu X, Ingram RL, and D'Costa A (1995). Moderate caloric restriction alters the subcellular distribution of somatostatin mRNA and increases growth hormone pulse amplitude in aged animals. *Neuroendocrinology* 61: 601–608.
132. Redman LM, Martin CK, Williamson DA, and Ravussin E (2008). Effect of caloric restriction in non-obese humans on physiological, psychological and behavioral outcomes. *Physiol Behav* 94: 643–648.
133. Liu H, Bravata DM, Olkin I, Nayak S, Roberts B, Garber AM, and Hoffman AR (2007). Systematic review: The safety and efficacy of growth hormone in the healthy elderly. *Ann Intern Med* 146: 104–115.
134. Bartke A (2008). Growth Hormone and Aging: A Challenging Controversy. Clin Interv Aging 3: 659–665.
135. Shechter M, Ginsberg S, Scheinowitz M, Feinberg MS, and Laron Z (2007). Obese adults with primary growth hormone resistance (Laron Syndrome) have normal endothelial function. *Growth Horm IGF Res* 17: 165–170.
136. Bonafe M, Barbieri M, Marchegiani F, Olivieri F, Ragno E, Giampieri C et al. (2003). Polymorphic variants of insulin-like growth factor I (IGF-I) receptor and phosphoinositide 3-kinase genes affect IGF-I plasma levels and human longevity: Cues for an evolutionarily conserved mechanism of life span control. *J Clin Endocrinol Metab* 88(7): 3299–3304.
137. Kojima T, Kamei H, Aizu T, Arai Y, Takayama M, Nakazawa S et al. (2004). Association analysis between longevity in the Japanese population and polymorphic variants of genes involved in insulin and insulin-like growth factor 1 signaling pathways. *Exp Gerontol* 39: 1595–1598.
138. Suh Y, Atzmon G, Cho M-O, Hwang D, Liu B, Leahy DJ et al. (2008). Functionally significant insulin-like growth factor I receptor mutations in centenarians. *Proc Natl Acad Sci U S A* 2008 March 4; 105(9): 3438–3442.
139. McElwee J, Schuster E, Blanc E, Piper M, Thomas J, Patel D et al. (2007). Evolutionary conservation of regulated longevity assurance mechanisms. *Genome Biol* 8(7): R132.
140. Longo V and Finch C (2003). Evolutionary Medicine: From Dwarf Model Systems to Healthy Centenarians? *Science* 299(5611): 1342–1346.
141. Barke A (2003). Can growth hormone (GH) accelerate aging? Evidence from GH-transgenic mice. *Neuroendocrinology* 78: 210–216.
142. Atzmon G, Pollin TI, Crandall J, Tanner K, Schechter CB, Soherer PE, et al. (2008). Adiponectin levels and genotype: a potential regulator of life span in humans. *J Gerontol A Biol Sci Med Sci* 63: 447–453.

Exploring Mechanisms of Aging Retardation by Caloric Restriction: Studies in Model Organisms and Mammals

Rozalyn M. Anderson, Ricki J. Colman, and Richard Weindruch

Abstract There has been expanding interest in research on aging and in the identification of underlying mechanisms of the aging process. The promise that by understanding aging we may also understand the factors that lead to age-associated disease has been held for some time and now, through the use of model organisms, this potential is being realized. A key subgroup within this gamut are investigations that seek to understand the impact of nutrition and diet on aging, and foremost among these are studies of caloric restriction (CR). Herein, we discuss a number of commonly used laboratory organisms, describe the methodology employed to study aging and the impact of nutrition, briefly discuss the main findings from these studies and present candidate factors emerging from these studies that may play a mechanistic role in the retardation of aging by CR.

Keywords Caloric restriction · Longevity · Aging · Model organisms · Mice · Nonhuman primates · Mitochondrial metabolism · Metabolic reprogramming

Introduction

A fundamental issue in the study of aging is the choice of organism; this represents a trade-off between ease of manipulation, cost, time and translatability to human aging (Table 1). Species that are best characterized are fast growing, short-lived, and their genetic manipulation is well understood and readily accomplished. For obvious reasons, the simplicity and ease of use of these organisms places them far ahead in terms of: (i) what is known about the biology of aging and, (ii) the factors that may be involved in the aging process. At the other end of the spectrum are

R.M. Anderson (✉)
Wisconsin National Primate Research Center, Department of Medicine, University of Wisconsin, Geriqtic Research, Education, and Clinical Center, University of Wisconsin-Madison, Madison, WI, USA
e-mail: rmanderson5@wisc.edu

N.S. Wolf (ed.), *The Comparative Biology of Aging*,
DOI 10.1007/978-90-481-3465-6_4,

Table 1 Summary of attributes and disadvantages of model organisms and mammalian species used to study the aging process

Species	Yeast	Worms	Flies	Rodents	Primates
Lifespan	Divisions/days replicative/ chronological	Weeks	Weeks	Years	Decades
Ease of genetic manipulation	+++	+++	+++	+	–
Cost	Low	Low	Low	High	Very high
Translatability	+	+	+	++	+++

species more closely related to humans, with longer lifespans and more complex physiology. These species hold the most potential to contribute to our understanding of human aging; however, they are also the least well characterized genetically and are the most expensive to study.

Single-Celled Organism

Saccharomyces cerevisiae

Although a unicellular eukaryote, the simple yeast has been a remarkable source of insight into numerous cellular functions. Of particular interest here are the advances in our understanding of the retardation of aging by CR. In yeast there are two approaches to study aging and longevity: the first, known as *replicative aging*, measures the ability of an individual yeast cell to bud and produce daughters, and the second, known as *chronological aging*, measures the ability of a cell to remain viable in stasis. Replicative aging is thought to have parallels with aging of mitotic tissues in higher organisms while chronological aging may have more in common with aging in post-mitotic tissues. Many of the pathways identified as being important in yeast longevity are associated either with metabolism, the stress response or genomic instability [1, 2]. It is worth noting that some yeast-specific elements of aging may not be so relevant in other species. First, yeast, unlike many of the other model organisms commonly used to study mechanisms of CR, are facultative anaerobes. Second, yeast possess a mechanism of survival in times of stress whereby meiotic division of a diploid species produces highly robust and resilient spores. Finally, a mechanism for replicative aging has been described and involves recombination at the highly repetitive ribosomal DNA locus, resulting in toxic extrachromosomal rDNA circles that are asymmetrically apportioned to the mother cell during budding [3]. However, the yeast response to changes in nutrient conditions, the key stimulus at the core of the mechanism of CR, is likely to be translatable to other species.

Methodology

The individual yeast cell can achieve a finite number of divisions before senescence. The replicative lifespan is the number of divisions attainable and is highly strain dependent. In the process of budding, a daughter cell is produced from the mother cell roughly every 2 h for yeast grown in optimal conditions. To measure replicative aging, previously unbudded cells (virgin daughters) are selected and each round of budding monitored by microscopy. The daughters produced are counted and removed to avoid contamination of the initial starter cell. CR is implemented by either reducing the glucose or amino acid components in media [4, 5]. Progressive reduction in either component results in increasing extension of lifespan. Growth in conditions of reduced glucose is the most prevalent method used to apply CR in the study of replicative aging.

To measure chronological aging, yeast are grown to saturation (maximum cell density, with cells in stationary phase) and maintained in the absence of replication. The yeast cells remain metabolically active and when glucose becomes exhausted from the media, metabolism is redirected from fermentation to respiration, a process known as the diauxic shift. Although highly strain dependent, the median chronological lifespan for wild type strains is about one week, just a few days following the shift [6, 7]. At the appropriate time points, survival is measured by determining the ability of cells to re-enter the cell cycle and form colonies. CR is applied in this system by growing the initial culture in media with glucose concentration reduced from 2 to 0.5%, and has been shown to extend lifespan [8]. An alternative approach in the study of chronological aging involves transfer of cells to water once stationary phase has been reached [9]. This measure greatly extends lifespan and although it has been referred to as extreme CR, it is likely to have more in common with starvation/sporulation pathways [6], than with pathways that respond to nutrient fluctuation. The lifespan extension induced by this approach is further augmented by genetic manipulations that cause a relative increase in genes involved in the stress response and in sporulation [10].

Findings and Candidate Mechanisms

Replicative aging: As the yeast age, the time between budding extends and the yeast cell becomes larger and pitted with bud scars. Because the cell surface of the new bud is generated de novo, it is possible to separate old yeast from young yeast by labeling the initial culture. This approach has been used to demonstrate that aging is associated with a shift in glucose metabolism from glycolysis toward the glyoxylate pathway, gluconeogenesis and energy storage [11]. CR implemented by reduction of the glucose concentration in the media from 2 to 0.5% defers this shift. It is noteworthy that many of the genes involved are usually repressed in the presence of glucose and their increased expression with age represents deregulated gene expression. Genetic mutants that induce or impede this transition have decreased and increased lifespan, respectively [11], consistent with deregulation of glucose signaling being a factor in yeast aging.

CR is associated with a shift in metabolism towards respiration and lifespan extension by CR requires functional mitochondria [12]. The involvement of mitochondria in the response to CR is confirmed by transcriptional profiling, where 31% of genes altered by CR in wild type yeast encode mitochondrial proteins [13]. Factors identified in genetic studies as being important in CR differ depending on the extent of restriction applied, implying that experiments where the CR regimen involves reduction from 2 to 0.5% glucose may be distinct from those that reduce glucose levels to 0.05% [12, 14, 15]. Yeast are facultative anaerobes and can readily derive energy from fermentation. It is possible to generate yeast lacking functional mitochondria and these "petite" cells in some strain backgrounds have extended lifespan [16]. Respiratory deficient (ρ°) yeast have been used to test whether the shift toward increased respiration is necessary for lifespan extension by CR [14]. Although CR appears to be quite effective at extending lifespan in respiration deficient lines, this finding does not negate the importance of respiration in CR in wild type cells. Transcription profiling in ρ° cells has revealed a metabolic shift toward peroxisomal and metabolite-restoration (anaplerotic) pathways [17], indicating that ρ° cells are metabolically distinct from their respiring counterparts. The fact that these cells still respond to CR points to a multiplicity of mechanistic components in lifespan extension.

The role of the NAD-dependent histone deacetylase Sir2 [18–20] in extension of yeast replicative lifespan by CR remains controversial [7], and the extent of involvement of Sir2 in regulation of lifespan appears to be strain-specific [5, 21]. Sir2 is involved in gene silencing in yeast [22], where Sir2-dependent silencing at the highly repetitive ribosomal DNA locus prevents recombination [23] and genomic instability associated with extrachromosomal rDNA circles [3, 24]. Genetic mutants with increased stability at the rDNA locus show exceptional lifespan extension compared to wildtype cells when CR is implemented using the lower range of glucose concentration (0.05%); however, in cells on moderate CR (0.5% glucose) the same mutation has no effect [15]. These differences in methodology are likely to be the basis for much of the controversy surrounding the role of Sir2 in the mechanism of CR.

The metabolic pathway for regeneration of NAD has also been implicated in the mechanism of CR in extending replicative lifespan [25]. Additional copies of salvage pathway genes increase silencing at the rDNA locus [26]. In the NAD salvage pathway, Pnc1 converts nicotinamide, an inhibitory product of Sir2 activity, to non-inhibiting nicotinic acid [27]. Pnc1 is upregulated by stress activated transcription factors Msn2 and Msn4 [28] in response to CR and other regimens that extend replicative lifespan. These data demonstrate that CR induces an active cellular response (as opposed to a passive response) that promotes longevity. Nicotinamide prevents lifespan extension by CR at 0.5% glucose [29], and at 0.05% glucose does so in a manner that is Sirtuin independent [15]. Pnc1 is both nuclear and cytosolic and sequestered in peroxisomes following amino acid restriction [25]. It is not clear what is being regulated by the NAD salvage pathway in this organelle, but this overt altered distribution is not observed in glucose restricted yeast, indicating that the cellular response to CR may involve a metabolic compensatory shift that is tailored to the type of nutrient limitation.

The Tor (Target of Rapamycin) signaling pathway is a critical regulator of cell growth and metabolism [30, 31]. Inhibition of Tor signaling negatively impacts ribosome biogenesis, induces stress responsive transcription and extends replicative lifespan [28, 32]. Of particular importance here, Tor signaling has been implicated in the mechanism of lifespan extension by CR. Tor inhibition activates transcription factors Msn2 and Msn4 by promoting their nuclear localization, resulting in increased expression of stress responsive genes including Pnc1 [28]. In addition, nutrient limitation or treatment with rapamycin causes Tor to relocate from the nucleus to the cytosol away from the rDNA locus, reducing transcription from the rDNA promoter and possibly altering stability at that locus [33].

An intracellular pathway that communicates functional status of mitochondria to the nucleus has been identified in yeast and is called the retrograde response [34, 35]. This intra-organelle signaling response activates metabolic pathways including the glyoxlyate cycle, gluconeogenesis and peroxisomal pathways that provide an alternative mechanism of fuel utilization upon detection of mitochondrial dysfunction. Activation of the retrograde response pathway extends lifespan in yeast and is required for the increased lifespan of ρ° cells, but is not required for lifespan extension by CR [4]. Interestingly, there is crosstalk between the retrograde response pathway and the Tor signaling pathway where inhibition of Tor using rapamycin activates components of the retrograde response pathway through changes in subcellular localization [36]. This same mechanism appears to be important in the Tor regulation of Msn2 and Msn4, stress and nutrient limitation activated transcription factors that are required for replicative lifespan extension by CR [28].

Chronological aging: During chronological aging, yeast cells remain metabolically active and eventually exhaust glucose from the media inducing the diauxic shift – a redirect of metabolism from fermentation to respiration. This shift induces genes involved in the TCA and glyoxylate cycles and concomitantly reduces expression of genes involved in protein synthesis including ribosomal genes [37]. Important exceptions to this trend are genes encoding mitochondrial ribosomal genes that are induced upon glucose exhaustion. Differences in glucose metabolism or glucose sensing can only occur in pre-diauxic phase culture before glucose is exhausted from the media, indicating that the metabolic state of the yeast in this early stage is a key determinant of subsequent cell viability. Growth in low glucose increases respiration but decreases reactive oxygen species production relative to oxygen consumption [38]. Genetic or pharmacological inhibition of respiration in CR cultures negatively impacts survival, arguing that the shift to respiration is required for extension of chronological lifespan [8]. Interestingly, growth in non-fermentable carbon sources such as glycerol or ethanol, converts cells to obligate aerobes and extends chronological lifespan. Restriction from 2 to 0.5% does not further increase lifespan, indicating that under these conditions the shift to respiration is the determinant of increased lifespan and not reduction in nutrient availability [8].

A number of factors involved in regulation of chronological lifespan have been identified; including Sch9 glucose signaling kinase and homologue of AKT and S6K [39, 40], Msn2 and Msn4 stress/nutrient limitation activated transcription factors [41], and nutrient activated kinase Tor1 [42–44]. Inhibition of Tor by nutrient limitation or by rapamycin increases chronological lifespan in yeast in a manner that is dependent on Msn2 and Msn4 [43, 44]. Reduced Tor signaling causes an increase in respiration and up-regulates mitochondrial gene expression [42]. Moreover, extension of lifespan is dependent both on respiration and on the presence of glucose, indicating that under normal conditions Tor plays a role in glucose dependent inhibition of respiration in yeast.

Invertebrate Animals

Caenorhabditis elegans

The nematode *C. elegans* has been widely used as a model organism for studying aging and numerous single gene mutations have been identified that extend lifespan [45, 46]. Foremost among these are genes involved in the insulin-signaling pathway, many of the components of which had previously been associated with the process of dauer formation [47]. In their native environment, these tiny (1 mm) worms feed on microorganisms in soil and in laboratory conditions are generally grown in monoxenic nematode growth media (i.e., supplemented with bacteria). When food is scarce, animals early in juvenile development (L2 larval stage) can enter an alternative stage known as dauer diapause. Dauers are highly resistant to stress and can survive months while conditions are unfavorable. Once conditions improve and nutrients become available, normal development resumes as the dauer enters the L4 larval stage.

One of the earliest long-lived mutants identified was *daf-2* [48], a mutant of the DAF2 insulin-like receptor involved in dauer formation. Reduced signaling through this pathway alleviates repression of the forkhead transcription factor DAF-16 resulting in altered expression of metabolic and stress resistance pathway genes [49]. Aging of this organism has been well characterized [50] and under typical growth conditions, bacterial proliferation has been identified as a critical cause of death. Lifespan is extended when worms are grown in UV killed bacteria [51] or when antibiotics are used to prevent bacterial growth [50]. The difficulty in determining exact caloric intake has led to the consensus that restriction of food intake in *C. elegans* is best referred to as dietary restriction (DR) rather than CR.

Methodology

Median lifespan for wild type animals is 2–3 weeks [47]. Live worms are detected by their response to mechanical stimulus. DR may be implemented by using reduced concentrations of bacteria (from 10^{10} to 10^8 cells/ml) in the monoxenic media and results in 60% extended lifespan [52]. Traditionally worms are cultured on agar plates and synchronized larvae are generated to conduct a survival analysis. Greater bacterial density can be achieved on plates than in liquid culture and accurate control of available food is not feasible. In liquid culture, bacterial concentration can be regulated and lifespan increases with decreasing concentration of bacteria up to a point where the worms likely become starved [53]. An alternative approach is to grow the animals in chemically defined or semi-defined axenic (no bacteria) media, which generally results in a 50–80% increase in lifespan [54, 55]. Worms grown in axenic media show phenotypes associated with DR such as reduced size, delayed maturation, and reduced fecundity in terms of brood size but longer duration of fertility. A third approach in the investigation of worm DR utilizes animals with defective pharyngeal pumping (*eat* gene mutants) which have reduced feeding rates and ~50% increase in lifespan [56].

The toxicity of bacteria to the aged worm is emphasized in studies of adult onset dietary deprivation. Complete removal of the bacterial component of the diet of post-reproductive adult worms results in a 40–50% increase in lifespan [57, 58]. This regimen of dietary deprivation is independent of the insulin/IGF pathway and also of *Sir2.1*. Consumption of bacteria declines with age in worms suggesting that lifespan effect of removal of bacteria may be not be simply due to the impact on diet. Lifespan extension by dietary deprivation is suppressed by a diffusible component of a bacterial food source [59]. These data suggest that the extension of lifespan by this method may be due in part to alleviation of bacterial toxicity for aged worms.

Important points of difference between experiments include the bacterial strain used when the worms are maintained on agar plates with a lawn of bacteria, whether the worms are grown on plates or in suspension, whether bacteria are dead, alive or treated with antibiotics to prevent bacterial growth, and whether the worms are sterile or fecund, as the ability to produce eggs also influences longevity. These differences become significant for transcriptional profiling, epistatic and metabolic analysis experiments.

Findings and Candidate Mechanisms

Surprisingly little is known about the mechanistic basis for lifespan extension by DR in worms [45, 55, 60]. When worms are grown in axenic media there is an increase in oxygen consumption and heat production [54] indicative of a change in metabolism. This increase in respiration is also observed in *eat-2* mutants [53]. Growth in axenic media induces the expression of genes involved in the glyoxylate pathway, glyceronoegenesis and possibly glyceroneogenesis indicating that metabolite replenishing pathways are activated [61]. There are conflicting reports regarding the influence of DR on respiration as a function of bacterial concentration in liquid culture, although this may be due to differences in the range of bacterial concentrations tested [53, 62].

A role for altered mitochondria energy metabolism is suggested in the mechanism of lifespan extension by bacterial reduction because inhibitors of the electron transport chain prevent lifespan extension by this method [62]. Inhibition of glycolysis by chemical (2-deoxy-glucose) or RNA interference induces respiration and extends lifespan [63]. Transcriptional analysis of worms with impaired glucose metabolism reveals an increase in expression of genes involved in energy metabolism and fatty acid oxidation pointing to a critical role of mitochondrial bioenergetics in mediating the longevity of these animals. Reactive oxygen species formation is increased as a result of impaired glucose metabolism and this increase is required both for lifespan extension and increased stress resistance [63]. DR by bacterial dilution activates SKN-1, a transcription factor involved in the oxidative stress response [62] further linking increased respiration to activation of stress pathways.

DR implemented by either bacterial dilution or by axenic media act independently of the insulin/IGF signaling pathway. Neither regimen requires the forkhead transcription factor DAF-16 [62, 64]. Lifespan extension of *eat-2* mutants, the genetic mimic of DR, is also independent of *daf-16* and the effect on lifespan is additive with *daf-2* mutants [56]. More recently, a separate study of adult onset DR using the method of bacterial dilution on solid media identifies DAF-16 as a mediator of lifespan extension. In this case, the forkhead transcription factor is positively regulated by AMP-activated protein kinase (AMPK), which itself can extend lifespan when expressed in the active form [65]. The apparent conflict between data sets regarding the involvement of DAF-16 is most likely a result of different methodology in applying DR. Obviously these differences become critical when interpreting the data and extrapolating the importance of the findings to higher organisms.

Additional copies of *Sir-2.1*, the closest SIR2 homologue in worm, extends lifespan [66] by a mechanism that requires the forkhead transcription factor Daf-16 previously identified in the insulin signaling pathway [67]. Sir-2.1 is not required for lifespan extension by *daf-2* mutants but is required for lifespan extension in the genetic mimic of DR, the *eat-2* mutant [68]. The interaction between DAF-16 and SIR-2.1 involves 14-3-3 proteins, and activation of DAF-16 by this mechanism acts in parallel DAF-2 indicating a convergence of longevity pathways at the forkhead transcription factor [68, 69]. More recently the forkhead transcription factor PHA-4 has been implicated in the mechanism of DR. Extra copies of pha-4 extend lifespan and pha-4 is required for lifespan extension by DR but not for that observed in mutants of the insulin-signaling pathway [70]. It will be interesting to see if PHA-4 is sensitive to SIR-2.1 activation. The *C. elegans* homologue of Tor is LET-363 and inhibition of worm Tor extends lifespan [71]. The Tor signaling pathway is highly conserved and as part of the nutrient sensing function of this pathway, Tor interacts with a regulatory protein known as raptor. The worm homologue of raptor, DAF-15, has been previously identified as part of the dauer formation pathway and is regulated by DAF-16 [72]. The extension of lifespan by Tor inhibition is not additive with *eat-2* indicating that Tor may be involved in DR [73].

Drosophila melanogaster

The fruit fly has long been a favorite model for the study of complex biological phenomena including aging [74]. The lifespan of the wild type fly is ~1 month and adulthood is rapidly achieved within a day of emergence from the pupa. One of the advantages of *Drosophila* for aging studies is the shear numbers of genetically identical individuals that can be obtained and monitored; this adds enormous statistical power and allows comparative mortality analysis to be performed in addition to survival curves. The fruit fly is an obligate aerobe and is considerably more complex than the previous two models described above. *Drosophila* possess highly differentiated tissues and organs and a well defined endocrine system. Furthermore, the animals can fly, show learning and memory, and are capable of a variety of sensory and motor driven behaviors [74]. Another difference with this organism is that it is dioecious, that is males and females are distinct individuals, permitting the study of gender specific effects [75]. As was the case in *C. elegans*, the term DR is also favored in *Drosophila* studies.

Methodology

As for yeast and worms, DR is implemented by using media in which the caloric composition is reduced. The principal components of the standard media used to grow *Drosophila* are yeast and sugar and these are generally provided on a solid cornmeal, charcoal, or agar-based medium. *Drosophila* media differs considerably among investigators [76, 77]. Although DR studies have demonstrated that reduction of either the sugar or yeast component of the diet can extend lifespan, yeast dilution is more effective in this regard [78, 79]. Replacement of the yeast component of the diet with amino acids in the form of caesin shows a similar impact on lifespan as that observed in yeast dilution experiments; maximal longevity is attained when dietary amino acids are restricted but not absent [80]. For many of the DR studies in non-mammalian systems it is not plausible to measure precisely how much food each individual consumes in a survival study. This aspect has emerged as rather important in the *Drosophila* studies. DR elicits robust compensatory changes in food consumption, so that actual caloric intake is not proportional to the energetic content of the diet media [81]. However, direct measurement of calorie assimilation reveals that calorie intake is reduced on all diets that extend lifespan [79].

A further consideration for *Drosophila* longevity studies is the effect of reproduction and sexual activity on lifespan. Female lifespan is extended when their reproduction is repressed [82]. When nutrient availability is high, increased female egg production promotes mating and sexual activity shortens lifespan [83]. Mixed gender groups on control diets may have shortened lifespan compared to the DR group due to behavioral differences in addition to differences in nutrient availability. In single sex housing, female lifespan extension in response to DR is greater than that of males and the optimal degree of restriction for maximal lifespan extension is not equivalent among males and females [75].

Findings and Candidate Mechanisms

Genome wide transcriptional analysis of pooled whole flies indicates that 23% of transcripts detected are altered with age [84]. Genes involved in oxidative phosphorylation were among the most significant groups identified as being sensitive to aging with strong declines in expression observed. Furthermore, activity of mitochondrial respiratory complexes also declines with age [85]. In flies on CR, altered mitochondrial metabolism is suggested by the finding that the production of reactive oxygen species is reduced [86]. Functional analysis of gene expression profiles point to a relationship between mitochondrial dysfunction, the stress response and aging and suggest that muscle may be particularly sensitive to aging in *Drosophila* [87]. Investigation of isolated mitochondria from flight muscle from CR animals indicates that although the mitochondrial density is not altered by CR, mitochondrial morphology and activity are significantly different [88].

Several studies have employed genetic manipulation to discover factors that regulate longevity in flies, though substantially fewer have investigated the roles of those candidate genes in lifespan extension by CR. Mutation of the deacetylase dRpd3 extends lifespan in male and female flies and the effect is not additive with CR, suggesting a common mechanism [89]. In flies that are mutant for dRpd3 or on CR, levels of dSir2 are increased suggesting a role for the NAD dependent deacetylase in the regulation of longevity. Consistent with this, increased expression of dSir2 extends lifespan and decreased expression prevents extension of lifespan by CR [90]. Resveratrol and other activators of Sirtuins (the collective name for all members of the Sir2 family) extend lifespan in yeast, worms and flies suggesting evolutionary conservation of the role of these proteins in metazoan longevity [91].

Disruption of ecdysone steroid hormone signaling in *Drosophila* extends lifespan [92]. In the absence of this hormone, the ecdysone receptor complex interacts with transcriptional repressors dRpd3 and dSin3 [93]. Reduction of dRpd3 extends lifespan in flies [89] and reduction in dSin3 causes up-regulation of genes involved in the oxidative metabolism of fatty acid to acetyl-CoA and genes involved in mitochondrial oxidative phosphorylation [94]. The similarity between the ecdysone receptor complex and the nuclear hormone complexes in mammalian systems suggest that the mammalian counterparts may play a role in aging [95].

The forkhead transcription factor dFoxo is a mediator of insulin signaling in *Drosophilia* similar to its *C. elegans* counterpart Daf-16 [96]. Overexpression of dFoxo in the fat body extends lifespan [97, 98] and, similar to its *C. elegans* counterpart, does not appear to play a role in CR [99, 100] although is has been suggested that the response to CR is modified in the absence of dFoxo [99].

The nutrient sensitive kinase dTOR has not been tested in context of CR but has been implicated in Drosphila growth [101] and regulation of lifespan [102].

Reduction in dTOR activity is associated with increased dFoxo activity [103] demonstrating a connection between metabolism and longevity. It will be interesting to see if dTOR is involved in the mechanism of CR in *Drosophila* and whether the recently identified role for mammalian TOR in regulation of mitochondrial respiration [104] is conserved in the fruit fly.

Mammals

Mice and Rats

The overwhelming majority of work in CR from its inception with the founding report of McCay et al. in 1935 [105] through the late 1980s [106] involved studies in mice and rats. As discussed above, this situation has changed due to the ongoing decade-long explosion of DR/CR studies in yeast, worms and flies. In this section, we first discuss four CR methods. First, CR carried out in a conventional way (i.e., lowering caloric intake by ~30%) in both genetically normal and manipulated animals. We then evaluate three increasingly popular methods to achieve a "CR-like state". These are methionine restriction (MR), every other day feeding (EOD) and the so-called "CR mimetics". The latter are drugs or nutrients which trigger the CR molecular/cellular phenotype in otherwise normally fed animals. We then highlight certain findings deemed to be of special relevance and summarize current thinking on underlying mechanisms provided by the rodent work in combination with that discussed above.

Methodology

CR methods have been reviewed both many years ago [106] and recently [107]. Long ago, it was our experience that a striking inverse correlation exists between caloric intake and maximum lifespan in mice (Fig. 1) [108, 109]. Successful CR in the conventional sense involves reducing caloric intake while avoiding malnutrition. There exist some early data to suggest that CR diets enriched in protein, vitamins and mineral content may increase longevity to a greater extent than unenriched diets [109–111]; however, it is important to underscore that the topic of the optimal composition of CR diets remains overtly under-investigated.

The vast majority of rodent CR studies (~95%) have started the diet early in life (3–16 weeks of age) of laboratory rodents. The topic of middle-age (or later) onset CR has been far less investigated, which is unfortunate as this approach is most germane to human application. We were the first to show the onset of CR could be delayed until near middle age (12 months), yet that mice on CR thereafter live longer and have a lower tumor incidence [112]. Interestingly, only 4 weeks of CR imposed on old mice (34 months) resulted in a "younger" hepatic gene expression profile than that of age-matched controls [113]. Other work suggests benefits of adult-onset CR

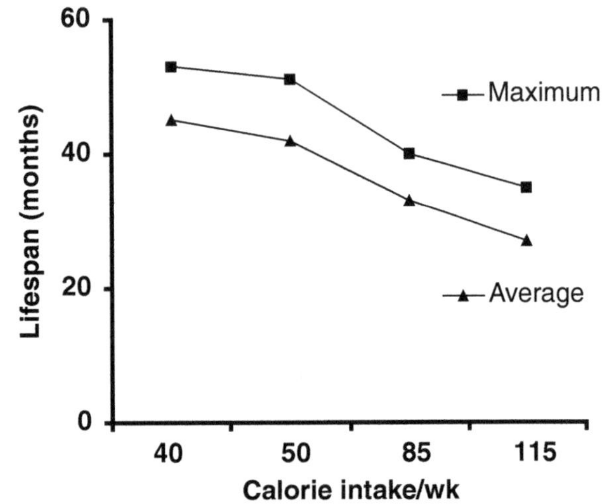

Fig. 1 The inverse linear relationship between calorie intake and lifespan extension in mice points to the involvement of metabolic regulators in the mechanism of CR. "Maximum" lifespan is the average of the longest-lived 10%. Adapted from [109]

in muscle. Male rats restricted at 17 months and studied at 32 months showed fewer signs of sarcopenia than controls [114]. Similarly, a recent study in rats found that 5 months of CR started at 21 months using protein-enriched diets slowed several age-associated biochemical and functional changes in two muscles [115]. Clearly, CR in rodents started at or beyond middle age can slow aspects of the aging process.

An alternate approach to understanding aging involves the study of transgenic and mutant animals that display increased longevity compared to wildtype littermates [116]. In general, data on age-related pathology and biological changes are lacking in many of these models and the extension of lifespan in some cases is background specific [116, 117]. In all of the genetic manipulated mouse models, the increase average lifespan is modest compared with CR, and effect on maximal lifespan is small if there is any effect at all [117]. A comparison of gene expression profiles of livers from long-lived dwarf mice or CR animals demonstrated a profound effect on insulin/IGF signaling in the dwarf mice that was not observed to the same extent in CR animals [118]. Tissues are highly specialized in mammalian systems and play distinct roles as energy consumers and regulators of metabolism. As a result many transcription factors and co-activators have distinct tissue-specific gene targets and binding partners that, accordingly, orchestrate responses to nutritional or other stimuli. The inference is that altered expression of a key regulator may be desirable in one tissue and detrimental in another. One of the main considerations in using transgenic or mutant strains is that the manipulation itself may significantly alter the normal metabolic state resulting in a model that is not reflective of the wild-type condition and, as such, caution must be exercised in extrapolation of these data to normal aging in wildtype strains.

Interest in EOD feeding (also called intermittent fasting) has increased markedly over the past decade and has been the subject of informative reviews [119, 120]. Without going into great detail, it is our conclusion that EOD may provide a diverse set of clinically relevant benefits without altering the aging process or maximal lifespan in an overt way. Interestingly, an intermittent feeding regimen in a short-lived fish (the three spined stickleback) results in a *shorter* lifespan [121]. In most rodent studies, EOD involves a mild weight loss such that it could be seen as a slight form of CR. In addition, periods of fasting likely also contribute in a major way to any effects. Even if the aging process proves not to be slowed by EOD diets, the effects being observed in EOD-fed rodents are striking, including improved functional recovery after spinal cord injury [122] and protection against age-related aortic sclerosis [123].

The gerontological field of MR began with a 1994 report by Richie et al. [124], showing that F344 rats fed such a diet (MR = 0.17% w/w methioine vs. 0.86% in the control diet) lived 44% longer (maximum lifespan) while weighing 43% less than controls. However, the report did not provide data on food intake. Accordingly, due to the large body weight difference it is difficult to dissect MR-induced lifespan extension from that caused by CR. Subsequent work has improved understanding of MR. Miller et al. [125], observed that CB6F1 mice on MR show a much higher rate of early deaths but an increased maximum lifespan. The body weights of MR mice were ~65% that of controls during most of adulthood; however, rigorous food intake data were not provided but that which were did not suggest major differences between groups. It was also reported that indicators of ocular and immunologic aging stayed "younger longer" with MR. Other investigators report that male Wistar rats on MR display lower rates of mitochondrial reactive oxygen species (ROS) production and ROS-induced damage in heart and liver [126]; however, clear data on food intakes and body weights were not provided. Overall, it appears that MR induces a metabolic state associated with aging retardation. However, there appears to be a need for a closer examination of food intakes in this model and more thought on whether it is most appropriate to express food intake per gram of animal (vs. per animal) in MR studies. This situation is strikingly reminiscent of the prolonged debate on how best to express metabolic rate data in the context of CR which produces animals differing not only in overall weight but in body composition.

A newer and highly exciting area of inquiry concerns the search for CR mimetics. Although the effects of CR are robust, it is difficult for nearly all people to adhere to the regimen long-term, sparking great interest in identifying compounds with the ability to mimic the effect of CR while eating normally [127]. Many studies have reported beneficial effects of the components of red wine on cardiovascular heath [128], thus an active area of research has focused on the ability of these compounds to mimic CR. A compound receiving much attention is resveratrol, a polyphenol found in red wine. Resveratrol has been suggested to extend lifespan of model organisms in a mechanism dependent on activation of sirtuin genes [91, 129], though some studies did not produce results supporting this notion [130, 131]. Conflicting data can also be found in mouse studies: high doses of resveratrol prevent early mortality caused by a high fat diet by activating the Sirt1 enzyme [132], but low doses of resveratrol do not appear to act through an increase in Sirt1 protein levels or activity [133]. However, the latter study did show that resveratrol markedly mimicked the gene expression profile of CR in heart, brain and muscle and prevented age-related declines in cardiac function, suggesting that it may slow aspects of the aging process in mammals. The search for CR mimetics, an important topic with major public health implications, is off to an exciting start.

Findings and Candidate Mechanisms

The pace of gerontological research on CR has accelerated overtly over the past 40 years as evidenced by 10 publications in 1970, 90 in 1990 and ~250 in 2008. In view of this situation we confine our comments to what we see as the most appealing mechanistic explanations for the retardation of aging and diseases by CR. We propose that CR induces an active response to altered nutrient availability and a reprogramming of energy metabolism that is a primary event in the mechanism of CR. Support for this hypothesis is detailed elsewhere including data from model organisms [134]. Other recent reviews also discuss the role of metabolism in the mechanism of CR as well as presenting alternate perspectives [135–137].

For the past decade we have conducted gene expression profiling to determine the transcriptional signature of aging and CR. Our investigations of CR-dependent changes in transcription profiles in aged mice have revealed two types of transcriptional effects: first, the prevention of age-associated changes by CR and second, the shifts in gene expression by CR that are not age-dependent, i.e. those that are indicative of CR-induced transcriptional reprogramming. These studies demonstrated that CR not only opposed age-associated changes in gene expression but also revealed CR-induced shifts in expression of gene not affected by aging in muscle [138], brain [139], and heart [140]. For example, in heart, aging is associated with expression changes indicative of increased structural protein turnover and neurodegeneration as well as a shift from fatty acid to glucose metabolism (as illustrated in [140]). Importantly, CR was associated with increased expression of genes involved in energy metabolism and decreased expression of pro-inflammatory genes [140]. These data point to mitochondrial alterations as a critical feature of CR, a concept that is supported by work in model organisms [134].

Our study of epididymal adipose tissue from 1-year-old mice, revealed that long-term CR had profound effects on the morphology, adiposity and transcriptional profile [141]. It showed that metabolic reprogramming is a prominent feature of the CR effect in white adipose tissue, with increased expression of genes involved in the glycolytic pathway, the lipolytic pathway, amino acid metabolism and importantly, mitochondrial metabolism. In addition to the striking activation of energy metabolism, there was a marked reduction in the expression of genes involved in inflammation [142]. It has become clear that adipose tissue is a source of hormones and inflammatory factors that influence metabolic homeostasis and systemic inflammation [143–145]. The lowered expression of pro-inflammatory genes in white adipose tissue by CR may provide an explanation for its ability to delay or prevent a broad spectrum of age-associated diseases driven by systemic inflammation.

A candidate factor that could contribute importantly to the metabolic reprogramming induced with CR in some tissues is the transcriptional co-activator PGC-1α (peroxisome proliferator activated receptor gamma co-activator 1 alpha). PGC-1α is a key regulator of genes involved in mitochondrial metabolism, including the nuclear encoded genes involved in the electron transport system [146, 147]. In addition to regulating the genes involved in energy metabolism, PGC-1α also influences the balance between carbohydrate and fat metabolism through co-activation of the peroxisome proliferator activated receptor (PPAR) nuclear receptor family of transcription factors [148, 149]. The PPARs have been linked to obesity and metabolic regulation and play a central role in the cross-talk between glucose and lipid homeostasis [150], and in liver, 19% of the transcriptional changes induced by CR are dependent on PPARα [151]. PGC-1α protein levels are elevated in epididymal white adipose tissue of 1 year old CR animals [152], indicating that it may be part of a regulated metabolic response to CR. We recently identified a novel mechanism of PGC-1α regulation involving SIRT1 NAD-dependent deacetylase and the nutrient sensitive kinase GSK3βa (Glycogen synthase kinase 3 beta). This mechanism permits temporal regulation of mitochondrial function through alterations in PGC-1α stability, localization and activity, and it is common to CR and the stress response [152]. Although a role for mitochondria in the mechanism of CR is conserved among studies, a homologue for PGC-1α has not been identified outside of mammalian systems. This evolutionary distinction may be crucial in determining factors that can influence human aging and underscores the importance of mammalian studies in the pursuit of potential targets that impact the aging process.

Dogs

It is noteworthy that a CR trial has been successfully conducted in Laborador retrievers [153]. The CR began at 6 weeks of age and involved a 25% lowering of the intake of the same diet fed to controls. CR increased the median lifespan by 1.8 years, marginally increased maximum lifespan and delayed the occurrence of osteoarthritis (a major disease of aging in this model). In sum, based on these and several other measures, CR induces physiological changes in dogs which parallel those seen in rodents and increases healthspan and lifespan.

Non-human Primate *Macaca mulatta*

While CR in small mammals has been shown repeatedly and under varying conditions to extend median and maximal lifespan, the efficacy of CR in a primate species is still under investigation. The marked similarities between human and nonhuman primates in almost all aspects of their anatomy, physiology and behavior make nonhuman primates uniquely useful for providing insights into the human condition. Rhesus macaques (*Macaca mulatta*), in contrast to the models previously described, have an average lifespan of ~27 years in captivity and a maximal lifespan ~40 years. Obviously an investigation encompassing this relatively long lifespan, while promising considerable insight that is readily translatable to human aging, necessitates a prolonged duration of study and incurs considerable cost. Furthermore, in most macaque colonies the animals are not genetically identical although there may be a subset of related animals within the cohort. This heterogeneity can present a challenge due to differences between individuals in terms of body composition, metabolism and behavior that can present difficulties in the interpretation of data gleaned over extended time periods.

Methodology

In order to address the efficacy of CR in a primate species, three major studies of varying design utilizing rhesus macaques were initiated in the late 1980s. Two of the rhesus studies, those at the National Institute on Aging (NIA) and ours at the Wisconsin National Primate Research Center (WNPRC), are ongoing and have the broad overall goal of determining the ability of CR to delay aging and extend maximal lifespan, while the third study, performed at the University of Maryland (UMD) focused more specifically on obesity and glucoregulatory function.

The NIA study began in 1987 with 60 male rhesus macaques of both Chinese and Indian origin. Three age groups were described: 20 were juveniles (1–2 years of age, 10 control, 10 CR), 20 were adolescents (3–5 years of age, 10 control, 10 CR), and the remaining 20 were considered old (>15 years of age, 10 control, 10 CR) [154]. In 1992, 60 female rhesus macaques were added to the study. As with the males, females were added in three age groups: juvenile (1–2 years of age, 10 control, 10 CR), adult (6–14 years of age, 10 control, 10 CR), and old (16–21 years of age, 10 control, 10 CR). This study utilizes a paradigm of 30% CR [155, 156]. All animals are fed the same semi-purified, nutritionally fortified diet containing 15% protein and 5% fat. Animals are fed twice per day and do not have continual access to food. Control animals are fed approximately ad libitum, while CR animals are fed 30% less then the standard amount for their age and body weight based on National Research Council food tables [157]. In order to reach the 30% restriction level, CR animals' food allotments were reduced by 10% per month over a three-month period [154, 155]. In line with the broad goals of this study, the NIA study includes a wide array of physiological measures to evaluate the effects of CR.

Unlike the NIA study, the WNPRC study was designed strictly to test the effects of moderate CR on adult rhesus monkeys of Indian origin. The WNPRC study began in 1989 with 30 adult male rhesus monkeys ranging in age from 8 to 14 years [158]. In 1994, 30 females (8–14 years of age) and an additional 16 males (6–14 years of age) were added [159]. Within the groups, animals were evenly randomized based on age and body weight into control and CR conditions. All animals are fed a semi-purified, lactalbumin-based diet containing 15% protein and 10% fat in the morning and receive a piece of food enrichment of approximately 100 calories (e.g. an apple) in the afternoon. This study utilizes a paradigm of 30% CR based on individualized food intake levels that were quantitated daily over a 3–6 month baseline period. At the end of the baseline period, the individualized food allotments were reduced by 10% per month for three months to reach the goal of a 30% CR. Animals in the CR group are fed a diet enriched by 30% in vitamins and minerals. Control animals have free access to food for between 6 and 10 h per day and generally have at least 20 g of food remaining each day. Food intake is quantified daily for all animals [158, 159]. Similar to the NIA study, the WNPRC study includes an extensive panel of physiological measurements to evaluate the efficacy of the CR paradigm.

The third study, performed at UMD, grew out of investigations into glucoregulatory function. This study utilized a paradigm similar to CR termed weight stabilization. Utilizing this approach, animals' food intakes are titrated to allow maintenance of a pre-determined body weight. This particular study included 8 adult male rhesus monkeys (12–19 years of age at study onset) subjected to a weight stabilization protocol that required a caloric reduction of approximately 35%. One hundred and nine (21 females, 96 males; 4–29 years of age at study onset) rhesus monkeys were used as the comparative control group. A key difference with this study, which focused more specifically on glucoregulatory function, was the emergence of 22 hyperinsulinemic animals and 20 diabetic animals within the control group [160]. In contrast to the NIA and WNPRC studies, in the UMD study a large portion of the control animals were obese.

Findings and Candidate Mechanisms

Although the three studies use varying methods, results are generally consistent among the studies. The most striking effects of CR in rhesus macaques involve body composition and glucoregulatory function. CR lowered body weights, decreased fat mass and improved glucoregulatory function at UMD [160], WNPRC [161–163] and the NIA [164, 165]. Both the NIA and WNPRC studies have shown lower bone mass in the CR animals that can be accounted for by lower body mass [166, 167] and maintenance of DHEAS levels with CR. Among other results, the NIA study has shown a reduction in body temperature with CR [168]. The WNPRC study has shown improved cardiovascular profiles in the restricted animals, including reduced levels of C-reactive protein (CRP) and decreased levels of triglyceride and phospholipids associated with low density lipoproteins [169]. Furthermore, we observe the attenuation of sarcopenia in CR animals [170]. All things considered, the real test of the efficacy of CR in this species is its ability to delay the onset of age-related diseases and extend maximal lifespan. To date in our study, it appears as if age-related diseases (i.e., diabetes, cardiovascular disease) are delayed or prevented by CR, but CR's ability to increase maximal lifespan remains unknown.

Humans

Although reduced caloric intake is prevalent in numerous global populations, the effects of malnutrition generally overshadow any benefits of CR. An apparent exception is the relatively large proportion of centenarians in the Okinawan population who reportedly eat fewer calories than individuals on mainland Japan as a whole [171–173]. Increased longevity among these people has been taken as evidence that reduced caloric intake increases average lifespan. In population studies in the USA, individuals reporting lower calorie intakes show increased resistance to developing several cancers and Parkinson's disease [174]. In contrast, there is increasing evidence that overindulgence and the resulting obesity accelerate the onset of numerous disorders previously associated with aging, including diabetes, hypertension, atherosclerosis and cancer [175, 176]. But these data do not provide direct information on the effects of a carefully monitored practice of CR. Maximal lifespan is the gold standard for determining whether an intervention slows primary aging. Although data have been generated from human subjects on CR as described below [177, 178], it seems extremely unlikely that information regarding the effect of CR on maximal lifespan in humans will become available in the foreseeable future.

Methodology

The direct practice of CR is ongoing in two distinct groups. The NIA is currently funding a multicenter study (CALERIE) of the effects CR in women and men. The primary goal of the CALERIE study is to determine whether humans develop the same adaptive responses to CR, such as decreases in the levels of growth factors, inflammatory cytokines, oxidative damage and metabolic rate, that occur in rodents. A second group on CR are members of the Calorie Restriction Society, who are practicing long-term CR with optimal nutrition.

Findings and Candidate Mechanisms

A wealth of data on the impact of short-term (6 months and 1 year long) CR has emerged from the NIH-funded CALERIE studies. The first study of overweight men and women on a 25% CR diet (n=12) for 6 months demonstrated reductions in body weight (~10%), core body temperature, and fasting levels of insulin compared to controls (n=10) [179]. Weight loss was reflected in a 24% reduction in body fat and 27% reduction in visceral fat [180]. Favorable changes in serum risk factors for cardiovascular disease were observed in CR individuals [181]. An impact of short-term, 25% CR on mitochondrial energy metabolism in humans is implied by the increased biogenesis observed in skeletal muscle [182]. A second study of 1 year 20% CR (n=18) demonstrated improved glucose tolerance and insulin action in overweight but healthy people [183]. In addition, substantial improvements in risk factors for cardiovascular disease were observed in agreement with the earlier study [184]. CR resulted in a reduction in weight (~10%), and fat mass comprised 77% of that weight loss [185]. However, the extended period of CR revealed problems with feasibility for long term studies due to lack of adherence to the diet – participants attained ~11% CR rather than the 20% objective for the study. It will be of tremendous interest to determine whether CR is as effective in non-overweight individuals.

The study of long term CR is made possible through the cooperation of individual that maintain a strict self-imposed restricted diet. Long-term practitioners of CR (on average 6 years) have reduced circulating levels of triglycerides, fasting glucose and fasting insulin compared to age and socioeconomic matched controls. Furthermore, blood pressure is lower in CR individuals and favorable lipoprotein profiles associated with reduced risk of atherosclerosis are observed [186, 187]. Clearly these early human studies are very promising and support continued exploration of the mechanism of aging retardation in other species and a means to understand human longevity.

Outcomes and Conclusion

The promise of slowing of human aging brings renewed focus to the biology of aging and the elucidation of the mechanistic basis of aging retardation by CR. While improved health and increased longevity are enormously attractive to most people, the idea of maintaining a reduced-calorie diet indefinitely is sufficiently unattractive to prevent most from undertaking the regimen. As a result, the development of nutraceuticals or drugs that mimic the effects of CR without requiring a severe dietary regimen is an exciting and active area of inquiry [133, 188, 189]. Adding further optimism is the synergistic interplay between the discoveries being rapidly generated in model organisms and their translation into mammalian models. Our most recent findings in our WNPRC aging and CR study indicate that adult-onset moderate CR delays the onset of age-associated pathologies and promotes survival in a primate species [190]. The conserved effect of CR on "health-span" and survival in non-human primates confirms the translatability of CR research as a means to explore aging and disease onset. The CR model has clearly evolved from a gerontological curiosity to likely being a major factor in shaping public health in the decades ahead.

References

1. Jazwinski SM (2005). Yeast longevity and aging – the mitochondrial connection. *Mech Ageing Dev* 126: 243–248.
2. Sinclair DA (2005). Toward a unified theory of caloric restriction and longevity regulation. *Mech Ageing Dev* 126: 987–1002.
3. Sinclair DA and Guarente L (1997). Extrachromosomal rDNA circles – a cause of aging in yeast. *Cell* 91: 1033–1042.
4. Jiang JC, Jaruga E, Repnevskaya MV, and Jazwinski SM (2000). An intervention resembling caloric restriction prolongs life span and retards aging in yeast. *FASEB J* 14: 2135–2137.
5. Lin SJ, Defossez PA, and Guarente L (2000). Requirement of NAD and SIR2 for life-span extension by calorie restriction in *Saccharomyces cerevisiae*. *Science* 289: 2126–2128.

6. Fabrizio P, Li L, and Longo VD (2005). Analysis of gene expression profile in yeast aging chronologically. *Mech Ageing Dev* 126: 11–16.
7. Longo VD and Kennedy BK (2006). Sirtuins in aging and age-related disease. *Cell* 126: 257–268.
8. Smith DL Jr., McClure JM, Matecic M, and Smith JS (2007). Calorie restriction extends the chronological lifespan of *Saccharomyces cerevisiae* independently of the Sirtuins. *Aging Cell* 6: 649–662.
9. Fabrizio P and Longo VD (2007). The chronological life span of *Saccharomyces cerevisiae. Methods Mol Biol* 371: 89–95.
10. Fabrizio P, Gattazzo C, Battistella L, Wei M, Cheng C, McGrew K, and Longo VD (2005). Sir2 blocks extreme life-span extension. *Cell* 123: 655–667.
11. Lin SS, Manchester JK, and Gordon JI (2001). Enhanced gluconeogenesis and increased energy storage as hallmarks of aging in *Saccharomyces cerevisiae. J Biol Chem* 276: 36000–36007.
12. Lin SJ, Kaeberlein M, Andalis AA, Sturtz LA, Defossez PA, Culotta VC, Fink GR, and Guarente L (2002). Calorie restriction extends *Saccharomyces cerevisiae* lifespan by increasing respiration. *Nature* 418: 344–348.
13. Lee YL and Lee CK (2008). Transcriptional response according to strength of calorie restriction in *Saccharomyces cerevisiae. Mol Cells* 26: 299–307.
14. Kaeberlein M, Hu D, Kerr EO, Tsuchiya M, Westman EA, Dang N, Fields S, and Kennedy BK (2005). Increased life span due to calorie restriction in respiratory-deficient yeast. *PLoS Genet* 1: e69.
15. Tsuchiya M, Dang N, Kerr EO, Hu D, Steffen KK, Oakes JA, Kennedy BK, and Kaeberlein M (2006). Sirtuin-independent effects of nicotinamide on lifespan extension from calorie restriction in yeast. *Aging Cell* 5: 505–514.
16. Kirchman PA, Kim S, Lai CY, and Jazwinski SM (1999). Interorganelle signaling is a determinant of longevity in *Saccharomyces cerevisiae. Genetics* 152: 179–190.
17. Epstein CB, Waddle JA, Hale WT, Dave V, Thornton J, Macatee TL, Garner HR, and Butow RA (2001). Genome-wide responses to mitochondrial dysfunction. *Mol Biol Cell* 12: 297–308.
18. Imai S, Armstrong CM, Kaeberlein M, and Guarente L (2000). Transcriptional silencing and longevity protein Sir2 is an NAD- dependent histone deacetylase. *Nature* 403: 795–800.
19. Landry J, Sutton A, Tafrov ST, Heller RC, Stebbins J, Pillus L, and Sternglanz R (2000). The silencing protein SIR2 and its homologs are NAD-dependent protein deacetylases. *Proc Natl Acad Sci U S A* 97: 5807–5811.
20. Tanner KG, Landry J, Sternglanz R, and Denu JM (2000). Silent information regulator 2 family of NAD- dependent histone/protein deacetylases generates a unique product, 1-O-acetyl-ADP-ribose. *Proc Natl Acad Sci U S A* 97: 14178–14182.
21. Kaeberlein M, Kirkland KT, Fields S, and Kennedy BK (2004). Sir2-independent life span extension by calorie restriction in yeast. *PLoS Biol* 2: E296.
22. Smith JS and Boeke JD (1997). An unusual form of transcriptional silencing in yeast ribosomal DNA. *Genes Dev* 11: 241–254.
23. Gottlieb S and Esposito RE (1989). A new role for a yeast transcriptional silencer gene, SIR2, in regulation of recombination in ribosomal DNA. *Cell* 56: 771–776.
24. Kaeberlein M, McVey M, and Guarente L (1999). The SIR2/3/4 complex and SIR2 alone promote longevity in *Saccharomyces cerevisiae* by two different mechanisms. *Genes Dev* 13: 2570–2580.
25. Anderson RM, Bitterman KJ, Wood JG, Medvedik O, and Sinclair DA (2003). Nicotinamide and PNC1 govern lifespan extension by calorie restriction in *Saccharomyces cerevisiae. Nature* 423: 181–185.
26. Anderson RM, Bitterman KJ, Wood JG, Medvedik O, Cohen H, Lin SS, Manchester JK, Gordon JI, and Sinclair DA (2002). Manipulation of a nuclear NAD+ salvage pathway delays aging without altering steady-state NAD+ levels. *J Biol Chem* 277: 18881–18890.

27. Bitterman KJ, Anderson RM, Cohen HY, Latorre-Esteves M, and Sinclair DA (2002). Inhibition of silencing and accelerated aging by nicotinamide, a putative negative regulator of yeast sir2 and human SIRT1. *J Biol Chem* 277: 45099–45107.
28. Medvedik O, Lamming DW, Kim KD, and Sinclair DA (2007). MSN2 and MSN4 link calorie restriction and TOR to sirtuin-mediated lifespan extension in *Saccharomyces cerevisiae*. *PLoS Biol* 5: e261.
29. Lamming DW, Latorre-Esteves M, Medvedik O, Wong SN, Tsang FA, Wang C, Lin SJ, and Sinclair DA (2005). HST2 mediates SIR2-independent life-span extension by calorie restriction. *Science* 309: 1861–1864.
30. Martin DE and Hall MN (2005). The expanding TOR signaling network. *Curr Opin Cell Biol* 17: 158–166.
31. Wullschleger S, Loewith R, and Hall MN (2006). TOR signaling in growth and metabolism. *Cell* 124: 471–484.
32. Kaeberlein M, Powers RW, 3rd, Steffen KK, Westman EA, Hu D, Dang N, Kerr EO, Kirkland KT, Fields S, and Kennedy BK (2005). Regulation of yeast replicative life span by TOR and Sch9 in response to nutrients. *Science* 310: 1193–1196.
33. Li H, Tsang CK, Watkins M, Bertram PG, and Zheng XF (2006). Nutrient regulates Tor1 nuclear localization and association with rDNA promoter. *Nature* 442: 1058–1061.
34. Butow RA and Avadhani NG (2004). Mitochondrial signaling: the retrograde response. *Mol Cell* 14: 1–15.
35. Jazwinski SM (2000). Metabolic control and ageing. *Trends Genet* 16: 506–511.
36. Komeili A, Wedaman KP, O'Shea EK, and Powers T (2000). Mechanism of metabolic control. Target of rapamycin signaling links nitrogen quality to the activity of the Rtg1 and Rtg3 transcription factors. *J Cell Biol* 151: 863–878.
37. DeRisi JL, Iyer VR, and Brown PO (1997). Exploring the metabolic and genetic control of gene expression on a genomic scale. *Science* 278: 680–686.
38. Barros MH, Bandy B, Tahara EB, and Kowaltowski AJ (2004). Higher respiratory activity decreases mitochondrial reactive oxygen release and increases life span in *Saccharomyces cerevisiae*. *J Biol Chem* 279: 49883–49888.
39. Fabrizio P, Pozza F, Pletcher SD, Gendron CM, and Longo VD (2001). Regulation of longevity and stress resistance by Sch9 in yeast. *Science* 292: 288–290.
40. Urban J, Soulard A, Huber A, Lippman S, Mukhopadhyay D, Deloche O, Wanke V, Anrather D, Ammerer G, Riezman H et al. (2007). Sch9 is a major target of TORC1 in *Saccharomyces cerevisiae*. *Mol Cell* 26: 663–674.
41. Fabrizio P, Liou LL, Moy VN, Diaspro A, Valentine JS, Gralla EB, and Longo VD (2003). SOD2 functions downstream of Sch9 to extend longevity in yeast. *Genetics* 163: 35–46.
42. Bonawitz ND, Chatenay-Lapointe M, Pan Y, and Shadel GS (2007). Reduced TOR signaling extends chronological life span via increased respiration and upregulation of mitochondrial gene expression. *Cell Metab* 5: 265–277.
43. Powers RW 3rd, Kaeberlein M, Caldwell SD, Kennedy BK, and Fields S (2006). Extension of chronological life span in yeast by decreased TOR pathway signaling. *Genes Dev* 20: 174–184.
44. Wei M, Fabrizio P, Hu J, Ge H, Cheng C, Li L, and Longo VD (2008). Life span extension by calorie restriction depends on Rim15 and transcription factors downstream of Ras/PKA, Tor, and Sch9. *PLoS Genet* 4: e13.
45. Houthoofd K and Vanfleteren JR (2007). Public and private mechanisms of life extension in *Caenorhabditis elegans*. *Mol Genet Genomics* 277: 601–617.
46. Kenyon C (2005). The plasticity of aging: insights from long-lived mutants. *Cell* 120: 449–460.
47. Braeckman BP, Houthoofd K, and Vanfleteren JR (2001). Insulin-like signaling, metabolism, stress resistance and aging in *Caenorhabditis elegans*. *Mech Ageing Dev* 122: 673–693.
48. Kenyon C, Chang J, Gensch E, Rudner A, and Tabtiang R:AC (1993). elegans mutant that lives twice as long as wild type. *Nature* 366: 461–464.

49. Murphy CT, McCarroll SA, Bargmann CI, Fraser A, Kamath RS, Ahringer J, Li H, and Kenyon C (2003). Genes that act downstream of DAF-16 to influence the lifespan of *Caenorhabditis elegans*. *Nature* 424: 277–283.
50. Garigan D, Hsu AL, Fraser AG, Kamath RS, Ahringer J, and Kenyon C (2002). Genetic analysis of tissue aging in *Caenorhabditis elegans*: a role for heat-shock factor and bacterial proliferation. *Genetics* 161: 1101–1112.
51. Gems D and Riddle DL (2000). Genetic, behavioral and environmental determinants of male longevity in *Caenorhabditis elegans*. *Genetics* 154: 1597–1610.
52. Klass MR (1977). Aging in the nematode *Caenorhabditis elegans*: major biological and environmental factors influencing life span. *Mech Ageing Dev* 6: 413–429.
53. Houthoofd K, Braeckman BP, Lenaerts I, Brys K, De Vreese A, Van Eygen S, and Vanfleteren JR (2002). No reduction of metabolic rate in food restricted *Caenorhabditis elegans*. *Exp Gerontol* 37: 1359–1369.
54. Houthoofd K, Braeckman BP, Lenaerts I, Brys K, De Vreese A, Van Eygen S, and Vanfleteren JR (2002). Axenic growth up-regulates mass-specific metabolic rate, stress resistance, and extends life span in *Caenorhabditis elegans*. *Exp Gerontol* 37: 1371–1378.
55. Walker G, Houthoofd K, Vanfleteren JR, and Gems D (2005). Dietary restriction in *C. elegans*: from rate-of-living effects to nutrient sensing pathways. *Mech Ageing Dev* 126: 929–937.
56. Lakowski B and Hekimi S (1998). The genetics of caloric restriction in *Caenorhabditis elegans*. *Proc Natl Acad Sci U S A* 95: 13091–13096.
57. Kaeberlein TL, Smith ED, Tsuchiya M, Welton KL, Thomas JH, Fields S, Kennedy BK, and Kaeberlein M (2006). Lifespan extension in *Caenorhabditis elegans* by complete removal of food. *Aging Cell* 5: 487–494.
58. Lee GD, Wilson MA, Zhu M, Wolkow CA, de Cabo R, Ingram DK, and Zou S (2006). Dietary deprivation extends lifespan in *Caenorhabditis elegans*. *Aging Cell* 5: 515–524.
59. Smith ED, Kaeberlein TL, Lydum BT, Sager J, Welton KL, Kennedy BK, and Kaeberlein M (2008). Age- and calorie-independent life span extension from dietary restriction by bacterial deprivation in *Caenorhabditis elegans*. *BMC Dev Biol* 8: 49.
60. Wolff S and Dillin A (2006). The trifecta of aging in *Caenorhabditis elegans*. *Exp Gerontol* 41: 894–903.
61. Castelein N, Hoogewijs D, De Vreese A, Braeckman BP, and Vanfleteren JR (2008). Dietary restriction by growth in axenic medium induces discrete changes in the transcriptional output of genes involved in energy metabolism in *Caenorhabditis elegans*. *Biotechnol J* 3: 803–812.
62. Bishop NA and Guarente L (2007). Two neurons mediate diet-restriction-induced longevity in *C. elegans*. *Nature* 447: 545–549.
63. Schulz TJ, Zarse K, Voigt A, Urban N, Birringer M, and Ristow M (2007). Glucose restriction extends *Caenorhabditis elegans* life span by inducing mitochondrial respiration and increasing oxidative stress. *Cell Metab* 6: 280–293.
64. Houthoofd K, Braeckman BP, Johnson TE, and Vanfleteren JR (2003). Life extension via dietary restriction is independent of the Ins/IGF-1 signalling pathway in *Caenorhabditis elegans*. *Exp Gerontol* 38: 947–954.
65. Greer EL, Dowlatshahi D, Banko MR, Villen J, Hoang K, Blanchard D, Gygi SP, and Brunet A (2007). An AMPK-FOXO pathway mediates longevity induced by a novel method of dietary restriction in *C. elegans*. *Curr Biol* 17: 1646–1656.
66. Tissenbaum HA and Guarente L (2001). Increased dosage of a sir-2 gene extends lifespan in *Caenorhabditis elegans*. *Nature* 410: 227–230.
67. Lin K, Dorman JB, Rodan A, and Kenyon C (1997). daf-16: An HNF-3/forkhead family member that can function to double the life-span of *Caenorhabditis elegans*. *Science* 278: 1319–1322.
68. Wang Y, Oh SW, Deplancke B, Luo J, Walhout AJ, and Tissenbaum HA (2006). *C. elegans* 14-3-3 proteins regulate life span and interact with SIR-2.1 and DAF-16/FOXO. *Mech Ageing Dev* 127: 741–747.

69. Berdichevsky A, Viswanathan M, Horvitz HR, and Guarente L (2006). *C. elegans* SIR-2.1 interacts with 14-3-3 proteins to activate DAF-16 and extend life span. *Cell* 125: 1165–1177.
70. Panowski SH, Wolff S, Aguilaniu H, Durieux J, and Dillin A (2007). PHA-4/Foxa mediates diet-restriction-induced longevity of *C. elegans. Nature* 447: 550–555.
71. Vellai T, Takacs-Vellai K, Zhang Y, Kovacs AL, Orosz L, and Muller F (2003). Genetics: influence of TOR kinase on lifespan in *C. elegans. Nature* 426: 620.
72. Jia K, Chen D, and Riddle DL (2004). The TOR pathway interacts with the insulin signaling pathway to regulate *C. elegans* larval development, metabolism and life span. *Development* 131: 3897–3906.
73. Hansen M, Taubert S, Crawford D, Libina N, Lee SJ, and Kenyon C (2007). Lifespan extension by conditions that inhibit translation in *Caenorhabditis elegans. Aging Cell* 6: 95–110.
74. Helfand SL and Rogina B (2003). Molecular genetics of aging in the fly: is this the end of the beginning? *Bioessays* 25: 134–141.
75. Magwere T, Chapman T, and Partridge L (2004). Sex differences in the effect of dietary restriction on life span and mortality rates in female and male *Drosophila melanogaster. J Gerontol A Biol Sci Med Sci* 59: 3–9.
76. Bass TM, Grandison RC, Wong R, Martinez P, Partridge L, and Piper MD (2007). Optimization of dietary restriction protocols in *Drosophila. J Gerontol A Biol Sci Med Sci* 62: 1071–1081.
77. Tatar M (2007). Diet restriction in *Drosophila melanogaster*. Design and analysis. *Interdiscip Top Gerontol* 35: 115–136.
78. Mair W, Piper MD, and Partridge L (2005). Calories do not explain extension of life span by dietary restriction in *Drosophila. PLoS Biol* 3: e223.
79. Min KJ, Flatt T, Kulaots I, and Tatar M (2007). Counting calories in *Drosophila* diet restriction. *Exp Gerontol* 42: 247–251.
80. Min KJ and Tatar M (2006). Restriction of amino acids extends lifespan in *Drosophila melanogaster. Mech Ageing Dev* 127: 643–646.
81. Carvalho GB, Kapahi P, and Benzer S (2005). Compensatory ingestion upon dietary restriction in *Drosophila melanogaster. Nat Methods* 2: 813–815.
82. Chapman T and Partridge L (1996). Female fitness in *Drosophila melanogaster*: an interaction between the effect of nutrition and of encounter rate with males. *Proc Biol Sci* 263: 755–759.
83. Piper MD and Partridge L (2007). Dietary restriction in Drosophila: delayed aging or experimental artefact? *PLoS Genet* 3: e57.
84. Pletcher SD, Macdonald SJ, Marguerie R, Certa U, Stearns SC, Goldstein DB, and Partridge L (2002). Genome-wide transcript profiles in aging and calorically restricted *Drosophila melanogaster. Curr Biol* 12: 712–723.
85. Morrow G and Tanguay RM (2008). Mitochondria and ageing in *Drosophila. Biotechnol J* 3: 728–739.
86. Miwa S, Riyahi K, Partridge L, and Brand MD (2004). Lack of correlation between mitochondrial reactive oxygen species production and life span in *Drosophila. Ann N Y Acad Sci* 1019: 388–391.
87. Girardot F, Lasbleiz C, Monnier V, and Tricoire H (2006). Specific age-related signatures in *Drosophila* body parts transcriptome. *BMC Genomics* 7: 69.
88. Magwere T, Goodall S, Skepper J, Mair W, Brand MD, and Partridge L (2006). The effect of dietary restriction on mitochondrial protein density and flight muscle mitochondrial morphology in *Drosophila. J Gerontol A Biol Sci Med Sci* 61: 36–47.
89. Rogina B, Helfand SL, and Frankel S (2002). Longevity regulation by *Drosophila* Rpd3 deacetylase and caloric restriction. *Science* 298: 1745.
90. Rogina B and Helfand SL (2004). Sir2 mediates longevity in the fly through a pathway related to calorie restriction. *Proc Natl Acad Sci U S A* 101: 15998–16003.

91. Wood JG, Rogina B, Lavu S, Howitz K, Helfand SL, Tatar M, and Sinclair D (2004). Sirtuin activators mimic caloric restriction and delay ageing in metazoans. *Nature* 430: 686–689.
92. Simon AF, Shih C, Mack A, and Benzer S (2003). Steroid control of longevity in *Drosophila melanogaster*. *Science* 299: 1407–1410.
93. Tsai CC, Kao HY, Yao TP, McKeown M, and Evans RM (1999). SMRTER, a *Drosophila* nuclear receptor coregulator, reveals that EcR-mediated repression is critical for development. *Mol Cell* 4: 175–186.
94. Pile LA, Spellman PT, Katzenberger RJ, and Wassarman DA (2003). The SIN3 deacetylase complex represses genes encoding mitochondrial proteins: implications for the regulation of energy metabolism. *J Biol Chem* 278: 37840–37848.
95. Tatar M, Bartke A, and Antebi A (2003). The endocrine regulation of aging by insulin-like signals. *Science* 299: 1346–1351.
96. Junger MA, Rintelen F, Stocker H, Wasserman JD, Vegh M, Radimerski T, Greenberg ME, and Hafen E (2003). The *Drosophila* forkhead transcription factor FOXO mediates the reduction in cell number associated with reduced insulin signaling. *J Biol* 2: 20.
97. Giannakou ME, Goss M, Junger MA, Hafen E, Leevers SJ, and Partridge L (2004). Long-lived *Drosophila* with overexpressed dFOXO in adult fat body. *Science* 305: 361.
98. Hwangbo DS, Gersham B, Tu MP, Palmer M, and Tatar M (2004). *Drosophila* dFOXO controls lifespan and regulates insulin signalling in brain and fat body. *Nature* 429: 562–566.
99. Giannakou ME, Goss M, and Partridge L (2008). Role of dFOXO in lifespan extension by dietary restriction in *Drosophila melanogaster*: not required, but its activity modulates the response. *Aging Cell* 7: 187–198.
100. Min KJ, Yamamoto R, Buch S, Pankratz M, and Tatar M (2008). *Drosophila* lifespan control by dietary restriction independent of insulin-like signaling. *Aging Cell* 7: 199–206.
101. Colombani J, Raisin S, Pantalacci S, Radimerski T, Montagne J, and Leopold P (2003). A nutrient sensor mechanism controls *Drosophila* growth. *Cell* 114: 739–749.
102. Kapahi P, Zid BM, Harper T, Koslover D, Sapin V, and Benzer S (2004). Regulation of lifespan in *Drosophila* by modulation of genes in the TOR signaling pathway. *Curr Biol* 14: 885–890.
103. Luong N, Davies CR, Wessells RJ, Graham SM, King MT, Veech R, Bodmer R, and Oldham SM (2006). Activated FOXO-mediated insulin resistance is blocked by reduction of TOR activity. *Cell Metab* 4: 133–142.
104. Schieke SM, Phillips D, McCoy JP, Jr., Aponte AM, Shen RF, Balaban RS, and Finkel T (2006). The mammalian target of rapamycin (mTOR) pathway regulates mitochondrial oxygen consumption and oxidative capacity. *J Biol Chem* 281: 27643–27652.
105. McCay CM, Crowell MF, and Maynard LA (1935). The effect of retarded growth upon the length of life span and upon the ultimate body size. *Nutrition* 5: 155–171; discussion 172.
106. Weindruch RH and Walford RL (1988). *The Retardation of Aging and Disease by Dietary Restriction*. Springfield, IL: Charles C Thomas.
107. Piper MD and Bartke A (2008). Diet and aging. *Cell Metab* 8: 99–104.
108. Weindruch R and Sohal RS (1997). Seminars in medicine of the Beth Israel Deaconess Medical Center. Caloric intake and aging. *N Engl J Med* 337: 986–994.
109. Weindruch R, Walford RL, Fligiel S, and Guthrie D (1986). The retardation of aging in mice by dietary restriction: longevity, cancer, immunity and lifetime energy intake. *J Nutr* 116: 641–654.
110. Ross MH (1961). Length of life and nutrition in the rat. *J Nutr* 75: 197–210.
111. Ross MH and Bras G (1973). Influence of protein under- and overnutrition on spontaneous tumor prevalence in the rat. *J Nutr* 103: 944–963.
112. Weindruch R, Gottesman SR, and Walford RL (1982). Modification of age-related immune decline in mice dietarily restricted from or after midadulthood. *Proc Natl Acad Sci U S A* 79: 898–902.

113. Cao SX, Dhahbi JM, Mote PL, and Spindler SR (2001). Genomic profiling of short- and long-term caloric restriction effects in the liver of aging mice. *Proc Natl Acad Sci U S A* 98: 10630–10635.
114. Aspnes LE, Lee CM, Weindruch R, Chung SS, Roecker EB, and Aiken JM (1997). Caloric restriction reduces fiber loss and mitochondrial abnormalities in aged rat muscle. *FASEB J* 11: 573–581.
115. Zangarelli A, Chanseaume E, Morio B, Brugere C, Mosoni L, Rousset P, Giraudet C, Patrac V, Gachon P, Boirie Y et al. (2006). Synergistic effects of caloric restriction with maintained protein intake on skeletal muscle performance in 21-month-old rats: a mitochondria-mediated pathway. *FASEB J* 20: 2439–2450.
116. Liang H, Masoro EJ, Nelson JF, Strong R, McMahan CA, and Richardson A (2003). Genetic mouse models of extended lifespan. *Exp Gerontol* 38: 1353–1364.
117. Barger JL, Walford RL, and Weindruch R (2003). The retardation of aging by caloric restriction: its significance in the transgenic era. *Exp Gerontol* 38: 1343–1351.
118. Swindell WR (2007). Gene expression profiling of long-lived dwarf mice: longevity-associated genes and relationships with diet, gender and aging. *BMC Genomics* 8: 353.
119. Martin B, Mattson MP, and Maudsley S (2006). Caloric restriction and intermittent fasting: two potential diets for successful brain aging. *Ageing Res Rev* 5: 332–353.
120. Varady KA and Hellerstein MK (2007). Alternate-day fasting and chronic disease prevention: a review of human and animal trials. *Am J Clin Nutr* 86: 7–13.
121. Inness CL and Metcalfe NB (2008). The impact of dietary restriction, intermittent feeding and compensatory growth on reproductive investment and lifespan in a short-lived fish. *Proc Biol Sci* 275: 1703–1708.
122. Plunet WT, Streijger F, Lam CK, Lee JH, Liu J, and Tetzlaff W (2008). Dietary restriction started after spinal cord injury improves functional recovery. *Exp Neurol* 213: 28–35.
123. Castello L, Froio T, Cavallini G, Biasi F, Sapino A, Leonarduzzi G, Bergamini E, Poli G, and Chiarpotto E (2005). Calorie restriction protects against age-related rat aorta sclerosis. *FASEB J* 19: 1863–1865.
124. Richie JP Jr., Leutzinger Y, Parthasarathy S, Malloy V, Orentreich N, and Zimmerman JA (1994). Methionine restriction increases blood glutathione and longevity in F344 rats. *FASEB J* 8: 1302–1307.
125. Miller RA, Buehner G, Chang Y, Harper JM, Sigler R, and Smith-Wheelock M (2005). Methionine-deficient diet extends mouse lifespan, slows immune and lens aging, alters glucose, T4, IGF-I and insulin levels, and increases hepatocyte MIF levels and stress resistance. *Aging Cell* 4: 119–125.
126. Sanz A, Caro P, Ayala V, Portero-Otin M, Pamplona R, and Barja G (2006). Methionine restriction decreases mitochondrial oxygen radical generation and leak as well as oxidative damage to mitochondrial DNA and proteins. *FASEB J* 20: 1064–1073.
127. Ingram DK, Zhu M, Mamczarz J, Zou S, Lane MA, and Roth GS (2006). deCabo R: calorie restriction mimetics: an emerging research field. *Aging Cell* 5: 97–108.
128. Opie LH and Lecour S (2007). The red wine hypothesis: from concepts to protective signalling molecules. *Eur Heart J* 28: 1683–1693.
129. Howitz KT, Bitterman KJ, Cohen HY, Lamming DW, Lavu S, Wood JG, Zipkin RE, Chung P, Kisielewski A, Zhang LL et al. (2003). Small molecule activators of sirtuins extend *Saccharomyces cerevisiae* lifespan. *Nature* 425: 191–196.
130. Bass TM, Weinkove D, Houthoofd K, Gems D, and Partridge L (2007). Effects of resveratrol on lifespan in *Drosophila melanogaster* and *Caenorhabditis elegans*. *Mech Ageing Dev* 128: 546–552.
131. Kaeberlein M, McDonagh T, Heltweg B, Hixon J, Westman EA, Caldwell SD, Napper A, Curtis R, DiStefano PS, Fields S et al. (2005). Substrate-specific activation of sirtuins by resveratrol. *J Biol Chem* 280: 17038–17045.

132. Baur JA, Pearson KJ, Price NL, Jamieson HA, Lerin C, Kalra A, Prabhu VV, Allard JS, Lopez-Lluch G, Lewis K et al. (2006). Resveratrol improves health and survival of mice on a high-calorie diet. *Nature* 444: 337–342.
133. Barger JL, Kayo T, Vann JM, Arias EB, Wang J, Hacker TA, Wang Y, Raederstorff D, Morrow JD, Leeuwenburgh C et al. (2008). A low dose of dietary resveratrol partially mimics caloric restriction and retards aging parameters in mice. *PLoS ONE* 3: e2264.
134. Anderson RM and Weindruch R (2007). Metabolic reprogramming in dietary restriction. *Interdiscip Top Gerontol* 35: 18–38.
135. Guarente L (2008). Mitochondria – a nexus for aging, calorie restriction, and sirtuins? *Cell* 132: 171–176.
136. Kennedy BK, Steffen KK, and Kaeberlein M (2007). Ruminations on dietary restriction and aging. *Cell Mol Life Sci* 64: 1323–1328.
137. Weindruch R, Colman RJ, Perez V, and Richardson AG (2008). How does caloric restriction increase the longevity of mammals? In *Molecular Biology of Aging*. (Guarente L, Partridge, L, Wallace, D, eds). New York: Cold Spring Harbor Laboratory Press.
138. Lee CK, Klopp RG, Weindruch R, and Prolla TA (1999). Gene expression profile of aging and its retardation by caloric restriction. *Science* 285: 1390–1393.
139. Lee CK, Weindruch R, and Prolla TA (2000). Gene-expression profile of the ageing brain in mice. *Nat Genet* 25: 294–297.
140. Lee CK, Allison DB, Brand J, Weindruch R, and Prolla TA (2002). Transcriptional profiles associated with aging and middle age-onset caloric restriction in mouse hearts. *Proc Natl Acad Sci U S A* 99: 14988–14993.
141. Higami Y, Pugh TD, Page GP, Allison DB, Prolla TA, and Weindruch R (2004). Adipose tissue energy metabolism: altered gene expression profile of mice subjected to long-term caloric restriction. *FASEB J* 18: 415–417.
142. Higami Y, Barger JL, Page GP, Allison DB, Smith SR, Prolla TA, and Weindruch R (2006). Energy restriction lowers the expression of genes linked to inflammation, the cytoskeleton, the extracellular matrix, and angiogenesis in mouse adipose tissue. *J Nutr* 136: 343–352.
143. Mohamed-Ali V, Pinkney JH, and Coppack SW (1998). Adipose tissue as an endocrine and paracrine organ. *Int J Obes Relat Metab Disord* 22: 1145–1158.
144. Kershaw EE and Flier JS (2004). Adipose tissue as an endocrine organ. *J Clin Endocrinol Metab* 89: 2548–2556.
145. Ahima RS (2006). Adipose tissue as an endocrine organ. *Obesity (Silver Spring)* 14(Suppl 5): 242S–249S.
146. Finck BN and Kelly DP (2006). PGC-1 coactivators: inducible regulators of energy metabolism in health and disease. *J Clin Invest* 116: 615–622.
147. Scarpulla RC (2006). Nuclear control of respiratory gene expression in mammalian cells. *J Cell Biochem* 97: 673–683.
148. Corton JC and Brown-Borg HM (2005). Peroxisome proliferator-activated receptor gamma coactivator 1 in caloric restriction and other models of longevity. *J Gerontol A Biol Sci Med Sci* 60: 1494–1509.
149. Puigserver P and Spiegelman BM (2003). Peroxisome proliferator-activated receptor-gamma coactivator 1 alpha (PGC-1 alpha): transcriptional coactivator and metabolic regulator. *Endocr Rev* 24: 78–90.
150. Evans RM, Barish GD, and Wang YX (2004). PPARs and the complex journey to obesity. *Nat Med* 10: 355–361.
151. Corton JC, Apte U, Anderson SP, Limaye P, Yoon L, Latendresse J, Dunn C, Everitt JI, Voss KA, Swanson C et al. (2004). Mimetics of caloric restriction include agonists of lipid-activated nuclear receptors. *J Biol Chem* 279: 46204–46212.
152. Anderson RM, Barger JL, Edwards MG, Braun KH, O'Connor CE, Prolla TA, and Weindruch R (2008). Dynamic regulation of PGC-1alpha localization and turnover implicates mitochondrial adaptation in calorie restriction and the stress response. *Aging Cell* 7: 101–111.

153. Lawler DF, Larson BT, Ballam JM, Smith GK, Biery DN, Evans RH, Greeley EH, Segre M, Stowe HD, and Kealy RD (2008). Diet restriction and ageing in the dog: major observations over two decades. *Br J Nutr* 99: 793–805.
154. Ingram DK, Cutler RG, Weindruch R, Renquist DM, Knapka JJ, April M, Belcher CT, Clark MA, Hatcherson CD, Marriott BM et al. (1990). Dietary restriction and aging: the initiation of a primate study. *J Gerontol* 45: B148–B163.
155. Lane MA, Ingram DK, Cutler RG, Knapka JJ, Barnard DE, and Roth GS (1992). Dietary restriction in nonhuman primates: progress report on the NIA study. *Ann N Y Acad Sci* 673: 36–45.
156. Mattison JA, Lane MA, Roth GS, and Ingram DK (2003). Calorie restriction in rhesus monkeys. *Exp Gerontol* 38: 35–46.
157. National Research Council CoAN, Agricultural Board (1978). Nutrient requirements of nonhuman primates, Washington, DC.
158. Kemnitz JW, Weindruch R, Roecker EB, Crawford K, Kaufman PL, and Ershler WB (1993). Dietary restriction of adult male rhesus monkeys: design, methodology, and preliminary findings from the first year of study. *J Gerontol* 48: B17–B26.
159. Ramsey JJ, Colman RJ, Binkley NC, Christensen JD, Gresl TA, Kemnitz JW, and Weindruch R (2000). Dietary restriction and aging in rhesus monkeys: the University of Wisconsin study. *Exp Gerontol* 35: 1131–1149.
160. Bodkin NL, Alexander TM, Ortmeyer HK, Johnson E, and Hansen BC (2003). Mortality and morbidity in laboratory-maintained Rhesus monkeys and effects of long-term dietary restriction. *J Gerontol A Biol Sci Med Sci* 58: 212–219.
161. Colman RJ, Ramsey JJ, Roecker EB, Havighurst T, Hudson JC, and Kemnitz JW (1999). Body fat distribution with long-term dietary restriction in adult male rhesus macaques. *J Gerontol A Biol Sci Med Sci* 54: B283–B290.
162. Gresl TA, Colman RJ, Roecker EB, Havighurst TC, Huang Z, Allison DB, Bergman RN, and Kemnitz JW (2001). Dietary restriction and glucose regulation in aging rhesus monkeys: a follow-up report at 8.5 yr. *Am J Physiol Endocrinol Metab* 281: E757–E765.
163. Kemnitz JW, Roecker EB, Weindruch R, Elson DF, Baum ST, and Bergman RN (1994). Dietary restriction increases insulin sensitivity and lowers blood glucose in rhesus monkeys. *Am J Physiol* 266: E540–E547.
164. Lane MA, Ball SS, Ingram DK, Cutler RG, Engel J, Read V, and Roth GS (1995). Diet restriction in rhesus monkeys lowers fasting and glucose-stimulated glucoregulatory end points. *Am J Physiol* 268: E941–E948.
165. Mattison JA, Roth GS, Lane MA, and Ingram DK (2007). Dietary restriction in aging nonhuman primates. *Interdiscip Top Gerontol* 35: 137–158.
166. Black A, Allison DB, Shapses SA, Tilmont EM, Handy AM, Ingram DK, Roth GS, and Lane MA (2001). Calorie restriction and skeletal mass in rhesus monkeys (*Macaca mulatta*): evidence for an effect mediated through changes in body size. *J Gerontol A Biol Sci Med Sci* 56: B98–B107.
167. Colman RJ, Lane MA, Binkley N, Wegner FH, and Kemnitz JW (1999). Skeletal effects of aging in male rhesus monkeys. *Bone* 24: 17–23.
168. Lane MA, Baer DJ, Rumpler WV, Weindruch R, Ingram DK, Tilmont EM, Cutler RG, and Roth GS (1996). Calorie restriction lowers body temperature in rhesus monkeys, consistent with a postulated anti-aging mechanism in rodents. *Proc Natl Acad Sci U S A* 93: 4159–4164.
169. Edwards IJ, Rudel LL, Terry JG, Kemnitz JW, Weindruch R, and Cefalu WT (1998). Caloric restriction in rhesus monkeys reduces low density lipoprotein interaction with arterial proteoglycans. *J Gerontol A Biol Sci Med Sci* 53: B443–B448.
170. Colman RJ, Beasley TM, Allison DB, and Weindruch R (2008). Attenuation of sarcopenia by dietary restriction in rhesus monkeys. *J Gerontol A Biol Sci Med Sci* 63: 556–559.
171. Chan YC, Suzuki M, and Yamamoto S (1997). Dietary, anthropometric, hematological and biochemical assessment of the nutritional status of centenarians and elderly people in Okinawa, Japan. *J Am Coll Nutr* 16: 229–235.

172. Kagawa Y (1978). Impact of Westernization on the nutrition of Japanese: changes in physique, cancer, longevity and centenarians. *Prev Med* 7: 205–217.
173. Suzuki M, Wilcox BJ, and Wilcox CD (2001). Implications from and for food cultures for cardiovascular disease: longevity. *Asia Pac J Clin Nutr* 10: 165–171.
174. Weindruch R (2003). Caloric restriction, gene expression, and aging. *Alzheimer Dis Assoc Disord* 17(Suppl 2): S58–S59.
175. Das M, Gabriely I, and Barzilai N (2004). Caloric restriction, body fat and ageing in experimental models. *Obes Rev* 5: 13–19.
176. Wisse BE (2004). The inflammatory syndrome: the role of adipose tissue cytokines in metabolic disorders linked to obesity. *J Am Soc Nephrol* 15: 2792–2800.
177. Heilbronn LK and Ravussin E (2003). Calorie restriction and aging: review of the literature and implications for studies in humans. *Am J Clin Nutr* 78: 361–369.
178. Holloszy JO and Fontana L (2007). Caloric restriction in humans. *Exp Gerontol* 42: 709–712.
179. Heilbronn LK, de Jonge L, Frisard MI, DeLany JP, Larson-Meyer DE, Rood J, Nguyen T, Martin CK, Volaufova J, Most MM et al. (2006). Effect of 6-month calorie restriction on biomarkers of longevity, metabolic adaptation, and oxidative stress in overweight individuals: a randomized controlled trial. *J Am Med Assoc* 295: 1539–1548.
180. Redman LM, Heilbronn LK, Martin CK, Alfonso A, Smith SR, and Ravussin E (2007). Effect of calorie restriction with or without exercise on body composition and fat distribution. *J Clin Endocrinol Metab* 92: 865–872.
181. Lefevre M, Redman LM, Heilbronn LK, Smith JV, Martin CK, Rood JC, Greenway FL, Williamson DA, Smith SR, and Ravussin E (2009). Caloric restriction alone and with exercise improves CVD risk in healthy non-obese individuals. *Atherosclerosis* 203(1): 206–213.
182. Civitarese AE, Carling S, Heilbronn LK, Hulver MH, Ukropcova B, Deutsch WA, Smith SR, and Ravussin E (2007). Calorie restriction increases muscle mitochondrial biogenesis in healthy humans. *PLoS Med* 4: e76.
183. Weiss EP, Racette SB, Villareal DT, Fontana L, Steger-May K, Schechtman KB, Klein S, and Holloszy JO (2006). Improvements in glucose tolerance and insulin action induced by increasing energy expenditure or decreasing energy intake: a randomized controlled trial. *Am J Clin Nutr* 84: 1033–1042.
184. Fontana L, Villareal DT, Weiss EP, Racette SB, Steger-May K, Klein S, and Holloszy JO (2007). Calorie restriction or exercise: effects on coronary heart disease risk factors. A randomized, controlled trial. *Am J Physiol Endocrinol Metab* 293: E197–E202.
185. Racette SB, Weiss EP, Villareal DT, Arif H, Steger-May K, Schechtman KB, Fontana L, Klein S, and Holloszy JO (2006). One year of caloric restriction in humans: feasibility and effects on body composition and abdominal adipose tissue. *J Gerontol A Biol Sci Med Sci* 61: 943–950.
186. Fontana L, Meyer TE, Klein S, and Holloszy JO (2004). Long-term calorie restriction is highly effective in reducing the risk for atherosclerosis in humans. *Proc Natl Acad Sci U S A* 101: 6659–6663.
187. Meyer TE, Kovacs SJ, Ehsani AA, Klein S, Holloszy JO, and Fontana L (2006). Long-term caloric restriction ameliorates the decline in diastolic function in humans. *J Am Coll Cardiol* 47: 398–402.
188. Barger JL, Kayo T, Pugh TD, Prolla TA, and Weindruch R (2008). Short-term consumption of a resveratrol-containing nutracuetical mixture mimics gene expression of long-term caloric restriction in mouse heart. *Exp Gerontol* 43(9): 859-866.
189. Pearson KJ, Baur JA, Lewis KN, Peshkin L, Price NL, Labinskyy N, Swindell WR, Kamara D, Minor RK, Perez E et al. (2008). Resveratrol delays age-related deterioration and mimics transcriptional aspects of dietary restriction without extending life span. *Cell Metab* 8: 157–168.
190. Colman RJ, Anderson RM, Johnson SC, Kastman EK, Kosmatka KJ, Beasley TM, Allison DB, Cruzen C, Simmons HA, Kemnitz JW, and Weindruch R (2009). Caloric Restriction Delays Disease Onset and Mortality in Rhesus Monkeys. *Science* 325: 201–204.

Cell Replication Rates In Vivo and In Vitro and Wound Healing as Affected by Animal Age, Diet, and Species

Norman S. Wolf

Abstract This chapter deals with the loss of cell replication capacity, or in some cases loss of a stimulus to carry out replication, that occurs with aging in animals of several species. The effect of dietary conditions, hormonal stimulus, and metabolic status are considered. The overall conclusion is that an age-related slowing of steady state or stimulated cell replication occurs in several tissues, whether tested in vivo or in vitro, and that this is also apparent in non-first intention wound healing. A brief summary of non-mammalian cell, tissue, and body part healing and replacement is included.

Keywords Human · Dog · Monkey · Mouse · Rat · Anuran · Froglet · Wound healing · Regeneration · Age · Caloric restriction · Bone · Dermis · Fracture · Cell replication · Telomere

Introduction

The rate of cell division decreases as a function of age and this has been linked to cellular, stromal, and circulating factors. When cell replication is tested in vivo all 3 factors may contribute. The latter two conditions can be eliminated by determining the replicative capacity of young and old donor cells in vitro, although other in situ conditions are lost. The apoptotic loss of useful cells, and the retention of replication-deficient senescent cells that may influence neighboring normal cells must be included in this rather complex mixture of aging changes that affect cell replication. The internal cell changes, such as oxidative damage to DNA, mitochondrial efficiency changes, and cell membrane content have been examined in aging cells and tissues. The capacity to heal non-first intention wounds is decreased with aging in several species. In previous studies my laboratory has noted age-related

N.S. Wolf (✉)
Department of Pathology, University of Washington, Seattle, WA 91895, USA
e-mail: normwolf@u.washington.edu

N.S. Wolf (ed.), *The Comparative Biology of Aging*,
DOI 10.1007/978-90-481-3465-6_5, © Springer Science+Business Media B.V. 2010

differences in cell turnover rates in vivo and cell replication capacity in vitro for several organs, as well as the delayed healing of wounds in old versus young animals. We also reported that long term caloric restriction (CR) had a dual effect on the cell replication in vivo and on wound healing in mice in most of the tissues that were examined, and that this depended upon nutritional availability. This information is provided in some detail below. Studies by many other laboratories are included in this chapter. Several studies have reported on the delayed wound or burn site healing in old human tissues, as well as in rodents and monkeys, and hormonal effects on this. The cause(s) for these age-related changes are probably multi-factorial, ranging from oxidative, inflammatory, and immune damage to nuclear and mitochondrial DNA that then contains altered message and metabolism, to protein cross-linkage, and cell membrane content, or a combination of all of these. Wound healing rate most certainly is dependent upon blood and plasma circulation and the delivery of hormones, cytokines and oxygen. I have included in the material that follows a description of events found in many species, noting the conditions under which these occur and attempting to explore the possible underlying molecular mechanisms. Finally, I have added reports on organ regeneration in non-mammals that suggest the possibility of increasing the wound healing capacity of humans.

Cell Replication and Wound Healing Activity Decrease with Old Age

Cell replication rates differ in aged versus young animal tissues under steady state ad libitum (AL) and caloric restriction (CR) conditions. For most tissues the in vivo cell replication rate of a CR animal depends upon the recent availability of sufficient energy, but the animal's age is always a factor.

Several studies in our laboratory have shown a slowing of cell replication in the aged mouse in several organs under steady state conditions. In addition, these studies showed the influence of caloric restriction in tissues of aged mice and in rhesus monkeys [1, 2–7].

In Vivo Studies of Cell Growth Capacity

In vivo studies carried out after 2 weeks continuous exposure to 2 mg/kg/h of BrdU uptake from subcutaneously placed mini-osmotic pumps indicated a reduced turnover of cells in several organs with age. In Fig. 1 those organs of B6D2F1 mice in which the pumps had been in place for 2 weeks were: bone marrow endothelial cells, hepatic cells, dermal fibroblasts, kidney epithelial cells, and pancreatic epithelial cells. A reduction in cell turnover with age was seen in all instances.

Shown in Fig. 2 are the results of 2 weeks administration of BrdU in young or old B6D2F1 mice with or without long term CR status and with or without placing the CR mice on AL feeding (referred to as refeeding, RF) for 1 week preceding the

Fig. 1 Cell replication rates for several organs and tissues as percent of cells undergoing at least one cell division in vivo, indicated by BrdU uptake over a 2 week period [4]

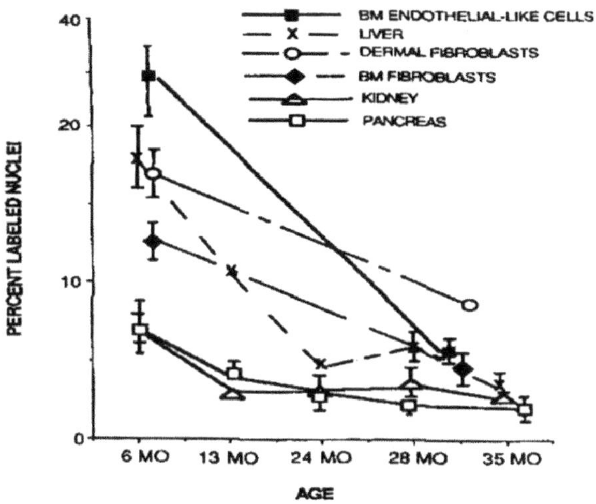

measurement of BrdU positive cells in multiple tissue sections of (A) Hepatocytes, (B) Kidney tubular cells, (C) Pancreatic acinar cells. It is notable that young and old AL fed mice differed significantly and that in most instances a change to AL feeding before measurement was necessary to increase the rate of cell turnover in the old CR mice. Under this condition cellular replicative capacity was preserved by long-term 40% caloric restriction (CR) that was begun from the fourth month of life and then continued onward. A difference in the rate of cell turnover was already present by midlife in AL diet animals, as seen in a comparison of 6 month old and 10 month old mice, but was most marked in the comparisons between AL and CR 28 month old mice. Although the "refeeding", i.e. placing the CR animals on AL diet, was for 4–8 weeks in this study, there also was a marked positive cell growth and wound closure effect with refeeding beginning as little as 48 h before skin wounding [3]. Apparently, energy reserves were not sufficient in the old CR mouse to allow the cell replication to return to young mouse level, thus the need for a brief period of refeeding. A different interpretation of this finding could be that upon refeeding hormone levels and/or a metabolic level adjustment occurred, rather than simply energy availability representing the primary factor.

The value of CR was not as marked when the BrdU uptake of dermal fibroblasts was measured, although refeeding of CR mice was still positive in cell turnover in this study group ($p < 0.05$), as illustrated in Fig. 3. As was found in the internal organs, unstimulated turnover in the 6 month old mouse dermal fibroblasts was actually less in the CR than the AL groups without refeeding (left hand panel). However, after refeeding CR diet animals an AL diet for 4 weeks the dermal fibroblasts from the 28 month old B6D2F1 mice significantly exceeded their ad libitum and CR only fed counterparts ($p = 0.05$). It is notable that 8 weeks of refeeding did not result in this strong change. It can be assumed that the mice were returned to a steady state AL status during this longer period of AL diet.

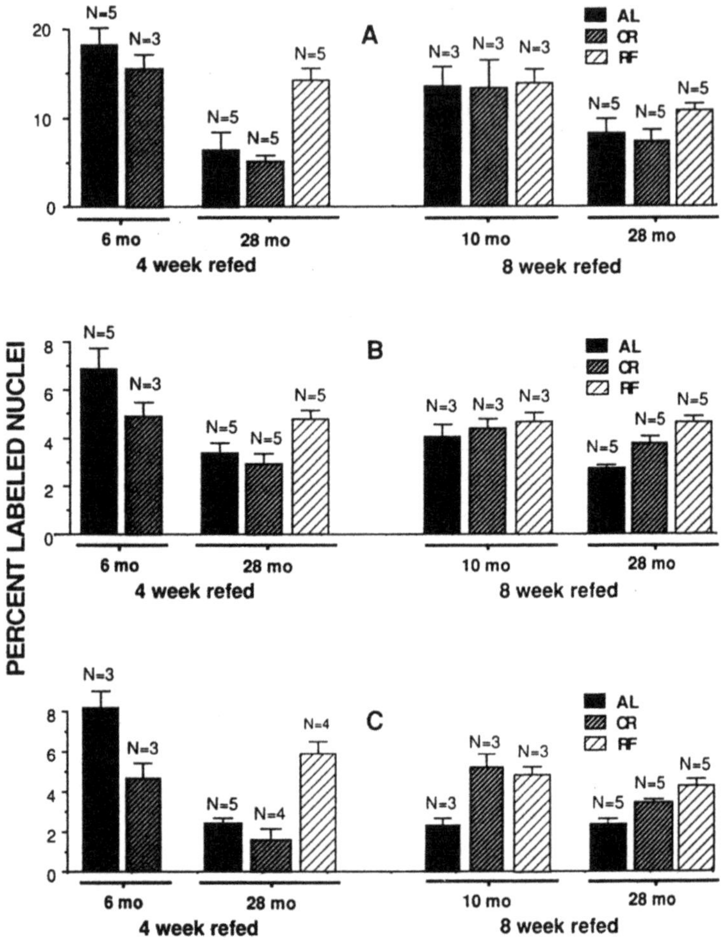

Fig. 2 BrdU uptake over a 2 week period in young and old AL or CR mice [4]

The external surface of the lens of the eye consists of a single layer of epithelial cells that divide continuously in the germinative zone and migrate from there to the end of the epithelial sheet coverage at the lens equator, from which point they migrate internally to form the lens fiber cells and then, with the resorption of all internal organelles, the clear lens fibers. Thus, there is a continuing replacement of the lens surface cells (excepting the small region of non-dividing cells of the anterior Central Zone). This in the mouse requires approximately 6 weeks for all surface cells to have undergone at least one cell division, as judged by all lens surface epithelial cells containing the provided BrdU. In 30–33 month old B6D2F1 mice this rate of cell turnover is reduced by over 50% from the young animal's state. A CR diet begun at 4 months of age and continued throughout life eliminates this slow-down

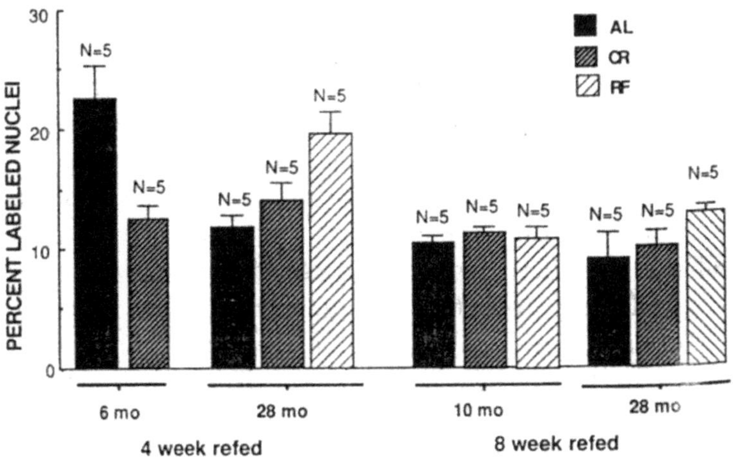

Fig. 3 BrdU uptake-measured dermal fibroblast replication over a 2 week period after a previous 4 or 8 weeks of refeeding in young and old B6D2F1 mice. Shown are CR mice, refed CR mice, and AL controls [4]

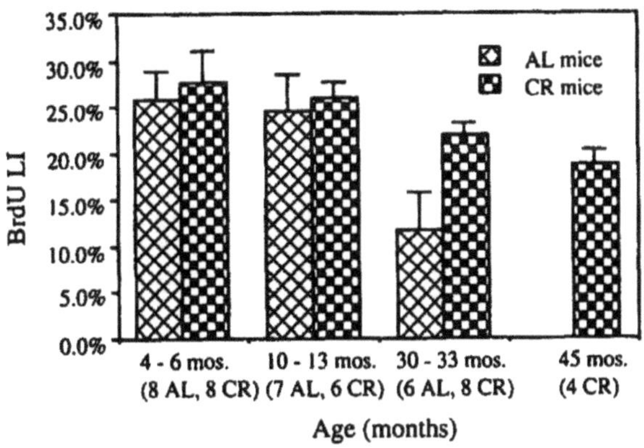

Fig. 4 Steady state in vivo replication rates (BrdU cell index) in surface lens epithelial cells in young and old AL and CR mice. Thirty to Thirty three month comparison AL to CR $p < 0.01$, 30–33 month AL comparison to 45 month CR $p < 0.01$ [2]

and at 45 months of age (when only the CR mice are still alive) the turnover rate is still not significantly different from that present in the 30–33 month old CR group and is higher than in 33 month old AL mice, as seen in Fig. 4 from Li et al. [2]. Note that in this in vivo study refeeding was not necessary for the value of CR to be apparent.

Cell Division Capacity In Vitro

Cell replication in vivo, as measured by BrdU uptake, represents the animal's ongoing cell turnover directly, as shown above. It is also possible to measure the maximum capacity of cells to divide in vitro over a specific period of time, noting this as percent of large clones developed that represent a higher number of clonal cell divisions, and to compare these measurements for age of the donor. This system removes the influence of donor blood supply, hormone/cytokine levels, and stromal cell influence. Shown in Fig. 5 are the relative capacities for large clone production (as percent of total clones) from young and old kidney epithelial cells read as doublings within a 7 day period of in vitro growth in DMEM with 10% fetal calf serum and 20% O_2 atmosphere. This can be expressed as the capacity to form large clones (more than 4–5 doublings) for several cell types in relation to donor age. The results of these studies demonstrated the loss of replication capacity with age and the advantage of the long term (from 16 weeks of age onward) CR state (Fig. 5). Results for dermal fibroblasts were similar (not shown here). These differences were also present in stromal cells obtained from the bone marrow of young and old mice (Fig. 6), and in lens epithelial cells (Fig. 7).

Resistance to oxidative damage was apparent in lens epithelial cells from CR mice exposed to H_2O_2 for 1 h in vitro in serum free medium. In this study lens cells from old B6D2F1 mice on standard AL diet were severely reduced in survival by H_2O_2 exposure. However, with the H_2O_2 treatment the old mice on long term CR diet showed a post-H_2O_2 survival level for lens cells equal to that of AL

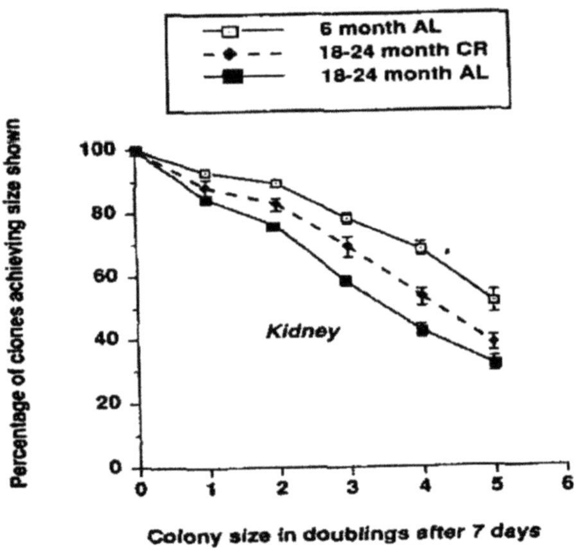

Fig. 5 In vitro colony size (growth capacity) for kidney epithelial cells form young and old AL or CR B6D2 mice [5]

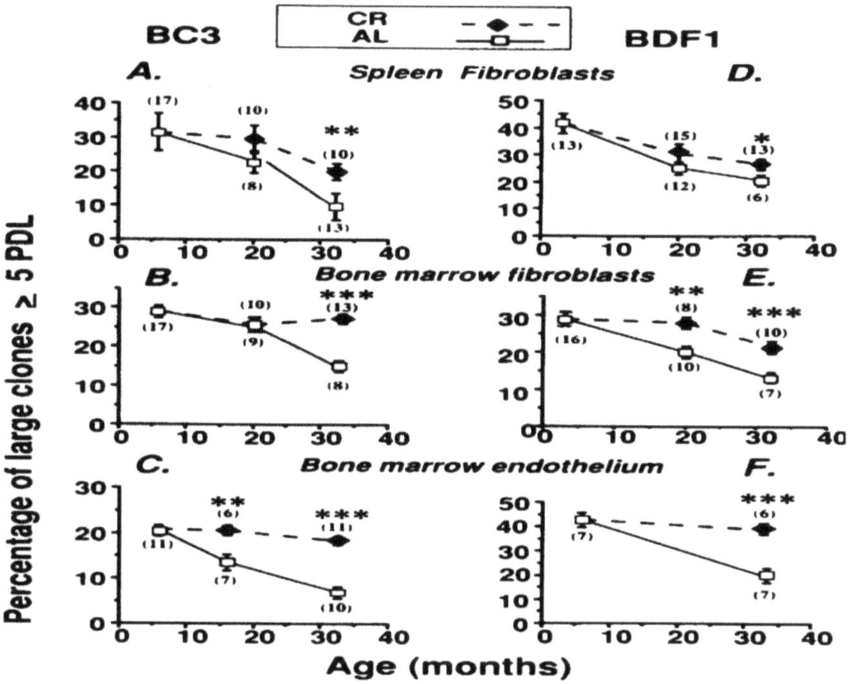

Fig. 6 In vitro large clone forming capacity of spleen stroma cells, B.M. fibroblasts, and B.M. endothelial cells from 2 different strains, C57BL6xC3HF1 and C57BL/6xDBA2F1. CR mice dark markers, AL mice open markers [5]

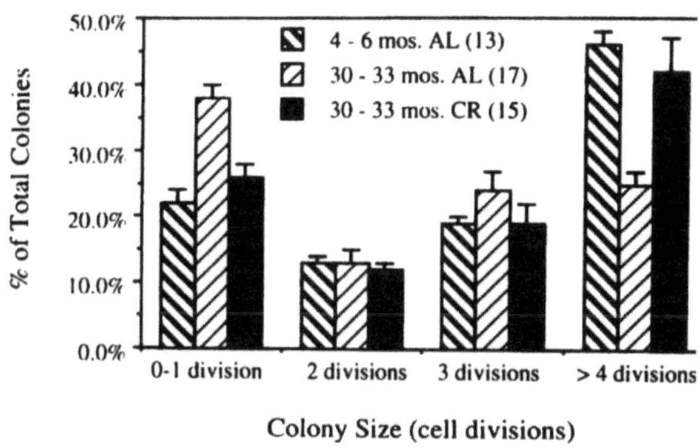

Fig. 7 The effect of age and diet on in vitro clone size formation by lens epithelial cells, B6D2F1 mice [2]

young mouse cells (Fig. 8). This study pointed to the protection against oxidative damage acquired by CR conditions for lens epithelium that is known to be sensitive to oxidative damage and the consequent development of cataracts. It is noted that refeeding was not necessary to demonstrate the protection from H_2O_2 exposure in

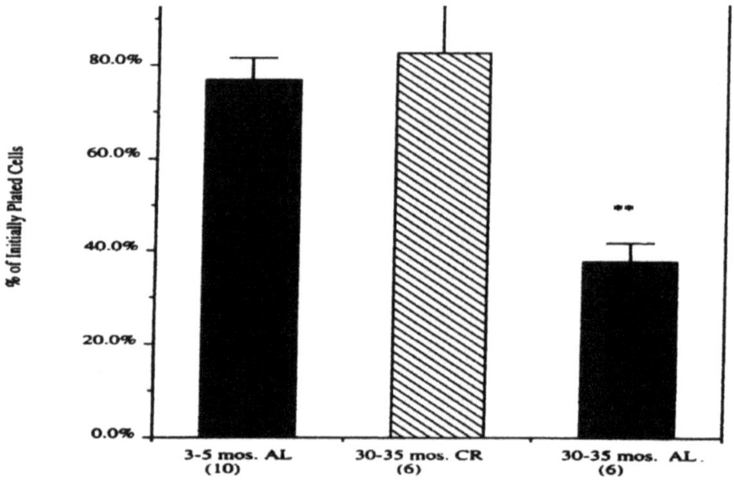

Fig. 8 The resistance to exposure to a pro-oxidant, H_2O_2, was also reduced in lens cells from old B6D2F1 mice and rescued by long term CR. The cells were exposed on the plates without serum for 1 h to 40 μM of H_2O_2, then FCS at 10% was resumed and growth allowed to proceed. Percent of viable cells present 48 h post-treatment is reported. $p < 0.01$ for AL old mouse cells versus young mouse cells or CR old mouse cells (Li et al. 1998, *Exp Cell Res* 239: 254–263)

vitro where sufficient energy and hormonal/cytokine support was supplied by the DMEM/10% FCS medium after the exposure to the H_2O_2. However, lens epithelial cells do not seem to require refeeding after CR even in in vivo studies in order to express superior replication capacity (Fig. 4).

Loss of cell in vitro replicative capacity with age, as shown in Fig. 9, was also apparent in dogs by the size of clones derived from dermal fibroblasts from the inner thigh when several members of the chihuahua, yorkshire, shetland sheepdog, and boston terrier breeds were grouped together for measurements made in young and old small dogs. A high percentage of large clones indicated high growth capacity in the young that declined with aging for the age groupings of 2–12 months, 6–8 years and >12 year old donors ($p < 0.01$). A separate study (Fig. 10) that included small and medium size dogs and 2 giant breeds (Irish Wolfhounds and Great Danes) demonstrated a significantly reduced in vitro dermal fibroblast growth capacity for those very large breeds, regardless of age of the donor. Medium size breeds included in the study, such as German Shepherds and Dobermans, did not differ from the small breeds that were also included in these measurements [8]. The 2 breeds weighing well over 100 pounds (Irish Wolfhound and Great Dane) are represented at far right in Fig. 10 and designated (Large II). Mean life span for those giant breeds is 7–8 years. In an interesting parallel, both reduction of life spans and the early advent and lifetime prevalence of age-related cataracts (ARC) varied in a direct fashion with increasing breed size when groupings of small, medium size, and giant breeds, 74 in total, were examined (submitted for publication by Urfer, Greer and Wolf). ARC is an at least partially oxidative damage driven pathology that includes the

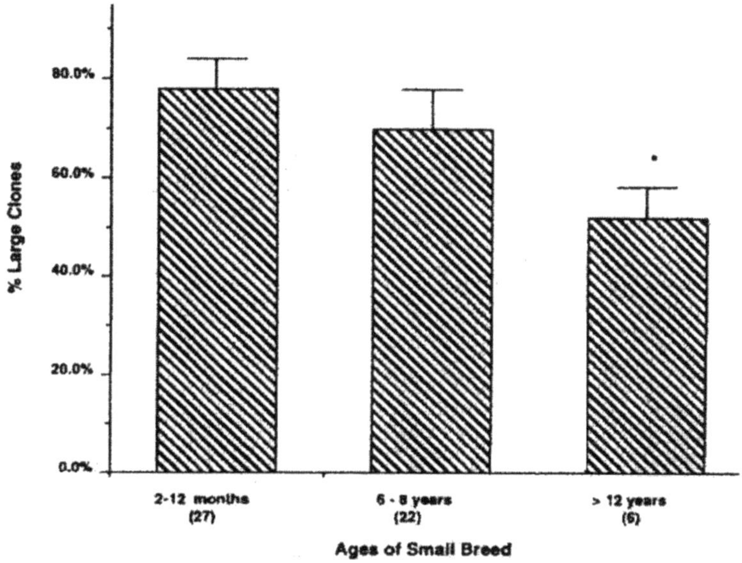

Fig. 9 Dermal fibroblast large clone forming capacity in vitro in young, middle aged and old small dogs. The old dog cell growth response differed at the $p > 0.01$ level from both young and middle-aged animals [8]

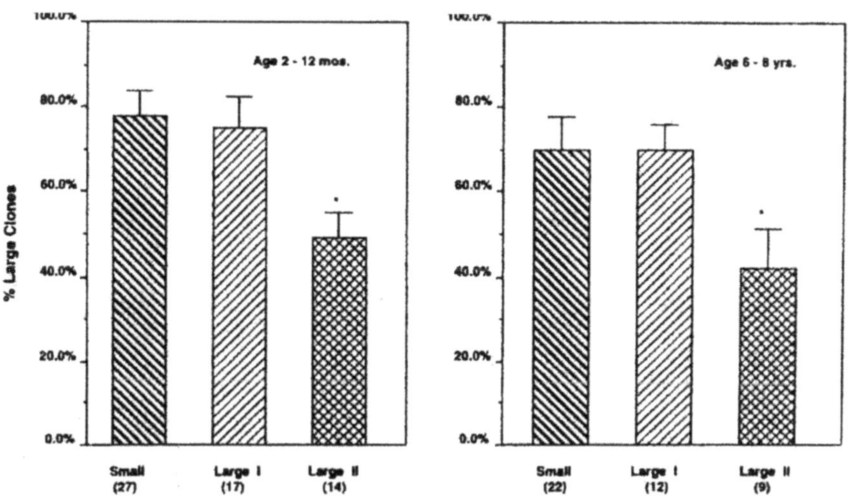

Fig. 10 Dermal fibroblasts, large clone percentages among all clones formed in age groupings of small, medium size and giant (Large II) dog breeds. Two to twelve month old dogs shown on the *left panel*, older dogs (6–8 years) set on *right panel*. Giant dogs (Large II) differed from the other sizes in either age grouping $p < 0.01$ [8]

loss of lens cell replication activity both in vivo and in vitro, as noted above for mice. Taken together, these results suggest that lifespan, an age-related pathology, ARC, and cell replication capacity in dogs are related to breed size and to animal's age. Further it appears that a defect in clonal cell growth capacity is seen early in life in the extreme case of gigantic size of dogs. This decrease in clonal cell growth capacity also can be seen early in life for giant mice (GH transgenic over-expressers) [6], and size in early life is a predictor of life span in wild type mice [9].

Cell Replication Decline with Aging or Senescence Among Several Species

Mice and dogs are not the only species that demonstrate a reduction in cell replication in vivo and in vitro. The replication rates of cells from young versus old animals have been studied in several mammalian species: mouse, rat, dog, monkey and human. The rate of cell turnover both in vivo and in vitro, and in response to wounding or to oxidative stress, has almost uniformly been reported to favor young versus old subjects. Middle aged animals, when included in these studies, occupied an intermediate status. It is important in assessing all of these studies to consider whether only the replication rate alone is involved, or whether apoptosis may remove a portion of the total population under study, and at what stage or time of the study. This caveat also would apply to studies in vitro when reporting the percent of cells that have gone through division, or when an increase in cell numbers is presented, especially after a toxic insult. Cristofalo and coworkers [10] have reviewed the evidence for replicative senescence tested in vitro and questioned the results of Martin et al. [11] that indicated that human dermal fibroblasts lost replicative capacity with increasing age of donor. Our own work with canine skin biopsies seems to support the earlier Martin study. Although less extensive in sample number it had a short in vitro expansion period and, thus a small number of previous cell replications that preceded the clone size measurement, suggesting that it accurately represented the corresponding in vivo status [8]. Using a similar clone size measurement, accompanied by a measurement of the presence of beta galactosidase that supported the clone size finding, the capacity for cell replication in vitro by rhesus monkey dermal fibroblasts was seen to decline in aged monkeys [12].

There are clearly genetic and epigenetic factors and agents that affect cell replication. Among these are the status of cellular chromatin arrangement (heterochromatin versus euchromatin), cell energy status (ATP availability), DNA deletions and mutations, status of gene promoters and suppressers, cell receptor status, telomere length, and various external stimuli: stromal environment, hormonal, or cytokine in nature. Further, the availability of stem or precursor cell reserves may be an important factor in maintaining numbers of functional cells. When cell types are tested in vitro for replication activity their inherent capacity is determined separately from conditions such as blood supply and cytokines availability. However, these conditions are present and important for cell growth in vivo in and differences in their presence in the young adult versus the old animal is a part of the process of aging change. The conclusion is that the one measurement clarifies conditions for the other and that both approaches should be used wherever possible. In mammals and other vertebrates the capacity for cell replication is present in many tissues. While replication in the adult mammal and birds had been thought to be nonexistent in the brain and heart, recent advances in stem cell research has shown that under proper conditions these tissues are not entirely excluded [13, 14]. It is also noted that telomere shortening with animal age has been reported in yeast, rats, dogs, horses, primates, and wild caught birds ([15–17, 1, 18–20] (see "Telomeres and telomerase: Inter-species comparisons of genetic, mechanistic and functional aging changes" by Gomes et al., this volume)).

Telomere Shortening with Age

Telomere shortening usually results eventually in cessation of cell replication in vivo and in vitro. Telomere shortening with aging has been found in many species and in tissues of rats, cats, dogs, baboons, humans and birds [21–25]. Surprisingly, in the long-lived bird, the Storm Petrel, telomeres were found to lengthen in old birds versus hatchlings [19]. Hausmann also found greater telomerase activity in long-lived bird species than in short lived species over lifespan, and considerable variance that depended upon the organ or tissue being examined [18]. This telomere length variability among bird species was also found by Hall [26]. Brummendorf [27] found a rapid telomere shortening in circulating blood lymphocytes and granulocytes as cats progressed in age from kittens to older cats, with lymphocyte telomeres shortening faster those of than granulocytes. Katepallii found a similar telomere shortening with age in horses [20]. Our own studies in rat lens cells indicated a significant 21% shortening with age of the donor, with a significant retention of length by long term CR in Brown Norway rats [1]. Nevertheless, the telomeres of mice and rats are quite long and it is difficult to explain the effect of telomere shortening as a cause for the reduced replication capacity of cells from these species. A summarization of telomere silencing regulation by the conditions at chromosome ends in several species is discussed by Ottaviani [28], and a consideration of the possibilities and limitations of use of telomere measurements in aging research is provided by Nakagawa [29]. Gallardo [30] has discussed the routes and means of telomerase biogenesis as currently understood. Telomere shortening to a critical length is not the only cause of cessation of cell replication but can be a major one. In general, telomere shortening is associated not only with age but also with internal and external origin stress [31, 32].

A decline in immune cell replication is seen in aging in several species and this is detailed in the chapter in this book by Nikolich-Zugich and Cicin-Sai. Effros [24, 33] has reviewed information on the definite decline of CD 8 T cells with aging. The capacity of T-cells to undergo replication is reduced with age and this is related to many factors associated with the cell surface and also include telomere shortening [34]. Katepalli [20] found telomere length in circulating white cells in horses to correlate with both T cell activity and total serum IgG, which were reduced with aging. The role of telomeres in immune aging has been summarized by Effros and co-workers [24]. The importance of telomeres and telomerase in aging is fully covered in the chapter on telomeres and telomerase in this book by Shay and co-authors.

DNA Damage in Old Cells That May Result in Senescence

Several recent studies have noted that cells from aging animals contained evidence of DNA damage. Busuttil [35] reported evidence of DNA mutations in a tissue specific manner and transcriptional "noise" in cardiomyocytes of old mice, as well as their increased sensitivity to H_2O_2 exposure in vitro. In baboons it was found that telomeres shortened with age in skin cells and that senescent cells there displayed activated ATM kinase, heterochomatinized nuclei and elevated p16 [23]. Human senescent cells apparently maintained their senescent state by continuous ATM, p53, p21 signaling [36, 37]. Xu [38] found that base excision repair decreased with age in multiple rat tissues. In both mice and rats DNA polymerase beta was decreased in brains of the old animals, and AP endonuclease in the spermatogenic cells of old mice. Given the numerous studies listing delayed or totally failed cell cycling and also specific gene expression markers of senescence, it seems well established that DNA damage is a marker for and probably one cause of senescence as well as carcinogenesis [39]. Singh [40], using the comet assay, found an increased number of double strand breaks in the kidney cells of old mice versus young mice. Our initial and ongoing studies indicate the increasing presence of 8-OHG in lens fiber cells in the lenses of old mice when compared to young mice.

Laun [41] suggested that DNA damage such as occurs in the old mother cell and last daughter cells in yeasts may represent the DNA damage that causes aging in higher animals. Gerson 2006 [42] suggested that DNA mutations inflicted on stem cells in humans and mice was likely to be the cause of subsequent cancer development. Ottaviani [28] discussed the changes in the subtelomeric region of the DNA strand in several species and the possible relationship of these changes to aging and cancer. Eruslimsky [43] noted the importance of cell senescence in aging, in particular in the cardiac system, and the insults that may damage DNA. Cristofalo [10] has discussed the multiple conditions and agents that may be involved in the senescence of cells in vivo and in vitro and possible flaws in related in vitro measurements. Laursen 2003 [44] showed the relationship of RecQ helicases and DNA topoisomerase III in processing homologous recombination in both yeast and mammals and the importance of these enzymes in DNA recombination, DNA replication, and cell cycle control. Chomyn [45] pointed out the accumulations of mtDNA mutations with age and suggested that mitochondrial DNA damage can lead to programmed cell death. Campisi [46, 47] and also Hayflick [48] have discussed cell senescence in multiple species and the affect of senescent cells on nearby cells and on the related subsequent development of cellular aging changes and cancer. Accumulated DNA damage can contribute to this changed state. The role of extensive cell divisions in bring on senescence and aging was proposed by Faragher 1998 [49], and by de Magealhaes [50].

Stem Cell Depletion with Aging

The reduction of stem or progenitor cells in several tissues as been shown to be an accompaniment of aging in rodents and humans. This can be one factor in the diminishment of wound healing capacity with aging, referred to below. However, Conboy has shown that the replication rate of hepatocytes and skeletal muscle satellite cells from old mouse muscle and hepatic cell replication rate was restored to normal when old mice were parabiosed with young mice, indicating the importance of circulating factors. Notch signaling and proliferation of muscle satellite cells were restored in this manner. In hepatic cells of old mice the expression of the cEBP-alpha complex accompanied hepatocyte proliferation [51]. Lees [52] examined 34 month old vs. 3 month old rats and reported a decreased differentiation of old rat skeletal muscle precursor cells into myotubes, and a decrease in muscle-specific protein, other than Myo-D. P27 expression was also lower in the old rats, as were myosin heavy chain and muscle creatine kinase. Torella found that cardiac stem cells of aged wild type mice had reduced telomerase when compared those of young mice, accompanied by telomere shortening and DNA strand uncapping, as well as increased DNA damage and cell death, reduced nuclear phospo-Akt and increased p27, p53, p16, and p19. Transgenic mice with increased IGF-1expression showed less of these age-related changes in cardiac stem cells, compensating in vivo by increasing cell turnover and without the losses in both phosphorylated Akt and telomerase that was seen in the old wild type mice [53]. A subset of pluripotent or unipotent stem cells or precursor cells that are in a resting state has been shown for hematopoietic stem cells [54]. Also, there is a reduced growth of old mouse hematopoietic organ stromal cells when grown in vitro [7, 55] reported that with older age in transgenic mice with limited telomerase activity and shortened telomeres there was a reduced capacity for cell renewal that correlated with a genetic defect in stem or precursor cells in bone marrow, intestines and testes. Dystra and de Haan [56] noted strain differences in the age effect on transplantability, accumulation of DNA damage and of ROS in mouse hematopoietic stem cells (while many strains showed losses with aging demonstrated both and vitro and in vivo, the C57BL/6 strain was not so affected). More recently the evidence for neural stem cells in the brain has accumulated. Wu [57] has provided evidence for the loss of hippocampus stem/progenitor cells even by middle age in mice and the retention of both numbers and turnover by forced exercise. Belluardo [58] documented the loss of stem/precursor cells from the subventricular zone of the hippocampus with age in rats. Interestingly, in relation to the findings noted under wound healing, the loss of replicating cells in the hippocampus with aging could be largely prevented with a constant release of pelleted estrogen (E2) in male mice [59]. The reduction in stem cell function with aging has been commented upon by many authors and is summarized in Gazit et al. [60]. Recently, the studies of Bergman et al. in the Frisen laboratory [61] have provided evidence that up to 50% of cardiomyocytes are replaced during the human lifetime, with the turnover ratio diminishing from approximately 1% per annum at age 25–0.45% at age 75.

Wound Healing in Young Versus Old Animals

Faragher [49] early on suggested that senescent cells might contribute to a number of pathological or debilitating conditions in old animals. Elsewhere in this volume Nikolich-Zugich and Cicin-Sain report on the decline of the immune system with aging. Putting infection aside, there is considerable evidence that wounds, in particular those that are not linearly apposed by sutures and that, therefore, heal by second intention, do so more slowly in old than in young animals. Healing of non-infected wounds involves at least four factors: (1) the rate of replication and orientation of the primary cell type (such as epithelial cells in the skin), (2) the rate and pattern of growth of supportive stromal tissue that supports the growth and orientation of the replacement primary cell type, (3) the growth of vascular tissue that provides nutrition for the primary and supportive cell types, and (4) the hormonal and cytokine stimulus (or repression) of growth and orientation of both stromal and primary cell types, and which may be produced by either of these cell types locally or which may arrive via the fluids and blood supply to the wounded area. An major factor, especially in surface wounds, is the orientation of the wound edges. A single straight incision with wound edges in apposition heals much more rapidly and with much less, if any, effect by the age of the individual than does a wound in which the edges are not in apposition and distanced from each other (an avulsive wound or the deliberate removal of a circumscribed area of tissue). In the latter case cell migration as well as cell replication are emphasized and the age of the individual becomes a factor in the healing rate.

Human Clinical Studies in Age-Related Wound Healing

Brem [62] suggested that delayed healing of skin wounds and ulcers in older human patients was due to reduced circulation and lower cytokine levels, but that properly treated wounds would heal at the same rate in young and old patients. In a study on old humans with distal limb and venous stasis ulcerative wounds he reported that, in spite of slower healing, the success rate for wound closure was similar to that of younger patients (nevertheless note that the healing was slower in the old subjects). Such studies are clinical success oriented, of course, and do not speak to the specific differences and the causes thereof between young and old subjects. Thomas [63] has emphasized the importance of co-morbidity in reports assessing the effects of aging in humans on the rate and success of wound healing. Seaman 2000 [64] has pointed out the importance of many of these factors in human wound healing. In spite of the difficulties of adequate controls, it is generally conceded that wounds and fractures in children heal more rapidly, even with the incorporation of pieces of tissue (such as fragmented bone) than similar wounds and fractures in adults [65]. Pittman [66] found delayed wound healing in the elderly human femoral neck. Yamada measured bone density in human humeri and found significant differences between young middle aged and old groups, as well as between the sexes. In bone tissue Kloss, [67] determined the increased fragility of maxillo-facial bone in elderly humans and the related imbalance of hormones and osteoblast/osteoclast activity with aging. He measured loss of maxillofacial bone in the aged humans and emphasized the differences between post-menopausal bone loss and senile osteopenia. Vernon-Roberts [68] studied human low spinal disc damage and found that transdiscal spinal disc tears and rim lesions increased with patient age. Calleja-Agius [69] described skin collagen loss, reduced skin elasticity, and slower wound healing, in post-menopausal women. Sherman [70] reported from a large study group that undesired outcomes from arthroscopic surgery was most common in humans over 50 years of age with no difference between the sexes, although this may have represented a less precise approach to sex differences. Rattan 2007 [71] studied human skin fibroblasts under stress in vitro. He found a reduction in several cell characteristics with increasing age of the donor, but that these changes could be prevented by a previous moderate level of heat shock or glyoxal, a hormetic effect that also increased the rate of wound closure.

Mimura 2006 [72] found the ex-vivo central corneal cells from old humans retained replication competence with donor age increase, but with lesser beginning numbers, and noted the senescence-related SA-beta-Gal presence only in the old donor cells. England [73] found that the healing of mucosal wounds experimentally placed in the hard palates of young and old human volunteers healed at different rates and significantly more slowly in the older group. In addition, they healed more slowly in the older women than older men. This sex difference may have been related to the estrogen reduction on the older female group [74]. Although a related study [75] was in mice, it is relevant that skin wounds in ovariectomized female mice treated with either raloxifene or tamoxifen healed more rapidly than similar incisions in control untreated mice, and with reduced inflammation and presence of inflammatory cytokines. Also, ovariectomy in rats resulted in slower alveolar healing after tooth extraction [76]. Thus, it appears that estrogen levels may well determine in part the rate of wound healing in females (and see [77]).

Age-Related Wound Healing in Laboratory Species

Since studies in humans cannot be designed to effectively exclude differences in the genetic, biological health status, and co-morbidity factors, studies in genetically identical but age differing animal models are more effectively controlled. Our own experience with ad libitum and CR fed young and old mice has been referred to above. We found that the healing of circular dermal depth skin wounds was significantly slowed in old versus young B6D2F1 mice, and that CR preserved the young animal healing rate in the old mouse, but only when ad libitum refeeding was commenced in the old CR animal at least 48 h before the wounding [3]. Under the refed CR condition there was an increase in replication and migration of fibroblast and endothelial cells at the healing wound's margins. In our related study it was found that refeeding was necessary to effect an increase in collagen message and in dermal fibroblast contractile ability on Type I collagen gels by CR mouse donors over old AL and old non-refed CR mice [3]. The refeeding restored the capacity of fibroblasts from old donors to that of young donors [3]. Ballas and Davidson [78] reported that 5th passage old rat skin fibroblasts provided a greater gel contraction and suggested that this was due to matrix metalloproteinase 2 production in the fibroblasts from the old donors. They suggested, therefore, that proteolysis might have a role in delayed wound healing in old animals. Reiser [79] reported slower accumulation of fibroblasts in a sponge implant model in both aged AL and CR rats. They found that in the subcutaneously implanted sponges, the aged ad libitum fed animals had delayed collagen deposition and that CR rats did also. This study did not include a refed cohort. Hollinger 2008 [80] reported delayed bone fracture healing in 24 month old rats, but significant increased healing rate when rhPDGF in a tri-calcium phosphate collagen mixture was topically applied. Sekine [81] used BrdU uptake measurement of cell turnover in 3 or 6 week versus 36 week old rat temporo-madibular condyle fractures. Condyles on only one side were fractured. Labeling increased in both condyles following the fracture to one side only. The labeling index of both injured and non-injured condyles decreased with age and as compared to controls. The degree of unrepaired osteolysis in 26 month old mice compared to 1 month old mice after implant was 17 times greater. Kaar [82] examined osteolysis induced by inert particle placed in mouse calvarium and found that osteolysis was increased as age of subject increased. Benatti [83] recorded peridontal healing to be delayed in old rats. Jarvinen [84] found that repair of skeletal muscle after inflicted injury was superior in young as apposed to old rats, with a greater capillary growth response at the injury site in the young animals. Thus, one may concluded that bone healing is delayed with advanced age in both rodents and humans.

Brem [85] found that both age and diabetes reduce skin healing in db/db mice. Neither condition, acting alone, affected the biomechanical properties of the wound, but when both were present both breaking point and wound stiffness were significantly affected. Kennedy [86] found slower wound contraction with age in rats, rabbits, and guinea pigs was attributable to the age of the animal. Keylock [87] carried out a study in 3 month old and 18 month old BalbC mice in which treadmill exercised or sedentary mice were skin wounded and re-wounded 4 months later. Wound healing was significantly faster in the young mice compared to old controls, with reductions in TNF-alpha, keratinocyte chemoattractant, and monocyte chemoattarctant at the wound site in the old mice. Relatedly, FGF subtype messages were found to be more highly expressed in full thickness skin healing in 2 month old than in 9 month old mice Komi-Kuramochi [88]. Schatteman [89] found that bone marrow cells from of old donors added at a skin wound site actually delayed skin wound healing and decreased local healing vessel size in both normal and diabetic mice.

Wagner [90] studied neonatal vs. adult healing in rat skin, finding that the neonates developed less TGF and FGF subtypes and less PDGF at wound site than the adult animals and that they healed more quickly and with less scar formation. Biondo-Smones [91] also found somewhat more cell and tissue responses in old versus young rats for healing of full thickness abdominal incisions, although the healing was complete in both groups. Bondono-Simoes [92] and also Quirinia [93] found decreased healing with age of both incision and ischemic flap wounds in old rats. It was also found that slower healing in the aged animal was largely due to local ischemia. Kluger [94] saw significantly more rapid healing that was by regeneration rather than by scar formation in young rats as opposed to old with inflicted splenic wounds. Biondo-Simoes 2006 [91] found a more rapid (but not more complete) liver regeneration following wounding in young rats than in middle aged rats. Kumari [95] recorded larger brain infarctions produced in db/db diabetic mice after hypoxic-anoxic damage. This study compared diabetic mice to non-diabetics and did not include old mice. Nevertheless, it suggests that under some conditions co-existing pathologies can be at the base of inferior age-related responses. Khodr [96] using vacuum induced foot pad blisters and healing rates in young and old rats showed paradoxical results with an anti-ROS treatment depending upon whether the antioxidant was locally injected early or late after the induction of the damage. Use of hyperbaric O_2 in the healing skin wounds had a greater positive effect in the old rats. However, Stoop [97], using young and 30 month old rats, found that age does not affect deposition or strength of collagen in intestinal anastomoses. Roth reported slower healing of skin wounds in old rhesus monkeys [98].

Metabolic Factors and Cytokine Levels Affect Wound Healing

Levels of many cytokines and hormones affect cell replication rate and wound healing. Spindler [99] used both Fisher 344 and BNF1 rats for explants of patellar tendons and found that age decreased both cellularity and DNA uptake, and that PDGF was effective in increasing both only in the young donor tendons. Reiser [79] used polyvinyl sponges implanted subcutaneously in young and old F344 rats and found lower collagen deposition but greater collagen crosslinking in the old. CR reduced the rate and extent of collagen deposition (see also Fig. 3 this chapter and reference to Reed et al. [3]); Schmidmaler [100] used local application of IGF1 or TGF beta at fracture sites in 5 month old rats and found increased fracture healing. Sobin [101] examined wounds in young and old rats and found that initial PAS positivity in the newly formed vessels of old rats is reduced more quickly post-wounding than that of young animals, suggesting a less complete healing acitivty. Wagner [90] Noted the more rapid skin wound healing in 3 day old newborns versus adult rats, and that the young contained much lower expression of TGFalpha, TFGbeta 1,2 and 3, IGF1, PDGF A, and bFGF than those seen in adult wounds. It is well known that ulcers and wounds in diabetic animals close or heal more slowly. While infection often plays a role in wound healing, levels of some hormones or cytokines, or the reduced response to them, are often involved. The effect of estrogen and/or those hormones lacking in menopausal humans or ovariectomized rats on wound healing or fracture healing have been referred to above. Gilliver reviewed wound healing in humans and found that estrogens encouraged, while testosterone inhibited, wound healing [74]. However, England [102] found that high testosterone level was a stimulant to wound healing in older women. Wound healing hormone and cytokine effects have been studied in mice, rats and humans with and without the diabetic state [103–108]. The diabetic state involves a large number of hormonal, growth factor, and gene level alterations, however the general statement may be made that the diabetic state delays healing of wounds and ulcers in all mammalian species studied so far and that this is more severe in the aged.

Conclusions on Wound Healing and Age of the Animal

Thus, various age related studies have looked at the replicative behavior of the cells from young versus old animals in vivo or in vitro; at the changes in cell DNA, including oxidative and undefined stress, and at intentional and accidental wound repair, burns, and special metabolic status. Specific genes and transcription factors, DNA damage, factors produced by tissue stroma that affect cells locally, and the levels of circulating hormones and cytokines are all involved to some degree in the reduction of cell turnover capacity or limited activity with advancing age in multiple species of animals. The overall conclusions that can be drawn from the many studies referred to above are that (1) wounded tissues in old humans or laboratory rodents heal more slowly than in the young, (2) CR increases the rate of wound healing in old rodents, but only under conditions in which adequate energy supply is present (as in vitro, or in refeeding in vivo), (3) females with lower estrogen levels than those present in the young intact subject (post-menopause or post-ovariectomy status) have slower healing rates, (4) previous exercise increases the rate of wound healing, (5) some hormone and some cytokine levels are lower in old subject wounds than those in the young, and contribute to delayed cell turnover or wound repair, (6) cell replication and wound healing studies in rodents have similar results to those in humans and are, therefore, generally indicative of expected results in humans, (7) Co-morbidity must always be considered in comparing cell replication rate of wound healing studies in all species. Infection and diabetes slow wound healing in both humans and rodents, further abetted by advanced age. The disadvantage in healing of fractures caused by advanced age in rats and mice correlates with the well-known slow healing of hip fractures in older humans. Many of the factors that are active in stimulation or repression of cell replication have related counterparts in several species (Table 1).

It is recognized that not all research confirms these conclusions. Therefore, they represent generalized conclusions.

Table 1 Agents, conditions that affect cell replication, or wound healing in the aged animal

Species	Agent/Condition	Cell Replic or Wound Healing	Special Condition
Yeast	Accumulation of rDNA circles	⇓	Limitation of lifespan, total replication.
	Replication rate control	⇓	Low glucose slows or prevents replication.
	CR	⇑	Extends life span and total turnover, but ⇓ rate of cell replications
Mouse and Rat	Physical activity	⇑	
	SirT1 status	⇑?	
	Caloric Restriction	⇑	Refeeding needed for in vivo increase in cell turnover.
	Estrogens	⇑	
	Testosterone	⇑	
	Telomere shortening	⇓	
Dog	Telomere shortening	⇓	Breed differences
N.H. Primate	CR	⇑	
Human	Physical activity	⇑	
	testosterone	⇑?	
	estrogens	⇑	
	growth hormone	⇑	

It is recognized that not all research confirms these conclusions. Therefore, they represent generalized conclusions.

Cell Replication and Wound Healing in Non-mammalian Species

Cell replication and organ regeneration in fish and amphibians is of interest in this chapter only as it approaches the pathways of cell replication and wound healing that may be related to the (minimal) tissue regeneration seen in mammals (a notable exception being the mammalian liver). There is a considerable literature in this area and it is reviewed here selectively for the above purpose. It is known that in the developing chick embryo whole organ structures can be regenerated [109]. In anurans, especially late development frog tadpoles, and in many adult urodels that include newts and salamanders, whole limb regeneration occurs after amputation. Further, in some fishes, such as the adult zebra fish, some body portions, such as the caudal fin, can be completely regenerated. These regenerations require the formation of a blastema cone at the wound site after the amputation, a form of epimorphosis in which undifferentiated cells develop into the future regenerated organ, whereas other forms of regeneration in non-vertebrates such as the hydra appears to result from rearrangement of preexisting regional cells (morphallaxis) [110]. Still another program is seen in planarians, in which regeneration appears to come from a group of committed stem cells [60, 111, 112]. In most of these studies, especially in those vertebrates referred to above, there appears to be a suspension of fibrosis, i.e., scar formation, that allows the blastema to generate cells that will regenerate the entire limb, lens, retina, or even heart [21, 113, 114]. This process in limb regeneration requires regeneration of bone, muscle, nerve and skin. Should the requirements for the genetic and epigenetic processing and the techniques for duplicating that are seen in urodel regeneration be developed in mammals, the field of wound healing and organ repair would take on a new and rewarding status, a la Star Trek. This is not completely unrealistic, as major advances are being made in the field. In addition, it is notable that this rush of replication and differentiation in the lower forms of animals has not been reported to result in cancer.

A large number of genes have been identified that speak to understanding the regenerative process, and also to its complexity. Among these in the axolotil are NvGas6, a heat shock response protein [115], trans-retinoic acid present in different concentrations at tip and posterior ends of the blastema in axoloti [116], HSP 70 [117], and in the regenerating zebra fish fin MsxB and C, FGF20A, and the TGF related ligand Activin-beta [118]. Blocking p53 expression in axolotis results in suppressing the p53 targets Mdm2 and Gadd45. Further, the axolotis p53 gene product contained multiple amino acid changes from the human form that is often seen in human tumors [119]. In the froglet genes important for regeneration of tendon and dermis are active, the scleraxis and Dermo-1 genes, respectively [110]. These studies await further development. At present, it seems apparent that at the point of tissue replacement fibrosis must be restrained and that several growth and differentiation factors are active, varying between species, and usually present at the site of blastema formation and regeneration. In comparing wound repair in these non-mammalian species to that in the neonate human an apparent repression of fibrosis (or failure to stimulate it early on) is one characteristic of the restorative healing mode, which in the human neonate leaves no scar. It is encouraging that the genes affecting this non-mammalian regrowth and multi-cell type differentiation have recognizable homologues in mammals.

References

1. Pendergrass WR, Penn PE, Li J, and Wolf NS (2001 Aug). Age-related telomere shortening occurs in lens epithelium from old rats and is slowed by caloric restriction. *Exp Eye Res* 73(2): 221–228.
2. Li Y, Yan Q, and Wolf NS (1997 Jan). Long-term caloric restriction delays age-related decline in proliferation capacity of murine lens epithelial cells in vitro and in vivo. *Invest Ophthalmol Vis Sci* 38(1): 100–107.
3. Reed MJ, Penn PE, Li Y, Birnbaum R, Vernon RB, Johnson TS, et al. (1996 Jul 31). Enhanced cell proliferation and biosynthesis mediate improved wound repair in refed, caloric-restricted mice. *Mech Ageing Dev* 89(1): 21–43.
4. Wolf NS, Penn PE, Jiang D, Fei RG, and Pendergrass WR (1995 Apr). Caloric restriction: conservation of in vivo cellular replicative capacity accompanies life-span extension in mice. *Exp Cell Res* 217(2): 317–323.
5. Pendergrass WR, Li Y, Jiang D, Fei RG, and Wolf NS (1995 Apr). Caloric restriction: conservation of cellular replicative capacity in vitro accompanies life-span extension in mice. *Exp Cell Res* 217(2): 309–316.

6. Pendergrass WR, Li Y, Jiang D, and Wolf NS (1993 July). Decrease in cellular replicative potential in "giant" mice transfected with the bovine growth hormone gene correlates to shortened life span. *J Cell Physiol* 156(1): 96–103.
7. Jiang D, Fei RG, Pendergrass WR, and Wolf NS (1992 Nov). An age-related reduction in the replicative capacity of two murine hematopoietic stroma cell types. *Exp Hematol* 20(10): 1216–1222.
8. Li Y, Deeb B, Pendergrass W, and Wolf N (1996 Nov). Cellular proliferative capacity and life span in small and large dogs. *J Gerontol A Biol Sci Med Sci* 51(6): B403–B408.
9. Miller RA, Chrisp C, and Atchley W (2000 Sep). Differential longevity in mouse stocks selected for early life growth trajectory. *J Gerontol A Biol Sci Med Sci* 55(9): B455–B461.
10. Cristofalo VJ, Lorenzini A, Allen RG, Torres C, and Tresini M (2004, Oct–Nov). Replicative senescence: a critical review. *Mech Ageing Dev* 125(10–11): 827–848.
11. Martin GM (1977 Nov). Cellular aging – clonal senescence. A review (Part I). *Am J Pathol* 89(2): 484–511.
12. Pendergrass WR, Lane MA, Bodkin NL, Hansen BC, Ingram DK, Roth GS, et al. (1999 July). Cellular proliferation potential during aging and caloric restriction in rhesus monkeys (Macaca mulatta). *J Cell Physiol* 180(1): 123–130.
13. Carlen M, Meletis K, Goritz C, Darsalia V, Evergren E, Tanigaki K, et al. (2009 Mar). Forebrain ependymal cells are Notch-dependent and generate neuroblasts and astrocytes after stroke. *Nat Neurosci* 12(3): 259–267.
14. Bergmann O (2009 Apr 3). Evidence for cardiomyocyte renewal in humans. *Science* 2009(324): 98–102.
15. Wang X and Baumann P (2008 Aug 22). Chromosome fusions following telomere loss are mediated by single-strand annealing. *Mol Cell* 31(4): 463–473.
16. Cadile CD, Kitchell BE, Newman RG, Biller BJ, and Hetler ER (2007 Dec). Telomere length in normal and neoplastic canine tissues. *Am J Vet Res* 68(12): 1386–1391.
17. McKevitt TP, Nasir L, Devlin P, and Argyle DJ (2002, June). Telomere lengths in dogs decrease with increasing donor age. *J Nutr* 132(6 Suppl 2): 1604S–1606S.
18. Haussmann MF, Winkler DW, Huntington CE, Nisbet IC, and Vleck CM (2004 June). Telomerase expression is differentially regulated in birds of differing life span. *Ann N Y Acad Sci* 1019: 186–190.
19. Haussmann MF and Mauck RA (2008 Jan). Telomeres and longevity: testing an evolutionary hypothesis. *Mol Biol Evol* 25(1): 220–228.
20. Katepalli MP, Adams AA, Lear TL, and Horohov DW (2008). The effect of age and telomere length on immune function in the horse. *Dev Comp Immunol* 32(12): 1409–1415.
21. Espejel S, Klatt P, Menissier-de Murcia J, Martin-Caballero J, Flores JM, Taccioli G, et al. (2004 Nov 22). Impact of telomerase ablation on organismal viability, aging, and tumorigenesis in mice lacking the DNA repair proteins PARP-1, Ku86, or DNA-PKcs. *J Cell Biol* 167(4): 627–638.
22. Colitz CM, Davidson MG, and Mc GM (1999 Dec). Telomerase activity in lens epithelial cells of normal and cataractous lenses. *Exp Eye Res* 69(6): 641–649.
23. Jeyapalan JC, Ferreira M, Sedivy JM, and Herbig U (2007 Jan). Accumulation of senescent cells in mitotic tissue of aging primates. *Mech Ageing Dev* 128(1): 36–44.
24. Effros RB (2009, Feb 19). Kleemeier award lecture 2008 – The Canary in the coal mine: Telomeres and human healthspan. *J Gerontol A Biol Sci Med Sci*: Series A; 64A(5): 511–515.
25. Greenwood MJ and Lansdorp PM (2003, Nov–Dec). Telomeres, telomerase, and hematopoietic stem cell biology. *Arch Med Res* 34(6): 489–495.
26. Hall ME, Nasir L, Daunt F, Gault EA, Croxall JP, Wanless S, et al. (2004 Aug 7). Telomere loss in relation to age and early environment in long-lived birds. *Proc Biol Sci* 271(1548): 1571–1576.
27. Brummendorf TH, Mak J, Sabo KM, Baerlocher GM, Dietz K, Abkowitz JL, et al. (2002 Oct). Longitudinal studies of telomere length in feline blood cells: implications for hematopoietic stem cell turnover in vivo. *Exp Hematol* 30(10): 1147–1152.

28. Ottaviani A, Gilson E, and Magdinier F (2008 Jan). Telomeric position effect: from the yeast paradigm to human pathologies? *Biochimie* 90(1): 93–107.
29. Nakagawa S, Gemmell NJ, and Burke T (2004 Sep). Measuring vertebrate telomeres: applications and limitations. *Mol Ecol* 13(9): 2523–2533.
30. Gallardo F and Chartrand P (2008, Oct–Dec). Telomerase biogenesis: the long road before getting to the end. *RNA Biol* 5(4): 212–215.
31. von Zglinicki T (1998 Nov). Telomeres: influencing the rate of aging. *Ann N Y Acad Sci* 20(854): 318–327.
32. von Zglinicki T and Martin-Ruiz CM (2005 Mar). Telomeres as biomarkers for ageing and age-related diseases. *Curr Mol Med* 5(2): 197–203.
33. Effros RB (2004 Oct). Replicative senescence of CD8 T cells: potential effects on cancer immune surveillance and immunotherapy. *Cancer Immunol Immunother* 53(10): 925–933.
34. Miller RA, Berger SB, Burke DT, Galecki A, Garcia GG, Harper JM, et al. (2005 Jun). T cells in aging mice: genetic, developmental, and biochemical analyses. *Immunol Rev* 205: 94–103.
35. Busuttil R, Bahar R, and Vijg J (2007 Apr 14). Genome dynamics and transcriptional deregulation in aging. *Neuroscience* 145(4): 1341–1347.
36. Herbig U, Ferreira M, Condel L, Carey D, and Sedivy JM (2006 Mar 3). Cellular senescence in aging primates. *Science* 311(5765): 1257.
37. Herbig U, Jobling WA, Chen BP, Chen DJ, and Sedivy JM (2004 May 21). Telomere shortening triggers senescence of human cells through a pathway involving ATM, p53, and p21(CIP1), but not p16(INK4a). *Mol Cell* 14(4): 501–513.
38. Xu G, Herzig M, Rotrekl V, and Walter CA (2008, Jul–Aug). Base excision repair, aging and health span. *Mech Ageing Dev* 129(7–8): 366–382.
39. Vijg J. (2008, Jul–Aug). The role of DNA damage and repair in aging: new approaches to an old problem. *Mech Ageing Dev.* 129(7–8): 498–502.
40. Singh NP, Ogburn CE, Wolf NS, van Belle G, and Martin GM (2001). DNA double-strand breaks in mouse kidney cells with age. *Biogerontology* 2(4): 261–270.
41. Laun P, Bruschi CV, Dickinson JR, Rinnerthaler M, Heeren G, Schwimbersky R, et al. (2007). Yeast mother cell-specific ageing, genetic (in)stability, and the somatic mutation theory of ageing. *Nucleic Acids Res* 35(22): 7514–7526.
42. Gerson SL, Reese J, and Kenyon J (2006). DNA repair in stem cell maintenance and conversion to cancer stem cells. *Ernst Schering Found Symp Proc* 5: 231–244.
43. Erusalimsky JD and Kurz DJ (2005, Aug–Sep). Cellular senescence in vivo: its relevance in ageing and cardiovascular disease. *Exp Gerontol* 40(8–9): 634–642.
44. Laursen LV, Bjergbaek L, Murray JM, and Andersen AH (2003). RecQ helicases and topoisomerase III in cancer and aging. *Biogerontology* 4(5): 275–287.
45. Chomyn A and Attardi G (2003 May 9). MtDNA mutations in aging and apoptosis. *Biochem Biophys Res Commun* 304(3): 519–529.
46. Campisi J (2000, Jan–Feb). Cancer, aging and cellular senescence. *In Vivo* 14(1): 183–188.
47. Campisi J and Sedivy J (2009 Feb). How does proliferative homeostasis change with age? What causes it and how does it contribute to aging? *J Gerontol A Biol Sci Med Sci* 64(2): 164–166.
48. Hayflick L (1998, Nov–Dec). How and why we age. *Exp Gerontol* 33(7–8): 639–653.
49. Faragher RG and Kipling D (1998 Dec). How might replicative senescence contribute to human ageing? *Bioessays* 20(12): 985–991.
50. de Magalhaes JP and Faragher RG (2008 June). Cell divisions and mammalian aging: integrative biology insights from genes that regulate longevity. *Bioessays* 30(6): 567–578.
51. Conboy IM, Conboy MJ, Wagers AJ, Girma ER, Weissman IL, and Rando TA (2005 Feb 17). Rejuvenation of aged progenitor cells by exposure to a young systemic environment. *Nature* 433(7027): 760–764.
52. Lees SJ, Rathbone CR, and Booth FW (2006 Feb). Age-associated decrease in muscle precursor cell differentiation. *Am J Physiol Cell Physiol* 290(2): C609–C615.

53. Torella D, Rota M, Nurzynska D, Musso E, Monsen A, Shiraishi I, et al. (2004 Mar 5). Cardiac stem cell and myocyte aging, heart failure, and insulin-like growth factor-1 overexpression. *Circ Res* 94(4): 514–524.
54. Pietrzyk ME, Wolf NS, and Priestley GV (1989 July). Cycling patterns of hemopoietic stem cell subpopulations in young and old BDF1 mice. *Mech Ageing Dev* 49(1): 79–86.
55. Hao LY, Armanios M, Strong MA, Karim B, Feldser DM, Huso D, et al. (2005 Dec 16). Short telomeres, even in the presence of telomerase, limit tissue renewal capacity. *Cell* 123(6): 1121–1131.
56. Dykstra B and de Haan G (2008 Jan). Hematopoietic stem cell aging and self-renewal. *Cell Tissue Res* 331(1): 91–101.
57. Wu CW, Chang YT, Yu L, Chen HI, Jen CJ, Wu SY, et al. (2008 Nov). Exercise enhances the proliferation of neural stem cells and neurite growth and survival of neuronal progenitor cells in dentate gyrus of middle-aged mice. *J Appl Physiol* 105(5): 1585–1594.
58. Belluardo N, Mudo G, Bonomo A, Di Liberto V, Frinchi M, and Fuxe K (2008 Feb). Nicotine-induced fibroblast growth factor-2 restores the age-related decline of precursor cell proliferation in the subventricular zone of rat brain. *Brain Res* 8(1193): 12–24.
59. Saravia F, Beauquis J, Pietranera L, and De Nicola AF (2007 June). Neuroprotective effects of estradiol in hippocampal neurons and glia of middle age mice. *Psychoneuroendocrinology* 32(5): 480–492.
60. Gazit R, Weissman IL, and Rossi DJ (2008 Oct). Hematopoietic stem cells and the aging hematopoietic system. *Semin Hematol* 45(4): 218–224.
61. Bergmann O, Bhardwaj RD, Bernard S, Zdunek S, Barnabé-Heider F, Walsh S, Zupicich J, Alkass K, Buchholz BA, Druid H, Jovinge S, and Frisén J (2009, Apr 5). Evidence for cardiomyocyte renewal in humans. *Science* 324: 98–102.
62. Brem H, Tomic-Canic M, Tarnovskaya A, Ehrlich HP, Baskin-Bey E, Gill K, et al. (2003). Healing of elderly patients with diabetic foot ulcers, venous stasis ulcers, and pressure ulcers. *Surg Technol Int* 11: 161–167.
63. Thomas DR (2001). Age-related changes in wound healing. *Drugs Aging* 18(8): 607–620.
64. Seaman S (2000, Jan). Considerations for the global assessment and treatment of patients with recalcitrant wounds. *Ostomy Wound Manage* 46(1A Suppl): 10S–29S, quiz 30S–1S.
65. Bartlett CS 3rd, Weiner LS, and Yang EC (1997 July). Treatment of type II and type III open tibia fractures in children. *J Orthop Trauma* 11(5): 357–362.
66. Pittman J (2007, Jul–Aug). Effect of aging on wound healing: current concepts. *J Wound Ostomy Continence Nurs* 34(4): 412–415, quiz 6–7.
67. Kloss FR and Gassner R (2006 Feb). Bone and aging: effects on the maxillofacial skeleton. *Exp Gerontol* 41(2): 123–129.
68. Vernon-Roberts B, Moore RJ, and Fraser RD (2007 Dec 1). The natural history of age-related disc degeneration: the pathology and sequelae of tears. *Spine* 32(25): 2797–2804.
69. Calleja-Agius J, Muscat-Baron Y, and Brincat MP (2007 Jun). Skin ageing. *Menopause Int* 13(2): 60–64.
70. Sherman OH, Fox JM, Snyder SJ, Del Pizzo W, Friedman MJ, Ferkel RD, et al. (1986 Feb). Arthroscopy – "no-problem surgery". An analysis of complications in two thousand six hundred and forty cases. *J Bone Joint Surg Am* 68(2): 256–265.
71. Rattan SI, Sejersen H, Fernandes RA, and Luo W (2007 Nov). Stress-mediated hormetic modulation of aging, wound healing, and angiogenesis in human cells. *Ann N Y Acad Sci* 1119: 112–121.
72. Mimura T and Joyce NC (2006 Apr). Replication competence and senescence in central and peripheral human corneal endothelium. *Invest Ophthalmol Vis Sci* 47(4): 1387–1396.
73. Engeland CG, Bosch JA, Cacioppo JT, and Marucha PT (2006, Dec). Mucosal wound healing: the roles of age and sex. *Arch Surg* 141(12): 1193–1197, discussion 8.
74. Gilliver SC, Ashworth JJ, and Ashcroft GS (2007, Jan–Feb). The hormonal regulation of cutaneous wound healing. *Clin Dermatol* 25(1): 56–62.
75. Hardman MJ, Emmerson E, Campbell L, and Ashcroft GS (2008 Feb). Selective estrogen receptor modulators accelerate cutaneous wound healing in ovariectomized female mice. *Endocrinology* 149(2): 551–557.

76. Pereira MC, Zecchin KG, Campagnoli EB, and Jorge J (2007 Nov). Ovariectomy delays alveolar wound healing after molar extractions in rats. *J Oral Maxillofac Surg* 65(11): 2248–2253.
77. Meyer RA Jr, Tsahakis PJ, Martin DF, Banks DM, Harrow ME, and Kiebzak GM (2001 May). Age and ovariectomy impair both the normalization of mechanical properties and the accretion of mineral by the fracture callus in rats. *J Orthop Res* 19(3): 428–435.
78. Ballas CB and Davidson JM (2001, May–Jun). Delayed wound healing in aged rats is associated with increased collagen gel remodeling and contraction by skin fibroblasts, not with differences in apoptotic or myofibroblast cell populations. *Wound Repair Regen* 9(3): 223–237.
79. Reiser K, McGee C, Rucker R, and McDonald R (1995 Jan). Effects of aging and caloric restriction on extracellular matrix biosynthesis in a model of injury repair in rats. *J Gerontol A Biol Sci Med Sci* 50A(1): B40–B47.
80. Hollinger JO, Onikepe AO, MacKrell J, Einhorn T, Bradica G, Lynch S, et al. (2008 Jan). Accelerated fracture healing in the geriatric, osteoporotic rat with recombinant human platelet-derived growth factor-BB and an injectable beta-tricalcium phosphate/collagen matrix. *J Orthop Res* 26(1): 83–90.
81. Sekine J, Sano K, and Inokuchi T (1995, Nov). Effect of aging on the rat condylar fracture model evaluated by bromodeoxyuridine immunohistochemistry. *J Oral Maxillofac Surg* 53(11): 1317–1321, discussion 22–23.
82. Kaar SG, Ragab AA, Kaye SJ, Kilic BA, Jinno T, Goldberg VM, et al. (2001 Mar). Rapid repair of titanium particle-induced osteolysis is dramatically reduced in aged mice. *J Orthop Res* 19(2): 171–178.
83. Benatti BB, Neto JB, Casati MZ, Sallum EA, Sallum AW, and Nociti FH, Jr (2006 Aug). Periodontal healing may be affected by aging: a histologic study in rats. *J Periodontal Res* 41(4): 329–333.
84. Jarvinen M, Aho AJ, Lehto M, and Toivonen H (1983 Feb). Age dependent repair of muscle rupture. A histological and microangiographical study in rats. *Acta Orthop Scand* 54(1): 64–74.
85. Brem H, Tomic-Canic M, Entero H, Hanflik AM, Wang VM, Fallon JT, et al. (2007 June). The synergism of age and db/db genotype impairs wound healing. *Exp Gerontol* 42(6): 523–531.
86. Kennedy DF and Cliff WJ (1979 Apr). A systematic study of wound contraction in mammalian skin. *Pathology* 11(2): 207–222.
87. Keylock KT, Vieira VJ, Wallig MA, DiPietro LA, Schrementi M, and Woods JA (2008 Jan). Exercise accelerates cutaneous wound healing and decreases wound inflammation in aged mice. *Am J Physiol Regul Integr Comp Physiol* 294(1): R179–R184.
88. Komi-Kuramochi A, Kawano M, Oda Y, Asada M, Suzuki M, Oki J, et al. (2005 Aug). Expression of fibroblast growth factors and their receptors during full-thickness skin wound healing in young and aged mice. *J Endocrinol* 186(2): 273–289.
89. Schatteman GC and Ma N (2006 Mar). Old bone marrow cells inhibit skin wound vascularization. *Stem Cells* 24(3): 717–721.
90. Wagner W and Wehrmann M (2007, Nov–Dec). Differential cytokine activity and morphology during wound healing in the neonatal and adult rat skin. *J Cell Mol Med* 11(6): 1342–1351.
91. Biondo-Simoes Mde L, Matias JE, Montibeller GR, Siqueira LC, Nunes Eda S, and Grassi CA (2006, Jul–Aug). Effect of aging on liver regeneration in rats. *Acta Cir Bras* 21(4): 197–202.
92. Biondo-Simoes Mde L, Terranova O, Ioshii SO, Borsato KS, Weingartner J, Nogueira G, et al. (2005, Mar–Apr). Effects of aging on abdominal wall healing in rats. *Acta Cir Bras* 20(2): 124–133.
93. Quirinia A and Viidik A (1996 Oct 25). The impact of ischemia on wound healing is increased in old age but can be countered by hyperbaric oxygen therapy. *Mech Ageing Dev* 91(2): 131–144.

94. Kluger Y, Rabau M, Rub R, Weinbroum A, Chaushu G, Ben-Avraham R, et al. (1999 Aug). Comparative study of splenic wound healing in young and adult rats. *J Trauma* 47(2): 261–264.
95. Kumari R, Willing LB, Krady JK, Vannucci SJ, and Simpson IA (2007 Apr). Impaired wound healing after cerebral hypoxia-ischemia in the diabetic mouse. *J Cereb Blood Flow Metab* 27(4): 710–718.
96. Khodr B and Khalil Z (2001 Jan 1). Modulation of inflammation by reactive oxygen species: implications for aging and tissue repair. *Free Radic Biol Med* 30(1): 1–8.
97. Stoop MJ, Dirksen R, and Hendriks T (1996 Jan). Advanced age alone does not suppress anastomotic healing in the intestine. *Surgery* 119(1): 15–19.
98. Roth GS, Kowatch MA, Hengemihle J, Ingram DK, Spangler EL, Johnson LK, et al. (1997 Mar). Effect of age and caloric restriction on cutaneous wound closure in rats and monkeys. *J Gerontol A Biol Sci Med Sci* 52(2): B98–B102.
99. Spindler KP, Nanney LB, and Davidson JM (1995). Proliferative responses to platelet-derived growth factor in young and old rat patellar tendon. *Connect Tissue Res* 31(2): 171–177.
100. Schmidmaier G, Wildemann B, Bail H, Lucke M, Fuchs T, Stemberger A, et al. (2001 Apr). Local application of growth factors (insulin-like growth factor-1 and transforming growth factor-beta1) from a biodegradable poly(D,L-lactide) coating of osteosynthetic implants accelerates fracture healing in rats. *Bone* 28(4): 341–350.
101. Sobin SS, Bernick S, and Ballard KW (1992 July). Acute wound repair in an aged animal: a model for accelerated aging of the microvasculature? *J Gerontol* 47(4): B121–B125.
102. Engeland CG, Sabzehei B, and Marucha PT (2008, Dec 11). Sex hormones and mucosal wound healing. *Brain Behav Immun* 23(5): 629–635.
103. Shishodia S, Singh T, and Chaturvedi MM (2007). Modulation of transcription factors by curcumin. *Adv Exp Med Biol* 595: 127–148.
104. Shanley LJ, McCaig CD, Forrester JV, and Zhao M (2004 Apr). Insulin, not leptin, promotes in vitro cell migration to heal monolayer wounds in human corneal epithelium. *Invest Ophthalmol Vis Sci* 45(4): 1088–1094.
105. Bitar MS (1997 Aug). Insulin-like growth factor-1 reverses diabetes-induced wound healing impairment in rats. *Horm Metab Res* 29(8): 383–386.
106. Doyle JW, Smith RM, and Roth TP (1997 Jan). The effect of hyperglycemia and insulin on the replication of cultured human microvascular endothelial cells. *Horm Metab Res* 29(1): 43–45.
107. Atalay M, Oksala N, Lappalainen J, Laaksonen DE, Sen CK, and Roy S (2009 Feb). Heat shock proteins in diabetes and wound healing. *Curr Protein Pept Sci* 10(1): 85–95.
108. Xu KP, Li Y, Ljubimov AV, and Yu FS (2009, Feb 2). High glucose suppresses EGFR-PI3K-AKT signaling pathway and attenuates corneal epithelial wound healing. *Diabetes* 58(5): 1077–1088.
109. Ferretti P, Zhang F, and O'Neill P (2003 Feb). Changes in spinal cord regenerative ability through phylogenesis and development: lessons to be learnt. *Dev Dyn* 226(2): 245–256.
110. Satoh A, Nakada Y, Suzuki M, Tamura K, and Ide H (2006 Apr). Analysis of scleraxis and dermo-1 genes in a regenerating limb of Xenopus laevis. *Dev Dyn* 235(4): 1065–1073.
111. Rossi L, Salvetti A, Batistoni R, Deri P, and Gremigni V (2008 Jan). Planarians, a tale of stem cells. *Cell Mol Life Sci* 65(1): 16–23.
112. Sanchez Alvarado A and Kang H (2005 Jun 10). Multicellularity, stem cells, and the neoblasts of the planarian Schmidtea mediterranea. *Exp Cell Res* 306(2): 299–308.
113. Straube WL and Tanaka EM (2006 Oct). Reversibility of the differentiated state: regeneration in amphibians. *Artif Organs* 30(10): 743–755.
114. Araki M (2007 Feb). Regeneration of the amphibian retina: role of tissue interaction and related signaling molecules on RPE transdifferentiation. *Dev Growth Differ* 49(2): 109–120.
115. Beug S, Vascotto SG, and Tsilfidis C (2006 Mar). Newt orthologue of growth arrest-specific 6 (NvGas6) is implicated in stress response during newt forelimb regeneration. *Dev Dyn* 235(3): 711–722.

116. Scadding SR and Maden M (1994 Apr). Retinoic acid gradients during limb regeneration. *Dev Biol* 162(2): 608–617.
117. Levesque M, Guimond JC, Pilote M, Leclerc S, Moldovan F, and Roy S (2005 Aug). Expression of heat-shock protein 70 during limb development and regeneration in the axolotl. *Dev Dyn* 233(4): 1525–1534.
118. Jazwinska A, Badakov R, and Keating MT (2007 Aug 21). Activin-betaA signaling is required for zebrafish fin regeneration. *Curr Biol* 17(16): 1390–1395.
119. Villiard E, Brinkmann H, Moiseeva O, Mallette FA, Ferbeyre G, and Roy S (2007). Urodele p53 tolerates amino acid changes found in p53 variants linked to human cancer. *BMC Evol Biol* 7: 180.

Sirtuin Function in Longevity

Daniel L. Smith Jr. and Jeffrey S. Smith

Abstract The use of model organisms has provided great insight into the process of aging, particularly during the last decade of research. Of the longevity factors identified, few have received more attention than Sir2 and the Sirtuins. Since the original discovery of Sir2 as a transcriptional silencing factor in yeast, these NAD^+-dependent protein deacetylases have been identified across organisms from yeast to humans and implicated in the fundamental regulation of genomic stability, metabolism, and aging. In this chapter, the historical context of Sirtuin function in yeast is discussed against the backdrop of the catalytic activity that contributes to its prominence in so many research areas. The role as a longevity factor and mediator of the calorie restriction response are discussed in detail for yeast and expanded to worms, flies and mice. Finally, a brief description of Sirtuin activators is included. Over a decade after its identification as a longevity factor, Sirtuin research is poised to answer the critical question of translation research and provide a clear answer to the role these proteins may play in mediating lifespan and disease in mammals, particularly in the response to calorie restriction. No matter the final answer, Sirtuin biology has built a strong foundation on which additional research will likely provide additional interesting results for the next decade and beyond.

Keywords Sirtuin · Aging · Longevity · SIR2 · Histone deacetylase · Metabolism · NAD^+ · Silencing · Caloric restriction · Lifespan

Much debate and speculation has accompanied the proposal that the fundamental regulation of aging may be shared across organisms, from budding yeast to humans. Truly, the notion of a shared aging mechanism is the underlying premise for the usc of modcl organisms to pursuc thc most basic biological proccsscs of aging. If successful, identification of the molecular regulators of lifespan would provide useful targets for pharmacological or nutritional interventions to modulate the aging

J.S. Smith (✉)
Department of Biochemistry and Molecular Genetics, University of Virginia Health System, School of Medicine, Charlottesville, VA 22908, USA
e-mail: jss5y@virginia.edu

N.S. Wolf (ed.), *The Comparative Biology of Aging*,
DOI 10.1007/978-90-481-3465-6_6,

process and prevent age-related diseases. The role of the Sirtuins in age modulation serves as an example of the translational approach to understanding the molecular basis of aging.

The Sirtuins comprise a large family of phylogenetically conserved protein deacetylases whose enzymatic activity requires the electron carrier, NAD^+. The founding member of this protein family is Sir2 from the budding yeast, *Saccharomyces cerevisiae*, and the name "Sirtuin" refers to Sir2-like proteins that have been identified from all kingdoms of life [1]. The genomes of eukaryotes tend to encode multiple family members. *S. cerevisiae* has a total of 5 sirtuins that consist of Sir2, Hst1, Hst2, Hst3, and Hst4 [2, 3], and humans have a total of seven (SIRT1 through SIRT7) [4]. In bacterial and Archea species, there tend to be only one or two sirtuins. For example, *Salmonella typhimurium* only has one called CobB [5]. The conservation of these proteins underlies their importance in regulating multiple cellular processes related to genetic stability, metabolism, and longevity, which will be the focus of this chapter. Much of the groundwork for the sirtuin field was laid with the yeast system, so the first section of the chapter will discuss the use of yeast in determining the function of Sir2 and its importance in aging.

Historical Introduction on Sir2 as a Silencing Factor in Yeast

SIR2 (*S*ilent *I*nformation *R*egulator 2) was originally identified as a gene involved in proper mating type control of yeast haploid cells back in the late 1970s and early 1980s [6, 7]. Haploid mating type in *S. cerevisiae* is dictated by the *MAT* locus, which can exist either as *MAT*a (encoding a-specific genes) or *MAT*α (encoding α-specific genes). *MAT*a and *MAT*α cells can mate to form an a/α diploid. Transcriptionally silenced versions of these mating-type cassettes, known as *HML*α and *HMR*a are present on opposite arms of chromosome III, and are used as templates for homologous recombination with the *MAT* locus in a process known as mating type switching [8]. *SIR2* was found to be required for silencing of the *HML*α and *HMR*a cassettes, hence the silencing information regulator name [9]. When *SIR2* is mutated, *HML* and *HMR* are both expressed, which results in a mating defect, the classic phenotype of a haploid *sir2* mutant [9]. Additional *SIR* genes (*SIR1*, *SIR3*, and *SIR4*) that function with *SIR2* in silencing *HML* and *HMR* were identified during the same time period [9], but unlike Sir2, the silencing proteins encoded by these genes are not highly conserved in other organisms and do not have any known enzymatic activity. Sir2, Sir3, and Sir4 interact to form the SIR silencing complex that is recruited to *HML* and *HMR* by flanking cis-acting DNA sequences called the E and I silencers [10]. These silenced loci are generally considered a primitive form of heterochromatin that is epigenetically propagated from generation to generation.

Two additional forms of Sir2-dependent silencing have been described in *S. cerevisiae*. The first, telomere position effect (TPE) is the silencing of genes in close proximity to telomeres [11]. The second form, ribosomal DNA (rDNA) silencing, is the Sir2-dependent repression of Pol II-transcribed genes within the rDNA locus [12, 13]. Unlike *HM* silencing and TPE, rDNA silencing does not require Sir1, Sir3, or Sir4 [13]. Within the nucleolus, Sir2 is the catalytic subunit of another multi-protein histone deacetylase complex called RENT [14, 15]. In addition to silenced reporter genes, there are endogenous Pol II-transcribed genes such as *TAR1*, encoding a mitochondrial protein, embedded within the rDNA that are under Sir2 control [16]. Recently, non-coding RNAs have even been identified as being transcribed from the intergenic spacer regions between each Pol I transcription unit [17, 18]. The silencing of the non-coding RNAs is possibly related to another key function of Sir2 at the rDNA, which is to suppress homologous recombination between rDNA repeats. In the absence of *SIR2*, one of the intergenic promoters (E-pro) becomes activated, and transcription from this promoter somehow forces the dissociation of cohesin from the intergenic spacer [17, 19]. The loss of cohesin then triggers an increase in rDNA recombination frequency [19], which is a key player in the role of Sir2 in longevity – discussed in more detail below.

Sir2 as a Longevity Factor in Yeast

A great deal of attention was shifted to the aging process when Leonard Guarente's lab pulled a specific allele of *SIR4* (*SIR4-42*) out of a genetic screen in yeast for mutants that enhance stress resistance and extend the replicative lifespan (RLS) of this organism [20]. RLS is defined as the number of times a mother cell divides (buds) before senescing. This was originally characterized as a phenotype by Mortimer in the 1950s [21]. Using similar microscopic techniques to what persists today, the asymmetrically dividing mother can be visually monitored for daughter bud formation. These daughter cells are sequentially manipulated away from the mother until the total replicative capacity of the single mother is determined. The average RLS usually ranges from ~20 to 40 divisions depending on the strain background [22], and as these cells age they become sterile and unable to mate due to a loss of silencing at the *HM* loci [23]. It was shown that the *SIR4-42* mutation causes a C-terminal truncation of Sir4 that forces the SIR complex found normally at the telomeres and *HM* loci to relocalize to the rDNA in the nucleolus [24]. The increased concentration of Sir2 in the nucleolus presumably results in stronger rDNA silencing and improved suppression of rDNA recombination. The implications of this finding were solidified with the demonstration that one of the causes of replicative aging in yeast is the accumulation of extrachromosomal rDNA circles (ERCs) that are produced as a result of rDNA recombination [25]. These DNA episomes can replicate during each S phase because of the autonomous replicating sequence (ARS) origin of DNA replication in each rDNA repeat. However, they are asymmetrically inherited by mother cells because of a septin-dependent diffusion barrier between the mother and daughter and the anchoring of the ERCs to the nuclear basket of the nuclear pore complexes (NPCs) [26]. The result is an exponential accumulation of the ERCs in old mother cells after each division that eventually kills the mother through an uncharacterized mechanism [25]. Artificial induction of ERC production has been shown to decrease lifespan [25]; however, some mutants with increased ERC levels still exhibit normal or increased lifespans [27]. By redistributing Sir2 from telomeres and *HM* loci to the rDNA, the *SIR4-42* mutation therefore delays the formation of ERCs and eventual replicative senescence. Subsequent studies directly demonstrated that simply integrating an extra copy of *SIR2* into the genome of a WT strain was sufficient to extend the RLS, while deleting *SIR2* resulted in a very short RLS that correlated with accumulation of ERCs [28]. Importantly, the short lifespan of a *sir2*Δ mutant could be partially rescued by combination with a *fob1*Δ mutation [28], which greatly reduces the frequency of rDNA recombination and extends lifespan [29, 30]. This asymmetric ERC distribution model for yeast *SIR2*-dependent replicative aging remains the predominant idea in the field, but there are potentially other Sir2-mediated processes that could also play a role, including the asymmetric inheritance of oxidatively damaged proteins to mother cells, a process that breaks down in *sir2*Δ mutants [31]. These carbonylated proteins form aggregates with Hsp104, which accumulate in *sir2*Δ mutant mother cells. Hsp104 over expression can partially suppress the short RLS of the *sir2*Δ mutant [32].

There are two types of aging in budding yeast, replicative (discussed above) and chronological. The chronological lifespan (CLS) is most simply defined by the number days that non-dividing yeast cells remain viable. These are typically cultures of cells that have depleted nutrients in liquid culture during the growth phase and have subsequently entered a "quiescent" stationary phase (G_0). These cells are highly sensitive to oxidative damage, and are typically believed to be a model for aging in largely non-mitotic tissues such as neurons [33]. Unlike RLS, deleting *SIR2* does not shorten CLS and over-expressing *SIR2* does not extend CLS [34]. Surprisingly, *sir2*Δ mutants instead have an extended CLS [35, 34]. Deleting the *SIR2* homolog, *HST3*, shortens both RLS and CLS [34, 36], which could be related to its role (along with Hst4) in maintaining genome stability [37]. Hst3 (along with Hst4) is a histone H3 K56 deacetylase [37], but has also been hypothesized to regulate acetyl CoA production due to the requirement of *HST3* and *HST4* for allowing the growth of yeast media containing propionate or acetate [38]. In *Salmonella*, the Hst3 homolog CobB, has been shown to regulate acetyl CoA synthetase by deacetylating a lysine residue in its active site [39], an activity shared by mammalian SIRT1 and SIRT3 [40, 41].

Catalytic Activity of Sirtuins

As mentioned above, the sirtuins are predominantly NAD^+-dependent protein deacetylases, but this was not determined until 2000. In 1993, silent chromatin at *HML* and *HMR* was known to be hypoacetylated on H3 and H4 in a *SIR2* and *SIR3* dependent manner [42]. In the same study, *SIR2* over expression resulted in deacetylation of bulk histone H2B, H3, and H4, suggesting that Sir2 was somehow promoting deacetylation [42]. The sequence of Sir2 and the other sirtuins did not resemble any other known histone deacetylases, and it was not until the CobB protein of *Salmonella* was found to be a Sir2 homolog that metabolized an NAD^+ intermediate was the link with NAD^+ suggested [5]. CobB and Sir2 were both shown to have relatively weak mono-ADP ribosyltransferase activity that used NAD$^+$ as a substrate [1, 43]. Furthermore, the Boeke lab also isolated the NAD$^+$ salvage pathway enzyme Npt1 from a genetic screen for rDNA silencing factors that implicated NAD$^+$ in Sir2-dependent silencing [44, 45]. The Guarente and Sternglanz labs both determined that rather than transferring ADP-ribose to a histone tail, the sirtuins removed an acetyl group from the tails when NAD$^+$ was added to the reaction mix, indicating these enzymes were actually a new class of histone deacetylase, now known as the class III HDACs, differentiating them from the class I and class II HDACs that do not use utilize NAD$^+$ [46, 47].

The sirtuin-mediated HDAC reaction is quite interesting. For every acetyl group removed from a lysine residue on the histone tail, one molecule of NAD$^+$ is consumed [48, 49]. The acetyl group on the target lysine is transferred to the ADP-ribose moiety of NAD$^+$ at the 2′ OH group. In the process, the glycosidic bond between the ADP-ribose and nicotinamide moieties is cleaved, releasing one molecule of 2′O-acetyl-ADP ribose and one molecule of nicotinamide (Fig. 1) [48, 49]. Multiple sirtuins have now been shown to have this activity, and a few have also been shown to have robust ADP-ribosyltransferase activity, including the TbSIR2RP1 protein from *Trypanosoma brucei* [50]. Mammalian SIRT4 does not

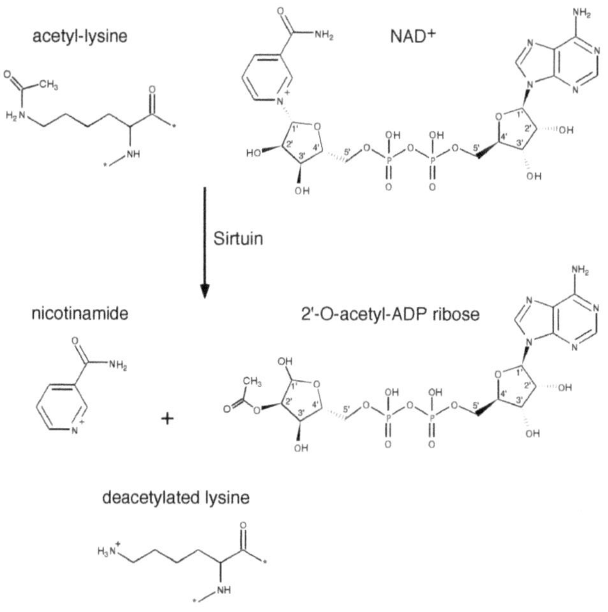

Fig. 1 The sirtuin-mediated protein deacetylation reaction. The substrates are a target protein with an acetylated lysine residue, and nicotinamide adenine dinucleotide (NAD$^+$). In a tightly coupled reaction, the glycosidic bond between the nicotinamide moiety and the ADP-ribose moiety of NAD$^+$ is broken, releasing free nicotinamide. The acetyl group is transferred from the acetylated lysine to the 2′ OH group of the ADP-ribose, yielding the deacetylated protein and 2′-O-acetyl-ADP ribose

have detectable deacetylase activity, but can ADP-ribosylate histones and glutamate dehydrogenase in the mitochondria of pancreatic β cells, where it regulates insulin secretion [51, 52]. SIRT6 has been shown to ADP-ribosylate itself [53], although it is not clear if this activity is physiologically relevant, because the human version of SIRT6 was recently shown to also be a histone H3 lys9-specific histone deacetylase that modulates telomere chromatin [54].

Calorie Restriction and Sir2 in Yeast Lifespan Extension

Calorie restriction is a dietary regimen that extends both the average and maximum lifespan of multiple eukaryotic organisms ranging from yeast to mammals. In the yeast system, CR is typically defined by a reduction of glucose concentration in the growth medium from the standard 2% down to 0.5%, which results in extension of both RLS and CLS [35, 55, 34]. More severe decreases in glucose concentration down to 0.1 or 0.05%, now often referred to as "extreme CR", cause an even greater extension of RLS and CLS [35, 56, 57, 34]. Since Sir2 couples the consumption of NAD^+ with the deacetylation of histones and functions in promoting longevity, it was logical to propose that Sir2 could provide a critical link between the cellular energy status and the regulation of longevity. The Guarente lab initially reported that *SIR2* was required for RLS extension induced by a *cdc25-1* mutation, which reduces signalling through the RAS/PKA pathway, therefore mimicking the response to reduced glucose [55]. Later studies demonstrated that *SIR2* was also required for the RLS extension caused by an actual reduction in media glucose concentration [58, 59]. The proposed underlying mechanism is that CR growth conditions result in the activation of Sir2, such that rDNA recombination is suppressed, therefore delaying the accumulation of ERCs in old mother cells.

The above model for Sir2-mediated lifespan extension in yeast is simple, straightforward, and has greatly influenced the direction of aging research in higher eukaryotes. However, the model has also been the subject of debate over the last several years. For example, in one study from the Jazwinski lab, *SIR2* was not required for the extension of RLS caused by CR under an extreme CR condition (0.1% glucose) [56]. The Kennedy and Fields labs then demonstrated that in a relatively long-lived strain background (BY4742), *SIR2* was dispensable for the CR effect if rDNA recombination was blocked by a *fob1*Δ mutation [60]. This challenged the model of Sir2 acting through the suppression of rDNA recombination during CR. One of the Sir2 homologs in yeast, Hst2, has been reported by the Sinclair lab to partially substitute for Sir2 in suppressing rDNA recombination during CR [61], although an independent report observed extension of RLS in a *sir2*Δ *hst2*Δ double mutant, or even when all five yeast sirtuins were deleted [62, 36]. It has been suggested that some of the variation could be due to differences in strain backgrounds, media preparation, and assay techniques [63]. Unfortunately, this remains an unresolved issue. CR has been shown to prevent oxidative damage of proteins and other cellular components in yeast and higher eukaryotes [64, 65], so it remains possible that some of the effect of CR on lifespan in the absence of Sir2 and the other sirtuins is through the suppression of this damage [31]. Studies with chronological lifespan, an aging process that is unaffected by ERC production, have clearly revealed that Sir2 and the other sirtuins are not required for lifespan extension caused by either "moderate" or "extreme" CR [35, 66, 34]. More work is therefore needed to understand how Sir2 functions in yeast lifespan regulation, especially in regards to calorie restriction and the identification of other non-histone targets for deacetylation.

NAD^+ Biosynthesis and the Activation of Sir2

As discussed above, the sirtuins are protein deacetylases that consume NAD^+, implying that they could be regulated by fluctuations in the intracellular concentrations of NAD^+ and related precursors or byproducts in yeast and other organisms (see overview of NAD^+ biosynthesis pathways in Fig. 2). The nicotinamide generated by sirtuins and other NAD^+-consuming enzymes is a strong non-competitive inhibitor of the deacetylation reaction [47], and product inhibition during the reaction has been demonstrated in vitro [67]. NAM can also be efficiently imported into cells such that 5 mM NAM in the growth media completely inhibits Sir2-mediated gene silencing, causing hyper recombination in the rDNA, and shortened RLS [68]. In yeast cells, the released nicotinamide moiety is converted to nicotinic acid by the nicotinamidase Pnc1 [69]. Nicotinic acid is not a sirtuin inhibitor, so Pnc1 promotes the deacetylation reaction, at least partially, by clearing the toxic NAM byproduct [58, 67]. Indeed, over expression of Pnc1 has been shown to extend the replicative lifespan of yeast, [58], and organismal lifespan of *Drosophila* [70]. It also enhances adult *C. elegans* survival during oxidative stress [71]. Expression of the *PNC1* gene is upregulated by moderately stressful growth conditions, including calorie restriction, and is required for CR-mediated extension of RLS [58, 67].

Nicotinic acid that is produced by Pnc1 or imported into the cell is converted to nicotinic acid mononucleotide (NaMN) by the conserved nicotinic acid phosphoribosyltransferase, Npt1 [72–74]. This enzyme is often concentrated in the nucleus [72, 74], and is the rate-limiting step of the Preiss-Handler pathway [75], which functions as a salvage pathway for NAD^+ synthesis. NaMN is converted to NaAD by the highly conserved nicotinic acid/nicotinamide adenylyltransferases, Nma1 and Nma2 [72]. NAD^+ is then generated by NAD synthetase, Qns1. NAD^+ can also be synthesized de novo from tryptophan using the Bna1 through Bna6 proteins, which eventually produce NaMN that then merges with the Preiss-Handler pathway to produce NAD^+. Interestingly, over expression of Npt1, Nma1, or Nma2 strengthens rDNA silencing and extends RLS, but over expression of the de novo pathway Bna proteins has no effect [72]. Deletion of *NPT1* reduces the intracellular NAD^+ concentration by 2–3 fold [55, 45], inhibits rDNA silencing, and shortens RLS, while deletion of *BNA1* has little effect [76, 74]. It has therefore been proposed that a flux of NAD^+ biosynthesis by Pnc1 and the Preiss-Handler pathway in the nucleus is necessary to sustain high Sir2 activity, and therefore longevity [72, 74]. While not

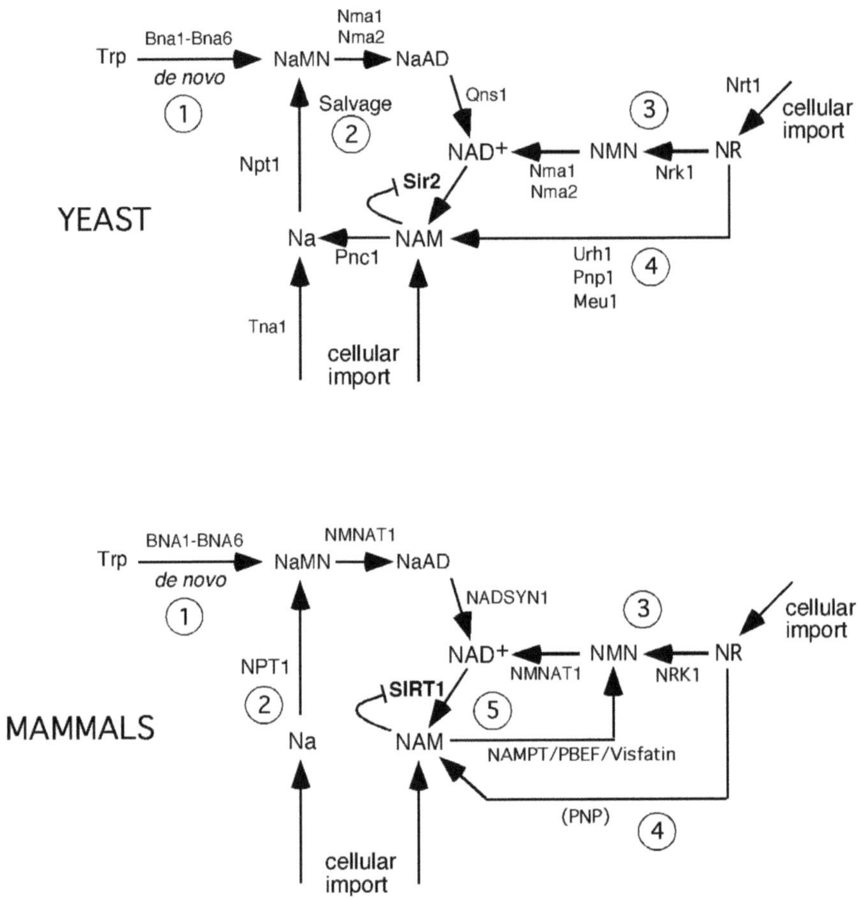

Fig. 2 A comparison of the NAD^+ biosynthesis and salvage pathways in yeast and mammals. The de novo NAD^+ synthesis pathway (*1*) is conserved between yeast and mammals. The Preiss-Handler pathway (*2*) is part of the nicotinamide (NAM) salvage pathway in yeast, and is a separate pathway in mammals. The nicotinamide riboside (NR) pathway involving phosphorylation of NR by Nrk1 (*3*) is conserved. A new pathway involving degradation of NR into NAM by a set of hydrolases and phosphorylases in yeast (*4*), is likely also conserved in mammals (probably carried out by purine nucleoside phosphorylase, PNP). Mammals have a nicotinamide phosphoribosyltransferase (NAMPT/PBEF/visfatin) that converts NAM into nicotinamide mononucleotide (NMN). This pathways does not exist in yeast and other lower eukaryotes, who convert NAM into nicotinic acid (Na) for utilization by the Preiss-Handler pathway using a nicotinamidase called Pnc1. Na, NAM, and NR are imported into the cell from the growth medium. Other abbreviations: tryptophan (Trp); nicotinic acid mononucleotide (NaMN); deamido adenine dinucleotide (NaAD). Nrt1 is an NR transporter, and Tna1 is a nicotinic acid transporter

upregulated by CR like Pnc1, Npt1 is also required for CR-mediated extension of RLS [55]. In general, it can be concluded from yeast studies that maintaining high NAD^+ concentrations and low nicotinamide concentrations are important for Sir2 activity and longevity.

In mammals, the nicotinamidase activity of Pnc1 has been replaced by the nicotinamide phosphoribosyltransferase, NAMPRT, which is also known as PBEF and visfatin [77]. Instead of salvaging NAM through a nicotinic acid intermediate like yeast and invertebrates do, the NAMPRT of mammals more directly converts NAM into nicotinamide mononucleotide, which the nicotinic acid/nicotinamide adenylyltransferases then convert to NAD^+ [78]. Over expression of NAMPRT raises the NAD^+ concentration of mammalian cells and appears to enhance the activity of mouse SIRT1 (Sir2α), although its effects on lifespan have not been tested [77]. Interestingly, NAMPRT was recently shown to be a systemic protein in mouse plasma that regulates insulin production by pancreatic β cells through the production of nicotinamide mononucleotide, which is also present in high concentrations in plasma [79]. Importantly, the maintenance of efficient NAD^+ biosynthesis promotes glucose-stimulated insulin secretion from β cells [79], perhaps through stimulation of SIRT1 (see relationship between SIRT1 and diabetes below).

Sirtuins in the Regulation of *C. elegans* and *Drosophila* Lifespan

Following the initial reports of *SIR2*-dependent replicative lifespan regulation in yeast [28], other more complex organisms were investigated for similar genes that might mediate lifespan. The first non-yeast sirtuin lifespan results were in *C. elegans*, which has 4 genes with sequence similarity to yeast *SIR2* that are denoted *sir-2.1*, *sir-2.2*, *sir-2.3* and *sir-2.4*. Strains carrying duplications of chromosomal regions that covered each of the 4 worm sirtuin genes were tested for lifespan effect. Only a duplication containing the *sir-2.1* locus, the most similar gene to yeast *SIR2* of the four, resulted in an extension of lifespan [80]. Additional lifespan experiments that used restricted fragments expressing just the *sir-2.1* locus were also found to increase lifespan by up to 50% [80]. The authors proposed a role for SIR-2.1 in the DAF-16 insulin-like signalling pathway to respond to nutrient status [80]. While protein levels were not assayed in this study, it is assumed that there is either an increased amount of SIR-2.1 deacetylase activity or protein level [80]. Opposite the over-expression data, a *sir-2.1* null mutation had no overt lifespan effect [81]. Similarly, independent studies determined that a *sir-2.1(ok434)* deletion or *sir-2.1(pk1640::Tc1)* transposon insertion mutation only slightly reduced median and maximum lifespan [82, 83]. Such a small lifespan effect (control $= 17.4 \pm 0.3$ days; *sir-2.1(ok434)* $= 16.7 \pm 0.2$ days [$p = 0.0449$]) may not be a complete surprise, as one of the major mechanisms of *SIR2*-dependent replicative lifespan regulation in yeast (ERC accumulation) has not yet been reported in other organisms. Additional experiments with the *sir-2.1(ok434)* mutant demonstrated an increased sensitivity to a variety of stresses, suggesting a stress resistance role for the native protein [83], which is consistent with the increased oxidative stress resistance of worms over-expressing the nicotinamidase Pnc1 described above [71]. Specific deacetylation targets for the *C. elegans* sirtuins have not been seriously explored, although SIR-2.1 does physically interact with 14-3-3 proteins [83], which also interact with DAF-16 (the FOXO protein of *C. elegans*). DAF-16 is required for the lifespan extension caused by over-expression of *sir-2.1* [80], and deacetylation of FOXO proteins in mammals results in activation of stress-response genes and promotes cell survival [84].

The role that *sir-2.1* plays in CR-mediated lifespan extension in *C. elegans* remains unclear. CR or dietary restriction (DR) is accomplished in worms by a variety of manipulations. The most common technique involves basic food dilution. Alternatively, a genetic "mimic" of CR, the *eat-2(ad465)* mutant, has been used [83], that is believed to reduce food ingestion and/or absorption [85]. The *eat-2(ad465)* mutant lifespan extension is partially blocked by the *sir-2.1(ok434)* mutation, initially suggesting that *sir-2.1* mediates the CR lifespan effect in worms [83]. A third method of DR consists of complete removal of the *E. coli* food source [86], and this "extreme" form of DR causes a significant increase in lifespan that is independent of *daf-2*, *daf-16*, and *sir-2.1* [86]. Additionally, lifespan of the *eat-2(ad465)* mutant, which mimics CR, was further extended by complete removal of *E. coli* and this extension was likewise independent of *sir-2.1* [86]. These authors argue that the *eat-2(ad465)* mutation and DR by *E. coli* removal work through a similar mechanism that is independent of *sir-2.1*. Similar results showing *sir-2.1*-independent lifespan extension were obtained in another study that also used removal of bacteria as a tool for DR, and in this case was called "dietary deprivation" (DD) [87]. As with CR in yeast, it remains to be determined if the complete removal of food is actually caloric restriction or a form of starvation. Studies using 2-deoxy-D-glucose (2-DG) to disrupt glucose metabolism and mimic CR failed to show a *sir-2.1* dependence [88]. Finally, the Kenyon lab reported that *sir-2.1* was not required for lifespan increases in *C. elegans* induced by an *eat-2(ad1116)* mutation, DR by diet manipulation, or genetic manipulations that inhibit protein translation [89]. Therefore, just as in yeast, the role of *sir-2.1* in mediating the lifespan effect of CR in *C. elegans* remains an open debate.

In *D. melanogaster*, five genes have been identified with similarity to yeast *SIR2*, with *dSir2* reported as the closest match [90]. The results of genetic *dSir2* manipulations have been variable. In one study the results were similar to those with *C. elegans sir-2.1*, such that *dSir2* mutations did not significantly shorten lifespan [91]. However, in a separate study using a P element excision strain, in which a null allele was generated, the *dSir2* mutant was found to have a short lifespan [90], similar to the yeast system. Rogina et al. subsequently showed that over expression of *dSir2* increased lifespan [92]. The dSir2 protein has been implicated in position effect variegation [90, 93], and polycomb-mediated silencing [94], although the silencing effects are unrelated to longevity [95]. As with *C. elegans*, the in vivo deacetylation targets of dSir2 remain unidentified. Even so, dSir2 appears to be involved in the lifespan extension caused by dietary restriction, and mRNA levels of dSir2 are reported to increase under such restricted dietary conditions [96]. Furthermore, a reduction in dSir2 levels blocks the extension of lifespan that normally occurs with DR [92].

Mammalian Sirtuins in Lifespan Regulation and Age-Related Disease

SIRT1. There are 7 sirtuin genes (*SIRT1-7*) in mammals, with *SIRT1* showing the greatest sequence similarity to yeast *SIR2*. Most early work on mammalian sirtuins therefore focused on *SIRT1*. Homozygous *SIRT1* KO mice are often embryonic lethal, and the pups that are born have reduced body weights. Individuals that survive to adulthood are sterile due to a defect in gametogenesis [97]. Therefore, any assessment of a role for SIRT1 in controlling aging or mediating CR effects using KO mice could be confounded by the developmental abnormalities. Initial analysis concluded that the lack of SIRT1 does not affect any of the general metabolic changes that occur under a CR diet, such as reduced blood glucose, triglycerides, and insulin-like growth factors, but does block the typical increase in physical activity induced by CR [98]. However, the KO mice are hyperphagic and have an overactive metabolism [99]. The mechanism for the effect of CR-mediated activity is unknown, but could be related to their already hypermetabolic condition. So far, there have been no reports of *SIRT1* KO mice having any significant effect on adult lifespan. Furthermore, studies with human centenarians have so far not found any significant association between common sequence variations in *SIRT1* and lifespan [100].

Despite a lack of clear lifespan effects caused by manipulations of *SIRT1*, there are multiple examples of expression changes of *SIRT1* in relation to age, nutrient condition (CR), or stress state. Measurements of SIRT1 protein levels in mouse embryonic fibroblasts (MEFs) derived from mice with normal, accelerated, or delayed aging all showed a decrease with serial passage of cells, but the decrease was more rapid in the premature aging strain [101]. It is interesting to note that spontaneous immortalization of the passaged MEFs resulted in the restoration of SIRT1 levels [101]. CR causes an increase in SIRT1 expression in rat brain, fat, kidney, and liver [102]. Nutrient withdrawal (or overnight fasting) in mice increases SIRT1 expression in skeletal muscle, liver and heart [103]. Furthermore, nutrient withdrawal in rat PC12 cells increases *SIRT1* mRNA and protein levels and its interactions with both p53 and FOXO3a [103], both of which are deacetylation targets of SIRT1, and improve stress resistance and cell survival when deacetylated [84]. Over expression of SIRT1 in PC12 cells also reveals interactions with PGC1-α. SIRT1 facilitates PGC1-α deacetylation in vitro and in vivo, allowing the cells to respond to metabolic conditions appropriately by adjusting mitochondrial biogenesis [104].

The over-expression of *SIR2* in yeast and *C. elegans* clearly extends lifespan [28, 80]. Mice with an extra copy of *SIRT1* have been generated using a transgenic "knock-in" approach [105], but the effect of the transgene on lifespan has not been reported. Despite the lack of information on lifespan, the *SIRT1-KI* mice do share some (but not all) metabolic phenotypes with normal mice that are calorie restricted [105], suggesting that *SIRT1* could possibly mediate some of the effects of CR. Among these phenotypes is an improvement in glucose tolerance that is similar to the phenotype observed in transgenic "BESTO" mice specifically overexpressing SIRT1 in pancreatic β cells [106]. Decreased glucose tolerance and insulin resistance are hallmarks of type 2 diabetes, a common age-associated disease in the human population. SIRT1 over expression in β cells appears to improve insulin secretion during high glucose conditions by repressing transcription of uncoupling protein 2 (UCP2), which leads to mitochondrial ATP production [106]. In pancreatic β cells, the increased ATP enhances the response of the insulin secretion machinery by closing K_{ATP} channels [107]. Additionally, SIRT1 has been proposed to regulate insulin signalling by deacetylation of insulin receptor substrate 2 (IRS2) and repression of protein tyrosine phosphatase 1B (PTP1) [108, 109]. The identification of a link between SIRT1 and diabetes has led to the development of small molecule activators of SIRT1 deacetylation activity for possible use as type 2 diabetes therapeutics [110]. Resveratrol, a polyphenolic compound found in red wine, was the first of these compounds identified [111]. However, resveratrol, is a relatively poor activator that has substrate-specific effects on activity [112, 113]. Consistent with a model for SIRT1 activation being effective against diabetes, moderate SIRT1 over expression in mice, meant to mimic a gain of function, was recently shown to decrease energy expenditure and improve glucose tolerance in models of insulin resistance and diabetes [114].

A common theme with SIRT1 is that over expression or activation of this protein promotes cell survival, especially in the context of organs subject to age-associated disease like the heart and nervous system. SIRT1 is upregulated in the heart during heart failure or stresses such as paraquot injection or pressure overload [115]. *SIRT1* expression in the heart is also higher in old compared to young monkeys. Moderate over expression of SIRT1 in the heart of a transgenic mouse was protective against the age-dependent changes that normally occur in the organ, such as cardiomyocyte hypertrophy, fibrosis, and myocyte apoptosis [115]. Furthermore, moderate *SIRT1* over expression protected the heart against oxidative damage caused by paraquot treatment. Importantly, these protective effects were specific to transgenic lines that had lower to moderate over expression levels (up to 7.5-fold). Lines with higher over expression (12.5-fold) actually showed worse outcomes than normal mice and had increased oxidative stress levels [115]. There is likely going to be an optimal range of SIRT1 activity in various tissues that is pro-survival. However, SIRT1 is not always a pro-survival molecule. This was clearly shown in the case of SIRT1-mediated deacetylation of NF-κB, where SIRT1 activity potentiates apoptosis in lung cancer cell lines when they are stimulated with TNFα [116]. A list of SIRT1 targets relevant to aging or age-associated disease is shown in Table 1.

SIRT2. The mammalian SIRT2 protein is most closely related to the yeast Hst2 protein, and like Hst2, is predominantly localized to the cytoplasm [138, 145]. Despite this localization pattern, SIRT2 can act as a histone H4-K16 deacetylase during the G2/M transition of the cell cycle [146], suggesting a role for SIRT2 in mitosis. Indeed cells overexpressing SIRT2 have a prolonged mitotic phase of the cell cycle due to control of mitotic exit [147], revealing a possible role for the enzyme as part of a mitotic checkpoint. Furthermore, SIRT2 interacts with and deacetylates α-tubulin [138], where in oligodendrocytes it controls mitosis and differentiation [148]. Oligodendrocytes are involved in the insulation of axons in the

Table 1 Deacetylation targets of Sirtuins[a]

Organism	Sirtuin	Substrate[b]	References
S. cerevisiae	Sir2	Histones	[46, 48]
	Hst3	H3 K56	[37]
C. elegans	SIR-2.1	Unknown	
D. melanogaster	dSir2	Histones	[117]
Human	SIRT1	AceCS1, Atg5, Atg7, Atg8, BCL6, β-catenin, FOXO1, FOXO3a, FOXO4, HIC1, histone H1 (K26), histone H3 (K9,K14), histone H4 (K16), H2A.z, Ku70, LXR, MEF2, MyoD, NF-κB, p300/CBP, p53, p73, PCAF, PGC-1α, Rb, TAFI68	[118, 102, 119, 120, 40, 121–127, 103, 104, 128, 129, 130, 131, 132, 133, 134, 116, 135]
	SIRT2	α-tubulin, FOXO1, FOXO3a, histone H3 (K14), histone H4 (K16), p53	[136–140]
	SIRT3	AceCS2, GDH, histone H4 (K16)	[40, 141, 41]
	SIRT4	GDH, BSA, histones	[52, 51]
	SIRT5	cytochrome c, histone H4	[138, 142]
	SIRT6	Histone H3 (K9)	[54]
	SIRT7	p53	[143]

[a] Adapted and modified from [144].
[b] Targets with putative effects on aging or age-associated disease.

central nervous system. Interestingly, SIRT2 and SIRT1 have both been implicated in neurodegenerative age-related diseases such as Alzheimer's, amyotrophic lateral sclerosis (ALS), and Parkinson's. Over expression or activation of SIRT1 is protective against neurodegeneration in cell-based or mouse models of Alzheimer's and ALS [149]. In contrast, inhibition of SIRT2 reduces α-synuclein-dependent neuronal defects in a cellular model of Parkinson's [150]. It is possible that their differential effects on degeneration are due to their different deacetylation targets and cellular localization. Both of these sirtuins deacetylate FOXO proteins in response to a variety of stresses, including oxidative stress [151, 137, 140]. However, the responses are complex, suggesting an intricate crosstalk between the sirtuins and FOXO-dependent pathways [152]. The common theme between these disorders is the accumulation of specific protein aggregates that are cytotoxic and highly carbonylated (i.e. α-synuclein for Parkinson's disease) [153]. Therefore, an intriguing link back to the yeast system is that *S. cerevisiae* Sir2 is involved in the asymmetric inheritance of oxidatively damaged (carbonylated) proteins, which form large aggregates in old mother cells [32]. The effect of Hst2 on the inheritance of these aggregates has not been reported.

SIRT3, SIRT4, SIRT5. The mitochondrial sirtuins, SIRT3, SIRT4, and SIRT5 have not been directly implicated in aging, but mounting evidence indicates that they are involved in energy metabolism, which again has important implications in age-related diseases. Mitochondrial SIRT3 is the best characterized of the three, and has been shown to deacetylate acetyl CoA synthetase (AceCS2), which activates the enzyme to convert acetate into acetyl-CoA [41]. In yeast, acetyl-CoA synthetase in the nucleus (Acs2) produces the acetyl-CoA that is used as a substrate for histone acetylation by HATs [154]. The same could be true for AceCS2 in the mitochondria. In the absence of mouse SIRT3, a large number of mitochondrial proteins become hyperacetylated, suggesting SIRT3 is a major protein deacetylase in this compartment [155]. In contrast, mice knocked out for SIRT4 or SIRT5 show no increase in overall mitochondrial protein acetylation [155]. As stated in an earlier section, SIRT4 is a mono-ADP ribosyltransferase, and has been shown to ADP ribosylate glutamate dehydrogenise, inactivating the enzyme [51]. CR reduces the expression level of SIRT4, promoting the conversion of glutamate into α-ketoglutarate for entry into the TCA cycle. In pancreatic β cells, this promotes amino acid-stimulated insulin secretion, providing another link between CR, Sirtuins, and an age-associated disease (diabetes) [51]. The aging-related targets for SIRT5 remain uncharacterized. Interestingly, a variable number of tandem repeats (VNTR) polymorphism occurs in exon 5 of the *SIRT3* gene that harbors enhancer activity and is linked to males older than 90, suggesting that under-expression of *SIRT3* may be detrimental for longevity [156].

SIRT6. Although SIRT1 knockout mice do not appear to have a significant aging phenotype, homozygous SIRT6 KO mice do have a premature aging-like phenotype that includes lymphopenia, loss of subcutaneous fat, lordokyphosis, and severe metabolic defects. These problems begin at about 2 weeks of age, and the mice eventually die at about 4 weeks [157]. These mice also have a problem with genomic instability, initially thought to be caused by a defect in DNA repair [157]. However, more recent work has determined that SIRT6 is a histone H3 K9 deacetylase, and that knocking down SIRT6 expression levels in cell culture results in problems with telomere maintenance [54]. There is also an increase in telomere fusion events, a shared characteristic of cells from Werner's syndrome patients with mutations in the WRN helicase. Werner's syndrome is characterized by premature aging. The WRN protein associates with telomeres and functions in telomere maintenance and recombination [158, 159]. Interestingly, SIRT6 deacetylation of H3-K9 appears to promote the association of WRN with telomere chromatin, suggesting that SIRT6 and WRN may have some overlapping function in longevity regulation [54].

SIRT7. SIRT7 has been implicated in age-related heart disease. SIRT7 KO mice actually do have a short lifespan and suffer from degenerative cardiac hypertrophy as they age [143]. Unlike the SIRT6 KO mouse, the SIRT7 KOs do not have additional aging phenotypes that are typical of Werner's syndrome or progeria. Their premature death is most likely due to the heart disease. SIRT7 interacts with Pol I in the nucleolus of cells in culture. Over expression of SIRT7 stimulates rDNA transcription by Pol I, and knockdown of SIRT7 inhibits Pol I transcription [160]. One of the hallmarks of cardiac hypertrophy is an increase in rDNA transcription and the size of the nucleoli to compensate for the increased requirement for ribosomes in actively growing myocardiocytes [161]. At first glance these results do not fit with the SIRT7 KO mouse results, and could possibly be due to the effect of over-expression levels, especially if proper SIRT7 dosage is as important as SIRT1 dosage. The effects of SIRT7 in the KO study was suggested to be through deacetylation of p53, rather than through Pol I [143]. A summary of lifespan effects caused

Table 2 Lifespan effects by sirtuin manipulations

Organism	Genotype	Diet or mutation	Lifespan effect[a]	References
S. cerevisiae	*SIR2* OE	Normal	⇑⇑	[28]
	*sir2*Δ	Normal	⇓⇓	[28]
	*hst3*Δ	Normal	⇓	[36]
C. elegans	*sir-2.1* OE	Normal	⇑	[80, 71, 82]
	–	Normal	⇔	[162, 81]
	–	Normal	⇓	[86, 82, 83]
	–	Normal	⇔⇓	[87]
	–	CR	⇔	[83]
	–	Food deprivation	⇑⇑	[86]
	–	*eat-2*	⇑⇑	[86]
	–	Dietary deprivation	⇑⇑	[87]
	–	*eat-2(ad1116)*	⇑⇑	[89]
	–	CR	⇑⇑	[89]
	–	2DG	⇑⇑	[88]
D. melanogaster	*dSir2* OE	Normal	⇑	[92]
	–	Normal	⇔	[91]
	–	Normal	⇓	[90, 92]
	–	Normal	⇓⇓	[163]
		Normal		
	–	CR	⇔	[92]
M. musculus	*SIRT1 -/-*	Normal	⇓⇓⇓[b]	[97]
	SIRT6 -/-	Normal	⇓⇓	[157]
	SIRT7 -/-	Normal	⇓⇓	[143]

[a] Upward arrow means extended lifespan, downward arrows indicate shortened lifespan, and sideways arrows indicate no effect.
[b] Large amount of embryonic lethality.

by manipulations of the various sirtuins in mammals and other organisms is shown in Table 2.

Resveratrol as an Activator of Sirtuins

Given the strong links between sirtuins and longevity, or the prevention of age-associated diseases, the development of a small molecule drug that could activate a specific sirtuin is highly desirable. Resveratrol was the first sirtuin activator identified and was reported to activate yeast Sir2 and human SIRT1 in vitro, and to extend yeast replicative lifespan when provided to cells exogenously [111]. As described above, resveratrol has substrate specific effects in vitro [112, 113], and the extension of lifespan in yeast has been difficult to confirm [113]. The effect of resveratrol on *C. elegans* and *Drosophila* lifespan has also been inconsistent in different studies. Some studies demonstrated lifespan extension [82, 81], while others did not or showed variable results [162]. Despite these discrepancies, resveratrol clearly induces physiological changes in mammals that are related to the effects of caloric restriction or over expression of SIRT1 [164]. This does not necessarily mean that resveratrol exclusively works through the activation of Sirtuins. Resveratrol has antioxidant properties and has effects on multiple protein targets, including the AMP-activated kinase (AMPK) [165]. High concentrations of resveratrol in the diet have been shown to prevent pathologies due to a high calorie diet in mice, but did not extend lifespan [166]. More potent Sirtuin activators have been developed [110]. It will be interesting to learn whether any of these compounds have positive effects on lifespan, in addition to their effects on age-related disease, such as diabetes [110].

Acknowledgments We would like to thank Marty Mayo and members of the Smith lab for helpful discussions in the course of writing this work. J.S.S. is a member of the University of Virginia Institute on Aging and is funded by the NIH/National Institutes of Aging (NIA) and General Medical Science (NIGMS). D.L.S. is funded by NIH training grant DK062710-05 to the University of Alabama-Birmingham Obesity Training Program.

References

1. Frye RA (1999). Characterization of five human cDNAs with homology to the yeast *SIR2* gene: Sir2-like proteins (Sirtuins) metabolize NAD and may have protein ADP-ribosyltransferase activity. *Biochem Biophys Res Comm* 260: 273–279.
2. Brachmann CB, Sherman JM, Devine SE, Cameron EE, Pillus L, and Boeke JD (1995). The *SIR2* gene family, conserved from bacteria to humans, functions in silencing, cell cycle progression, and chromosome stability. *Genes Dev* 9: 2888–2902.
3. Derbyshire MK, Weinstock KG, and Strathern JN (1996). *HST1*, a new member of the *SIR2* family of genes. *Yeast* 12: 631–640.
4. Frye RA (2000). Phylogenetic classification of prokaryotic and eukaryotic Sir2-like proteins. *Biochem Biophys Res Commun* 273: 793–798.
5. Tsang AW and Escalante-Semerena JC (1998). Cobb, a new member of the SIR2 family of eucaryotic regulatory proteins, is required to compensate for the lack of nicotinate mononucleotide:5,6-dimethylbenzimidazole phosphoribosyltransferase activity in *cobT* mutants during cobalamin biosynthesis in *Salmonella typhimurium* LT2. *J Biol Chem* 273: 31788–31794.
6. Klar AJ, Fogel S, and Macleod K (1979). *MAR1*-a regulator of the *HM*a and *HM*α loci in *SACCHAROMYCES CEREVISIAE*. *Genetics* 93: 37–50.
7. Rine J, Strathern JN, Hicks JB, and Herskowitz I (1979). A suppressor of mating-type locus mutations in *Saccharomyces cerevisiae*: evidence for and identification of cryptic mating-type loci. *Genetics* 93: 877–901.
8. Haber JE (1998). Mating-type gene switching in *Saccharomyces cerevisiae*. *Annu Rev Genet* 32: 561–599.
9. Rine J and Herskowitz I (1987). Four genes responsible for a position effect on expression from *HML* and *HMR* in *Saccharomyces cerevisiae*. *Genetics* 116: 9–22.
10. Rusche LN, Kirchmaier AL, and Rine J (2003). The establishment, inheritance, and function of silenced chromatin in *Saccharomyces cerevisiae*. *Annu Rev Biochem* 72: 481–516.
11. Gottschling DE, Aparicio OM, Billington BL, and Zakian VA (1990). Position effect at *S. cerevisiae* telomeres: reversible repression of Pol II transcription. *Cell* 63: 751–762.
12. Bryk M, Banerjee M, Murphy M, Knudsen KE, Garfinkel DJ, and Curcio MJ (1997). Transcriptional silencing of Ty1 elements in the *RDN1* locus of yeast. *Genes Dev* 11: 255–269.

13. Smith JS and Boeke JD (1997). An unusual form of transcriptional silencing in yeast ribosomal DNA. *Genes Dev* 11: 241–254.
14. Shou W, Seol JH, Shevchenko A, Baskerville C, Moazed D, Chen ZW, Jang J, Charbonneau H, and Deshaies RJ (1999). Exit from mitosis is triggered by Tem1-dependent release of the protein phosphatase Cdc14 from nucleolar RENT complex. *Cell* 97: 233–244.
15. Straight AF, Shou W, Dowd GJ, Turck CW, Deshaies RJ, Johnson AD, and Moazed D (1999). Net1, a Sir2-associated nucleolar protein required for rDNA silencing and nucleolar integrity. *Cell* 97: 245–256.
16. Coelho PS, Bryan AC, Kumar A, Shadel GS, and Snyder M (2002). A novel mitochondrial protein, Tar1p, is encoded on the antisense strand of the nuclear 25S rDNA. *Genes Dev* 16: 2755–2760.
17. Kobayashi T and Ganley AR (2005). Recombination regulation by transcription-induced cohesin dissociation in rDNA repeats. *Science* 309: 1581–1584.
18. Li C, Mueller JE, and Bryk M (2006). Sir2 represses endogenous polymerase II transcription units in the ribosomal DNA nontranscribed spacer. *Mol Biol Cell* 17: 3848–3859.
19. Kobayashi T, Horiuchi T, Tongaonkar P, Vu L, and Nomura M (2004). SIR2 regulates recombination between different rDNA repeats, but not recombination within individual rRNA genes in yeast. *Cell* 117: 441–453.
20. Kennedy BK, Nicanor R, Austriaco J, Zhang J, and Guarente L (1995). Mutation in the silencing gene *SIR4* can delay aging in *S. cerevisiae*. *Cell* 80: 485–496.
21. Mortimer RK and Johnston JR (1959). Life span of individual yeast cells. *Nature* 183: 1751–1752.
22. Kaeberlein M, Kirkland KT, Fields S, and Kennedy BK (2005a). Genes determining yeast replicative life span in a long-lived genetic background. *Mech Ageing Dev* 126: 491–504.
23. Smeal T, Claus J, Kennedy B, Cole F, and Guarente L (1996). Loss of transcriptional silencing causes sterility in old mother cells of *S. cerevisiae*. *Cell* 84: 633–642.
24. Kennedy BK, Gotta M, Sinclair DA, Mills K, Mcnabb DS, Murthy M, Pak SM, Laroche T, Gasser SM, and Guarente L (1997). Redistribution of silencing proteins from telomeres to the nucleolus is associated with extension of life span in *S. cerevisiae*. *Cell* 89: 381–391.
25. Sinclair DA and Guarente L (1997). Extrachromosomal rDNA circles-A cause of aging in yeast. *Cell* 91: 1033–1042.
26. Shcheprova Z, Baldi S, Frei SB, Gonnet G, and Barral Y (2008). A mechanism for asymmetric segregation of age during yeast budding. *Nature* 454: 728–734.
27. Conrad-Webb H and Butow RA (1995). A polymerase switch in the synthesis of rRNA in *Saccharomyces cerevisiae*. *Mol Cell Biol* 15: 2420–2428.
28. Kaeberlein M, Mcvey M, and Guarente L (1999). The SIR2/3/4 complex and SIR2 alone promote longevity in *Saccharomyces cerevisiae* by two different mechanisms. *Genes Dev* 13: 2570–2580.
29. Defossez PA, Prusty R, Kaeberlein M, Lin SJ, Ferrigno P, Silver PA, Keil RL, and Guarente L (1999). Elimination of replication block protein Fob1 extends the life span of yeast mother cells. *Mol Cell* 3: 447–455.
30. Kobayashi T and Horiuchi T (1996). A yeast gene product, Fob1 protein, required for both replication fork blocking and recombinational hotspot activities. *Genes Cells* 1: 465–474.
31. Aguilaniu H, Gustafsson L, Rigoulet M, and Nystrom T (2003). Asymmetric inheritance of oxidatively damaged proteins during cytokinesis. *Science* 299: 1751–1753.
32. Erjavec N, Larsson L, Grantham J, and Nystrom T (2007). Accelerated aging and failure to segregate damaged proteins in Sir2 mutants can be suppressed by overproducing the protein aggregation-remodeling factor Hsp104p. *Genes Dev* 21: 2410–2421.
33. Chen Q, Ding Q, and Keller JN (2005b). The stationary phase model of aging in yeast for the study of oxidative stress and age-related neurodegeneration. *Biogerontology* 6: 1–13.
34. Smith DL Jr, Mcclure JM, Matecic M, and Smith JS (2007). Calorie restriction extends the chronological lifespan of *Saccharomyces cerevisiae* independently of the Sirtuins. *Aging Cell* 6: 649–662.

35. Fabrizio P, Gattazzo C, Battistella L, Wei M, Cheng C, Mcgrew K, and Longo VD (2005). Sir2 blocks extreme life-span extension. *Cell* 123: 655–667.
36. Tsuchiya M, Dang N, Kerr EO, Hu D, Steffen KK, Oakes JA, Kennedy BK, and Kaeberlein M (2006). Sirtuin-independent effects of nicotinamide on lifespan extension from calorie restriction in yeast. *Aging Cell* 5: 505–514.
37. Celic I, Masumoto H, Griffith WP, Meluh P, Cotter RJ, Boeke JD, and Verreault A (2006). The sirtuins Hst3 and Hst4p preserve genome integrity by controlling histone H3 lysine 56 deacetylation. *Curr Biol* 16: 1280–1289.
38. Starai VJ, Takahashi H, Boeke JD, and Escalante-Semerena JC (2003). Short-chain fatty acid activation by acyl-coenzyme A synthetases requires SIR2 protein function in *Salmonella enterica* and *Saccharomyces cerevisiae*. *Genetics* 163: 545–555.
39. Starai VJ, Celic I, Cole RN, Boeke JD, and Escalante-Semerena JC (2002). Sir2-dependent activation of acetyl-CoA synthetase by deacetylation of active lysine. *Science* 298: 2390–2392.
40. Hallows WC, Lee S, and Denu JM (2006). Sirtuins deacetylate and activate mammalian acetyl-CoA synthetases. *Proc Natl Acad Sci U S A* 103: 10230–10235.
41. Schwer B, Bunkenborg J, Verdin RO, Andersen JS, and Verdin E (2006). Reversible lysine acetylation controls the activity of the mitochondrial enzyme acetyl-CoA synthetase 2. *Proc Natl Acad Sci U S A* 103: 10224–10229.
42. Braunstein M, Rose AB, Holmes SG, Allis CD, and Broach JR (1993). Transcriptional silencing in yeast is associated with reduced nucleosome acetylation. *Genes Dev* 7: 592–604.
43. Tanny JC, Dowd GJ, Huang J, Hilz H, and Moazed D (1999). An enzymatic activity in the yeast Sir2 protein that is essential for gene silencing. *Cell* 99: 735–745.
44. Smith JS, Caputo E, and Boeke JD (1999). A genetic screen for ribosomal DNA silencing defects identifies multiple DNA replication and chromatin-modulating factors. *Mol Cell Biol* 19: 3184–3197.
45. Smith JS, Brachmann CB, Celic I, Kenna MA, Muhammad S, Starai VJ, Avalos JL, Escalante-Semerena JC, Grubmeyer C, Wolberger C, and Boeke JD (2000). A phylogenetically conserved NAD^+-dependent protein deacetylase activity in the Sir2 protein family. *Proc Natl Acad Sci U S A* 97: 6658–6663.
46. Imai S-I, Armstrong CM, Kaeberlein M, and Guarente L (2000). Transcriptional silencing and longevity protein Sir2 is an NAD-dependent histone deacetylase. *Nature* 403: 795–800.
47. Landry J, Slama JT, and Sternglanz R (2000a). Role of NAD^+ in the deacetylase activity of the SIR2-like proteins. *Biochem Biophys Res Commun* 278: 685–690.
48. Landry J, Sutton A, Tafrov ST, Heller RC, Stebbins J, Pillus L, and Sternglanz R (2000b). The silencing protein SIR2 and its homologs are NAD-dependent protein deacetylases. *Proc Natl Acad Sci U S A* 97: 5807–5811.
49. Tanny JC and Moazed D (2001). Coupling of histone deacetylation to NAD breakdown by the yeast silencing protein Sir2: Evidence for acetyl transfer from substrate to an NAD breakdown product. *Proc Natl Acad Sci U S A* 98: 415–420.
50. Garcia-Salcedo JA, Gijon P, Nolan DP, Tebabi P, and Pays E (2003). A chromosomal SIR2 homologue with both histone NAD-dependent ADP-ribosyltransferase and deacetylase activities is involved in DNA repair in *Trypanosoma brucei*. *EMBO J* 22: 5851–5862.
51. Haigis MC, Mostoslavsky R, Haigis KM, Fahie K, Christodoulou DC, Murphy AJ, Valenzuela DM, Yancopoulos GD, Karow M, Blander G, Wolberger C, Prolla TA, Weindruch R, Alt FW, and Guarente L (2006). SIRT4 inhibits glutamate dehydrogenase and opposes the effects of calorie restriction in pancreatic beta cells. *Cell* 126: 941–954.
52. Ahuja N, Schwer B, Carobbio S, Waltregny D, North BJ, Castronovo V, Maechler P, and Verdin E (2007). Regulation of insulin secretion by SIRT4, a mitochondrial ADP-ribosyltransferase. *J Biol Chem* 282: 33583–33592.
53. Liszt G, Ford E, Kurtev M, and Guarente L (2005). Mouse Sir2 homolog SIRT6 is a nuclear ADP-ribosyltransferase. *J Biol Chem* 280: 21313–21320.
54. Michishita E, Mccord RA, Berber E, Kioi M, Padilla-Nash H, Damian M, Cheung P, Kusumoto R, Kawahara TL, Barrett JC, Chang HY, Bohr VA, Ried T, Gozani O, and

Chua KF (2008). SIRT6 is a histone H3 lysine 9 deacetylase that modulates telomeric chromatin. *Nature* 452: 492–496.

55. Lin S-J, Defossez P-A, and Guarente L (2000). Requirement of NAD and SIR2 for life-span extension by calorie restriction in *Saccharomyces cerevisiae*. *Science* 289: 2126–2128.
56. Jiang JC, Wawryn J, Shantha Kumara HM, and Jazwinski SM (2002). Distinct roles of processes modulated by histone deacetylases Rpd3p, Hda1p, and Sir2p in life extension by caloric restriction in yeast. *Exp Gerontol* 37: 1023–1030.
57. Kaeberlein M, Andalis AA, Fink GR, and Guarente L (2002). High osmolarity extends life span in *Saccharomyces cerevisiae* by a mechanism related to calorie restriction. *Mol Cell Biol* 22: 8056–8066.
58. Anderson RM, Bitterman KJ, Wood JG, Medvedik O, and Sinclair DA (2003). Nicotinamide and PNC1 govern lifespan extension by calorie restriction in *Saccharomyces cerevisiae*. *Nature* 423: 181–185.
59. Lin SJ, Kaeberlein M, Andalis AA, Sturtz LA, Defossez PA, Culotta VC, Fink GR, and Guarente L (2002). Calorie restriction extends *Saccharomyces cerevisiae* lifespan by increasing respiration. *Nature* 418: 344–348.
60. Kaeberlein M, Kirkland KT, Fields S, and Kennedy BK (2004). Sir2-independent life span extension by calorie restriction in yeast. *PLoS Biol* 2: E296.
61. Lamming DW, Latorre-Esteves M, Medvedik O, Wong SN, Tsang FA, Wang C, Lin SJ, and Sinclair DA (2005). *HST2* mediates *SIR2*-independent life-span extension by calorie restriction. *Science* 309: 1861–1864, Epub 2005 Jul 28.
62. Kaeberlein M, Steffen KK, Hu D, Dang N, Kerr EO, Tsuchiya M, Fields S, and Kennedy BK (2006a). Comment on "HST2 mediates SIR2-independent life-span extension by calorie restriction". *Science* 312: 1312, author reply 1312.
63. Lin SJ and Guarente L (2006). Increased life span due to calorie restriction in respiratory-deficient yeast. *PLoS Genet* 2: e33, author reply e34.
64. Reverter-Branchat G, Cabiscol E, Tamarit J, and Ros J (2004). Oxidative damage to specific proteins in replicative and chronological-aged *Saccharomyces cerevisiae*: common targets and prevention by calorie restriction. *J Biol Chem* 279: 31983–31989.
65. Sohal RS and Weindruch R (1996). Oxidative stress, caloric restriction, and aging. *Science* 273: 59–63.
66. Murakami CJ, Burtner CR, Kennedy BK, and Kaeberlein M (2008). A method for high-throughput quantitative analysis of yeast chronological life span. *J Gerontol A Biol Sci Med Sci* 63: 113–121.
67. Gallo CM, Smith DL, Jr, and Smith JS (2004). Nicotinamide clearance by Pnc1 directly regulates Sir2-mediated silencing and longevity. *Mol Cell Biol* 24: 1301–1312.
68. Bitterman KJ, Anderson RM, Cohen HY, Latorre-Esteves M, and Sinclair DA (2002). Inhibition of silencing and accelerated aging by nicotinamide, a putative negative regulator of yeast sir2 and human SIRT1. *J Biol Chem* 277: 45099–45107.
69. Ghislain M, Talla E, and Francois JM (2002). Identification and functional analysis of the *Saccharomyces cerevisiae* nicotinamidase gene, *PNC1*. *Yeast* 19: 215–224.
70. Balan V, Miller GS, Kaplun L, Balan K, Chong ZZ, Li F, Kaplun A, Vanberkum MF, Arking R, Freeman DC, Maiese K, and Tzivion G (2008). Lifespan extension and neuronal cell protection by *Drosophila* nicotinamidasc. *J Biol Chem* 283: 27810–27819.
71. van der Horst A, Schavemaker JM, Pellis-van Berkel W, and Burgering BM (2007). The *Caenorhabditis elegans* nicotinamidase PNC-1 enhances survival. *Mech Ageing Dev* 128: 346–349.
72. Anderson RM, Bitterman KJ, Wood JG, Medvedik O, Cohen H, Lin SS, Manchester JK, Gordon JI, and Sinclair DA (2002). Manipulation of a nuclear NAD^+ salvage pathway delays aging without altering steady-state NAD^+ levels. *J Biol Chem* 277: 18881–18890.
73. Rajavel M, Lalo D, Gross JW, and Grubmeyer C (1998). Conversion of a cosubstrate to an inhibitor: phosphorylation mutants of nicotinic acid phosphoribosyltransferase. *Biochemistry* 37: 4181–4188.

74. Sandmeier JJ, Celic I, Boeke JD, and Smith JS (2002). Telomeric and rDNA silencing in *Saccharomyces cerevisiae* are dependent on a nuclear NAD^+ salvage pathway. *Genetics* 160: 877–889.
75. Imsande J (1964). A cross-linked control system. I. Properties of a triphosphate-dependent nicotinic acid mononucleotide pyrophosphorylase from *Bacillus subtilis*. *Biochim Biophys Acta* 85: 255–264.
76. Belenky P, Racette FG, Bogan KL, Mcclure JM, Smith JS, and Brenner C (2007). Nicotinamide riboside promotes Sir2 silencing and extends lifespan via Nrk and Urh1/Pnp1/Meu1 pathways to NAD^+. *Cell* 129: 473–484.
77. Revollo JR, Grimm AA, and Imai S (2004). The NAD biosynthesis pathway mediated by nicotinamide phosphoribosyltransferase regulates Sir2 activity in mammalian cells. *J Biol Chem* 279: 50754–50763.
78. Magni G, Amici A, Emanuelli M, Orsomando G, Raffaelli N, and Ruggieri S (2004). Enzymology of NAD^+ homeostasis in man. *Cell Mol Life Sci* 61: 19–34.
79. Revollo JR, Korner A, Mills KF, Satoh A, Wang T, Garten A, Dasgupta B, Sasaki Y, Wolberger C, Townsend RR, Milbrandt J, Kiess W, and Imai S (2007). Nampt/PBEF/Visfatin regulates insulin secretion in beta cells as a systemic NAD biosynthetic enzyme. *Cell Metab* 6: 363–375.
80. Tissenbaum HA and Guarente L (2001). Increased dosage of a *sir-2* gene extends lifespan in *Caenorhabditis elegans*. *Nature* 410: 227–230.
81. Wood JG, Rogina B, Lavu S, Howitz K, Helfand SL, Tatar M, and Sinclair D (2004). Sirtuin activators mimic caloric restriction and delay ageing in metazoans. *Nature* 430: 686–689.
82. Viswanathan M, Kim SK, Berdichevsky A, and Guarente L (2005). A role for SIR-2.1 regulation of ER stress response genes in determining *C. elegans* life span. *Dev Cell* 9: 605–615.
83. Wang Y and Tissenbaum HA (2006). Overlapping and distinct functions for a *Caenorhabditis elegans SIR2* and DAF-16/FOXO. *Mech Ageing Dev* 127: 48–56.
84. Giannakou ME and Partridge L (2004). The interaction between FOXO and SIRT1: tipping the balance towards survival. *Trends Cell Biol* 14: 408–412.
85. Lakowski B and Hekimi S (1998). The genetics of caloric restriction in *Caenorhabditis elegans*. *Proc Natl Acad Sci U S A* 95: 13091–13096.
86. Kaeberlein TL, Smith ED, Tsuchiya M, Welton KL, Thomas JH, Fields S, Kennedy BK, and Kaeberlein M (2006b). Lifespan extension in *Caenorhabditis elegans* by complete removal of food. *Aging Cell* 5: 487–494.
87. Lee GD, Wilson MA, Zhu M, Wolkow CA, de Cabo R, Ingram DK, and Zou S (2006). Dietary deprivation extends lifespan in *Caenorhabditis elegans*. *Aging Cell* 5: 515–524.
88. Schulz TJ, Zarse K, Voigt A, Urban N, Birringer M, and Ristow M (2007). Glucose restriction extends *Caenorhabditis elegans* life span by inducing mitochondrial respiration and increasing oxidative stress. *Cell Metab* 6: 280–293.
89. Hansen M, Taubert S, Crawford D, Libina N, Lee SJ, and Kenyon C (2007). Lifespan extension by conditions that inhibit translation in *Caenorhabditis elegans*. *Aging Cell* 6: 95–110.
90. Astrom SU, Cline TW, and Rine J (2003). The *Drosophila melanogaster* $sir2^+$ gene is nonessential and has only minor effects on position-effect variegation. *Genetics* 163: 931–937.
91. Newman BL, Lundblad JR, Chen Y, and Smolik SM (2002). A *Drosophila* homologue of Sir2 modifies position-effect variegation but does not affect life span. *Genetics* 162: 1675–1685.
92. Rogina B and Helfand SL (2004). Sir2 mediates longevity in the fly through a pathway related to calorie restriction. *Proc Natl Acad Sci U S A* 101: 15998–16003.
93. Rosenberg MI and Parkhurst SM (2002). Drosophila Sir2 is required for heterochromatic silencing and by euchromatic Hairy/E(Spl) bHLH repressors in segmentation and sex determination. *Cell* 109: 447–458.

94. Furuyama T, Banerjee R, Breen TR, and Harte PJ (2004). SIR2 is required for polycomb silencing and is associated with an E(Z) histone methyltransferase complex. *Curr Biol* 14: 1812–1821.
95. Frankel S and Rogina B (2005). *Drosophila* longevity is not affected by heterochromatin-mediated gene silencing. *Aging Cell* 4: 53–56.
96. Rogina B, Helfand SL, and Frankel S (2002). Longevity regulation by *Drosophila* Rpd3 deacetylase and caloric restriction. *Science* 298: 1745.
97. Mcburney MW, Yang X, Jardine K, Hixon M, Boekelheide K, Webb JR, Lansdorp PM, and Lemieux M (2003). The mammalian SIR2α protein has a role in embryogenesis and gametogenesis. *Mol Cell Biol* 23: 38–54.
98. Chen D, Steele AD, Lindquist S, and Guarente L (2005a). Increase in activity during calorie restriction requires Sirt1. *Science* 310: 1641.
99. Boily G, Seifert EL, Bevilacqua L, He XH, Sabourin G, Estey C, Moffat C, Crawford S, Saliba S, Jardine K, Xuan J, Evans M, Harper ME, and Mcburney MW (2008). SirT1 regulates energy metabolism and response to caloric restriction in mice. *PLoS One* 3: e1759.
100. Flachsbart F, Croucher PJ, Nikolaus S, Hampe J, Cordes C, Schreiber S, and Nebel A (2006). Sirtuin 1 (SIRT1) sequence variation is not associated with exceptional human longevity. *Exp Gerontol* 41: 98–102.
101. Sasaki T, Maier B, Bartke A, and Scrable H (2006). Progressive loss of SIRT1 with cell cycle withdrawal. *Aging Cell* 5: 413–422.
102. Cohen HY, Miller C, Bitterman KJ, Wall NR, Hekking B, Kessler B, Howitz KT, Gorospe M, de Cabo R, and Sinclair DA (2004). Calorie restriction promotes mammalian cell survival by inducing the SIRT1 deacetylase. *Science* 305: 390–392.
103. Nemoto S, Fergusson MM, and Finkel T (2004). Nutrient availability regulates SIRT1 through a forkhead-dependent pathway. *Science* 306: 2105–2108.
104. Nemoto S, Fergusson MM, and Finkel T (2005). SIRT1 functionally interacts with the metabolic regulator and transcriptional coactivator PGC-1α. *J Biol Chem* 280: 16456–16460.
105. Bordone L, Cohen D, Robinson A, Motta MC, van Veen E, Czopik A, Steele AD, Crowe H, Marmor S, Luo J, Gu W, and Guarente L (2007). *SIRT1* transgenic mice show phenotypes resembling calorie restriction. *Aging Cell* 6: 759–767.
106. Moynihan KA, Grimm AA, Plueger MM, Bernal-Mizrachi E, Ford E, Cras-Meneur C, Permutt MA, and Imai S (2005). Increased dosage of mammalian Sir2 in pancreatic beta cells enhances glucose-stimulated insulin secretion in mice. *Cell Metab* 2: 105–117.
107. Koster JC, Permutt MA, and Nichols CG (2005). Diabetes and insulin secretion: the ATP-sensitive K^+ channel (K ATP) connection. *Diabetes* 54: 3065–3072.
108. Sun C, Zhang F, Ge X, Yan T, Chen X, Shi X, and Zhai Q (2007). SIRT1 improves insulin sensitivity under insulin-resistant conditions by repressing PTP1B. *Cell Metab* 6: 307–319.
109. Zhang J (2007). The direct involvement of SirT1 in insulin-induced insulin receptor substrate-2 tyrosine phosphorylation. *J Biol Chem* 282: 34356–34364.
110. Milne JC, Lambert PD, Schenk S, Carney DP, Smith JJ, Gagne DJ, Jin L, Boss O, Perni RB, Vu CB, Bemis JE, Xie R, Disch JS, Ng PY, Nunes JJ, Lynch AV, Yang H, Galonek H, Israelian K, Choy W, Iffland A, Lavu S, Medvedik O, Sinclair DA, Olefsky JM, Jirousek MR, Elliott PJ, and Westphal CH (2007). Small molecule activators of SIRT1 as therapeutics for the treatment of type 2 diabetes. *Nature* 450: 712–716.
111. Howitz KT, Bitterman KJ, Cohen HY, Lamming DW, Lavu S, Wood JG, Zipkin RE, Chung P, Kisielewski A, Zhang LL, Scherer B, and Sinclair DA (2003). Small molecule activators of sirtuins extend *Saccharomyces cerevisiae* lifespan. *Nature* 425: 191–196.
112. Borra MT, Smith BC, and Denu JM (2005). Mechanism of human SIRT1 activation by resveratrol. *J Biol Chem* 280: 17187–17195.
113. Kaeberlein M, Mcdonagh T, Heltweg B, Hixon J, Westman EA, Caldwell SD, Napper A, Curtis R, Distefano PS, Fields S, Bedalov A, and Kennedy BK (2005b). Substrate-specific activation of sirtuins by resveratrol. *J Biol Chem* 280: 17038–17045.

114. Banks AS, Kon N, Knight C, Matsumoto M, Gutierrez-Juarez R, Rossetti L, Gu W, and Accili D (2008). SirT1 gain of function increases energy efficiency and prevents diabetes in mice. *Cell Metab* 8: 333–341.
115. Alcendor RR, Gao S, Zhai P, Zablocki D, Holle E, Yu X, Tian B, Wagner T, Vatner SF, and Sadoshima J (2007). Sirt1 regulates aging and resistance to oxidative stress in the heart. *Circ Res* 100: 1512–1521.
116. Yeung F, Hoberg JE, Ramsey CS, Keller MD, Jones DR, Frye RA, and Mayo MW (2004). Modulation of NF-κB-dependent transcription and cell survival by the SIRT1 deacetylase. *EMBO J* 23: 2369–2380.
117. Parsons XH, Garcia SN, Pillus L, and Kadonaga JT (2003). Histone deacetylation by Sir2 generates a transcriptionally repressed nucleoprotein complex. *Proc Natl Acad Sci U S A* 100: 1609–1614.
118. Chen IY, Lypowy J, Pain J, Sayed D, Grinberg S, Alcendor RR, Sadoshima J, and Abdellatif M (2006). Histone H2A.z is essential for cardiac myocyte hypertrophy but opposed by silent information regulator 2α. *J Biol Chem* 281: 19369–19377.
119. Dai JM, Wang ZY, Sun DC, Lin RX, and Wang SQ (2007). SIRT1 interacts with p73 and suppresses p73-dependent transcriptional activity. *J Cell Physiol* 210: 161–166.
120. Fulco M, Schiltz RL, Iezzi S, King MT, Zhao P, Kashiwaya Y, Hoffman E, Veech RL, and Sartorelli V (2003). Sir2 regulates skeletal muscle differentiation as a potential sensor of the redox state. *Mol Cell* 12: 51–62.
121. Heltweg B, Gatbonton T, Schuler AD, Posakony J, Li H, Goehle S, Kollipara R, Depinho RA, Gu Y, Simon JA, and Bedalov A (2006). Antitumor activity of a small-molecule inhibitor of human silent information regulator 2 enzymes. *Cancer Res* 66: 4368–4377.
122. Lee IH, Cao L, Mostoslavsky R, Lombard DB, Liu J, Bruns NE, Tsokos M, Alt FW, and Finkel T (2008). A role for the NAD-dependent deacetylase Sirt1 in the regulation of autophagy. *Proc Natl Acad Sci U S A* 105: 3374–3379.
123. Li X, Zhang S, Blander G, Tse JG, Krieger M, and Guarente L (2007b). SIRT1 deacetylates and positively regulates the nuclear receptor LXR. *Mol Cell* 28: 91–106.
124. Luo J, Nikolaev AY, Imai S, Chen D, Su F, Shiloh A, Guarente L, and Gu W (2001). Negative control of p53 by Sir2α promotes cell survival under stress. *Cell* 107: 137–148.
125. Mattagajasingh I, Kim CS, Naqvi A, Yamamori T, Hoffman TA, Jung SB, Dericco J, Kasuno K, and Irani K (2007). SIRT1 promotes endothelium-dependent vascular relaxation by activating endothelial nitric oxide synthase. *Proc Natl Acad Sci U S A* 104: 14855–14860.
126. Motta MC, Divecha N, Lemieux M, Kamel C, Chen D, Gu W, Bultsma Y, Mcburney M, and Guarente L (2004). Mammalian SIRT1 represses forkhead transcription factors. *Cell* 116: 551–563.
127. Muth V, Nadaud S, Grummt I, and Voit R (2001). Acetylation of TAF(I)68, a subunit of TIF-IB/SL1, activates RNA polymerase I transcription. *EMBO J* 20: 1353–1362.
128. Rodgers JT, Lerin C, Haas W, Gygi SP, Spiegelman BM, and Puigserver P (2005). Nutrient control of glucose homeostasis through a complex of PGC-1α and SIRT1. *Nature* 434: 113–118.
129. Stankovic-Valentin N, Deltour S, Seeler J, Pinte S, Vergoten G, Guerardel C, Dejean A, and Leprince D (2007). An acetylation/deacetylation-SUMOylation switch through a phylogenetically conserved psiKXEP motif in the tumor suppressor HIC1 regulates transcriptional repression activity. *Mol Cell Biol* 27: 2661–2675.
130. van der Horst A, Tertoolen LG, de Vries-Smits LM, Frye RA, Medema RH, and Burgering BM (2004). FOXO4 is acetylated upon peroxide stress and deacetylated by the longevity protein hSir2(SIRT1). *J Biol Chem* 279: 28873–28879.
131. Vaquero A, Scher M, Lee D, Erdjument-Bromage H, Tempst P, and Reinberg D (2004). Human SirT1 interacts with histone H1 and promotes formation of facultative heterochromatin. *Mol Cell* 16: 93–105.

132. Vaziri H, Dessain SK, Ng Eaton E, Imai SI, Frye RA, Pandita TK, Guarente L, and Weinberg RA (2001). hSIR2(SIRT1) functions as an NAD-dependent p53 deacetylase. *Cell* 107: 149–159.
133. Wong S and Weber JD (2007). Deacetylation of the retinoblastoma tumour suppressor protein by SIRT1. *Biochem J* 407: 451–460.
134. Yang Y, Hou H, Haller EM, Nicosia SV, and Bai W (2005). Suppression of FOXO1 activity by FHL2 through SIRT1-mediated deacetylation. *EMBO J* 24: 1021–1032.
135. Zhao X, Sternsdorf T, Bolger TA, Evans RM, and Yao TP (2005). Regulation of MEF2 by histone deacetylase 4- and SIRT1 deacetylase-mediated lysine modifications. *Mol Cell Biol* 25: 8456–8464.
136. Jin YH, Kim YJ, Kim DW, Baek KH, Kang BY, Yeo CY, and Lee KY (2008). Sirt2 interacts with 14-3-3 beta/gamma and down-regulates the activity of p53. *Biochem Biophys Res Commun* 368: 690–695.
137. Jing E, Gesta S, and Kahn CR (2007). SIRT2 regulates adipocyte differentiation through FoxO1 acetylation/deacetylation. *Cell Metab* 6: 105–114.
138. North BJ, Marshall BL, Borra MT, Denu JM, and Verdin E (2003). The human Sir2 ortholog, SIRT2, is an NAD+-dependent tubulin deacetylase. *Mol Cell* 11: 437–444.
139. Vaquero A, Sternglanz R, and Reinberg D (2007). NAD^+-dependent deacetylation of H4 lysine 16 by class III HDACs. *Oncogene* 26: 5505–5520.
140. Wang F, Nguyen M, Qin FX, and Tong Q (2007). SIRT2 deacetylates FOXO3a in response to oxidative stress and caloric restriction. *Aging Cell* 6: 505–514.
141. Scher MB, Vaquero A, and Reinberg D (2007). SirT3 is a nuclear NAD^+-dependent histone deacetylase that translocates to the mitochondria upon cellular stress. *Genes Dev* 21: 920–928.
142. Schlicker C, Gertz M, Papatheodorou P, Kachholz B, Becker CF, and Steegborn C (2008). Substrates and regulation mechanisms for the human mitochondrial sirtuins Sirt3 and Sirt5. *J Mol Biol* 382: 790–801.
143. Vakhrusheva O, Smolka C, Gajawada P, Kostin S, Boettger T, Kubin T, Braun T, and Bober E (2008). Sirt7 increases stress resistance of cardiomyocytes and prevents apoptosis and inflammatory cardiomyopathy in mice. *Circ Res* 102: 703–710.
144. Taylor DM, Maxwell MM, Luthi-Carter R, and Kazantsev AG (2008). Biological and potential therapeutic roles of sirtuin deacetylases. *Cell Mol Life Sci* 65: 4000–4018.
145. Perrod S, Cockell MM, Laroche T, Renauld H, Ducrest AL, Bonnard C, and Gasser SM (2001). A cytosolic NAD-dependent deacetylase, Hst2p, can modulate nucleolar and telomeric silencing in yeast. *EMBO J* 20: 197–209.
146. Vaquero A, Scher MB, Lee DH, Sutton A, Cheng HL, Alt FW, Serrano L, Sternglanz R, and Reinberg D (2006). SirT2 is a histone deacetylase with preference for histone H4 Lys 16 during mitosis. *Genes Dev* 20: 1256–1261.
147. Dryden SC, Nahhas FA, Nowak JE, Goustin AS, and Tainsky MA (2003). Role for human SIRT2 NAD-dependent deacetylase activity in control of mitotic exit in the cell cycle. *Mol Cell Biol* 23: 3173–3185.
148. Li W, Zhang B, Tang J, Cao Q, Wu Y, Wu C, Guo J, Ling EA, and Liang F (2007a). Sirtuin 2, a mammalian homolog of yeast silent information regulator-2 longevity regulator, is an oligodendroglial protein that decelerates cell differentiation through deacetylating α-tubulin. *J Neurosci* 27: 2606–2616.
149. Kim D, Nguyen MD, Dobbin MM, Fischer A, Sananbenesi F, Rodgers JT, Delalle I, Baur JA, Sui G, Armour SM, Puigserver P, Sinclair DA, and Tsai LH (2007). SIRT1 deacetylase protects against neurodegeneration in models for Alzheimer's disease and amyotrophic lateral sclerosis. *EMBO J* 26: 3169–3179.
150. Outeiro TF, Kontopoulos E, Altmann SM, Kufareva I, Strathearn KE, Amore AM, Volk CB, Maxwell MM, Rochet JC, Mclean PJ, Young AB, Abagyan R, Feany MB, Hyman BT, and Kazantsev AG (2007). Sirtuin 2 inhibitors rescue α-synuclein-mediated toxicity in models of Parkinson's disease. *Science* 317: 516–519.

151. Brunet A, Sweeney LB, Sturgill JF, Chua KF, Greer PL, Lin Y, Tran H, Ross SE, Mostoslavsky R, Cohen HY, Hu LS, Cheng HL, Jedrychowski MP, Gygi SP, Sinclair DA, Alt FW, and Greenberg ME (2004). Stress-dependent regulation of FOXO transcription factors by the SIRT1 deacetylase. *Science* 303: 2011–2015.
152. Gan L and Mucke L (2008). Paths of convergence: sirtuins in aging and neurodegeneration. *Neuron* 58: 10–14.
153. Nystrom T (2005). Role of oxidative carbonylation in protein quality control and senescence. *EMBO J* 24: 1311–1317.
154. Takahashi H, Mccaffery JM, Irizarry RA, and Boeke JD (2006). Nucleocytosolic acetyl-coenzyme A synthetase is required for histone acetylation and global transcription. *Mol Cell* 23: 207–217.
155. Lombard DB, Alt FW, Cheng HL, Bunkenborg J, Streeper RS, Mostoslavsky R, Kim J, Yancopoulos G, Valenzuela D, Murphy A, Yang Y, Chen Y, Hirschey MD, Bronson RT, Haigis M, Guarente LP, Farese RV, Jr, Weissman S, Verdin E, and Schwer B (2007). Mammalian Sir2 homolog SIRT3 regulates global mitochondrial lysine acetylation. *Mol Cell Biol* 27: 8807–8814.
156. Bellizzi D, Rose G, Cavalcante P, Covello G, Dato S, de Rango F, Greco V, Maggiolini M, Feraco E, Mari V, Franceschi C, Passarino G, and de Benedictis G (2005). A novel VNTR enhancer within the *SIRT3* gene, a human homologue of *SIR2*, is associated with survival at oldest ages. *Genomics* 85: 258–263.
157. Mostoslavsky R, Chua KF, Lombard DB, Pang WW, Fischer MR, Gellon L, Liu P, Mostoslavsky G, Franco S, Murphy MM, Mills KD, Patel P, Hsu JT, Hong AL, Ford E, Cheng HL, Kennedy C, Nunez N, Bronson R, Frendewey D, Auerbach W, Valenzuela D, Karow M, Hottiger MO, Hursting S, Barrett JC, Guarente L, Mulligan R, Demple B, Yancopoulos GD, and Alt FW (2006). Genomic instability and aging-like phenotype in the absence of mammalian SIRT6. *Cell* 124: 315–329.
158. Laud PR, Multani AS, Bailey SM, Wu L, Ma J, Kingsley C, Lebel M, Pathak S, Depinho RA, and Chang S (2005). Elevated telomere-telomere recombination in WRN-deficient, telomere dysfunctional cells promotes escape from senescence and engagement of the ALT pathway. *Genes Dev* 19: 2560–2570.
159. Opresko PL, Otterlei M, Graakjaer J, Bruheim P, Dawut L, Kolvraa S, May A, Seidman MM, and Bohr VA (2004). The Werner syndrome helicase and exonuclease cooperate to resolve telomeric D loops in a manner regulated by TRF1 and TRF2. *Mol Cell* 14: 763–774.
160. Ford E, Voit R, Liszt G, Magin C, Grummt I, and Guarente L (2006). Mammalian Sir2 homolog SIRT7 is an activator of RNA polymerase I transcription. *Genes Dev* 20: 1075–1080.
161. Diez J (1994). Current work in the cell biology of left ventricular hypertrophy. *Curr Opin Cardiol* 9: 512–519.
162. Bass TM, Weinkove D, Houthoofd K, Gems D, and Partridge L (2007). Effects of resveratrol on lifespan in *Drosophila melanogaster* and *Caenorhabditis elegans*. *Mech Ageing Dev* 128: 546–552.
163. Kusama S, Ueda R, Suda T, Nishihara S, and Matsuura ET (2006). Involvement of *Drosophila* Sir2-like genes in the regulation of life span. *Genes Genet Syst* 81: 341–348.
164. Baur JA and Sinclair DA (2006). Therapeutic potential of resveratrol: the in vivo evidence. *Nat Rev Drug Discov* 5: 493–506.
165. Harikumar KB and Aggarwal BB (2008). Resveratrol: a multitargeted agent for age-associated chronic diseases. *Cell Cycle* 7: 1020–1035.
166. Pearson KJ, Baur JA, Lewis KN, Peshkin L, Price NL, Labinskyy N, Swindell WR, Kamara D, Minor RK, Perez E, Jamieson HA, Zhang Y, Dunn SR, Sharma K, Pleshko N, Woollett LA, Csiszar A, Ikeno Y, Le Couteur D, Elliott PJ, Becker KG, Navas P, Ingram DK, Wolf NS, Ungvari Z, Sinclair DA, and de Cabo R (2008). Resveratrol delays age-related deterioration and mimics transcriptional aspects of dietary restriction without extending life span. *Cell Metab* 8: 157–168.

The Role of TOR Signaling in Aging

Matt Kaeberlein and Lara S. Shamieh

Abstract The target of rapamycin (TOR) kinase defines a highly conserved nutrient-response pathway that is known to modulate longevity in invertebrate organisms. Multiple mutations in this pathway that reduce TOR signaling have been reported to increase life span in different organisms, as has pharmacological inhibition of TOR. Multiple TOR-regulated processes are also known to play a role in longevity control, including autophagy, mRNA translation initiation, and mitochondrial metabolism. TOR signaling interacts with insulin/IGF-1 signaling via Akt kinase and maps genetically to the same longevity pathway as dietary restriction. Studies are underway to determine whether inhibition of TOR is sufficient to increase life span in mammals. TOR-inhibitors are clinically useful in humans and may prove beneficial against multiple age-associated diseases.

Keywords Target of rapamycin · mTOR · Autophagy · mRNA translation · Dietary restriction

Introduction and Overview

The target of rapamycin (TOR) kinase is an evolutionarily conserved sensor of nutrients and growth factors required for viability in eukaryotes. TOR was first identified as the molecular target of an anti-fungal compound (rapamycin) produced by the bacterium *Streptomyces hygroscopicus* [1]. Rapamycin was subsequently shown to inhibit the protein product of two yeast genes: TOR1 and TOR2 [2]. Since then, TOR proteins have been identified in a variety of different species, including humans.

M. Kaeberlein (✉)
Department of Pathology, University of Washington, Seattle, WA, USA
e-mail: kaeber@u.washington.edu

N.S. Wolf (ed.), *The Comparative Biology of Aging*,
DOI 10.1007/978-90-481-3465-6_7,

Table 1 Known components of TOR complex 1 and TOR complex 2

S. cervesisiae	*C. elegans*	*D. melanogaster*	Mammals
TORC1			
TOR1/TOR2	let-363	TOR-PA	mTOR
KOG1	daf-15	Raptor-PA	Raptor
LST8	C10H11.8	CG3004-PA	mLST8
TCO89	–	–	–
TORC2			
TOR2	let-363	TOR-PA	mTOR
AVO1	–	–	–
AVO2	–	–	–
BIT61	–	–	–
LST8	C10H11.8	CG3004-PA	mLST8
SLM1	–	–	–
SLM2	–	–	–
TSC11	F29C12.2	Rictor-PA	Rictor

Components of the TORC1 and TORC2 complexes in *C. elegans*, *D. melanogaster*, and mammals were identified based on orthology mapping from the yeast protein sequences available on the *Saccharomyces cerevisiae* Genome Database http://www.yeastgenome.org

In all species studied thus far, TOR has been found to act in two distinct complexes: TOR complex 1 (TORC1) and TOR complex 2 (TORC2) [3, 4]. TORC1 and TORC2 have different cellular functions and are composed of different constituent proteins (Table 1). Both complexes are essential, as loss of the TORC1 specific component, raptor, or the TORC2 specific component, rictor, leads to inviability in yeast and embryonic lethality in mice [5, 6]. Rapamycin specifically inhibits TORC1 activity, and it is currently thought that longevity control mediated by TOR (*discussed further below*) occurs exclusively by altering TORC1 activity. TORC2 is generally thought to be rapamycin insensitive and plays an important role in organization of the actin cytoskeleton [7, 8].

The upstream regulatory features of TORC1 activity are better understood than those of TORC2 (Fig. 1). TORC1 is activated by insulin and other growth factors via signaling through phosphatidyl 3-OH kinase and Akt [9]. Activation of TORC1 by Akt is mediated by inhibition of tuberous sclerosis complex 2 (TSC2), which is itself an inhibitor of the small GTPase Rheb [10–12]. The mechanism by which Rheb activates TORC1 is not known, but requires GTP-bound Rheb and may involve direct physical interaction [13, 14]. In addition to being regulated by growth factors, TORC1 is also activated by environmental nutrients and repressed by the energy sensing AMP-activated protein kinase (AMPK). These multiple inputs place TOR at a key regulatory nexus in responding to nutrients, growth cues, and cellular energy status.

Over the past few years, an important role has emerged for TOR in determining longevity and the progression of age-associated diseases. Inhibition of TOR-signaling has been associated with improved outcomes in animal models of

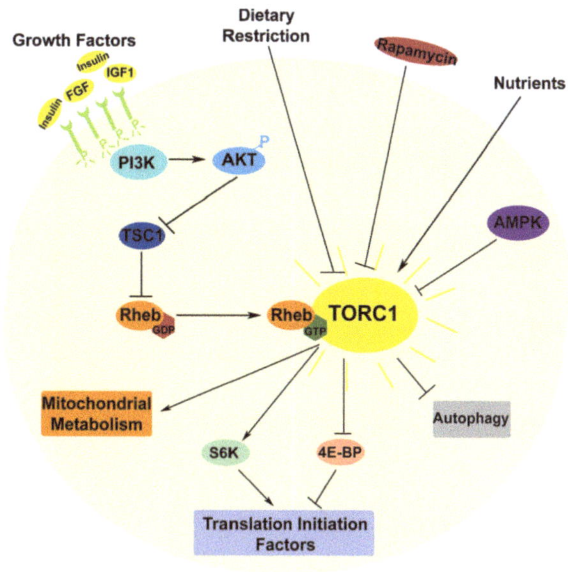

Fig. 1 The TORC1 signaling network. TOR complex 1 (TORC1) activity is promoted both by growth factors through Akt and directly by nutrient availability. Activation of AMP kinase reduces TORC1 signaling, as does dietary restriction. Downstream targets of TORC1 that have been implicated in longevity control include autophagy, mRNA translation, and mitochondrial metabolism

cancer, diabetes, cardiac disease, and neurodegeneration. More strikingly, several different genetic and pharmacological interventions that decrease TOR signaling have been found to increase life span in four different invertebrate and yeast aging models (Table 2). The remainder of this chapter describes the known links

Table 2 Multi-organism comparison of interventions that reduce TOR signaling and increase life span

Intervention	Yeast CLS	Yeast RLS	Worm	Fly	Mouse
Dietary restriction	√	√	√	√	√
TOR mutation/knock-down	√	√	√	√	
Raptor mutation/knock-down			√		
Activation of AMP kinase			√		
Activation of Tsc1/2				√	
Pharmacological inhibition of TORC1 (e.g. rapamycin)	√	√			
S6 Kinase mutation/knock-down	√	√	√	√	
Ribosomal protein mutation/knock-down		√	√		
Translation initiation factor mutation/knock-down		√	√		
Mutations reducing amino acid uptake	√	√	√		

Multiple interventions that reduce TORC1 signaling have been reported to increase life span in different model organisms. Data was derived from multiple sources [15–40].

between TOR signaling and aging and discusses potential mechanisms by which TOR activity might regulate longevity in divergent eukaryotic species.

TOR Signaling Modulates Aging in Invertebrate Organisms

It has been known for many years that nutrient availability is an important environmental determinant of longevity. Given the central role of TOR signaling in the cellular response to nutritional cues, it is not surprising that TOR signaling also plays a role in determining life span in at least three widely divergent species: the budding yeast *Saccharomyces cerevisiae*, the nematode *Caenorhabditis elegans*, and the fruit fly *Drosophila melanogaster*. This section describes the studies related to TOR and aging in invertebrate model organisms and in yeast.

Role of TOR in Yeast Aging

Two different paradigms for aging in the budding yeast have been described: replicative and chronological [41]. Replicative aging is defined by the number of daughter cells produced by a mother cell before senescence, and has been suggested as a model for aging in mitotic cells of multicellular organisms [42]. Chronological aging is defined by the length of time that yeast cells can maintain viability in a quiescent-like stationary phase, and is a model for aging of post-mitotic cells [43]. Nutrient availability affects life span in both yeast aging systems, and dietary restriction (DR) protocols have been described that involve reducing the glucose concentration of the media from 2 to 0.5% or lower [44, 31, 33, 35].

Similar to the case in multicellular eukaryotes, both TORC1 and TORC2 complexes are present in yeast [45]. Unlike multicellular organisms which generally have only a single ORF coding for the TOR kinase itself, yeast contain two ORFs that code for partially redundant TOR kinases: *TOR1* and *TOR2* [6]. *TOR2* is thought to function in both TORC1 and TORC2 and is essential for viability. *TOR1*, in contrast, is thought to function only in the rapamycin-sensitive *TORC1* complex and is not essential, presumably due to sufficient residual Tor2-dependent TORC1 activity. Yeast cells carrying a deletion of *TOR1* show reduced growth and altered sensitivity to rapamycin, indicating that *TORC1* activity is reduced in these mutants [46].

The importance of TOR signaling in yeast replicative life span determination was uncovered from an unbiased longevity screen of yeast single-gene deletion strains [25]. In this study, replicative life span was determined for 564 individual single-gene deletion mutants from the yeast ORF deletion collection. Of the 564 strains analyzed, 13 were found to live significantly longer than the parental wild type strain. Among these 13 replicatively long-lived strains was a mutant lacking *TOR1*, which codes for one of two partially redundant TOR kinases in yeast.

Deletion strains lacking individual components of the TOR signaling pathway were also identified as long-lived, including two ribosomal proteins of the large subunit (Rpl31a and Rpl6b) and Ure2, a regulator of glutamine metabolism. Treatment of cells with the TOR-inhibitor methionine sulfoximine was also shown to increase replicative life span [25]; methionine sulfoximine inhibits TOR indirectly, by reducing glutamine synthetase activity leading to lower cytoplasmic levels of glutamine [47].

An independent screen of the yeast ORF deletion collection for chronologically long-lived mutants, uncovered a similar role for TOR signaling in yeast chronological life span determination. Deletion of *TOR1* was found to significantly increase chronological life span, as did deletion of additional factors known to increase TOR activity by promoting uptake of amino acids from the environment [34]. Treatment of cells with rapamycin was also sufficient to increase chronological life span.

Role of TOR in Nematode Aging

In *C. elegans*, aging is typically assayed by measuring the survival of adult animals maintained on a nutrient-agar surface and fed an *E. coli* OP50 bacterial food source. Gene expression can be specifically knocked-down by RNAi, which is accomplished by replacing the OP50 food source with an *E. coli* clone expressing double stranded RNA corresponding to the ORF of interest. RNAi knock-down of either TOR (*let-363*) or raptor (*daf-15*) during adulthood significantly increases life span [23, 39]. Animals completely lacking TOR function from egg arrest in the third larval stage (L3), but are also long-lived [39]. Mutation of an intestinal peptide transporter (*pep-2*) that is thought to act upstream of TOR has also been shown to increase life span in *C. elegans* [48], as has deletion or RNAi knock-down of the S6 kinase homolog, *rsks-1* [22, 49].

In addition to increasing life span, knock-down of TOR signaling induces a variety of other phenotypes which may be related to its longevity effects. For example, RNAi knock-down of either TOR or raptor leads to increased fat storage and enhanced dauer formation, both hallmarks of mutants that promote longevity in the insulin/IGF-1-like signaling pathway [23, 39]. The relationship between TOR and insulin/IGF-1-like signaling in worms (and in other organisms) is complex, however, as TOR and raptor appear to behave differently with respect to this pathway; life span extension from knock-down of raptor requires the FOXO-family transcription factor *daf-16* [23], while knock-down of TOR increases life span independently of *daf-16* [22, 49]. *Daf-16* activity is repressed by insulin/IGF-1-like signaling, and mutations that increase life span in the insulin/IGF-1-like signaling pathway do so by activating *daf-16* [50, 51]. Expression of raptor is also repressed by *daf-16* [23], however, which further complicates genetic dissection of precisely how TORC1 interacts with insulin-like signaling to modulate longevity in *C. elegans*.

Role of TOR in Fly Aging

In flies, proteins functioning upstream of TORC1, downstream of TORC1, and in the TORC1 complex have been shown to modulate longevity. Similar to the case in mammals, dTsc1 and dTsc2 negatively regulate TORC1 activity in flies, and overexpression of either protein is sufficient to increase life span [27]. Expression of a dominant negative allele of TOR or S6K is also sufficient to increase fly life span [27].

Along with increased life span, inhibition of TOR signaling is associated with improved function in fly models of age-associated disease. Treatment of flies with rapamycin induces autophagy and enhances clearance of aggregation prone proteins, including expanded polyglutamine peptides and mutant forms of tau [52, 53]. Mutation of TOR confers protection against age-associated declines in fly heart function and protects against insulin resistance in animals with a hyperactive allele of the insulin responsive transcription factor, dFOXO [54].

The Relationship Between TOR and Dietary Restriction

DR, defined as a reduction in nutrient availability without malnutrition, is the only intervention known to increase life span in yeast, worms, flies, and mammals [55]. The conserved role of TOR as a nutrient-responsive signaling pathway, combined with the observation that decreased TOR activity increases life span in multiple organisms, suggests the possibility that TOR signaling mediates life span extension from DR [56]. Consistent with this idea, it is clear that DR reduces TOR activity in a variety of organisms, as evidenced by an induction of autophagy and reduced S6 kinase activity [57, 58, 3]. What remains to be determined is the degree to which the longevity and health benefits of dietary restriction can be directly attributed to TOR signaling and TOR-regulated targets.

Genetic studies in yeast are consistent with the idea that DR acts through TOR. In the replicative aging paradigm, deletion of *TOR1* increases the life span of wild type mother cells, but does not further increase the life span of cells subjected to DR [25]. Such non-additivity is generally interpreted to be consistent with the hypothesis that two interventions act to modulate longevity via a similar mechanism. Further supporting this idea, DR, deletion of *TOR1*, or deletion of *SCH9* (yeast S6K) each increase replicative life span additively with mutation of the replication fork block protein Fob1 and independent of the Sir2 protein deacetylase [44, 25, 59, 60]. In the yeast chronological aging paradigm, life span extension from DR is also thought to act via reduced TOR signaling, perhaps involving a TOR-mediated shift toward enhanced mitochondrial respiratory activity [61].

Similar to the case in yeast, inhibition of TOR signaling fails to further increase the life span of worms subjected to DR [22]. Interestingly, several recent studies have reported that induction of autophagy (via reduced TOR signaling) is required for the life span extension associated with DR in *C. elegans* [21, 62, 63]. The importance of autophagy and other TOR-regulated processes (see Section 4) in promoting longevity in response to DR is currently an area of intense interest.

Downstream Effectors of TOR Signaling and Their Relationship to Aging

Although invertebrate systems have demonstrated that reduced TOR signaling is sufficient to slow aging, the downstream mechanisms responsible for these effects remain to be precisely determined. TOR signaling is known to regulate several different processes that may contribute to its role in aging (Fig. 1), including (1) repressing autophagy, (2) repressing stress response pathways, (3) regulating mitochondrial metabolism, and (4) regulating translation by promoting ribosome function and translation initiation. Recent experimental evidence suggests that multiple TOR-regulated processes act in concert to promote health and longevity in response to reduced TOR signalling.

Autophagy

Autophagy, which literally means "self eating", is a degradative process through which cellular components are engulfed by cytoplasmic vesicles and transported to the lysozome/vacuole for degradation [64]. Autophagy is repressed by TOR signaling and is induced in response to starvation or treatment with TOR inhibitors, such as rapamycin. A decline in the autophagic response has been reported in aging mammals [65], and increased autophagy is required for life span extension in long-lived mutants with reduced insulin/IGF-1-like signaling in *C. elegans* [66]. Autophagy is thought to be of particular importance in protecting against age-associated neurodegenerative diseases caused by protein misfolding or aggregation [67, 68].

Several recent studies have uncovered an important role for autophagy in the response to DR. Autophagy is induced by DR in yeast, worms, and flies [69–71], and is reported to be required for life span extension from DR or TOR-inhibition in both worms and flies [21, 62, 69]. Autophagy also plays a protective role in neurodegenerative diseases associated with polyglutamine toxicity in worms [72, 73] and mice [74]. These findings are consistent with a model in which TOR-mediated induction of autophagy in response to DR plays a causal role in the observed effects on health and longevity; however, it remains to be determined whether induction of autophagy is sufficient to phenocopy any of the effects of DR.

Stress Response

The correlation between longevity and stress resistance has been well established in a variety of organisms. In yeast, TORC1, in concert with protein kinase A and S6K (Sch9), represses several stress-responsive transcription factors, including Msn2, Msn4, Rim15, and Gis1. As a consequence, reduced TOR signaling leads to induction a constitutive stress response. In the yeast chronological aging system, induction of this stress response seems to be particularly important for life span extension from DR or deletion of S6K [75], and it has been proposed that up-regulation of superoxide dismutase activity via TOR-regulated stress-response factors is necessary for chronological life span extension [76].

Unlike the case for chronological life span, the majority of data suggests that TOR-repressed stress factors have only a minimal role in modulation of replicative life span. One report suggested that Msn2/Msn4-dependent up-regulation of the nicotinamidase enzyme Pnc1 (and subsequent activation of Sir2) is important for replicative life span extension from reduced TOR signaling [77]. This model seems unlikely to be correct, however, as a prior report showed that triple deletion of Msn2, Msn4, and Rim15 modestly increases replicative life span and does not prevent life span extension from deletion of S6K [19]. In a separate report, double deletion of both Msn4 and Msn4 did not prevent life span extension from DR [31] and a series of studies has shown that neither Sir2 nor the other Sir2-family members is required for replicative life span extension from DR, deletion of *TOR1*, or deletion of S6K [78, 44, 25, 59, 37, 60]. Thus, while TOR signaling may indirectly interact with sirtuins to modulate longevity, the preponderance of genetic data suggests that these conserved longevity modifiers act largely via distinct genetic pathways.

Metabolic Effects and Mitochondrial Function

TOR signaling has long been known to be responsive to nutrient availability and abundance, but TOR signaling also regulates how those nutrients are obtained and the manner in which they are utilized once they enter the cell. The precise metabolic effects of altered TOR signaling are still relatively poorly understood; however, recent studies have suggested that mitochondrial function is both a regulator of TOR activity, as well as a downstream target that responds to TOR signaling.

In yeast, plentiful glucose leads to high TOR activity and ATP generation primarily by alcoholic fermentation. When glucose becomes limiting (such as under DR conditions), TOR activity is reduced, mitochondrial genes are induced, and yeast switch over to primarily respiratory metabolism. Interestingly, *TOR1* mutants show an induction of mitochondrial enzymes and increased oxygen consumption even under high glucose conditions [61]. Mutation of *TOR1* fails to increase chronological life span in respiratory-deficient cells, consistent with the hypothesis that a constitutive metabolic shift toward greater mitochondrial activity may be causally involved in the chronological life span extension associated with reduced TOR signaling [61]. A detailed description of the factors involved in TOR-mediated regulation of mitochondrial function in yeast has not been described and will be an important area of future research.

The retrograde response is a signaling pathway that transduces a mitochondrially-generated signal to regulate expression of nuclear encoded genes in response to mitochondrial dysfunction [79]. Retrograde signaling has been reported to regulate yeast replicative life span through the activity an Rtg2, a gene required for proper activation of retrograde response target genes [80]. Interestingly, TOR signaling has also been shown to regulate retrograde response target genes in yeast [81, 82]. Furthermore, mitochondrial isocitrate dehydrogenase is regulated by the retrograde response and deletion/RNAi knock-down of the gene coding for this enzyme increases life span in both yeast [25, 36] and worms [83]. Thus, it is reasonable to speculate that TOR signaling could influence longevity via altered expression of retrograde-response target genes, although there is not yet direct evidence supporting this hypothesis.

In mammals, the relationship between mTOR signaling and mitochondrial function is complex. Mitochondrial function can influence TORC1 activity and is influenced by TORC1 activity. mTOR is associated with mitochondria, suggesting a direct physical interaction [84]. In response to mitochondrial uncouplers or respiratory chain inhibitors, TORC1 activity is reduced [85]. Conversely, mitochondrial activity and oxygen consumption are reduced by rapamycin or knock-down of raptor [86]. Modulation of mitochondrial function by TORC1 appears to be mediated, at least in part, by regulating a physical interaction between the YY1 transcription factor and the peroxisome proliferator-activated receptor gamma coactivator 1α (PCG-1α) [87], which has been suggested to play a role in DR and in mediating beneficial effects of resveratrol in mice fed a high-fat diet [88]. Additional studies will be necessary to determine the relevance of TOR-mitochondrial interactions in aging and disease.

mRNA Translation

TOR signaling is a primary mechanism by which protein synthesis is modulated in response to nutrient availability and abundance of growth factors [89]. TOR signaling promotes mRNA translation via multiple inputs, including direct activation of S6K and repression of eukaryotic initiation factor 4E (eIF4E) binding proteins (4E-BP). Activation of S6K, in turn, promotes the activity of translation initiation factors, such as eukaryotic initiation factor 4B (eIF4B), and directly stimulates production of ribosomal proteins and ribosome biogenesis.

Based on recent studies, a compelling case can be made for regulation of mRNA translation as an important factor accounting for increased longevity in response to reduced TOR signaling or TOR-inhibition. One piece of evidence supporting this idea is the finding from multiple different laboratories that reduced S6 kinase activity is sufficient to increase life span in yeast, worms, and flies [20, 22, 78, 27, 49]. RNAi knock-down of several different translation initiation factors or ribosomal proteins has also been found to increase life span in worms and [16, 17, 22, 49, 38], in some cases, deletion of the yeast ortholog has been shown to increase replicative life span [36, 37].

The observation that modulation of longevity both by TOR signaling and by TOR-regulated protein synthesis factors is conserved between yeast and worms, which are evolutionarily separated by approximately 1.5 billion years, suggests a central role for this pathway in longevity determination [36]. These observations also suggest the hypothesis that DR, reduced TOR signaling, and reduce protein synthesis promote longevity in a linear pathway. Epistasis analysis from studies in *C. elegans* indicate that this model is likely to be overly simplistic, however, as TOR and DR map to a single epistasis group with respect to longevity, but S6K and translation initiation factors map to a separate epistasis group [21]. One explanation that has been proposed is that, while DR leads to reduced protein synthesis via reduced TOR activity, the life span extension observed from knock-down of S6K and other protein synthesis factors in well fed animals occurs via a different mechanism [21]. Further studies in yeast, worms, and flies will be required to sort out the complex interactions involved between different TOR-regulated targets and longevity under both fed and food restricted conditions.

TOR Signaling in Mammalian Aging

Aside from DR, TOR signaling is the only modifier of longevity shown to be conserved in both yeast aging paradigms, in worms, and in flies. The importance of TOR signaling in mammalian aging remains unknown. Since DR increases life span in rodents, a key test of the hypothesis that DR is mediated by reduced TOR signaling will be to determine whether inhibition of TOR is sufficient to increase life span in mice. Initial longevity experiments with mice fed a diet supplemented with rapamycin are underway as part of the National Institutes on Aging Interventions Testing Program [90] and will be of particular interest for the gerontological community.

Independent of longevity data in rodents, there is reason to be optimistic that reduced TOR signaling may be beneficial for a variety of age-associated diseases in mammals. For example, mice treated with rapamycin show resistance to cancer, neurodegeneration, and cardiac disease [91, 92]. Additionally, S6 kinase knock-out mice show phenotypes consistent with a genetic mimic of dietary restriction, including improved insulin sensitivity and resistance to age- and diet-induced obesity [93].

There is also emerging data that inhibition of TOR is likely to have beneficial health effects in humans. Rapamycin (Sirolimus) is used clinically as an immunosuppressant and to prevent coronary stent restenosis [94]. Rapamycin is also in clinical trials as an anti-cancer therapy [95]. It is noteworthy that reduced cancer incidence is a primary feature of DR in rodents, suggesting that rapamycin is mimicking at least some DR phenotypes in people. It will be of great interest to determine whether rapamycin is therapeutic toward other age-associated diseases, such as Alzheimer's disease and diabetes.

Conclusion

The target of rapamycin kinase is one the best current candidates for a therapeutically useful anti-aging target in humans. TOR signaling is likely to mediate at least some of the benefits associated with DR. Among known longevity interventions, only DR and TOR signaling are known to modulate longevity in both yeast aging paradigms, in worms, and in flies. The downstream targets of TOR signaling are highly conserved from yeast to humans, and several of these downstream targets have also been implicated in longevity control. TOR signaling is known to modulate a variety of age-associated diseases, including cancer, metabolic disease, neurodegeneration, and cardiac disease. Finally, clinically useful inhibitors of TOR signaling are already available and in use. It will be of great interest to discover whether inhibition of TOR is sufficient to increase life span and delay the onset of age-associated diseases in rodents and people.

Acknowledgments Studies related to this topic in the Kaeberlein lab are funded by a Glenn/AFAR Breakthroughs in Gerontology Award to M. K. and a grant from the Ellison Medical Foundation. L. S. S. is supported by NIH training grant P30 AG013280.

References

1. Vezina C, Kudelski A, and Sehgal SN (1975). Rapamycin (AY-22,989), a new antifungal antibiotic. I. Taxonomy of the producing streptomycete and isolation of the active principle. *J Antibiot (Tokyo)* 28: 721–726.
2. Heitman J, Movva NR, and Hall MN (1991). Targets for cell cycle arrest by the immunosuppressant rapamycin in yeast. *Science* 253: 905–909.
3. De Virgilio C and Loewith R (2006). The TOR signalling network from yeast to man. *Int J Biochem Cell Biol* 38: 1476–1481.
4. Martin DE and Hall MN (2005). The expanding TOR signaling network. *Curr Opin Cell Biol* 17: 158–166.
5. Guertin DA, Stevens DM, Thoreen CC, Burds AA, Kalaany NY, Moffat J, Brown M, Fitzgerald KJ, and Sabatini DM (2006). Ablation in mice of the mTORC components raptor, rictor, or mLST8 reveals that mTORC2 is required for signaling to Akt-FOXO and PKCalpha, but not S6K1. *Dev Cell* 11: 859–871.
6. Helliwell SB, Howald I, Barbet N, and Hall MN (1998). TOR2 is part of two related signaling pathways coordinating cell growth in Saccharomyces cerevisiae. *Genetics* 148: 99–112.
7. Jacinto E, Facchinetti V, Liu D, Soto N, Wei S, Jung SY, Huang Q, Qin J, and Su B (2006). SIN1/MIP1 maintains rictor-mTOR complex integrity and regulates Akt phosphorylation and substrate specificity. *Cell* 127: 125–137.
8. Sarbassov DD, Ali SM, Kim DH, Guertin DA, Latek RR, Erdjument-Bromage H, Tempst P, and Sabatini DM (2004). Rictor, a novel binding partner of mTOR, defines a rapamycin-insensitive and raptor-independent pathway that regulates the cytoskeleton. *Curr Biol* 14: 1296–1302.
9. Gingras AC, Svitkin Y, Belsham GJ, Pause A, and Sonenberg N (1996). Activation of the translational suppressor 4E-BP1 following infection with encephalomyocarditis virus and poliovirus. *Proc Natl Acad Sci USA* 93: 5578–5583.
10. Inoki K, Li Y, Xu T, and Guan KL (2003). Rheb GTPase is a direct target of TSC2 GAP activity and regulates mTOR signaling. *Genes Dev* 17: 1829–1834.
11. Inoki K, Li Y, Zhu T, Wu J, and Guan KL (2002). TSC2 is phosphorylated and inhibited by Akt and suppresses mTOR signalling. *Nat Cell Biol* 4: 648–657.
12. Manning BD, Tee AR, Logsdon MN, Blenis J, and Cantley LC (2002). Identification of the tuberous sclerosis complex-2 tumor suppressor gene product tuberin as a target of the phosphoinositide 3-kinase/akt pathway. *Mol Cell* 10: 151–162.
13. Long X, Muller F, and Avruch J (2004). TOR action in mammalian cells and in *Caenorhabditis elegans*. *Curr Top Microbiol Immunol* 279: 115–138.

14. Long X, Spycher C, Han ZS, Rose AM, Muller F, and Avruch J (2002). TOR deficiency in *C. elegans* causes developmental arrest and intestinal atrophy by inhibition of mRNA translation. *Curr Biol* 12: 1448–1461.
15. Apfeld J, O'Connor G, McDonagh T, DiStefano PS, and Curtis R (2004). The AMP-activated protein kinase AAK-2 links energy levels and insulin-like signals to lifespan in *C. elegans*. *Genes Dev* 18: 3004–3009.
16. Chen D, Pan KZ, Palter JE, and Kapahi P (2007). Longevity determined by developmental arrest genes in *Caenorhabditis elegans*. *Aging Cell* 6: 525–533.
17. Curran SP and Ruvkun G (2007). Lifespan regulation by evolutionarily conserved genes essential for viability. *PLoS Genet* 3: e56.
18. Curtis R, O'Connor G, and DiStefano PS (2006). Aging networks in Caenorhabditis elegans: AMP-activated protein kinase (aak-2) links multiple aging and metabolism pathways. *Aging Cell* 5: 119–126.
19. Fabrizio P, Pletcher SD, Minois N, Vaupel JW, and Longo VD (2004). Chronological aging-independent replicative life span regulation by Msn2/Msn4 and Sod2 in Saccharomyces cerevisiae. *FEBS Lett* 557: 136–142.
20. Fabrizio P, Pozza F, Pletcher SD, Gendron CM, and Longo VD (2001). Regulation of longevity and stress resistance by Sch9 in yeast. *Science* 292: 288–290.
21. Hansen M, Chandra A, Mitic LL, Onken B, Driscoll M, and Kenyon C (2008). A Role for Autophagy in the Extension of Lifespan by Dietary Restriction in *C. elegans*. *PLoS Genet* 4: e24.
22. Hansen M, Taubert S, Crawford D, Libina N, Lee SJ, and Kenyon C (2007). Lifespan extension by conditions that inhibit translation in Caenorhabditis elegans. *Aging Cell* 6: 95–110.
23. Jia K, Chen D, and Riddle DL (2004). The TOR pathway interacts with the insulin signaling pathway to regulate C. elegans larval development, metabolism and life span. *Development* 131: 3897–3906.
24. Jiang JC, Jaruga E, Repnevskaya MV, and Jazwinski SM (2000). An intervention resembling caloric restriction prolongs life span and retards aging in yeast. *Faseb J* 14: 2135–2137.
25. Kaeberlein M, Powers RW, 3rd, Steffen KK, Westman EA, Hu D, Dang N, Kerr EO, Kirkland KT, Fields S, and Kennedy BK (2005b). Regulation of yeast replicative life span by TOR and Sch9 in response to nutrients. *Science* 310: 1193–1196.
26. Kaeberlein TL, Smith ED, Tsuchiya M, Welton KL, Thomas JH, Fields S, Kennedy BK, and Kaeberlein M (2006b). Lifespan extension in Caenorhabditis elegans by complete removal of food. *Aging Cell* 5: 487–494.
27. Kapahi P, Zid BM, Harper T, Koslover D, Sapin V, and Benzer S (2004). Regulation of lifespan in Drosophila by modulation of genes in the TOR signaling pathway. *Curr Biol* 14: 885–890.
28. Klass MR (1977). Aging in the nematode *Caenorhabditis elegans*: major biological and environmental factors influencing life span. *Mech Ageing Dev* 6: 413–429.
29. Lakowski B and Hekimi S (1998). The genetics of caloric restriction in *Caenorhabditis elegans*. *Proc Natl Acad Sci USA* 95: 13091–13096.
30. Lee GD, Wilson MA, Zhu M, Wolkow CA, de Cabo R, Ingram DK, and Zou S (2006). Dietary deprivation extends lifespan in *Caenorhabditis elegans*. *Aging Cell* 5: 515–524.
31. Lin SJ, Defossez PA, and Guarente L (2000). Requirement of NAD and SIR2 for life-span extension by calorie restriction in *Saccharomyces cerevisiae*. *Science* 289: 2126–2128.
32. Mair W, Goymer P, Pletcher SD, and Partridge L (2003). Demography of dietary restriction and death in Drosophila. *Science* 301: 1731–1733.
33. Murakami CJ, Burtner CR, Kennedy BK, and Kaeberlein M (2008). A method for high-throughput quantitative analysis of yeast chronological life span. *J Gerontol A Biol Sci Med Sci* 63: 113–121.
34. Powers RW 3rd, Kaeberlein M, Caldwell SD, Kennedy BK, and Fields S (2006). Extension of chronological life span in yeast by decreased TOR pathway signaling. *Genes Dev* 20: 174–184.

35. Smith DL Jr., McClure JM, Matecic M, and Smith JS (2007). Calorie restriction extends the chronological lifespan of *Saccharomyces cerevisiae* independently of the Sirtuins. *Aging Cell* 6: 649–662.
36. Smith ED, Tsuchiya M, Fox LA, Dang N, Hu D, Kerr EO, Johnston ED, Tchao BN, Pak DN, Welton KL, et al. (2008). Quantitative evidence for conserved longevity pathways between divergent eukaryotic species. *Genome Res* 18: 564–570.
37. Steffan KK, MacKay VL, Kerr EO, Tsuchiya M, Hu D, Fox LA, Dang N, Johnston ED, Oakes JA, Tchao BN, et al. (2008). Yeast lifespan extension by depletion of 60S ribosomal subunits is mediated by Gcn4. *Cell* 133: 292–302.
38. Syntichaki P, Troulinaki K, and Tavernarakis N (2007). eIF4E function in somatic cells modulates ageing in *Caenorhabditis elegans*. *Nature* 445: 922–926.
39. Vellai T, Takacs-Vellai K, Zhang Y, Kovacs AL, Orosz L, and Muller F (2003). Genetics: influence of TOR kinase on lifespan in *C. elegans*. *Nature* 426: 620.
40. Weindruch RH and Walford RL (1988). *The Retardation of Aging and Disease by Dietary Restriction*. Springfield, IL: Thomas.
41. Kaeberlein M (2006). Longevity and aging in the budding yeast. In *Handbook of Models for Human Aging*. (PM Conn, ed.). Boston: Elsevier Press, pp. 109–120.
42. Mortimer RK and Johnston JR (1959). Life span of individual yeast cells. *Nature* 183: 1751–1752.
43. Fabrizio P and Longo VD (2003). The chronological life span of Saccharomyces cerevisiae. *Aging Cell* 2: 73–81.
44. Kaeberlein M, Kirkland KT, Fields S, and Kennedy BK (2004). Sir2-independent life span extension by calorie restriction in yeast. *PLoS Biol* 2: E296.
45. Loewith R, Jacinto E, Wullschleger S, Lorberg A, Crespo JL, Bonenfant D, Oppliger W, Jenoe P, and Hall MN (2002). Two TOR complexes, only one of which is rapamycin sensitive, have distinct roles in cell growth control. *Mol Cell* 10: 457–468.
46. Helliwell SB, Wagner P, Kunz J, Deuter-Reinhard M, Henriquez R, and Hall MN (1994). TOR1 and TOR2 are structurally and functionally similar but not identical phosphatidylinositol kinase homologues in yeast. *Mol Biol Cell* 5: 105–118.
47. Crespo JL, Powers T, Fowler B, and Hall MN (2002). The TOR-controlled transcription activators GLN3, RTG1, and RTG3 are regulated in response to intracellular levels of glutamine. *Proc Natl Acad Sci USA* 99: 6784–6789.
48. Meissner B, Boll M, Daniel H, and Baumeister R (2004). Deletion of the intestinal peptide transporter affects insulin and TOR signaling in *Caenorhabditis elegans*. *J Biol Chem* 279: 36739–36745.
49. Pan KZ, Palter JE, Rogers AN, Olsen A, Chen D, Lithgow GJ, and Kapahi P (2007). Inhibition of mRNA translation extends lifespan in *Caenorhabditis elegans*. *Aging Cell* 6: 111–119.
50. Lin K, Dorman JB, Rodan A, and Kenyon C (1997). daf-16: An HNF-3/forkhead family member that can function to double the life-span of *Caenorhabditis elegans*. *Science* 278: 1319–1322.
51. Ogg S, Paradis S, Gottlieb S, Patterson GI, Lee L, Tissenbaum HA, and Ruvkun G (1997). The Fork head transcription factor DAF-16 transduces insulin-like metabolic and longevity signals in *C. elegans*. *Nature* 389: 994–999.
52. Berger Z, Ravikumar B, Menzies FM, Oroz LG, Underwood BR, Pangalos MN, Schmitt I, Wullner U, Evert BO, O'Kane CJ, et al. (2006). Rapamycin alleviates toxicity of different aggregate-prone proteins. *Hum Mol Genet* 15: 433–442.
53. Ravikumar B, Vacher C, Berger Z, Davies JE, Luo S, Oroz LG, Scaravilli F, Easton DF, Duden R, O'Kane CJ, et al. (2004). Inhibition of mTOR induces autophagy and reduces toxicity of polyglutamine expansions in fly and mouse models of Huntington disease. *Nat Genet* 36: 585–595.
54. Luong N, Davies CR, Wessells RJ, Graham SM, King MT, Veech R, Bodmer R, and Oldham SM (2006). Activated FOXO-mediated insulin resistance is blocked by reduction of TOR activity. *Cell Metab* 4: 133–142.

55. Masoro EJ (2005). Overview of caloric restriction and ageing. *Mech Ageing Dev* 126: 913–922.
56. Kapahi P and Zid B (2004). TOR pathway: linking nutrient sensing to life span. *Sci Aging Knowledge Environ 2004* PE34.
57. Arsham AM and Neufeld TP (2006). Thinking globally and acting locally with TOR. *Curr Opin Cell Biol* 18: 589–597.
58. Bhaskar PT and Hay N (2007). The two TORCs and Akt. *Dev Cell* 12: 487–502.
59. Kaeberlein M, Steffen KK, Hu D, Dang N, Kerr EO, Tsuchiya M, Fields S, and Kennedy BK (2006a). Comment on HST2 mediates SIR2-independent life-span extension by calorie restriction. *Science* 312: 1312, author reply 1312.
60. Tsuchiya M, Dang N, Kerr EO, Hu D, Steffen KK, Oakes JA, Kennedy BK, and Kaeberlein M (2006). Sirtuin-independent effects of nicotinamide on lifespan extension from calorie restriction in yeast. *Aging Cell* 5: 505–514.
61. Bonawitz ND, Chatenay-Lapointe M, Pan Y, and Shadel GS (2007). Reduced TOR signaling extends chronological life span via increased respiration and upregulation of mitochondrial gene expression. *Cell Metab* 5: 265–277.
62. Jia K and Levine B (2007). Autophagy is required for dietary restriction-mediated life span extension in *C. elegans*. *Autophagy* 3: 597–599.
63. Toth ML, Sigmond T, Borsos E, Barna J, Erdelyi P, Takacs-Vellai K, Orosz L, Kovacs AL, Csikos G, Sass M, et al. (2008). Longevity pathways converge on autophagy genes to regulate life span in *Caenorhabditis elegans*. *Autophagy* 4: 330–338.
64. Klionsky DJ (2007). Autophagy: from phenomenology to molecular understanding in less than a decade. *Nat Rev Mol Cell Biol* 8: 931–937.
65. Cuervo AM and Dice JF (2000). Age-related decline in chaperone-mediated autophagy. *J Biol Chem* 275: 31505–31513.
66. Melendez A, Talloczy Z, Seaman M, Eskelinen EL, Hall DH, and Levine B (2003). Autophagy genes are essential for dauer development and life-span extension in *C. elegans*. *Science* 301: 1387–1391.
67. Martinez-Vicente M and Cuervo AM (2007). Autophagy and neurodegeneration: when the cleaning crew goes on strike. *Lancet Neurol* 6: 352–361.
68. Ventruti A and Cuervo AM (2007). Autophagy and neurodegeneration. *Curr Neurol Neurosci Rep* 7: 443–451.
69. Juhasz G, Erdi B, Sass M, and Neufeld TP (2007). Atg7-dependent autophagy promotes neuronal health, stress tolerance, and longevity but is dispensable for metamorphosis in Drosophila. *Genes Dev* 21: 3061–3066.
70. Morck C and Pilon M (2006). *C. elegans* feeding defective mutants have shorter body lengths and increased autophagy. *BMC Dev Biol* 6: 39.
71. Takeshige K, Baba M, Tsuboi S, Noda T, and Ohsumi Y (1992). Autophagy in yeast demonstrated with proteinase-deficient mutants and conditions for its induction. *J Cell Biol* 119: 301–311.
72. Jia K, Hart AC, and Levine B (2007). Autophagy genes protect against disease caused by polyglutamine expansion proteins in Caenorhabditis elegans. *Autophagy* 3: 21–25.
73. Khan LA, Yamanaka T, and Nukina N (2008). Genetic impairment of autophagy intensifies expanded polyglutamine toxicity in *Caenorhabditis elegans*. *Biochem Biophys Res Commun* 368: 729–735.
74. Yamamoto A, Cremona ML, and Rothman JE (2006). Autophagy-mediated clearance of huntingtin aggregates triggered by the insulin-signaling pathway. *J Cell Biol* 172: 719–731.
75. Wei M, Fabrizio P, Hu J, Ge H, Cheng C, Li L, and Longo VD (2008). Life span extension by calorie restriction depends on Rim15 and transcription factors downstream of Ras/PKA, Tor, and Sch9. *PLoS Genet* 4: e13.
76. Fabrizio P, Liou LL, Moy VN, Diaspro A, SelverstoneValentine J, Gralla EB, and Longo VD (2003). SOD2 functions downstream of Sch9 to extend longevity in yeast. *Genetics* 163: 35–46.

77. Medvedik O, Lamming DW, Kim KD, and Sinclair DA (2007). MSN2 and MSN4 link calorie restriction and TOR to sirtuin-mediated lifespan extension in *Saccharomyces cerevisiae*. *PLoS Biol* 5: e261.
78. Kaeberlein M, Hu D, Kerr EO, Tsuchiya M, Westman EA, Dang N, Fields S, and Kennedy BK (2005a). Increased Life Span due to Calorie Restriction in Respiratory-Deficient Yeast. *PLoS Genet* 1: e69.
79. Butow RA and Avadhani NG (2004). Mitochondrial signaling: the retrograde response. *Mol Cell* 14: 1–15.
80. Kirchman PA, Kim S, Lai CY, and Jazwinski SM (1999). Interorganelle signaling is a determinant of longevity in *Saccharomyces cerevisiae*. *Genetics* 152: 179–190.
81. Dilova I, Chen CY, and Powers T (2002). Mks1 in concert with TOR signaling negatively regulates RTG target gene expression in *S. cerevisiae*. *Curr Biol* 12: 389–395.
82. Powers T, Dilova I, Chen CY, and Wedaman K (2004). Yeast TOR signaling: a mechanism for metabolic regulation. *Curr Top Microbiol Immunol* 279: 39–51.
83. Hamilton B, Dong Y, Shindo M, Liu W, Odell I, Ruvkun G, and Lee SS (2005). A systematic RNAi screen for longevity genes in *C. elegans*. *Genes Dev* 19: 1544–1555.
84. Desai BN, Myers BR, and Schreiber SL (2002). FKBP12-rapamycin-associated protein associates with mitochondria and senses osmotic stress via mitochondrial dysfunction. *Proc Natl Acad Sci USA* 99: 4319–4324.
85. Kim DH, Sarbassov DD, Ali SM, King JE, Latek RR, Erdjument-Bromage H, Tempst P, and Sabatini DM (2002). mTOR interacts with raptor to form a nutrient-sensitive complex that signals to the cell growth machinery. *Cell* 110: 163–175.
86. Schieke SM, Phillips D, McCoy JP, Jr., Aponte AM, Shen RF, Balaban RS, and Finkel T (2006). The mammalian target of rapamycin (mTOR) pathway regulates mitochondrial oxygen consumption and oxidative capacity. *J Biol Chem* 281: 27643–27652.
87. Cunningham JT, Rodgers JT, Arlow DH, Vazquez F, Mootha VK, and Puigserver P (2007). mTOR controls mitochondrial oxidative function through a YY1-PGC-1alpha transcriptional complex. *Nature* 450: 736–740.
88. Baur JA, Pearson KJ, Price NL, Jamieson HA, Lerin C, Kalra A, Prabhu VV, Allard JS, Lopez-Lluch G, Lewis K, et al. (2006). Resveratrol improves health and survival of mice on a high-calorie diet. *Nature* 444: 337–342.
89. Kaeberlein M and Kennedy BK (2007). Protein translation, 2007. *Aging Cell* 6: 731–734.
90. Miller RA, Harrison DE, Astle CM, Floyd RA, Flurkey K, Hensley KL, Javors MA, Leeuwenburgh C, Nelson JF, Ongini E, et al. (2007). An Aging Interventions Testing Program: study design and interim report. *Aging Cell* 6: 565–575.
91. Gao XM, Wong G, Wang B, Kiriazis H, Moore XL, Su YD, Dart A, and Du XJ (2006). Inhibition of mTOR reduces chronic pressure-overload cardiac hypertrophy and fibrosis. *J Hypertens* 24: 1663–1670.
92. Wullschleger S, Loewith R, and Hall MN (2006). TOR signaling in growth and metabolism. *Cell* 124: 471–484.
93. Um SH, Frigerio F, Watanabe M, Picard F, Joaquin M, Sticker M, Fumagalli S, Allegrini PR, Kozma SC, Auwerx J, et al. (2004). Absence of S6K1 protects against age- and diet-induced obesity while enhancing insulin sensitivity. *Nature* 431: 200–205.
94. Cheng-Lai A and Frishman WH (2004). Sirolimus-eluting coronary stents: novel devices for the management of coronary artery disease. *Am J Ther* 11: 218–228.
95. Weil RJ (2008). Incorporating molecular tools into early-stage clinical trials. *PLoS Med* 5: e21.

Mitochondria, Oxidative Damage and Longevity: What Can Comparative Biology Teach Us?

Yun Shi, Rochelle Buffenstein, and Holly Van Remmen

Abstract The most studied theory of aging is the oxidative stress theory of aging, and evidence supporting or disputing the theory has come primarily from investigations using common model organisms such as *C. elegans, Drosophila*, and laboratory rodent models. However, studies using more non-traditional animal models offer an excellent opportunity to critically evaluate different aging hypotheses. The advantage of studying a broader spectrum of species is that one can significantly expand the amount of information obtained on a wide range of biological phenotypes/traits such as life span, body weight, and metabolic rate. In addition, the ultimate validity of a hypothesis/theory can be more critically tested using as many samples, in this case, species as possible. In this chapter we present evidence regarding different aspects of oxidative stress theory of aging with special emphasis on metabolic rate, reactive oxygen species generation, and oxidative damage to macromolecules. The purpose of the chapter is to initiate the integration of current knowledge and also to inspire readers to consider the advantages and power of using a comparative biology approach to study aging.

Keywords Mitochondria · Aging · Oxidative damage · Reactive oxygen species · Metabolic rate

Introduction

One of the most widely supported theories of aging is the oxidative stress theory of aging originally proposed by Harman more than 50 years ago [1]. This theory suggests that aging is a result of accumulation of oxidative cellular damage caused by free radicals produced during normal aerobic respiration. The underlying

H. Van Remmen (✉)
Department of Physiology, Cellular and Structural Biology, Barshop Institute for Longevity and Aging Studies, University of Texas Health Science Center at San Antonio; Audie Murphy Division, South Texas Veterans Health Care System, San Antonio, TX, USA
e-mail: vanremmen@uthscsa.edu

N.S. Wolf (ed.), *The Comparative Biology of Aging*,
DOI 10.1007/978-90-481-3465-6_8, © Springer Science+Business Media B.V. 2010

principle of this theory is that oxidative stress, i.e. an imbalance between reactive oxygen species (ROS) production and antioxidant defense/damage repair mechanisms is the driving force underlying aging (reviewed by Beckman and Ames [2] and Muller et al. [3]).

At the cellular level, a primary site for the generation of reactive oxygen species (ROS) resides in the mitochondria. Thus, a great of deal of research has focused on studying the role of mitochondrial ROS formation, mitochondrial function and oxidative damage in the context of aging [4, 3]. Many of these studies report an age-related increase in mitochondrial ROS production, a decline in mitochondrial function, and an accrual of oxidative damage to macromolecules [4]. However, reports using genetic manipulations and interventional approaches to alter oxidative stress in traditional experimental animal models, namely nematodes, fruit flies and mice have generated a plethora of equivocal and contradictory findings concerning the validity of this theory with significant evidence in support or opposition, and others that are inconclusive [3]. It should also be noted that the vast majority of these studies have utilized only a few commonly used laboratory animal species (mostly inbred mice or rats) that have been kept in captivity for generations. Indeed, most supporting evidence for the oxidative stress theory of aging is based on studies from only specific strains of particular species, which makes the generalization of conclusions difficult if not impossible.

Few studies have tested the ubiquity of this theory across the animal kingdom or exploited the natural variability in lifespan among organisms to test whether oxidative stress correlates with species longevity. One would expect that if the oxidative stress theory of aging had a solid foundation that it would be evident in phylogenetically diverse organisms throughout the animal kingdom (from single celled yeasts to mammals) and that markers of oxidative stress would correlate with maximum lifespan. Indeed, the use of comparative approaches to determine shared mechanisms, as well as different patterns among various species (especially in long-living organisms), is a powerful strategy to test different hypotheses of aging and is fundamental to the understanding of aging process. Uncovering the molecular, cellular and physiological mechanisms underlying the variation in species longevity, especially those of the exceptionally long-lived ones, will help to ultimately intervene with specific strategies to prolong lifespan and improve quality of life.

The goal of this chapter is to review the literature exploring the relationship between the role of mitochondrial function, mitochondrial ROS, oxidative damage and lifespan exploiting comparative biology approaches. It is written from a biology of aging point of view and by no means covers all evolutionary and ecological studies. Future prospective of each area of research will also be briefly proposed at the end of each section.

Metabolic Rate and Lifespan

The nature of the relationship between metabolic rate, body size and lifespan has long been of interest [5–7]. Since the time of Aristotle it has been documented that larger animals live longer than do smaller species. Maximum lifespan potential

(MLSP), is an important species characteristic that is determined from the maximum longevity record for that particular species. This trait is primarily based upon captive records, and may often be an underestimate of MLSP, since captive species may have different nutritional and exercise requirements and that these may be suboptimal in zoos and other captive settings. Nevertheless this species specific trait varies more than 100 fold in mammals and by more than 40,000 fold across the animal kingdom. MLSP, like almost every other biological trait is dependent upon the size of the organism, such that MLSP increases in a predictable manner as the species average body mass enlarges. Based upon the allometric equations (MLSP is proportional to mass $(kg)^{0.22}$ in mammals) for this relationship determined by Hulbert et al. [7], using existing mammalian longevity and mass records, it has been calculated that species that are twice as large as others will live ~16% longer than the smaller species.

Similarly, mass specific basal metabolic rate (BMR) scales allometrically with body mass, although in this instance, for every doubling of body mass there is a ~15–20% decline in mass specific BMR. In other words, small mammals have markedly higher energy requirements per gram of tissue than do large organisms, and this is attributed to their large surface area to volume ratio and greater heat exchange with the environment. The coupling of the relationships between both MLSP and BMR with body size was instrumental in formulating a causal relationship between metabolic rate and longevity, such that animals with high metabolic rates have shorter lives than do those with low metabolic rates. This theory was first proposed by Buffon in 1749 and formalized by Pearl in 1928 in the "rate of living theory" who proposed that organisms had a fixed amount of heart beats and energy available to them and thus longevity was inversely proportional to metabolic rate and that life-time energy expenditure (LEE – calculated from multiplying mass specific BMR by MLSP) is constant [6]. The rate of living theory predicts that longevity is inversely proportional to metabolic rate. In 1908, Rubner had (erroneously as it later turned out) noted that a gram of tissue from five chosen mammalian species, ranging in size from mice to horses, would expend the same amount of energy per life-time of the organism. That concept, coupled with prior knowledge of oxygen toxicity [8], led to a plausible mechanistic theory explaining why organisms age at different rates. The more recent free-radical/oxidative stress theory of aging [1] stems from both the rate of living theory and the free-radical theory of oxygen toxicity [9] and provides a mechanistic explanation for those early comparative observations.

Although there is a broad correlation between body size and both MLSP and BMR; when a more comprehensive assessment of >240 mammals is used, calculated LEE is not constant, but rather MLSP is negatively correlated with LEE. Furthermore BMR, the integral component upon which LEE is determined, is specifically measured at rest in post absorptive, non-growing, non-breeding healthy adults housed in their thermo neutral zone and is not an accurate indicator of total daily energy expenditure; it does not take into account the energy costs associated with daily activity, foraging, digestion, growth or reproduction. If average daily metabolic rates (ADMR) or field metabolic rates (FMR) are used instead of BMR, LEE among mammals declines with increasing body mass [7, 10]. There are also multiple exceptions to the presumed constant relationship between metabolic rate and lifespan posited by the rate of living theory. For example voluntary exercise and its associated increase in metabolic rate does not shorten life span of rats [11] or humans [12] and is generally thought to extend healthy lifespan. Dietary caloric restriction is well known to extend lifespan in a wide range of species [13], yet this process is not accompanied by attenuations in mass specific metabolic rate resulting in elevated LEE [14]. Also, there is no inverse relationship between lifespan and mass-specific metabolic rates of individual mice, dogs or flies [15–17]. In addition, although birds compared with mammals have essentially similar BMR and higher ADMR and FMR, birds are generally much longer-living than similar-sized mammals. Furthermore, within both groups of endotherms there are significant species differences in MLSP that cannot be explained by metabolic rate differences, for example while BMR of naked mole-rats is 75% of that of mice [18], their order of magnitude greater longevity would result in the highest mass specific LEE of any known mammal [19]. Intraspecific data also provide compelling exceptions to this theory such that within a species those individuals with the highest metabolic rates live the longest and this is attributed to a higher degree of mitochondrial uncoupling in the longer lived individuals [15].

Early evidence in support of the rate of living theory came from poikilothermic invertebrates. Loeb and Northrop [20] showed that flies kept at cooler temperatures lived considerably longer than their more active counterparts raised in hot environments. However, Arking et al. [21], in a well controlled study, showed that even within one species (*Drosophila*), long-lived strains do not simply have lower metabolic rates but rather their LEE potential is considerably greater than those of shorter-lived strains and proposed a qualitative genetic difference (e.g. mitochondrial efficacy) among strains of disparate longevity. Not surprisingly, given all these exceptions and current knowledge of the importance of high activity levels in healthy long lifespans, the rate of living hypothesis is no longer considered a viable explanation for the relationship between MLSP and metabolism [7, 22]. Its offshoot, the oxidative stress theory of aging, nevertheless remains one of the most widely accepted theories of aging and is based upon qualitative differences in mitochondrial function and efficiency evident in organisms with disparate longevity.

Mitochondrial Function and Free Radical Generation

Since most organisms adopted an aerobic lifestyle when oxygen appeared in the atmosphere about two billion years ago, the mitochondrion has evolved to become an important cellular component to provide energy. As an inevitable consequence of aerobic metabolism, reactive oxygen species are formed and can cause damage to adjacent structural and functional components. It is widely accepted that the functional declines that characterize the aging process are to a large extent related to the cumulative effects of oxidative damage to mitochondria. Accrual of damage within these organelles would hinder the ability of the mitochondria to provide energy and this alone may contribute to the aging phenotype. Species differences in free radical generation, mitochondrial efficacy with and without damage, and repair processes may contribute to their disparate longevity.

Mitochondrial Function. Aerobic metabolism relies on mitochondria to generate ATP for cellular energy utilization. Mitochondria utilize about 90% of the resting oxygen consumption [23] and approximately 80% is coupled to ATP synthesis, whereas the other 20% is uncoupled by the mitochondria by proton leak [24]. Different tissues consume different proportions of BMR due to distinct physiological functions and blood supply with kidney and heart in the higher end, liver and heart in the middle, and skeletal muscle, fat and skin in the lower end, at least in the case of rats [25, 24].

Aging has been associated with both a reduced capacity in cellular function and an impairment in mitochondrial function. One aspect of mitochondrial function that has been extensively investigated in a variety of organisms is changes in the activity of electron transport chain (ETC) complexes. A few studies have suggested that there are differential changes in electron transport chain activities with age, mostly in post-mitotic tissues such as heart and brain. Complexes I and IV show a selectively decreased enzymatic activities in electron transfer in isolated mitochondria from rat liver, brain, heart and kidney upon aging, whereas complexes II and III are minimally affected [26]. Kwong reported the activities of respiratory complexes I, II, III and IV from brain, heart, skeletal muscle, liver and kidney of young, middle aged and old C57Bl/6 mice. No common patterns in age-related changes of the various complex activities were observed, yet adverse effects of aging were more apparent in brain, heart and skeletal muscle [27]. Recently Choksi re-evaluated the age-related changes in ETC complexes more accurately utilizing specific inhibitors of each complex. He found that only complex I and V decreased activity in the heart mitochondria of old mice, and there were no age-associated differences in complex IV. Surprisingly he also reported age-related increases in complexes II and III activity [28] even though age-related oxidative modifications such as carbonylation, 4-hydroxy-nonenal adducts and nitrotyrosine modifications were identified in specific subunits of all five complexes using a proteomic approach. These disparate age-related responses of the various complexes despite augmented oxidative damage with age are puzzling and suggest that accrued oxidative damage is not the predominant cause of changes in ETC activity with age. The activity of complex I has also been reported to decrease with aging in dogs and rats, especially in skeletal muscle [29]. In rhesus monkeys, activities of brain mitochondrial complex I and IV were shown to negatively correlate with age, with no significant change in Complex II, III or V [30]. The lack of consistent and reproducible changes in ETC activities among different species and in different tissues points to the need for a more comprehensive analysis of age-related changes in transcriptional, translational, and post-translational regulations of the ETC complexes. In fact, down regulation of genes of electron transport chain at the transcriptional level has been regarded as one of the common aging signatures in Drosophila, mice and humans in multiple tissues [31].

Mitochondrial Free Radical Production. The mechanism by which mitochondria utilize proton-motive force to generate ATP also leads to free radical generation due to electron leakage from the electron transport chain. A free radical contains one or more unpaired electrons and is chemically very reactive. ROS is a term used to describe oxygen radicals as well as non-radical derivatives of oxygen. ROS production is initiated from the reaction of single electron reduction of oxygen which forms superoxide anion ($O_2^{\cdot -}$). $O_2^{\cdot -}$ then rapidly undergoes dismutation to form hydrogen peroxide (H_2O_2). Hydroxyl radicals (·OH) can form from $O_2^{\cdot -}$ and H_2O in the presence of iron or copper during Fenton reaction. These three reactive radical and nonradical oxygen species are the primary sources of the oxidative stress effectors. $O_2^{\cdot -}$ is considered to be formed primarily at complex I (NADH dehydrogenase) and complex III (ubiquinol: cytochrome c oxidoreductase) of the electron transport chain resides in the inner mitochondrial membrane [32]. The rate of $O_2^{\cdot -}$ production is dependent on two parameters, namely local O_2 concentration and the presence of reducing equivalents. The exact mechanisms of superoxide formation are still under intense investigation. In comparison to $O_2^{\cdot -}$, H_2O_2 is more membrane permeable and can diffuse through mitochondrial membranes and propagate oxidative damage. In addition H_2O_2 is easily detectable. In an isolated mitochondria preparation, specific substrates and inhibitors can be administered to donate electrons through respiratory chain to generate ATP (in the presence of ADP) and ROS. Oxygen consumption and mitochondrial membrane potential can be detected as well. Even though physiological concentrations of ROS have been suggested to serve as signaling molecules, excessive mitochondrial ROS is no doubt detrimental. Figure 1 shows a schematic model of mitochondrial ROS generation and antioxidants that scavenge ROS. Potential damage caused by ROS to cellular constituents (DNA, lipid and protein) is also illustrated.

Based on the free radical/oxidative stress theory of aging, one would predict that animals with longer lifespans would either produce less ROS or would have more effective defense/repair mechanisms against free radical damage. Along these lines, some comparative studies reported that mitochondrial ROS production is inversely correlated with MLSP among different species [33–35]. In other words, short lived animals reportedly produce more mitochondrial ROS than long lived ones. The majority of the supporting evidence came from two research groups (Sohal and Barja) using animals including flies, mammals (such as mouse, rat, guinea pig, rabbit and cow) and birds (such as canaries, parakeets, and pigeons) with MLSP ranging from weeks to over thirty years. Tables 1, 2, 3, and 4 list a summary of the literature regarding the relationship between mitochondrial ROS from different tissues and MLSP of various animals.

The rate of $O_2^{\cdot -}$ generation (in submitochondrial particles) and H_2O_2 (in isolated mitochondria) in mitochondria isolated from the multiple vital organs such as liver, kidney, heart and brain are reported to correlate inversely with MLSP among mammals [36, 37, 33]. Herrero et al. showed that mitochondrial ROS is also lower in long lived birds – pigeons (MLSP: 35 years) [41], canaries (MLSP: 24 years) and parakeets (MLSP: 21 years) [42] compared to similar sized rodents, however the ROS production among all the species in the study did not correlate with MLSP. These studies provide a partial explanation for the exceptional longevities for birds than mammals despite their very high metabolic rate.

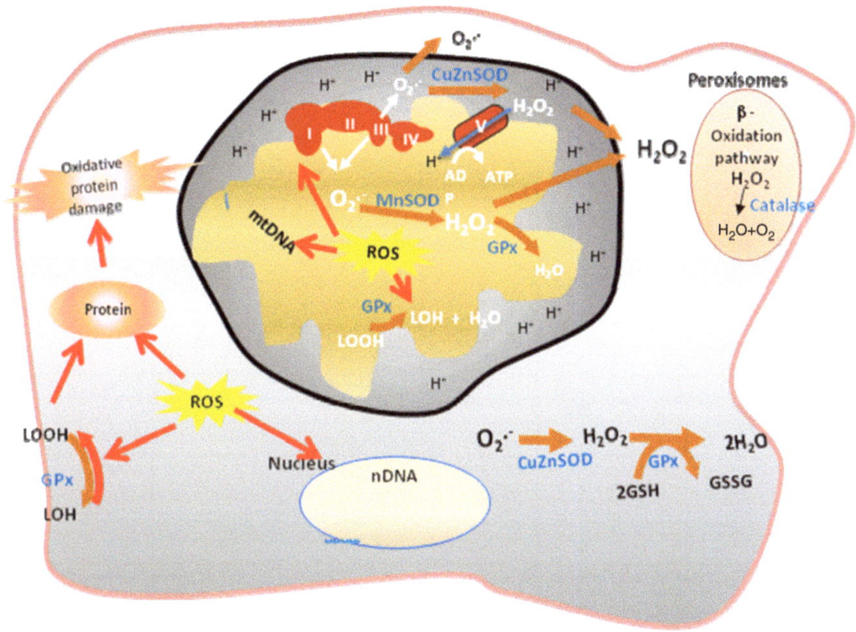

Fig. 1 Mitochondria-centered oxidative stress model. The diagram depicts a typical mammalian cell, with mitochondrial ROS, their sites of production, antioxidant enzymes, and potential damages to macromolecules. While there are other non-mitochondrial pathways that are capable of generating ROS such as peroxisomal beta-oxidation pathway shown here, the mitochondrion is emphasized here and enlarged in proportion to the cell

A similar finding was also reported in ectotherms between short-lived and long-lived snakes [38], in vascular systems between white-footed mouse and mouse [44], and in a long-lived bat, short-lived shrew and intermediate lived white-footed mouse comparison [43]. These data, although not corrected for phylogenetic differences among species, show that long-lived species generate less ROS than do the shortest living species, but do not show species differences in ROS production that correlate with longevity differences. Mitochondrial ROS generation occurs continuously throughout life, independent of the rate of mitochondria oxygen consumption, and is a species specific modality that may be influenced by the MLSP of each species [35, 41, 42, 45]. The exact components of the respiratory chain responsible for the lower mitochondrial ROS production in long-lived species are unclear, although there is some evidence predominantly implicating that complex I may be the determinant for the rate of ROS production [34, 41].

A major drawback of most of the earlier studies, as commented by Speakman [10] and Lambert [35] is the lack of control for an important potential covariant – body size. The observed correlation could simply reflect the well-known effects of body size on numerous biological traits including MLSP. In addition, most studies did not correct for phylogenetic interdependence. Many of the traits may simply be specific features associated with that particular phylogenetic group

Table 1 Summary of the relationship between liver mitochondrial ROS production and MLSP

Free radicals	Samples	Detection methods	Substrates	Species	MLSP (years)	Production (nmole free radicals/min/mg prot.)[a]	Association with MLSP	References
O_2^-	submitochondrial particals (SMPs)	SOD inhibitable reduction of ferricytochrome c in the presence of antimycin A and KCN	succinate + antimycin A	housefly[a] (flight muscle)	0.25	3.30 ± 0.36	The rate of O_2^- generation was inversely related to MLSP ($r = -0.92$).	[36]
				mouse	3.5	0.66 ± 0.05		
				rat	4.5	0.77 ± 0.06		
				rabbit	18	0.46 ± 0.05		
				pig	27	0.42 ± 0.06		
				cow	30	0.13 ± 0.01		
H_2O_2	isolated mitochondria	the oxidation of p-hydroxyphenylacetate (PHPA) by the enzymatic reduction of H_2O_2 by horseradish peroxidase (HRP)	succinate	mouse	3.5	58	Inverse relationship between succinate supported H_2O_2 production and MLSP was reported but no correlation in the presence of antimycin A or rotenone.	[37]
				rat	4.5	128		
				guinea pig	7.5	36		
				rabbit	18	38		
				pig	27	3		
				cow	30	4		

Table 1 (continued)

Free radicals	Samples	Detection methods	Substrates	Species	MLSP (years)	Production (nmole free radicals/min/mg prot.)[a]	Association with MLSP	References
H_2O_2	isolated mitochondria	coupled PHPA oxidation and reduction of H_2O_2 by HRP	succinate in the presence of SOD, antimycin A and rotenone	long lived colubrid snakes	>15	68	Long lived colubrid snakes have reduced H_2O_2 production than short-lived species	[38]
				short lived colubrid snakes	<10	96		

[a] data value is either from original data table or read from graph.

Table 2 Summary of the relationship between heart mitochondrial ROS production and MLSP

Free radicals	Samples	Detection methods	Substrates	Species	MLSP (years)	Production (nmole free radicals/min/mg prot.)[a]	Association with MLSP	References
O_2^-	SMPs	SOD inhibitable reduction of acetylated ferricytochrome c	succinate + antimycin A	mouse	3.5	4120 ± 220	Long lived peromyscus showed lower rate of mitochondrial O_2^- and H_2O_2 production than short lived mouse.	[39]
				white-footed mouse	8	2370 ± 60		
H_2O_2	isolated mitochondria	coupled oxidation of PHPA	succinate	mouse	3.5	2.32 ± 0.08	The rate of O_2^- and H_2O_2 generation were inversely correlated to MLSP.	[33]
				white-footed mouse	8	0.42 ± 0.02		
$O_2\cdot^-$	SMPs	SOD inhibitable reduction of acetylated ferricytochrome c	succinate + antimycin A	mouse	3.5	2.89		
				hamster	4	2.86		
				rat	4.5	1.71		
				guinea pig	7.5	1.61		
				rabbit	18	1.18		
				pig	27	1.36		
				cow	30	0.57		
H_2O_2	isolated mitochondria	coupled oxidation of PHPA	succinate	mouse	3.5	1.06		
				hamster	4	2.06		
				rat	4.5	1.00		
				guinea pig	7.5	1.72		
				rabbit	18	0.84		
				pig	27	0.19		
				cow	30	0.28		

Table 2 (continued)

Free radicals	Samples	Detection methods	Substrates	Species	MLSP (years)	Production (nmole free radicals/min/mg prot.)[a]	Association with MLSP	References
$O_2^{\,-}$	SMPs	SOD inhibitable reduction of acetylated ferricytochrome c	succinate + antimycin A	rat pigeon	4.5 35	2.13 1.43	significant lower rate of $O_2^{\,-}$ and H_2O_2 generation in pigeon than in rat	[40]
H_2O_2	isolated mitochondria	coupled oxidation of PHPA	succinate	rat pigeon	4.5 35	0.95 0.29	Compared with rat, pigeon has lower mitochondrial H_2O_2 production rate	[41]
H_2O_2	isolated mitochondria	Homovanillic acid (HVA)	Pyruvate/Malate + inhibitors; succinate + inhibitors	rat pigeon	4.5 35	1.33 ± 0.25 0.35 ± 0.08		
H_2O_2	isolated mitochondria	HVA	pyruvate/malate + inhibitors	mouse canary parakeet	3.5 24 21	1.12 0.23 0.58	Lower H_2O_2 generation in the two birds than mouse due to low oxygen consumption i the parakeet and a low free radical leak in the canary	[42]

Table 2 (continued)

Free radicals	Samples	Detection methods	Substrates	Species	MLSP (years)	Production (nmole free radicals/min/mg prot.)[a]	Association with MLSP	References
H_2O_2	isolated mitochondria	coupled oxidation of PHPA	NA	little brown bat white-footed mouse short-tailed shrew	34 8 1-2	0.050 ± 0.016 0.025 ± 0.016 0.119 ± 0.013	Reduced mitochondrial inefficiency in *M. lucifugus* relative to *B. vrevicauda* and *P leucopus* in brain and kidney	[43]
H_2O_2	isolated mitochondria	Amplex Red or PHPA coupled reduction of H_2O_2 in the presence of HRP	Pyruvate/Malate + inhibitors; succinate + inhibitors	mouse rat white-footed mouse naked mole-rat Damara mole-rat guinea pig baboon Brazilian free-tailed bat little brown bat ox quail pigeon	3.5 5 8 28 15 8 37.5 12 34 30 6 35	2.6 2.6 2.1 1.8 1.2 2.5 1.3 1.3 1.1 2.1 2.5 1.1	MLSP was negatively correlated with only succinate supported H_2O_2, statistical significance remains after correction for body mass and phylogeny	[35]

[a] data value is either from original data table or read from graph. When multiple substrates and/or inhibitors were used in the study, only succinate supported O_2^- and H_2O_2 production were listed above. Underline indicates the values for the free radicals production rate.

Table 3 Summary of the relationship between kidney mitochondrial ROS production and MLSP

Free radicals	Samples	Detection methods	Substrates	Species	MLSP (years)	Production (nmole free radicals/min/ mg prot.)[a]	Association with MLSP	References
O_2^-	SMPs	SOD inhibitable reduction of acetylated ferricytochrome c	succinate + antimycin A	mouse	3.5	1.25	The rate of O_2^- and H_2O_2 generation were inversely related to MLSP.	[45]
				hamster	4	0.64		
				rat	4.5	1.00		
				guinea pig	7.5	0.54		
				rabbit	18	0.29		
				pig	27	0.28		
				cow	30	0.18		
H_2O_2	isolated mitochondria	coupled oxidation of PHPA	succinate	mouse	3.5	0.88		
				hamster	4	0.9		
				rat	4.5	0.53		
				guinea pig	7.5	0.75		
				rabbit	18	NA		
				pig	27	0.23		
				cow	30	0.1		
O_2^-	SMPs	SOD inhibitable reduction of acetylated ferricytochrome c	succinate + antimycin A	rat	4.5	0.96	The rate of O_2^- and H_2O_2 generation were lower in pigeon mitochondria compared with rat.	[46]
				pigeon	35	0.48		

Table 3 (continued)

Free radicals	Samples	Detection methods	Substrates	Species	MLSP (years)	Production (nmole free radicals/min/ mg prot.)[a]	Association with MLSP	References
H_2O_2	isolated mitochondria	coupled oxidation of PHPA	succinate	rat	4.5	0.52		
				pigeon	35	0.4		
H_2O_2	isolated mitochondria	PHPA	NA	little brown bat	34	0.048 ± 0.010	Reduced mitochondrial inefficiency in *M. lucifugus* relative to *B. vrevicauda* and *P. leucopus* in kidney	[16]
				white-footed mouse	8	0.023 ± 0.010		
				short-tailed shrew	1–2	0.117 ± 0.009		

[a] data value is either from original data table or read from graph.

Table 4 Summary of the relationship between brain mitochondrial ROS and MLSP

Free radicals	Samples	Detection methods	Substrates	Species	MLSP (years)	Production (nmole free radicals/min/mg prot.)[a]	Association with MLSP	References
O_2^-	SMPs	SOD inhibitable reduction of acetylated ferricytochrome c	succinate + antimycin A	mouse white-footed mouse	3.5 8	680 ± 20 460 ± 10	lower rate of mitochondrial O_2^- and H_2O_2 production in long lived peromyscus than in mouse	[74]
H_2O_2	isolated mitochondria	coupled oxidation of PHPA	succinate	mouse white-footed mouse	3.5 8	0.64 ± 0.04 0.19 ± 0.02	The rate of O_2^- and H_2O_2 generation inversely correlated to MLSP	[46]
O_2^-	SMPs	SOD inhibitable reduction of acetylated ferricytochrome c	succinate + antimycin A	rat pigeon	4.5 35	0.96 0.56		
H_2O_2	isolated mitochondria	coupled oxidation of PHPA	succinate	rat pigeon	4.5 35	1.28 0.43		

Table 4 (continued)

Free radicals	Samples	Detection methods	Substrates	Species	MLSP (years)	Production (nmole free radicals/min/mg prot.)[a]	Association with MLSP	References
H_2O_2	isolated mitochondria	PHPA	NA	little brown bat white-footed mouse short-tailed shrew	34 8 1–2	0.085 ± 0.006 0.060 ± 0.006 0.202 ± 0.005	Reduced mitochondrial inefficiency in *M. lucifugus* relative to *B. vrevicauda* and *P. leucopus* in brain	[16]

[a] data value is either from original data table or read from graph.

(e.g., birds). Diverse phylogenetic history in distant related species such as rat and pigeon, or closely evolved species like mouse and rat raises the question whether the correlation is valid through more stringent statistical analyses. Lambert revisited the question whether mitochondrial ROS is a determinant of longevity after correction for body mass or phylogeny interdependence. Residual analysis, after correction for mass shows a significant negative correlation between succinate supported heart mitochondrial H_2O_2 and MLSP. However, this inverse relationship fell short for regression analysis after correcting for phylogeny [35].

Another limitation of earlier studies lies in the techniques used to detect mitochondrial ROS. Submitochondrial particles (SMPs) which invert mitochondrial inner membrane to detect superoxide production has eliminated or minimized the effect of superoxide dismutase (Manganese SOD, MnSOD) in the matrix of mitochondria. At physiological conditions O_2^- has a very short half-life; it spontaneously or enzymatically undergoes dismutation to form H_2O_2. So measurement of O_2^- production rate using SMPs will lead to overestimation of actual steady-state O_2^-. Different probes have various stoichiometries in reactions with ROS. Commonly used probes for ROS detection have been critically evaluated elsewhere [46]. Another complication is the fact that different substrates are used to stimulate respiration. Even when the same substrate was used; the variation in ROS generation rate is considerably big (see Tables 1, 2, 3, and 4). In addition, mitochondrial ROS determination has most commonly studied in state 4 respiration, i.e. without ADP present.

It is worth mentioning that the detected inverse relationship between mitochondrial ROS and MLSP observed in [35] was only evident when one substrate (succinate) that is known to induce a significant amount of reverse electron transfer in mitochondria was used [35]. This study suggests that the reverse electron transfer through complex I may be an indicator of longevity. However, the physiologic significance of reverse electron transfer is still uncertain and it is not known why no correlation was evident when multiple other substrates representative of forward electron transport through the various components of the ETC were employed. Moreover, it is important to determine whether this phenomenon would be observed in mitochondria from other vital tissues such as brain, liver, kidney or skeletal muscle. Mitochondria from different tissues of the same species have different phenotypes in physiological and pathological situations [47–49]. In addition, interspecies difference in mitochondrial organization may influence the efficiency of mitochondria in substrate utilization and transport as well as proton leak. Furthermore, whether mitochondrial ROS would cause similar extent of damage in intact cells containing endogenous antioxidants is still an open question and probably more meaningful.

One interesting discovery is that although basal metabolic rate and average daily metabolic rate are higher in birds than rodents, isolated heart mitochondria from parakeet, canary [42] and pigeon [41] consume significantly less O_2 per milligram mitochondria protein than mitochondria those from rodents, at least under the assay conditions used. Nevertheless, mitochondrial free radical leak (calculated as the rate of H_2O_2 production divided by two times the rate of oxygen consumption) is lower in canary and pigeon than in mouse or rat, respectively. Another study compared mitochondrial function in bat, shrew and white-footed mouse [43]. However, this study is somewhat problematic as the selection of animals is not based on either size or phylogeny; and whole organism O_2 consumption was used to calculate mean mitochondrial inefficiency rather than O_2 consumption from isolated mitochondria.

Mitochondrial ROS has been measured longitudinally with age and has been shown to increase with age in multiple species. It is possible that long-lived animals could have a slower age-dependent increase of ROS generation production than short-lived ones. This hypothesis can be tested by comparing mitochondrial ROS among species with varying MLSP at different ages. This approach has been undertaken by Csiszar et al in a study evaluating vascular aging between exceptionally long-lived naked mole-rats (NMR, MLSP > 28 years) and laboratory rats (MSLP: ~3 years) [50]. Cellular O_2^- and H_2O_2 production significantly increased with age in rat arteries, whereas they did not change substantially with age in NMR vessels either expressed as an absolute age or as a percentage of MLSP. This at least provides a partial explanation for the successful aging in NMR although in an earlier study that no relationship between vascular ROS and MLSP was found among four species including NMR. Sasaki et al. recently also documented that age-dependent change in O_2^- production rate in brain slices inversely correlates with MLSP among mouse, rat and pigeon [51]. It should be noted that these two studies measured cellular O_2^- other than previous reports that compared O_2^- only of mitochondria origin. This could potentially account for some differences observed.

As our knowledge of mitochondrial physiology increases, along with continued advances in technology of in vivo imaging [52] and the application of new spin trap agents together with electron paramagnetic resonance (EPR) offering more accurate estimation of free radicals generation in physiological situations, we can expect a increase in our understanding of this issue in the future.

Antioxidant Defense Systems

A complex arrangement of mechanisms has evolved to eliminate ROS and repair potential damage. The primary lines of defense include (1) enzymatic antioxidants such as superoxide dismutase (SOD), catalase (CAT), glutathione peroxidase (GPX); (2) hydrophilic scavengers such as glutathione (GSH), ascorbate; (3) lipophilic scavengers such as tocopherols, flavonoids, carotenoids, and ubiquinol; and (4) proteins involved in regenerating oxidized antioxidants and protein thiols such as GSH reductase, thioredoxin, and thioredoxin reductase.

The free radical theory of aging predicts that difference in longevity could be contributed by species-specific antioxidant capacity. In other words, superior antioxidant defense might protect against oxidative damage to cellular components in long-lived species more efficiently than in short-lived ones, resulting in the differences in MLSP. Yet the other side of this argument would be that as a compensatory mechanism, short-lived species adapt higher ROS generation to optimize antioxidant defense systems in order to battle the deleterious effect of ROS. In the latter case antioxidant defense would therefore serve as a marker for in vivo ROS production rather than a determinate for MLSP. Association between various antioxidants and MLSP has been attempted [53–55] with somewhat ambiguous results: CAT and GPX activities negatively correlated with MLSP; SOD when divided by basal metabolic rate showed negative correlation with MLSP. Sohal found that SOD and CAT activities were positively correlated whereas GSH concentration was negatively correlated with MLSP in six different mammalian species, and GPX correlated positively in the brain but negatively in the liver and heart [56](Reviewed by [57]. When birds, fish, amphibians were included for comparison for antioxidants activities and MLSP, strong negative correlations for certain antioxidants such as brain GPX or lung GSH-reductase were obtained, but no correlations between SOD, GSSG/GSH ratio and MLSP [58–60]. A study attempting to correlate antioxidants and longevity in five South American bat species did not find common patterns in blood or tissue antioxidants levels [61].

One difficulty in studying the role of antioxidants lies in the fact that many members of the antioxidant defense system share redundant roles and are inducible in response to elevated oxidative stress. In order to estimate the overall ability to defend against ROS, it is necessary to evaluate the activities of all members as well as both transcriptional and translational levels of each antioxidant/antioxidant enzyme. No consistent differences in known processes that remove radicals and repair the damage have been found to correlate with MLSP [35, 43, 62, 63]. Levels of antioxidant defense cannot account for lifespan difference across the board. This observation goes along with the negative lifespan altering effect from various dietary supplementation studies and genetic interventions to increase or decrease antioxidants in mice [64, 65].

Oxidative Damage to Macromolecules

The purpose of comparing the rate of mitochondrial free radical generation and antioxidant defense capacity is to estimate the net steady state level of oxidative damage, which is the ultimate proposed mediator of aging according to the oxidative stress theory of aging. In general, studies investigating oxidative damage fall into two categories: (1) comparison of damage in young adults of multiple species; and (2) comparison of damage in young versus old animals in a single model organism. The former approach has been applied in the majority of reported studies to compare a snap shot of oxidative damage at a given age of each species. The latter approach has not often been applied in models other than conventional lab species, even though convincing evidence from those studies has shown that aging is associated with oxidative stress and oxidative macromolecular damage in various tissues [66, 4]. The chronological comparison approach argues that the age-dependent accumulation of oxidative damage is an intrinsic factor determining the rate of aging. Most comparative interspecific studies compare oxidative damage in young individuals of each species. Although this is generally regarded to be of practical reasons, there remains a scientific rationale in that traits influencing rates of aging ought to be present throughout life and if these traits facilitate long life, even at an early age their impact should be evident. The few studies that utilized both approaches are obviously more powerful for testing the oxidative stress theory of aging [63, 67, 51]. In these studies comparisons were not only made between species of the same chronological age, but also between species of equivalent physiological age (equal in percentage of their MLSP). This is a more convincing approach for making conclusions with respect to aging since the parameters are normalized within each species.

One critical issue which has gained increased attention is the intrinsic differences in the composition of macromolecules, such as amino acids usage in protein sequence especially methionine, lysine, and cysteine, degree of unsaturation in fatty acids and G/C content in DNA sequence. These differences may affect the susceptibility of the macromolecules to oxidative stress even though the endogenous free radicals production and antioxidant defense and repair mechanisms differ among various species as well. It is also important to keep in mind that the extent of this age-associated increase in oxidative damage to macromolecules varies greatly among different tissues and according to detection methods. In addition, it is important to distinguish damage arising specifically from oxidative insults from other types of damage when testing free radical theory of aging among multiple species. Although the task is very challenging, the list of oxidative stress specific markers has grown substantially due to the recent advances in technology.

(a) *Oxidative Damage to DNA.* Among all molecular modifications by oxidative stress, damage to DNA is the most important due to the potential loss or alteration of genetic information, especially in post mitotic tissues. Oxidative damage to DNA includes adducts of base and sugar groups, single- and double-strand breaks in the backbone, and cross-links to other molecules. Guanine has the lowest oxidation potential among four nucleobases thus it is most easily oxidized. 8-oxo-7,8-dihydroguanine (8-oxoGua) and 8-oxo-7,8-dihydro-2′-deoxyguanosne (8-oxodG) products of base excision repair and nucleotide excision repair respectively reflect oxidative DNA damage. Their easy detection in urinary excretion as well as plasma and tissues by high-performance liquid chromatography has allowed assessment of in vivo DNA damage in different species. 5-hydroxymethyluracil (5-HMUra) is a 5′ OH-mediated thymine oxidation product and its urinary excretion should also represent oxidative DNA damage [68, 69].

In agreement with low mitochondrial ROS generation, long-lived mammals have significantly lower levels of 8-oxodG in their mitochondrial DNA (mtDNA) in the brain and heart [70]. 8-oxodG is also lower in the heart and brain mtDNA of long-lived birds compared to similar sized rodents [71]. Interestingly, no correlation was found between 8-oxodG in nuclear DNA (nDNA) and MLSP in both of the studies. It suggests first of all that mtDNA is more vulnerable to oxidative damage; and secondly nDNA is not a predictor for MLSP. However urinary excretion of modified bases/nucleosides 8-oxoGua, 8-oxodG and 5-HMUra, which are reflective of oxidative DNA damage on the level of the whole organism has been reported to correlate with specific metabolic rate in six mammalian species (mice, rats, rabbits, dogs, pigs, and humans). Among the three, 8-oxo-Gua is the only one inversely correlated with MLSP [72]. Similar results are obtained from an independent study in urinary samples from mice, rats, guinea pigs, cats, chimpanzees, and humans [73]. It suggests 8-oxo-Gua rather than 8-oxo-dG may be a general marker of oxidative damage.

Despite the fact that there are lower levels of 8-oxodG in mtDNA in the brain and heart in long-lived versus short-lived mammals [70], mtDNA mutations do accumulate with age in an individual species [74–76]. Wang et al. compared the rate of mtDNA mutation between mice and human and discovered that it is greater in short-lived mice than in human [77]. So far, studies that examine a broader range of species in respect to DNA mutations are lacking.

Investigations regarding DNA repair mechanisms among species with various MSLP have yielded a general agreement that long-lived animals have superior DNA repair capacities in comparison to short-lived ones [78, 79]. It is reasonable from the evolutionary point of view to suggest that short-lived animals will not invest as much in DNA repair if they are only going to live for a short period of time, while animals with a relatively long lifespan would evolve to enhance DNA repair mechanisms to prevent tumorigenesis in response to prolonged exposure to endogenous and exogenous free radicals. Vijg reported that mice accumulate mutations during aging at a much faster pace than humans do, reflecting the high capability of the genome quality control mechanisms in humans than mice [80].

(b) *Oxidative Damage to Lipids*. Lipids are the basic component of biological membranes which is essential for life. The existence of double bonds in unsaturated fatty acids makes lipids sensitive to oxidation. Oxidative damage to lipids can occur through direct reaction with ROS such as H_2O_2 or O_2^- or indirectly by reactive aldehydes. Oxidation of lipids leads to the formation of hydroperoxides and endoperoxides, which in turn can undergo fragmentation to yield a broad range of reactive intermediates, including alkanals, alkenals, hydroxyalkenals, malondiadehyde (MDA), and hydroxynonenal (HNE). These carbonyl compounds and their peroxide precursors are highly unstable and reactive. They are well suited to attack nucleophilic groups in proteins, an activity that results in irreversible chemical, structural, and possibly functional alterations. These modifications are collectively named as advance lipoxidation end products (ALEs) and can be used to as indicators of lipid peroxidation. Another widely used parameter for lipid peroxidation is $F_{(2)}$-isoprostanes. Isoprostanes (IsoPs) are produced in vivo independently of cyclooxygenase enzymes, primarily by free radical-induced peroxidation of arachidonic acid. Measurement of F_2-isoprostanes is very reliable to assess oxidative stress in vivo [81]. Lipofuscin, considered a biomarker of intralysosomal lipid peroxidation has been documented in various species including nematodes, fruit flies, rats, bees, monkeys, crayfish and humans. A non-invasive way to evaluate in vivo lipid oxidation is to measure exhaled hydrocarbon (ethane and pentane). Lipid peroxidation caused by oxidative stress induces an irreversible impairment of membrane fluidity and plasticity, and can disrupt membrane-bound proteins thereby leading to irreversible damage to cellular integrity.

Many studies have been interested in determining age-related changes in oxidative damages to lipids. The mammal species studied include mouse, rat, dog and human. Age-related increases in lipofusin, MDA-TBARS, lipid hydroperoxides, exhaled hydrocarbons, and F_2-isoprostances [82] in various tissues have been observed (summarized in [7]). Recently, it has been shown that membrane fatty acid composition, especially the content of omega-6 polyunsaturated fatty acid (PUFA) does vary in a systemic manner with body size in mammals and in birds. It raises the possibility that variations in membrane composition can explain part of the differences in the rate of aging in a variety of systems. Low levels of unsaturation in fatty acids in the plasma membrane is a trait of all the long-lived homeothermic vertebrates studied, relative to their short-lived counterparts (rats and mice), and this feature could be one of the main reasons behind the low rate of aging in these animals. This has led to the "homeoviscous-longevity adaptation" hypothesis [83]. The relationship of susceptibility of lipids to peroxidation (reflected by the calculation of peroxidation index) and membrane composition among vertebrate species has been comprehensively reviewed by Hulbert et al. [7].

The long-lived naked mole-rat (MSLP > 28 years) is an outlier in oxidative lipid damage profile based on the prediction of oxidative stress theory. Lipid peroxidation measured by MDA adduct and isoprostanes showed 2-fold and 10-fold higher levels respectively in naked mole-rat than in mice at young age groups [63]. This alone is contradictory to the oxidative stress theory of aging. In addition, the naked mole-rat does not show age-related accumulation of oxidative damage from lipid peroxidation as has been shown in mouse [67]. It argues once against accrued damage caused by oxidative stress is the sole driving force of aging process and suggests that the naked mole-rat may have superior resistance to age-related increases in oxidative stress burden. This might be an example of species-specific mechanisms that the naked mole-rat has evolved to tolerate high levels of peroxidation.

(c) *Oxidative Damage to Proteins.* Proteins carry out important biological functions and are the most abundant macromolecules present in cells. Oxidation of proteins by ROS or other reactive species, leads to fragmentation of polypeptide chain, oxidation of amino acid side chains, and generation of protein-protein crosslinks [84]. As a consequence changes in protein confirmation due to oxidative damage, changes in enzymatic activity, binding affinity, and/or recognition sites for other interacting proteins occur. Among all types of modifications, protein carbonyls occur in orders of magnitude greater than other kinds of protein oxidation [85]. The carbonyl content of proteins has become a general bio-marker for estimation of oxidative stress-mediated protein oxidation. Other methods include detecting products of specific amino acid modifications, such as dityrosine, nitrotyrosine or methionine sulfoxide, to name a few. In addition, oxidation of carbohydrates causes advanced glycoxidation end-products (AGEs) on proteins [86]. Change in overall protein thiol content is also considered another important indicator for the age-related alteration in protein structure and oxidant status [87]. Furthermore alterations of protein conformation reflected by changes in hydrophobicity and formation of aggregates have been used as markers of an age-related increase in protein oxidation [88]. Incomplete repair or removal of oxidized proteins result in accumulation of the "damaged" proteins and manifest with age, therefore exacerbate age-related decline in cellular function. Many studies using individual animal models have demonstrated that cellular protein carbonyl content increase with age, as reported in the brain of gerbils [89, 90], mouse liver [91] and mouse and rat plasma [92].

Based on oxidative stress theory of aging we would predict that long living animals will have less oxidative damage in the proteome due to either low ROS production or efficient protein degradation pathway. Sohal's group has demonstrated in houseflies that (1) protein carbonyl content increase with age after exposure to X-irradiation or hyperoxia; (2) the extent of protein carbonyl is negatively correlated with their lifespans [93, 94]. The white-footed mouse (*Peromyscus leucopus,* MSLP: >8 years) showed lower levels of protein damage than mouse in response to experimental oxidative stress [39]. Another study that included four mammals and one bird (mouse, rat, rabbit, pig and pigeon) found that MLSP was inversely correlated to the susceptibility to acute oxidative stress reflected by protein carbonyl formation [95]. However, more recently, a study investigating potential differences in protein modification and proteasome activity between long lived pigeon and short-lived rat revealed that skeletal muscle from pigeon showed significantly higher levels multiple protein oxidative modification products including protein carbonyls. Finally, pigeon samples also showed significantly lower levels of the peptidase activities of the proteasome. This has been interpreted as evidence against a correlation between low protein oxidation and longer lifespan [96]. The naked mole rat as well as certain long-lived bat species also contradicts the validity of the oxidative stress theory of aging by having a significantly higher degree of oxidative damage to proteins [63, 55] and a relatively long lifespan. Thus, the relationship between oxidative protein damage and MLSP remains inconclusive considering available information.

Despite the fact that studies of global protein oxidative modification fail to support oxidative stress theory of aging, the discovery of specific susceptibility of individual proteins to oxidative stress are instrumental. It is of great interest and importance to indentify those proteins that are easily or heavily modified and the functional consequences of those oxidative modifications.

Summary

Aging is universal among all organisms and still remains a mystery. A fundamental question is whether there is a single cause behind all aging phenomena or for multifaceted as is the case many other biological processes. The diversity in size, metabolism, physiology and lifespan in the animal kingdom offers an unique opportunity to study this question; the comparative approach underlying these type of studies is undeniably more powerful than approaches simply using conventional lab models. With the increasing awareness of the statistics for phylogenetic correction involved in these studies, more genomic and epigenomic information available and more unconventional species used in aging research, we can expect to see significant advances in the fundamental knowledge of aging, and in particular the role of mitochondria and oxidative damage in the aging process. By taking advantage of systemic integration of knowledge from different aspects of research, we hope to gain a deeper understanding of the mechanisms of aging and to use that knowledge to intervene in order to slow aging.

References

1. Harman D (1956). Aging: a theory based on free radical and radiation chemistry. *J Gerontol* 11(3): 298–300.
2. Beckman KB and Ames BN (1998). The free radical theory of aging matures. *Physiol Rev* 78(2): 547–581.
3. Muller FL, Lustgarten MS, Jang Y, Richardson A, and Van Remmen H (2007). Trends in oxidative aging theories. *Free Radic Biol Med* 43(4): 477–503.
4. Van Remmen H and Richardson A (2001). Oxidative damage to mitochondria and aging. *Exp Gerontol* 36(7): 957–968.
5. Rubner M (1908). *Das Problem der Lebensdauer und Seine Beziehungen Zu Wachstum und Ernarhrung Muenchen*. Germany: Oldenburg.
6. Pearl R (1928). *The Rate of Living*. London: University of London Press.
7. Hulbert AJ, Pamplona R, Buffenstein R, and Buttemer WA (2007). Life and death: metabolic rate, membrane composition, and life span of animals. *Physiol Rev* 87(4): 1175–1213.
8. Lavoisier AL. Considérations Générales sur la Nature des Acides 1778.
9. Gershman R, Gilbert DL, Nye SW, Dwyer P, and Fenn WO (1954). Oxygen poisoning and X-irradiation: A mechanism in common. Science 119: 623–626.
10. Speakman JR (2005b). Correlations between physiology and lifespan–two widely ignored problems with comparative studies. *Aging Cell* 4(4): 167–175.
11. Holloszy JO, Smith EK, Vining M, and Adams S (1985). Effect of voluntary exercise on longevity of rats. *J Appl Physiol* 59(3): 826–831.
12. Lee IM, Hsieh CC, and Paffenbarger RS, Jr. (1995). Exercise intensity and longevity in men. The harvard alumni health study. *JAMA* 273(15): 1179–1184.
13. Masoro EJ (2005). Overview of caloric restriction and ageing. *Mech Ageing Dev* 126(9): 913–922.
14. McCarter RJ and Palmer J (1992). Energy metabolism and aging: a lifelong study of Fischer 344 rats. *Am J Physiol* 263(3 Pt 1): E448–E452.
15. Speakman JR, Talbot DA, Selman C, Snart S, McLaren JS, Redman P, Krol E, Jackson DM, Johnson MS, and Brand MD (2004). Uncoupled and surviving: individual mice with high metabolism have greater mitochondrial uncoupling and live longer. *Aging Cell* 3(3): 87–95.
16. Speakman JR, van Acker A, and Harper EJ (2003). Age-related changes in the metabolism and body composition of three dog breeds and their relationship to life expectancy. *Aging Cell* 2(5): 265–275.
17. Hulbert AJ, Clancy DJ, Mair W, Braeckman BP, Gems D, and Partridge L (2004). Metabolic rate is not reduced by dietary-restriction or by lowered insulin/IGF-1 signalling and is not correlated with individual lifespan in Drosophila melanogaster. *Exp Gerontol* 39(8): 1137–1143.
18. Buffenstein R and Yahav S (1991). The effect of diet on microfaunal population and function in the caecum of a subterranean naked mole-rat, Heterocephalus glaber. *Br J Nutr* 65(2): 249–258.
19. O'Connor TP, Lee A, Jarvis JU, and Buffenstein R (2002). Prolonged longevity in naked mole-rats: age-related changes in metabolism, body composition and gastrointestinal function. *Comp Biochem Physiol A Mol Integr Physiol* 133(3): 835–842.
20. Loeb J and Northrop JH (1917). What determines the duration of life in metazoa?. *Proc Natl Acad Sci USA* 3(5): 382–386.
21. Arking R, Buck S, Wells RA, and Pretzlaff R (1988). Metabolic rates in genetically based long lived strains of Drosophila. *Exp Gerontol* 23(1): 59–76.
22. Speakman JR (2005a). Body size, energy metabolism and lifespan. *J Exp Biol* 208(Pt 9): 1717–1730.
23. Chance B, Sies H, and Boveris A (1979). Hydroperoxide metabolism in mammalian organs. *Physiol Rev* 59(3): 527–605.
24. Rolfe DF and Brown GC (1997). Cellular energy utilization and molecular origin of standard metabolic rate in mammals. *Physiol Rev* 77(3): 731–758.

25. Davies M (1961). On body size and tissue respiration. *J Cell Comp Physiol* 57: 135–147.
26. Navarro A and Boveris A (2004). Rat brain and liver mitochondria develop oxidative stress and lose enzymatic activities on aging. *Am J Physiol Regul Integr Comp Physiol* 287(5): R1244–R1249.
27. Kwong LK and Sohal RS (2000). Age-related changes in activities of mitochondrial electron transport complexes in various tissues of the mouse. *Arch Biochem Biophys* 373(1): 16–22.
28. Choksi KB, Nuss JE, Boylston WH, Rabek JP, and Papaconstantinou J (2007). Age-related increases in oxidatively damaged proteins of mouse kidney mitochondrial electron transport chain complexes. *Free Radic Biol Med* 43(10): 1423–1438.
29. Sugiyama S, Takasawa M, Hayakawa M, and Ozawa T (1993). Changes in skeletal muscle, heart and liver mitochondrial electron transport activities in rats and dogs of various ages. *Biochem Mol Biol Int* 30(5): 937–944.
30. Bowling AC, Mutisya EM, Walker LC, Price DL, Cork LC, and Beal MF (1993). Age-dependent impairment of mitochondrial function in primate brain. *J Neurochem* 60(5): 1964–1967.
31. Zahn JM, Sonu R, Vogel H, Crane E, Mazan-Mamczarz K, Rabkin R, Davis RW, Becker KG, Owen AB, and Kim SK (2006). Transcriptional profiling of aging in human muscle reveals a common aging signature. *PLoS Genet* 2(7): e115.
32. St-Pierre J, Buckingham JA, Roebuck SJ, and Brand MD (2002). Topology of superoxide production from different sites in the mitochondrial electron transport chain. *J Biol Chem* 277(47): 44784–44790.
33. Ku HH, Brunk UT, and Sohal RS (1993). Relationship between mitochondrial superoxide and hydrogen peroxide production and longevity of mammalian species. *Free Radic Biol Med* 15(6): 621–627.
34. Barja G and Herrero A (1998). Localization at complex I and mechanism of the higher free radical production of brain nonsynaptic mitochondria in the short-lived rat than in the longevous pigeon. *J Bioenerg Biomembr* 30(3): 235–243.
35. Lambert AJ, Boysen HM, Buckingham JA, Yang T, Podlutsky A, Austad SN, Kunz TH, Buffenstein R, and Brand MD (2007). Low rates of hydrogen peroxide production by isolated heart mitochondria associate with long maximum lifespan in vertebrate homeotherms. *Aging Cell* 6(5): 607–618.
36. Sohal RS, Svensson I, Sohal BH, and Brunk UT (1989). Superoxide anion radical production in different animal species. *Mech Ageing Dev* 49(2): 129–135.
37. Sohal RS, Svensson I, and Brunk UT (1990b). Hydrogen peroxide production by liver mitochondria in different species. *Mech Ageing Dev* 53(3): 209–215.
38. Robert KA, Brunet-Rossinni A, and Bronikowski AM (2007). Testing the 'free radical theory of aging' hypothesis: physiological differences in long-lived and short-lived colubrid snakes. *Aging Cell* 6(3): 395–404.
39. Sohal RS, Ku HH, and Agarwal S (1993b). Biochemical correlates of longevity in two closely related rodent species. *Biochem Biophys Res Commun* 196(1): 7–11.
40. Ku HH and Sohal RS (1993). Comparison of mitochondrial pro-oxidant generation and anti-oxidant defenses between rat and pigeon: possible basis of variation in longevity and metabolic potential. *Mech Ageing Dev* 72(1): 67–76.
41. Herrero A and Barja G (1997). Sites and mechanisms responsible for the low rate of free radical production of heart mitochondria in the long-lived pigeon. *Mech Ageing Dev* 98(2): 95–111.
42. Herrero A and Barja G (1998). H2O2 production of heart mitochondria and aging rate are slower in canaries and parakeets than in mice: sites of free radical generation and mechanisms involved. *Mech Ageing Dev* 103(2): 133–146.
43. Brunet-Rossinni AK (2004). Reduced free-radical production and extreme longevity in the little brown bat (Myotis lucifugus) versus two non-flying mammals. *Mech Ageing Dev* 125(1): 11–20.
44. Csiszar A, Labinskyy N, Zhao X, Hu F, Serpillon S, Huang Z, Ballabh P, Levy RJ, Hintze TH, Wolin MS, Austad SN, Podlutsky A, and Ungvari Z (2007b). Vascular superoxide and

hydrogen peroxide production and oxidative stress resistance in two closely related rodent species with disparate longevity. *Aging Cell* 6(6): 783–797.
45. Barja G, Cadenas S, Rojas C, Perez-Campo R, and Lopez-Torres M (1994b). Low mitochondrial free radical production per unit O_2 consumption can explain the simultaneous presence of high longevity and high aerobic metabolic rate in birds. *Free Radic Res* 21(5): 317–327.
46. Wardman P (2007). Fluorescent and luminescent probes for measurement of oxidative and nitrosative species in cells and tissues: progress, pitfalls, and prospects. *Free Radic Biol Med* 43(7): 995–1022.
47. Forman HJ and Kennedy J (1976). Dihydroorotate-dependent superoxide production in rat brain and liver. A function of the primary dehydrogenase. *Arch Biochem Biophys* 173(1): 219–224.
48. Panov A, Dikalov S, Shalbuyeva N, Hemendinger R, Greenamyre JT, and Rosenfeld J (2007). Species- and tissue-specific relationships between mitochondrial permeability transition and generation of ROS in brain and liver mitochondria of rats and mice. *Am J Physiol Cell Physiol* 292(2): C708–C718.
49. Benard G, Faustin B, Passerieux E, Galinier A, Rocher C, Bellance N, Delage JP, Casteilla L, Letellier T, and Rossignol R (2006). Physiological diversity of mitochondrial oxidative phosphorylation. *Am J Physiol Cell Physiol* 291(6): C1172–C1182.
50. Csiszar A, Labinskyy N, Orosz Z, Xiangmin Z, Buffenstein R, and Ungvari Z (2007a). Vascular aging in the longest-living rodent, the naked mole rat. *Am J Physiol Heart Circ Physiol* 293(2): H919–H927.
51. Sasaki T, Unno K, Tahara S, Shimada A, Chiba Y, Hoshino M, and Kaneko T (2008). Age-related increase of superoxide generation in the brains of mammals and birds. *Aging Cell* 7: 459–469.
52. Wang W, Fang H, Groom L, Cheng A, Zhang W, Liu J, Wang X, Li K, Han P, Zheng M, Yin J, Wang W, Mattson MP, Kao JP, Lakatta EG, Sheu SS, Ouyang K, Chen J, Dirksen RT, and Cheng H (2008). Superoxide flashes in single mitochondria. *Cell* 134(2): 279–290.
53. Tolmasoff JM, Ono T, and Cutler RG (1980). Superoxide dismutase: correlation with life-span and specific metabolic rate in primate species. *Proc Natl Acad Sci USA* 77(5): 2777–2781.
54. Cutler RG (1991). Antioxidants and aging. *Am J Clin Nutr* 53(1 Suppl): 373S–379S.
55. Buffenstein R (2008). Negligible senescence in the longest living rodent, the naked mole-rat: insights from a successfully aging species. *J Comp Physiol [B]* 178(4): 439–445.
56. Sohal RS, Sohal BH, and Brunk UT (1990a). Relationship between antioxidant defenses and longevity in different mammalian species. *Mech Ageing Dev* 53(3): 217–227.
57. Perez-Campo R, Lopez-Torres M, Cadenas S, Rojas C, and Barja G (1998). The rate of free radical production as a determinant of the rate of aging: evidence from the comparative approach. *J Comp Physiol [B]* 168(3): 149–158.
58. Lopez-Torres M, Perez-Campo R, Rojas C, Cadenas S, and Barja G (1993). Maximum life span in vertebrates: relationship with liver antioxidant enzymes, glutathione system, ascorbate, urate, sensitivity to peroxidation, true malondialdehyde, in vivo H2O2, and basal and maximum aerobic capacity. *Mech Ageing Dev* 70(3): 177–199.
59. Barja G, Cadenas S, Rojas C, Lopez-Torres M, and Perez-Campo R (1994a). A decrease of free radical production near critical targets as a cause of maximum longevity in animals. *Comp Biochem Physiol Biochem Mol Biol* 108(4): 501–512.
60. Perez-Campo R, Lopez-Torres M, Rojas C, Cadenas S, and Barja G (1994). Longevity and antioxidant enzymes, non-enzymatic antioxidants and oxidative stress in the vertebrate lung: a comparative study. *J Comp Physiol [B]* 163(8): 682–689.
61. Wilhelm Filho D, Althoff SL, Dafre AL, and Boveris A (2007). Antioxidant defenses, longevity and ecophysiology of South American bats. *Comp Biochem Physiol C Toxicol Pharmacol* 146(1–2): 214–220.
62. Andziak B, O'Connor TP, and Buffenstein R (2005). Antioxidants do not explain the disparate longevity between mice and the longest-living rodent, the naked mole-rat. *Mech Ageing Dev* 126(11): 1206–1212.

63. Andziak B, O'Connor TP, Qi W, DeWaal EM, Pierce A, Chaudhuri AR, Van Remmen H, and Buffenstein R (2006). High oxidative damage levels in the longest-living rodent, the naked mole-rat. *Aging Cell* 5(6): 463–471.
64. Meydani M, Lipman RD, Han SN, Wu D, Beharka A, Martin KR, Bronson R, Cao G, Smith D, and Meydani SN (1998). The effect of long-term dietary supplementation with antioxidants. *Ann NY Acad Sci* 854: 352–360.
65. Van Remmen H, Qi W, Sabia M, Freeman G, Estlack L, Yang H, Mao Guo Z, Huang TT, Strong R, Lee S, Epstein CJ, and Richardson A (2004). Multiple deficiencies in antioxidant enzymes in mice result in a compound increase in sensitivity to oxidative stress. *Free Radic Biol Med* 36(12): 1625–1634.
66. Hamilton ML, Van Remmen H, Drake JA, Yang H, Guo ZM, Kewitt K, Walter CA, and Richardson A (2001). Does oxidative damage to DNA increase with age?. *Proc Natl Acad Sci USA* 98(18): 10469–10474.
67. Andziak B and Buffenstein R (2006). Disparate patterns of age-related changes in lipid peroxidation in long-lived naked mole-rats and shorter-lived mice. *Aging Cell* 5(6): 525–532.
68. Bjelland S and Seeberg E (2003). Mutagenicity, toxicity and repair of DNA base damage induced by oxidation. *Mutat Res* 531(1-2): 37–80.
69. Cadet J, Douki T, and Ravanat JL (2008). Oxidatively generated damage to the guanine moiety of DNA: mechanistic aspects and formation in cells. *Acc Chem Res* 41(8): 1075–1083.
70. Barja G and Herrero A (2000). Oxidative damage to mitochondrial DNA is inversely related to maximum life span in the heart and brain of mammals. *FASEB J* 14(2): 312–318.
71. Herrero A and Barja G (1999). 8-oxo-deoxyguanosine levels in heart and brain mitochondrial and nuclear DNA of two mammals and three birds in relation to their different rates of aging. *Aging (Milano)* 11(5): 294–300.
72. Foksinski M, Rozalski R, Guz J, Ruszkowska B, Sztukowska P, Piwowarski M, Klungland A, and Olinski R (2004). Urinary excretion of DNA repair products correlates with metabolic rates as well as with maximum life spans of different mammalian species. *Free Radic Biol Med* 37(9): 1449–1454.
73. Svoboda P, Maekawa M, Kawai K, Tominaga T, Savela K, and Kasai H (2006). Urinary 8-hydroxyguanine may be a better marker of oxidative stress than 8-hydroxydeoxyguanosine in relation to the life spans of various species. *Antioxid Redox Signal* 8(5–6): 985–992.
74. Muscari C, Giaccari A, Stefanelli C, Viticchi C, Giordano E, Guarnieri C, and Caldarera CM (1996). Presence of a DNA-4236 bp deletion and 8-hydroxy-deoxyguanosine in mouse cardiac mitochondrial DNA during aging. *Aging (Milano)* 8(6): 429–433.
75. Ozawa T (1999). Mitochondrial genome mutation in cell death and aging. *J Bioenerg Biomembr* 31(4): 377–390.
76. Lee CM, Chung SS, Kaczkowski JM, Weindruch R, and Aiken JM (1993). Multiple mitochondrial DNA deletions associated with age in skeletal muscle of rhesus monkeys. *J Gerontol* 48(6): B201–B205.
77. Wang E, Wong A, and Cortopassi G (1997) The rate of mitochondrial mutagenesis is faster in mice than humans. *Mutat. Res.* 377(2): 157–166.
78. Hart RW, D'Ambrosio SM, Ng KJ, and Modak SP (1979). Longevity, stability and DNA repair. *Mech Ageing Dev* 9(3–4): 203–223.
79. Cortopassi GA and Wang E (1996). There is substantial agreement among interspecies estimates of DNA repair activity. *Mech Ageing Dev* 91(3): 211–218.
80. Vijg J (2000). Somatic mutations and aging: a re-evaluation. *Mutat Res* 447(1): 117–135.
81. Montuschi P, Barnes P, and Roberts LJ, 2nd (2007). Insights into oxidative stress: the isoprostanes. *Curr Med Chem* 14(6): 703–717.
82. Ward WF, Qi W, Van Remmen H, Zackert WE, Roberts LJ, 2nd, and Richardson A (2005). Effects of age and caloric restriction on lipid peroxidation: measurement of oxidative stress by F2-isoprostane levels. *J Gerontol A Biol Sci Med Sci* 60(7): 847–851.
83. Pamplona R, Barja G, and Portero-Otin M (2002). Membrane fatty acid unsaturation, protection against oxidative stress, and maximum life span: a homeoviscous-longevity adaptation?. *Ann NY Acad Sci* 959: 475–490.
84. Stadtman ER (2006). Protein oxidation and aging. *Free Radic Res* 40(12): 1250–1258.

85. Dean RT, Fu S, Stocker R, and Davies MJ (1997). Biochemistry and pathology of radical-mediated protein oxidation. *Biochem J* 324(Pt 1): 1–18.
86. Thorpe SR and Baynes JW (2003). Maillard reaction products in tissue proteins: new products and new perspectives. *Amino Acids* 25(3–4): 275–281.
87. Agarwal S and Sohal RS (1994). Aging and protein oxidative damage. *Mech Ageing Dev* 75(1): 11–19.
88. Chao CC, Ma YS, and Stadtman ER (1997). Modification of protein surface hydrophobicity and methionine oxidation by oxidative systems. *Proc Natl Acad Sci USA* 94(7): 2969–2974.
89. Carney JM, Starke-Reed PE, Oliver CN, Landum RW, Cheng MS, Wu JF, and Floyd RA (1991). Reversal of age-related increase in brain protein oxidation, decrease in enzyme activity, and loss in temporal and spatial memory by chronic administration of the spin-trapping compound N-tert-butyl-alpha-phenylnitrone. *Proc Natl Acad Sci USA* 88(9): 3633–3636.
90. Dubey A, Forster MJ, and Sohal RS (1995). Effect of the spin-trapping compound N-tert-butyl-alpha-phenylnitrone on protein oxidation and life span. *Arch Biochem Biophys* 324(2): 249–254.
91. Chaudhuri AR, de Waal EM, Pierce A, Van Remmen H, Ward WF, and Richardson A (2006). Detection of protein carbonyls in aging liver tissue: A fluorescence-based proteomic approach. *Mech Ageing Dev* 127(11): 849–861.
92. Jana CK, Das N, and Sohal RS (2002). Specificity of age-related carbonylation of plasma proteins in the mouse and rat. *Arch Biochem Biophys* 397(2): 433–439.
93. Agarwal S and Sohal RS (1993). Relationship between aging and susceptibility to protein oxidative damage. *Biochem Biophys Res Commun* 194(3): 1203–1206.
94. Sohal RS, Agarwal S, Dubey A, and Orr WC (1993a). Protein oxidative damage is associated with life expectancy of houseflies. *Proc Natl Acad Sci USA* 90(15): 7255–7259.
95. Agarwal S and Sohal RS (1996). Relationship between susceptibility to protein oxidation, aging, and maximum life span potential of different species. *Exp Gerontol* 31(3): 365–372.
96. Portero-Otin M, Requena JR, Bellmunt MJ, Ayala V, and Pamplona R (2004). Protein nonenzymatic modifications and proteasome activity in skeletal muscle from the short-lived rat and long-lived pigeon. *Exp Gerontol* 39(10): 1527–1535.

Comparative Genomics of Aging

Jan Vijg, Ana Maria Garcia, Brent Calder, and Martijn Dollé

Abstract Genomes are inherently unstable as a consequence of their role as substrate for evolutionary change. In somatic cells the accumulation of both mutations and epimutations are the inevitable outcome of errors made during DNA replication or the repair of DNA damage. Genome instability has often been considered as a universal cause of aging, with genome maintenance as the main determinant of species-specific life span. It has been very difficult to test this hypothesis directly because of a lack of good model systems allowing a direct comparison of the rate of spontaneous genome alteration in somatic tissues during aging. Here we review the results of a direct comparison of spontaneous DNA mutation frequencies in somatic tissue of mice and fruit flies using transgenic mutational reporter genes.

Keywords Aging · *Drosophila* · Mouse · Gene mutations · DNA repair · DNA damage

Introduction

The genetic information kept in the repository that we now call the genome is walking a tightrope between stability and change. Too much change threatens an organism's very existence while too little change compromises its evolvability, i.e., the capacity to mutate itself away from environmental challenges through selection. This balance between stability and plasticity is accomplished through genome maintenance, probably our most ancient longevity system. Indeed, without some form of genome maintenance the first replicating nucleic acids in a world exposed to high fluxes of damaging ultraviolet radiation would not have been able to survive long enough to multiply [1]. Errors generated during repair provide the germ line

J. Vijg (✉)
Department of Genetics, Albert Einstein College of Medicine, Bronx, NY 10461, USA
e-mail: jvijg@aecom.yu.edu

N.S. Wolf (ed.), *The Comparative Biology of Aging*,
DOI 10.1007/978-90-481-3465-6_9, © Springer Science+Business Media B.V. 2010

with the DNA sequence variation that is the substrate of evolution. In this sense, longevity of protocells and subsequent unicellular organisms was balanced against evolvability. In modern, multicellular organisms genome maintenance is still critical to their survival, which is evident from the large investment cells make in encoding proteins solely devoted to repairing damage in DNA. In some cases an entire protein molecule is sacrificed for the removal of a single lesion [2].

While also in multicellular organisms germ line genomes must retain some plasticity, their somatic genomes could in theory be equipped with perfect maintenance systems. However, evolutionary theory would not predict maximal somatic maintenance for long periods after the time of first reproduction [3]. According to the disposable soma theory, investments in somatic maintenance are inversely related to growth and reproduction [4]. According to this reasoning, organisms that turnover rapidly would not require as much somatic maintenance as organisms with extended life spans. Observed correlations between stress response and organismal life span tend to support this argument [5]. Interestingly, the rates of DNA sequence change (i.e., rates of molecular evolution) among different phylogenetic groups differ by a factor of 5, with the slowest rates in higher primates and much faster rates in rodents, sea urchins and *Drosophila* [6]. In what is called the hominid slowdown, lineage-specific rates of primate evolution – as derived from the rate of DNA sequence change in the germ line compared for both mitochondrial and nuclear DNA – decline from apes to monkeys and humans; humans, the longest-lived primates evolved the slowest [7]. This suggests that long-lived primates have better genome maintenance systems to preserve the germ line than short-lived organisms. Indeed, recent results of a comparison of DNA basepair substitution rate variation across mammals, comparing both mitochondrial and nuclear loci at both synonymous and non-synonymous sites, mitochondrial synonymous substitution rates were negatively correlated with maximum recorded lifespan [8].

While there is some correlative evidence suggesting that long-lived mammals have a higher capacity to repair damage in DNA than short-lived animals [9], there are multiple factors that confound such correlative studies. First, structure and organization of the genome differ greatly between organisms, requiring different levels of maintenance without necessarily any consequence for longevity. Second, it is difficult if not impossible to provide an objective measure for DNA repair capacity, since there are many different repair pathways and their utilization may differ from organism to organism [10]. Physiologically relevant DNA repair activities in vivo are also difficult to measure. Finally, how should we interpret a perceived lack of DNA repair capacity? Repair activity may be high but at the cost of many errors. This would not be obvious from enzymatic assessment or monitoring disappearance of specific types of induced DNA damage.

There are roughly two different modes by which a cell can respond to the thousands of DNA lesions induced in its DNA on a daily basis from such diverse sources as hydrolyis, reactive oxygen species (ROS) and environmental mutagens (Fig. 1). If the damage inflicted on the DNA is so severe that it is deemed irreparable – for example, after high levels of radiation – the cell becomes subject to the action of signaling pathways inducing apoptosis or cellular senescence, the irreversible cessation

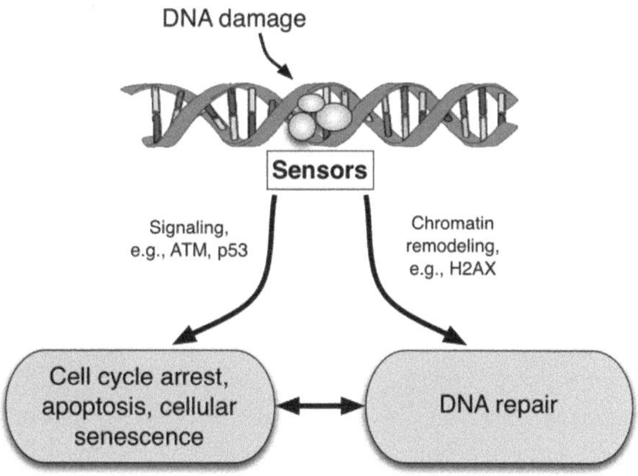

Fig. 1 The two major branches of genome maintenance. DNA repair *sensu stricto* aims to restore the original situation by removing the lesion. The complex of DNA damage signaling pathways assist in these repair activities or activates cellular responses that kill the cell or terminate its mitotic activity when it is beyond repair

of mitotic capacity [11]. Also temporary cell cycle arrest is utilized to provide more time for DNA repair. While such DNA damage response systems offer a short-term solution after exposure to high levels of genotoxic stress, on the somewhat longer term they may cause premature aging. Indeed, mice or humans with heritable defects in one or more genome maintenance systems, expected to lead to excessive levels of genotoxic stress, often show multiple symptoms of premature aging [12]. In such cases, aging phenotypes might be caused by accelerated loss of cells or a decreased regenerative capacity [13]. Under normal conditions cell loss and a general impairment of mitotic activity almost certainly contribute to aging. However, it is unlikely that normative aging is caused exclusively by cellular responses to DNA damage.

At relatively low levels, presumably the norm in wild type animals, DNA damage is efficiently repaired. In this case, i.e., during normative aging, adverse effects are not caused by the DNA damage directly, but by the mutations or epimutations that result from the inevitable errors made during the repair process or (in dividing cells) during replication of a damaged template. This failure to restore the correct DNA sequence or DNA histone modification patterns after removal of the lesion is irreversible, which is in striking contrast to changes in RNA or proteins. Mutations or epimutations are unavoidable and, as we shall see, accumulate with age; they are generally assumed to be the main cause of cancer in animals with renewable tissues.

Random mutations can affect cellular phenotypes by altering protein-coding sequences, but probably much more frequently by affecting gene regulatory patterns [14]. An increased burden of mutations (e.g., basepair substitutions, small and large deletions, translocations, copy number changes) and/or epimutations (changes in DNA methylation or histone modification patterns) in aged tissues could adversely affect regulated gene transcription, for example, through haploinsufficiency after deletions, position effects after translocations or derangements in chromatin looping, which can also result from deletions or from point mutations in nuclear matrix attachment regions. This would promote cancer, but could also cause functional decline of cell populations. What is the evidence that mutations and epimutations ever reach a stage where they could impact cellular function?

Somatic Mutations and Epimutations in Aging

While it was realized since Hermann Muller first pointed out, well before the discovery of the double helix, that mutations almost always have adverse effects [15], it was assumed that errors in the genome were too rare to have phenotypic effects, except by clonal outgrowth, i.e., in cancer [16]. However, we now know that spontaneous changes at the level of the genome are not only much more frequent than initially assumed, but also likely to have adverse phenotypic effects by affecting patterns of gene regulation.

Spontaneous mutation rates are far higher than previously estimated. Using microsatellite markers to study loss of heterozygosity events, several investigators reported instability at these loci not only in tumors, but also in normal histologically benign tissue [17]. These cases probably reflect clonal amplification of the cell that originally underwent the mutagenic event. Indeed, such clonally amplified mutations in human heritable disease genes are known to cause so-called segmental forms of the diseases [18]. However, there are many other cases indicating very high somatic mutation loads. For example, in both mouse and human brain, a significant fraction of cells, including neurons, were found to be aneuploid, with both loss and gain of chromosomes [19]. In mouse brain, a dramatic age-related increase in LOH, one of the most frequent forms of genome instability in mammalian cells [20], has been observed in neuronal progenitor cells [21].

Our own data using positive selection of mutant reporter genes recovered from genomic DNA from mouse tissues (see below) revealed a load of genome rearrangements in the mouse heart at old age (some deletions are millions of base pairs) of almost 40 per cell [22]. Results from Martin et al., using the HPRT selectable marker gene in human kidney tubular epithelial cells, indicate mutation frequencies of over 1 per 10,000 loci [23], corresponding to more than 1,000 mutations per cell when extrapolated to the genome overall. It should be kept in mind that in most selectable systems weak mutations, which could reduce cellular function without affecting viability too much, would go undetected.

Epigenomic alterations are of special concern because once established epigenetic states can drift compared to the more static DNA sequence [24]. Epigenetic changes are increasingly recognized as part of aging and age-related pathology [25]. We as well as others have found aging to be associated with a general hypomethylation, which is probably a consequence of less faithful maintenance of methylation patterns in repetitive elements [26]. However, hypermethylation, especially of promoter-associated CpG islands has been observed to increase with age in normal colonic tissue of patients with colorectal neoplasia [27]. Indeed, increased hypermethylation of tumor suppressor genes in normal, aged tissue likely contributes to the increased cancer risk at old age. The causes of such hypermethylation could be the same as DNA sequence alterations, i.e., errors in restoring normal patterns of methylation after DNA repair or replication [28].

Whatever the case, it is now clear that the overall mutation/epimutation load of cells in apparently normal tissues is substantial, and increases during aging [29]. The exact types of events, their cell and tissue-specificity and functional impact remain unknown. However, given the magnitude of the random genomic stress now being revealed, it seems unlikely that such (epi)genomic decay has no adverse effect other than increasing cancer risk. If random mutations and epimutations are a universal cause of aging, the rate and severity of such events may correlate with species-specific life span. As mentioned, comparative studies of germ line DNA mutation rate suggests that this is the case. However, there is virtually no comparative information on somatic mutation accumulation across species due to the lack of good model systems. Here we will discuss some results obtained by comparing *Drosophila melanogaster* and *Mus musculus*, using a similar, reporter gene-based model system.

Somatic Mutation Accumulation in Mice and Flies

Somatic mutagenesis is difficult to study in higher organisms, with most assays indirect and based on alterations in phenotypic characteristics, such as the mouse or *Drosophila* spot tests [30, 31]. Direct methods are available, but restricted to point mutations in restriction enzyme recognition sites [32]. In the past, we have generated transgenic mouse models harboring chromosomally integrated lacZ-plasmid constructs that can be recovered into *E. coli* for the subsequent quantification and sequence characterization of a broad range of spontaneous mutations [33]. The results with this system indicate that somatic mutations accumulate in virtually all organs and tissues albeit at different rates [34]. Also the types of mutations found to accumulate with age are very different among organs. For example, while many mutations in heart and liver were large genome rearrangements, e.g., deletions, inversions or translocations, sometimes involving millions of basepairs, virtually all mutations that had accumulated in the small intestine of old mice were point mutations, i.e., basepair substitutions or very small deletions or insertions [34].

A similar transgenic reporter model did not exist for invertebrates and information as to how the spontaneous mutation burden in somatic tissues of such organisms differs from those in mammals is absent. We recently generated several lines of *Drosophila melanogaster* harboring a lacZ-plasmid construct identical to the one in the mouse [35]. This system allows to directly compare somatic mutation frequencies and spectra as a function of age between a mammal and an insect (Fig. 2). Interestingly, the results for one transgenic line, i.e., line 11, harboring the lacZ-plasmid construct on chromosome 3 indicated a spontaneous mutant frequency in male flies of about 11×10^{-5} with a significantly higher mutant frequency of about 15×10^{-5} for females ($p < 0001$) [35]. This sexual dimorphism of spontaneous mutation frequency is not present in the mouse. (Of note, in this genetic background female flies live shorter than males.) Although some statistically significant variation as a function of the integration site was observed, neither the mutant frequency nor the mutation spectra were much different among the different lines. In all cases genome rearrangements in the fly were much more prominent than point mutations.

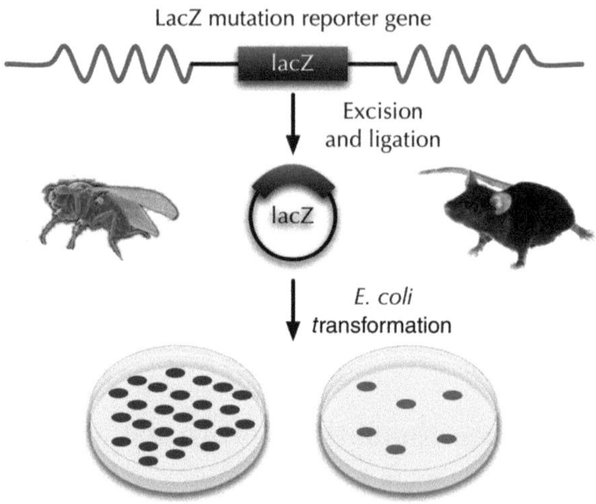

Fig. 2 Analysis of somatic mutation frequency and spectrum in transgenic mice and flies. A plasmid construct containing the lacZ reporter gene is integrated at a chromosomal location. After DNA extraction the lacZ-plasmids are excised and used to transform *E. coli* cells deleted for lacZ and harboring an inactivated gale gene. In this strain only plasmids with an inactivated lacZ gene allow a bacterial cell to survive. Hence, mutant lacZ genes are positively selected. The nature of the mutation (point mutation, deletion) can be identified by nucleotide sequencing and physical mapping of the breakpoints (in case of large rearrangements), since the exact position of the integrated reporter genes is known. For further details, see Garcia et al. [39]

Like in the mouse, also in the fly mutations accumulate with age. Lifespan in *Drosophila* is temperature-dependent and longevity decreases exponentially with increasing temperature between 12 and 30°C [36]. This may be caused by an increase in the rate of metabolic processes, presumably speeding up the ageing process. We found that the age-related increase in somatic mutations in the fly was much higher at higher temperatures (A. Garcia, submitted for publication).

When directly comparing the mutation frequency per locus between mouse tissues (which do not vary much from tissue to tissue) and whole fly tissue, a significantly higher somatic mutation burden in young flies as compared to young mice is evident. Indeed, our results indicate a 3- to 4-fold higher mutant frequency in *Drosophila* than in *Mus musculus* heart (Table 1). Because of the far larger size of the mouse genome, the total number of mutations per average cell is significantly higher in the mouse, i.e., about 5-fold. However, while the fly has a smaller genome than the mouse (about 16-fold), it is more compact in the sense that its gene density is much higher. Hence, random mutations should be more likely to have adverse effects in the fly genome than in that of the mouse.

Interestingly, many more spontaneous mutations in *Drosophila* appeared to be large genome rearrangements than in mice. Since genome rearrangements are much more likely to have adverse effects than point mutations (they can adversely affect

Table 1 Somatic mutation loads of young adult and old flies and mice

	Young		Old		
	lacZ mut. freq. ($\times 10^{-5}$)	Mutations per cell[e]	lacZ mut. freq. ($\times 10^{-5}$)	Mutations per cell[e]	Fold incr. with age
***Drosophila*[a]**					
All	15.6	10	25.2	16	1.6
Gene mutations[c]	4.5	4	6.6	6	1.5
Rearrangements[d]	11.1	5	18.6	9	1.7
Mouse heart[b]					
All	4.1	49	10.1	131	2.5
Gene mutations[c]	2.1	34	6.4	101	3.0
Rearrangements[d]	1.9	15	3.7	29	1.9
Ratio *Drosophila*/ mouse (inverse)					
All	3.8	0.2 (5)	2.5	0.1 (8.4)	
Gene mutations	2.1	0.1 (7.6)	1.0	0.1 (16)	
Rearrangements	5.7	0.3 (2.8)	5.0	0.3 (3.2)	

[a]Mean lacZ mutant frequencies of 1–7 day and 6 week old female *Drosophila* kept at 25°C ([35] and unpublished results).
[b]Mean lacZ mutant frequencies in heart of 3 and 32 months old male mice [22].
[c]Point mutations and small (intragene) deletions, inversions and insertions.
[d]Genome rearrangements (intergene deletions, inversions, insertions and translocations).
[e]Mutations per diploid genome, based on a 3303 bp target locus, a 3.24×10^8 bp diploid fly genome, and a 5.24×10^9 bp diploid mouse genome; rearrangements per cell are specifically adjusted for the twofold chance of one of the breakpoints falling within the reporter gene; "all" is the sum of "gene mutations" and "rearrangements".

the expression pattern of many genes), this would correlate with the life span difference. The high tolerance for rearrangements of the fly genome relative to that of the mouse may reflect differences in genome organization between these two species. For example, gene function in mammals may depend to a much greater extent on long-distance regulatory interactions among multiple genes than in *Drosophila*. Indeed, the increased biological complexity from yeast to mammals is not adequately reflected in expanded gene numbers, but may instead be due to increased regulatory intricacy [37]. In the mouse, sizable fractions of genome rearrangements could disrupt the many long-distance gene regulatory interactions and might be unsustainable. The evolution of more complex species with longer life spans and more numerous cell divisions most likely also required the evolution of more sophisticated mechanisms for replication and repair to prevent the deleterious effects of genome rearrangements.

Summary and Future Prospects

Functional decay caused by random (epi)genomic events is probably more prevalent than suspected in the past, when gene action was often considered in relative isolation. As revealed by early results of the ENCODE project, essentially all parts of the human genome are transcribed, much of it possibly for regulatory purposes [38]. Hence, the target for random alterations that adversely impact (but do not necessarily completely disrupt) cellular function is very large. Using model systems based on a lacZ reporter gene integrated in the genome we comparatively analyzed *Mus musculus* and *Drosophila melanogaster* for spontaneous somatic mutation frequencies and spectra. Because the lacZ reporter gene is not expressed it acts as a perfectly neutral target without distortion due to selection. The results indicate that in both species mutations accumulate with age irrespective of the mitotic state of the organ. In the mouse, mutations accumulate not only in spleen and small intestine, both organs of high mitotic activity, but also in the heart and liver, which are predominantly postmitotic organs. In flies, an organism mostly containing non-dividing cells, mutations also accumulate with age and they do so much faster at higher temperatures.

Although the spontaneous mutation frequency per locus in *Drosophila* is about 5-fold higher than in mouse tissues, the mutation load per amount of functional genome may be much higher because of the much higher gene density in the fly genome. Moreover, the spectrum of spontaneous mutations significantly differs between the two species, with genome rearrangements making up a much higher fraction in the fly somatic genome than in that of the mouse. We suggest that this difference reflects the difference in gene regulatory complexity, which requires more and more complex long-distance gene regulatory interactions. To extend and confirm these findings it would be of interest to generate additional model organisms harboring the same lacZ mutational reporter gene.

References

1. de Duve C (2005). The onset of selection. *Nature* 433: 581–582.
2. Friedberg EC, Walker GR, Siede W, Wood RD, Schultz RA, and Ellenberger T (2006). *DNA Repair and Mutagenesis*. Washington, DC: ASM Press.
3. Charlesworth B (2000). Fisher, Medawar, Hamilton and the evolution of aging. *Genetics* 156: 927–931.
4. Kirkwood TB (2005). Understanding the odd science of aging. *Cell* 120: 437–447.
5. Kapahi P, Boulton ME, and Kirkwood TB (1999). Positive correlation between mammalian life span and cellular resistance to stress. *Free Radic Biol Med* 26: 495–500.
6. Britten RJ (1986). Rates of DNA sequence evolution differ between taxonomic groups. *Science* 231: 1393–1398.
7. Goodman M (1985). Rates of molecular evolution: the hominoid slowdown. *Bioessays* 3: 9–14.
8. Welch JJ, Bininda-Emonds OR, and Bromham L (2008). Correlates of substitution rate variation in mammalian protein-coding sequences. *BMC Evol Biol* 8: 53.
9. Hart RW and Setlow RB (1974). Correlation between deoxyribonucleic acid excision-repair and life-span in a number of mammalian species. *Proc Natl Acad Sci U S A* 71: 2169–2173.
10. Hoeijmakers JH (2001). Genome maintenance mechanisms for preventing cancer. *Nature* 411: 366–374.
11. Campisi J (2003). Cancer and ageing: rival demons? *Nat Rev Cancer* 3: 339–349.
12. Hasty P, Campisi J, Hoeijmakers J, van Steeg H, and Vijg J (2003). Aging and genome maintenance: lessons from the mouse? *Science* 299: 1355–1359.
13. Sharpless NE and DePinho RA (2007). How stem cells age and why this makes us grow old. *Nat Rev Mol Cell Biol* 8: 703–713.
14. Bahar R, Hartmann CH, Rodriguez KA, Denny AD, Busuttil RA, Dollé MET, Calder RB, Chisholm GB, Pollock BH, Klein CA, and Vijg J (2006). Increased cell-to-cell variation in gene expression in aging mouse heart. *Nature* 441: 1011–1014.
15. Muller HJ (1927). Artificial transmutation of the gene. *Science* 66: 84–87.
16. Maynard Smith J (1962). Review lectures on senescence. I. The causes of ageing. *Proc Roy Soc B* 157: 115–127.
17. Arvanitis DA, Flouris GA, and Spandidos DA (2005). Genomic rearrangements on VCAM1, SELE, APEG1and AIF1 loci in atherosclerosis. *J Cell Mol Med* 9: 153–159.
18. Erickson RP (2003). Somatic gene mutation and human disease other than cancer. *Mutat Res* 543: 125–136.
19. Rehen SK, Yung YC, McCreight MP, Kaushal D, Yang AH, Almeida BS, Kingsbury MA, Cabral KM, McConnell MJ, Anliker B, Fontanoz M, and Chun J (2005). Constitutional aneuploidy in the normal human brain. *J Neurosci* 25: 2176–2180.
20. Shao C, Deng L, Henegariu O, Liang L, Raikwar N, Sahota A, Stambrook PJ, and Tischfield JA (1999). Mitotic recombination produces the majority of recessive fibroblast variants in heterozygous mice. *Proc Natl Acad Sci U S A* 96: 9230–9235.
21. Bailey KJ, Maslov AY, and Pruitt SC (2004). Accumulation of mutations and somatic selection in aging neural stem/progenitor cells. *Aging Cell* 3: 391–397.
22. Dollé ME and Vijg J (2002). Genome dynamics in aging mice. *Genome Res* 12: 1732–1738.
23. Martin GM, Ogburn CE, Colgin LM, Gown AM, Edland SD, and Monnat RJ, Jr. (1996). Somatic mutations are frequent and increase with age in human kidney epithelial cells. *Hum Mol Genet* 5: 215–221.
24. Fraga MF, Ballestar E, Paz MF, Ropero S, Setien F, Ballestar ML, Heine-Suner D, Cigudosa JC, Urioste M, Benitez J, Boix-Chornet M, Sanchez-Aguilera A, Ling C, Carlsson E, Poulsen P, Vaag A, Stephan Z, Spector TD, Wu YZ, Plass C, and Esteller M (2005). Epigenetic differences arise during the lifetime of monozygotic twins. *Proc Natl Acad Sci U S A* 102: 10604–10609.

25. Issa JP (2003). Age-related epigenetic changes and the immune system. *Clin Immunol* 109: 103–108.
26. Slagboom PE and Vijg J (1992). The dynamics of genome organization and expression during the aging process. *Ann N Y Acad Sci* 673: 58–69.
27. Issa JP, Ottaviano YL, Celano P, Hamilton SR, Davidson NE, and Baylin SB (1994). Methylation of the oestrogen receptor CpG Island links ageing and neoplasia in human colon. *Nat Genet* 7: 536–540.
28. Kastan MB, Gowans BJ, and Lieberman MW (1982). Methylation of deoxycytidine incorporated by excision-repair synthesis of DNA. *Cell* 30: 509–516.
29. Vijg J (2007). *Aging of the Genome*. New York: Oxford University Press.
30. Russell LB, Selby PB, von Halle E, Sheridan W, and Valcovic L (1981). Use of the mouse spot test in chemical mutagenesis: interpretation of past data and recommendations for future work. *Mutat Res* 86: 355–379.
31. Kaya B, Marcos R, Yanikoglu A, and Creus A (2004). Evaluation of the genotoxicity of four herbicides in the wing spot test of *Drosophila melanogaster* using two different strains. *Mutat Res* 557: 53–62.
32. Bielas JH and Loeb LA (2005). Quantification of random genomic mutations. *Nat Methods* 2: 285–290.
33. Boerrigter ME, Dolle ME, Martus HJ, Gossen JA, and Vijg J (1995). Plasmid-based transgenic mouse model for studying in vivo mutations. *Nature* 377: 657–659.
34. Vijg J and Dollé ME (2002). Large genome rearrangements as a primary cause of aging. *Mech Ageing Dev* 123: 907–915.
35. Garcia AM, Derventzi A, Busuttil R, Calder RB, Perez E, Jr., Chadwell L, Dolle ME, Lundell M, and Vijg J (2007b). A model system for analyzing somatic mutations in *Drosophila melanogaster*. *Nature Methods* 4: 401–403.
36. David JR (1988). Temperature. In *Drosophila as a Model Organism for Ageing Studies*. (FH Lints, MH Soliman, eds.). London: Blackie, pp. 33–45.
37. Levine M and Tjian R (2003). Transcription regulation and animal diversity. *Nature* 424: 147–151.
38. Birney E, Stamatoyannopoulos JA, Dutta A, Guigo R, Gingeras TR, Margulies EH, Weng Z, Snyder M, Dermitzakis ET, Thurman RE et al. (2007). Identification and analysis of functional elements in 1% of the human genome by the ENCODE pilot project. *Nature* 447: 799–816.
39. Garcia AM, Busuttil RA, Rodriguez A, Cabrera C, Lundell M, Dolle ME, and Vijg J (2007a). Detection and analysis of somatic mutations at a lacZ reporter locus in higher organisms: application to Mus musculus and *Drosophila melanogaster*. *Methods Mol Biol* 371: 267–287.

Changes in Lysosomes and Their Autophagic Function in Aging: The Comparative Biology of Lysosomal Function

Samantha J. Orenstein and Ana Maria Cuervo

Abstract The lysosome, the organelle with the greatest degradative capability in the cell, is an essential component of the systems responsible for cellular quality control. Lysosome malfunctioning alters cellular homeostasis and has been proposed to contribute to the accumulation of abnormal and damaged intracellular components in different human pathologies and aging organisms. In this chapter, we summarize the most recent advances in the characterization of the complex subset of molecular components that contribute to proper lysosomal functioning. We also provide a comparative analysis of the main properties and components of the lysosomal system in different species and review the evolutive changes of this essential catabolic pathway. A more complete characterization of the lysosomal system has recently revealed the importance of lysosomes in cellular physiology and has helped establish causal connections between impaired lysosomal function and certain diseases. In the last part of this chapter, we provide a brief summary of these connections with special emphasis on lysosomal changes in age-related disorders.

Keywords Chaperones · Lysosomes · Proteases · Protein degradation · Quality control

Introduction

Protein degradation, or the breakdown of proteins into their constitutive amino acids, is essential for cell survival and defense, to maintain protein and organelle homeostasis and to promote development, differentiation and growth [1–6]. Continuous protein turnover maintains the stability of the proteome by limiting the time that a given protein is inside the cell (half-life), which decreases the risk that

A.M. Cuervo (✉)
Department of Developmental and Molecular Biology, Marion Bessin Liver Research Center, Institute for Aging Research, Albert Einstein College of Medicine, Bronx, NY, USA
e-mail: amcuervo@aecom.yu.edu

N.S. Wolf (ed.), *The Comparative Biology of Aging*,
DOI 10.1007/978-90-481-3465-6_10,

the protein may become damaged or altered by the surrounding cellular environment [7, 8]. Protein degradation also functions as a quality control mechanism to eliminate abnormally synthesized proteins or proteins damaged during instances of oxidative stress, UV exposure, heat shock, etc. [1, 2, 9]. Degradation acts, under these conditions, to prevent protein aggregation, as partial unfolding of the damaged proteins favors abnormal interactions with other proteins and cellular components, often leading to protein aggregation and cellular toxicity [10–13].

The major components of the intracellular surveillance systems responsible for quality control are chaperones and the proteolytic systems. Cellular chaperones recognize specific areas on misfolded proteins (such as hydrophobic stretches), and with the aid of co-chaperones, attempt to refold the altered proteins to restore their proper functional conformation (Fig. 1) [11, 12]. However, if refolding is unsuccessful, the proteolytic systems, which make up the second line of defense and quality control, degrade the altered protein, thereby preventing further cellular damage [3, 12, 14].

Protein degradation also plays an important regulatory role inside cells, by allowing rapid changes in levels of intracellular proteins, which enables cells to adjust to

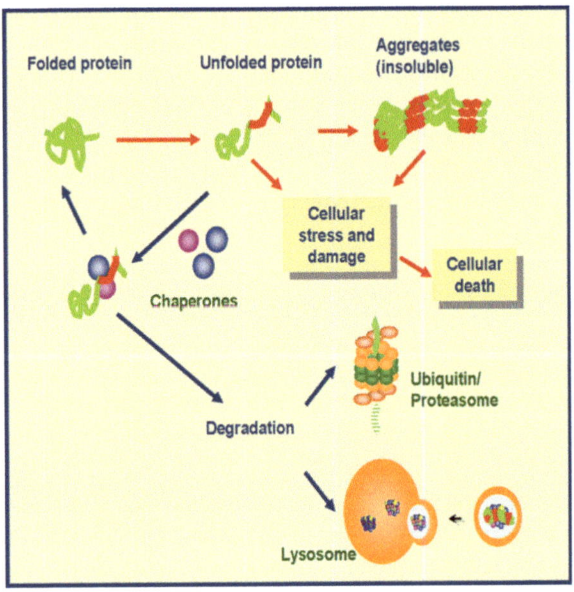

Fig. 1 Quality control inside cells. Two different components of cellular quality control are responsible for the stability of the cellular proteome, the chaperones and the proteolytic systems. Altered or damaged proteins are detected by cellular chaperones that assist them in their re-folding. However, if re-folding is not possible, the same chaperones target the altered proteins for degradation by the two main cellular proteolytic systems, the ubiquitin/proteasome system and the lysosomes. Failure of these quality control mechanisms leads to the cellular accumulation of toxic forms of the altered proteins and often to functional failure and cell death

the changing extracellular environment [3, 14, 15]. Thus, for example, when nutrients are scarce, the cellular energetic balance is maintained through the degradation of proteins no longer needed in order to provide the energy and amino acid building blocks necessary for the synthesis of proteins essential under these new conditions [2, 15, 16]. Lastly, conditions involving major structural changes, such as cellular differentiation, embryogenesis or tissue remodeling, also require active participation of the proteolytic systems [1, 17–19].

Protein degradation is a highly conserved process present and essential for the survival of all organisms from bacteria and yeast to complex multicellular organisms such as mammals. In light of the multiple functions in which proteolytic systems participate, it is easy to infer that malfunctioning of these systems has detrimental cellular consequences and underlies the basis for multiple pathologies. A decrease in the rate of intracellular degradation and functional decline of the main cellular proteolytic systems with age has been described in multiple organisms including worms, flies and almost all mammalian tissues. Reduced protein degradation has been proposed to contribute to the accumulation of altered intracellular components in aging tissues and the diminished resistance to stress of old organisms [20–24].

Two main proteolytic systems function to completely degrade proteins into their constitutive amino acids, the ubiquitin proteolytic system (UPS) and the lysosomes. The UPS primarily targets short lived proteins possessing important cellular regulatory functions, such as transcription factors, regulators of the cell cycle, and members of signaling cascades [9, 20, 25–28]. Substrate proteins are tagged by covalent linkage of multiple molecules of ubiquitin to lysine residues in their proteolytic core which are then selectively recognized by the regulatory subunits of the proteasome, the major protease of the UPS. Upon substrate recognition, the multisubunit regulatory complex removes the tag, unfolds the substrate protein and facilitates its access to the proteasome catalytic core [3, 4, 29]. Readers are directed to recent detailed reviews on the function and physiological relevance of the UPS [9, 20, 25–28]. The focus of this chapter will be the lysosomal system, particularly the role of this major proteolytic system in intracellular degradation or autophagy. We will review the characteristics and main components of this pathway, the consequences of its malfunctioning and the changes that this system undergoes with age. Although the essential components of the lysosomal system are highly phylogenetically conserved, we will also highlight in this chapter evolutive differences in the lysosomal system.

Lysosomes: Concept and Properties

Lysosomes are single membrane organelles able to degrade both intracellular and extracellular components [30]. In contrast to the UPS, lysosomes degrade mainly long-lived proteins, as well as entire organelles, thanks to their powerful enzymatic machinery. In fact, lysosomes contain a diverse variety of hydrolases – proteases, lipases, glycosidases and nucloteotidases – which allow the complete degradation of all types of macromolecules in the lysosomal lumen [30].

The lysosome is a highly conserved organelle present from yeast to mammals, although there are important distinctions in its characteristics and functions between these two (Table 1). Unlike multicellular organisms, which have multiple lysosomes completely devoted to degradation and recycling, the yeast *Saacharamyces cerevisiae,* one of the most commonly used experimental systems, contains a single vacuole which encompasses functions beyond mere protein degradation [31–33]. Like the lysosome, the vacuole contains a wide variety of hydrolases capable of breaking down all kinds of macromolecules, but in addition, it serves as a cellular reservoir for nutrients, a place for the containment and excretion of unwanted cellular substances, a means to maintain the cellular hydrostatic pressure and a mechanism to transport protons from the cytosol to the vacuole in order to maintain a stable cytoplasmic pH [34]. The vacuole is the site of accumulation of various amino acids, which can be utilized during times of food depletion [31, 34]. This unique storage function is not shared by its multicellular organism lysosomal counterparts. Furthermore, the yeast vacuole tightly controls intracellular calcium homeostasis by regulating both calcium transport into and out of the vacuole as well as phosphate and polyphosphate concentrations [35, 36].

Some of the functions of the yeast vacuole have diversified with the evolution toward other cellular processes such as exocytosis, and the lysosomal system changed from a single compartment occupying as much as 90% of the volume for certain cell types, to multiple smaller vesicles that preserve the plasticity characteristic of the vacuole. However, the yeast vacuole has not been replaced by lysosomes in all multicellular organisms. In fact, most mature plant cells have one or several vacuoles that conserve the multifunctional characteristic of the unicellular yeast vacuole. The plant vacuolar system retains the plasticity of the lysosomal system, and depending on the cellular conditions, some plant cells display a single large vacuole or multiple smaller vacuoles [37–39]. In plants the vacuole is essential to maintain turgor pressure against the cell wall through osmosis, which is also used for plant cellular elongation. In addition, similar to the lysosomal system, the plant vacuole is the main site for the degradation of proteins which control development, adaptation to environmental conditions, senescence, as well as defense [40].

Lysosomal Pathways for Protein Degradation

Lysosomes are the final degradative compartment for both extracellular components, through a process known as heterophagy, and intracellular components by autophagy. Both pathways are separately described in this section.

Table 1 Comparison of the properties of the most common lysosomal pathways in different species

Lysosomal system: components and pathways			Yeast	Invertebrate	Plants	Mammals
Lysosomes	Number		Single	Multiple	Single/ Multiple	Multiple
	Functions		– Degradation – Storage – Secretion – Regulation of cytosolic Ca^{+2}/pH	– Degradation – Recycling	– Degradation – Recycling – Maintenance of cell turgence	– Degradation – Recycling
Heterophagy	Phagocytosis	Cells	All cells	Specialized cells	Limited or non-existing	Specialized cells
		Function	Nutrition	Defense	?	Defense
	Pinocytosis	Absorptive	Low capability (no concentration of cargo) Cargo in suspension			
		Fluid phase	Higher capacity (partial concentration of cargo) Cargo binds non-specifically to the plasma membrane			
		Receptor mediated	The highest capacity (concentration of cargo) Cargo binds selectively to receptors at the plasma membrane			
Autophagy	Macroauto-phagy	Present	YES	YES	YES	YES
		Components	Conserved throughout species, although a single gene component in yeast often has several homologues in other species			
		Activation	Basal/ Inducible	Basal/ Inducible	Basal/ Inducible	Basal/ Inducible
		Functions	Energy source Quality control	Energy source, quality control, cellular defense, cellular remodeling, embryogenesis, etc		
	Microauto-phagy	Present	YES	YES	YES	YES
		Components	Some identified	Unknown, none conserved from yeast		
		Activation	Basal	Basal	Basal	Basal
		Functions	Organelle degradation	Continuous turnover of cytosolic components? Poorly characterized		

Table 1 (continued)

Lysosomal system: components and pathways		Yeast	Invertebrate	Plants	Mammals
Chaperone-mediated autophagy	Present	Not present			YES
	Components				Not conserved
	Activation				Basal/Inducible
	Functions				Energy Quality control Stress response
Vacuole-import degradation	Present	YES	Not present		
	Components	Not conserved			
	Activation	Inducible			
	Functions	Adaptation to refeeding			
Cytoplasm-to-vacuole targeting	Present	YES	Not present		
	Components	Not conserved			
	Activation	Basal			
	Functions	Vacuole biogenesis			

Heterophagy

Heterophagy describes the cellular internalization and degradation by the lysosome of exogenous materials and plasma membrane proteins. In this process, part of the plasma membrane invaginates and pinches off to form intracellular vesicles (endosomes or phagosomes) which then fuse with lysosomes to deliver the internalized cargo. The best characterized heterophagic or endocytic pathways are phagocytosis and pinocytosis, which differ in the type of cargo internalized and in the mechanisms and molecular components involved in this internalization (Fig. 2) [41–43]. Phagocytosis typically occurs in specialized cells such as macrophages and neutrophils, and involves the uptake of large particles such as apoptotic cells or bacteria [44]. This pathway involves a binding step, in which the substrate to be phagocyted interacts with cell surface receptors, followed by actin dependent cytoskeletal reorganization to form a phagosome around the substrate. Upon internalization the phagosome matures, through changes in the proteins that interact with its membrane, and it is targeted for fusion with the lysosome [45]. Phagocytosis in unicellular organisms, such as the protist or amoeba, is the main mechanism of cellular nutrition, however, in multicellular eukaryotes, phagocytosis has been preserved only in highly specialized cells that participate in non-specific host defense and immunity [46].

In contrast, pinocytosis, in its different forms, has been conserved through evolution in all cellular types to mediate the internalization and delivery to lysosomes of molecules and soluble components, rather than particulate structures as in

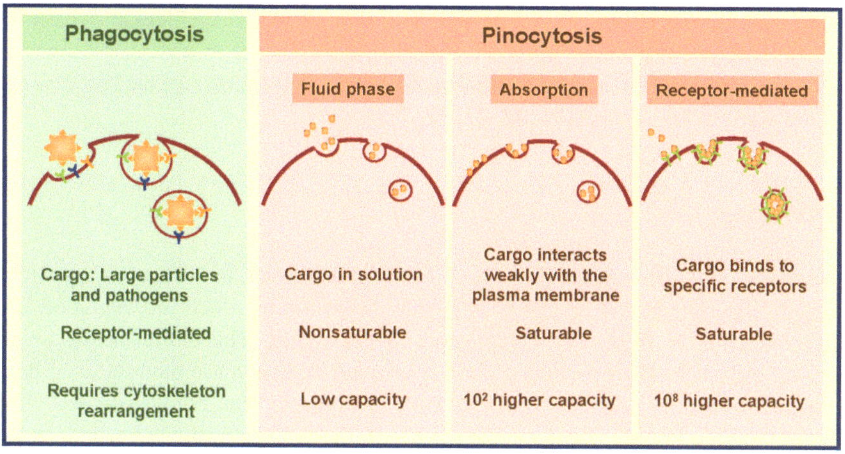

Fig. 2 Main types of endocytic degradation in lysosomes. Extracellular components and plasma membrane proteins can be targeted for degradation in lysosomes through endocytosis. There are two main types of endocytosis, phagocytosis or pinocytosis, depending on the size-nature of the internalized cargo and the requirement for major cytoskeleton reorganization during the internalization process. The three main types of pinocytosis are depicted on the right and the main characteristics that distinguish them are listed at the bottom

phagocytosis (Fig. 2) [42, 47–49]. The different types of pinocytosis – fluid phase, absorptive and receptor-mediated endocytosis – depend on the mechanism mediating cargo recognition and internalization. Whereas molecules are internalized in solution by fluid phase, they interact in a non-specific manner with the plasma membrane before internalization during absorptive endocytosis, leading this way to some degree of cargo concentration. Maximal concentration of cargo before internalization is attained through receptor-mediated endocytosis, where substrates interact selectively with specific receptors at the plasma membrane. Through lateral membrane movement, the receptor-cargo complexes concentrate in particular regions of the plasma membrane where internalization takes place. Depending on the type of pinocytosis, different structural proteins, such as clathrin and caveolin, associate to the invagination at the plasma membrane that pinches off to form clathrin-coated vesicles or caveolae [47, 50]. These caging proteins are then deassembled to release into the cytosol an early endosome, which matures into a late endosome and eventually fuses with the lysosome through similar mechanism as described for phagosomes [42]. An intermediate sorting step at the level of the late endosome often occurs to facilitate recycling of some components (i.e. receptors) back to the plasma membrane. All forms of pinocytosis occur in all prokaryotic and eukaryotic cells to allow continuous "sampling" of the extracellular environment and to capture nutrients and factors required for cell maintenance and development. In addition, in some specialized eukaryotic cells, endocytosis has been modified to fulfill particular cellular functions. For example in professional antigen presenting cells, endocytosis does not lead to complete degradation of cargo, but instead results in partial fragmentation into peptides that are then exposed at the plasma membrane through recycling. Similar is the case of endothelial cells covering the walls of blood vessels, where endocytic vesicles formed in the apical plasma membrane bypass the lysosomal system to fuse with the basal plasma membrane in a process known as transcytosis, which is used to transport nutrients from the lumen of the blood vessel to the surrounding tissues [41, 42, 47, 51].

Autophagy

Autophagy or "self-eating", the main focus of this chapter, is the process by which long-lived proteins and complete organelles from within a cell are degraded into their constitutive molecular components in lysosomes. The amino acids, lipids, sugars and nucleotides produced upon lysosomal digestion are then released back into the cytosol for further use in anabolic processes. This continuous recycling makes autophagy a very conservative process. Autophagy participates both in the continuous turnover of intracellular proteins and organelles necessary to maintain cellular homeostasis, through what is known as constitutive autophagy, and in the increased protein degradation in response to changes in the extracellular environment, known as inducible autophagy [1, 2, 52–54]. Basal and inducible autophagy co-exist in all cells, although their prevalence varies depending on cell type and cellular conditions.

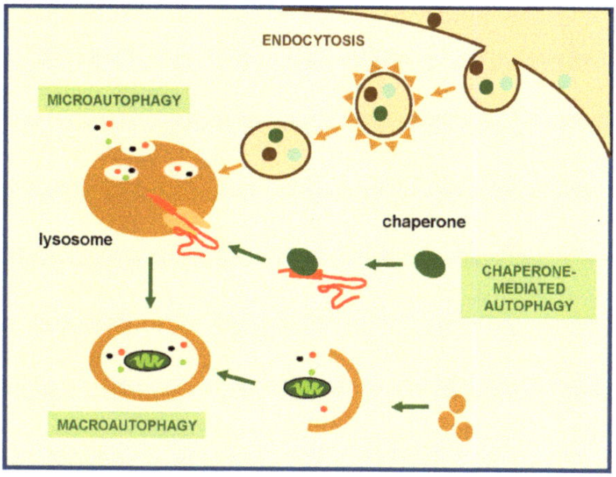

Fig. 3 Main types of autophagic degradation in lysosomes. Intracellular components can reach the lysosomal lumen for degradation by three different mechanisms in mammals: macroautophagy, microautophagy and chaperone-mediated autophagy. This last pathway has not been identified in yeast, where instead two additional mechanisms of delivery to the vacuole –*vid* and *cvt* – exist (see text for details). Endocytosis is shown in a different color to indicate its heterophagic nature

Three main types of autophagy have been described in mammalian cells, macroautophagy, microautophagy and chaperone-mediated autophagy (CMA) (Fig. 3) [15, 54]. Although alike in many respects, these autophagic pathways differ in the substrates degraded, the mechanism by which these substrates are trafficked to the lysosome, and the stimuli mediating their activation. Whereas macroautophagy and microautophagy are highly conserved forms of autophagy, present already in yeast and preserved in all mammalian cells, CMA, at least in its conventional form, appears late in evolution (essential components for this pathway are evolutionarily detected for the first time in birds) [21]. In contrast, yeast count on unique forms of autophagy that have not been conserved in other unicellular and multicellular organisms, such as the vacuole import and degradation pathway (*vid*) and the cytosol to vacuole transport (*cvt*) [15]. We provide next a brief description of each of these five types of autophagy, including a comparative analysis between those present only in particular species (Table 1). Readers interested in a more detailed description of the molecular mechanism and pathophysiology of each autophagic pathway are referred to recent comprehensive reviews [1, 2, 52–54].

Macroautophagy

Macroautophagy, first described in mammals [55], has since been observed in all known organisms, including yeast, where a detailed molecular characterization has been possible [56–58]. In this process, soluble proteins and entire organelles are sequestered "in bulk" within portions of the cytosol by a de novo formed double membrane (limiting membrane), which seals into a vesicle known as the autophagosome [59–61] (Fig. 3). The formation of the autophagosome involves a series of autophagy related proteins or Atgs that participate in the different steps of this process: nucleation, sequestration, elongation, sealing, autophagosome targeting and lysosomal fusion and degradation [62]. Atg proteins organize into different functional complexes. Two conjugation complexes regulate protein-to-protein conjugation and protein-to-lipid conjugation events required for autophagosome formation [63]. An initiation complex, whose major components Beclin-1 and VPS34, are shared with other complexes that regulate endocytosis, is required to activate autophagosome formation and elongation of the limiting membrane [64]. The fourth complex is a negative regulator of macroautophagy whose main component is the energy-sensing kinase TOR (target of rapamycin) [65, 66]. When nutrients are scarce this kinase is inactivated, thus releasing its negative effect on macroautophagy. Blockage of TOR by rapamycin treatment has the same activating effect on macroautophagy [67, 68]. Once the limiting membrane closes into an autophagosome, this double membrane vesicle fuses with the lysosome which provides the hydrolytic enzymes necessary for cargo degradation [69]. Homologues for most yeast Atg proteins have already been identified in mammals [70]. Interestingly, several mammalian variants are often found for each single yeast Atg. Further investigation is required to determine the reasons for this multiplicity of Atgs, but it is possible that yeast Atgs are multifunctional, and that these functions distribute among each of the variants with evolution. It is also possible that the multiplicity of Atg variants is a consequence of the different requirements of cell types and tissues in multicellular organisms. For example, three variants for Atg8 have been identified in mammals (LC3, GABA and GABARAB), and although the three are present in all cell types, their contribution to autophagosome formation varies in different tissues [71].

Macroautophagy is primarily induced upon nutrient deprivation in both mammals and yeast [68, 72–74] although recent studies have revealed the existence of basal macroautophagy as a continuous cellular process in most tissues. The high rates of cell death and tissue degeneration observed in mouse models conditionally knocked-out for essential autophagy genes in neural tissue [75, 76], heart [77] or liver [78] have underscored the importance of this basal form of macroautophagy in maintenance of cellular homeostasis. In contrast, the two main purposes of stress-induced macroautophagy are to preserve a positive cellular energetic balance during nutritional hardship by degrading dispensable intracellular components, and to get rid of any cellular component damaged by intra or extracellular stressors [1, 2, 52–54].

Macroautophagy is in most cases a non selective process by which large portions of the cytosol are sequestered "in bulk". However, recent studies support some level of selectivity in the macroautophagic degradation of organelles and particulate structures from the cytosol (i.e. protein aggregates). Selective forms of macroautophagy include mitophagy, ERphagy, pexophagy, aggregophagy, as well as others [79, 80].

Microautophagy

Microautophagy, also conserved from yeast to mammals, has been classically considered a constitutive form of autophagy which occurs even in the absence of typical autophagy inducers such as nutrient deprivation [81, 82]. Like macroautophagy, microautophagy involves the sequestration and degradation of whole cytosolic regions, but in this case, the engulfing membrane is the lysosomal membrane itself that invaginates or forms projections to trap both organelles and soluble proteins [82–85]. In contrast to the extensive molecular characterization of macroautophagy, little is known about microautophagy, in particular in mammals. This pathway was initially described in rat liver, based on the presence of multivesiculate lysosomes able to trap soluble cytosolic molecules (sugars, proteins, dyes) in a nonselective manner [82, 83]. Further studies in yeast revealed that full organelles, such a peroxisomes, could be directly engulfed by the vacuole upon formation of an organized membrane protrusion known as micropexophagic membrane apparatus (MIPA), which requires some of the macroautophagy Atgs but also unique microautophagy-related genes [84, 85]. Degradation of discrete nuclear regions by microautophagy has also been described in yeast. None of the unique genes identified to participate in these two processes, micropexophagy and piecemeal microautophagy of the nucleus, are conserved in mammals. Further studies are needed to determine if changes through evolution in these genes make it difficult to establish homology or if different cellular processes are responsible for peroxisome and nuclear clearance in higher order species [86, 87].

Chaperone-Mediated Autophagy (*CMA*)

Chaperone-mediated autophagy (CMA) is a selective form of autophagy for cytosolic proteins. The selectivity of this pathway stems from the fact that CMA substrate proteins all contain in their amino acid sequence a pentapeptide, biochemically related to KFERQ, which acts as a lysosomal targeting motif [88]. This motif is present in the sequence of about 30% of cytosolic proteins, but recent studies support that post-translational modifications in proteins (i.e. phosphorylation, acetylation, etc.) could generate a CMA-targeting motif in a protein missing one of the residues of the pentapeptide, thereby increasing the putative pool of CMA substrates. This pentapeptide sequence is recognized in the cytosol by the hsc70 chaperone complex, which targets substrates to the surface of the lysosome and helps them unfold to allow their translocation across the membrane [89, 90] (Fig. 3). Upon reaching the lysosomal membrane, the hsc70 complex interacts with the lysosomal associated membrane protein type 2A (LAMP-2A), which acts as a receptor for the CMA pathway [91]. Binding of substrate proteins to monomeric forms of this receptor drives its organization into a multimeric protein complex which mediates substrate translocation [92], with the assistance of a resident luminal form of hsc70 (lys-hsc70) that pulls the substrates into the lysosome for rapid degradation [93, 94].

Some level of basal CMA activity exists in all cells, but CMA is maximally activated by stressors such as oxidative stress or nutritional deprivation [95, 96]. Binding of substrate proteins to the lysosomal receptor is a rate limiting step in CMA [97, 98]. In fact, levels and dynamics of the CMA receptor, LAMP-2A, are tightly regulated and modulate changes in CMA activity [92, 99]. Translational upregulation of LAMP-2A is behind the activation of CMA during oxidative stress [100], whereas in conditions of nutrient deprivation, the increase in lysosomal levels of LAMP-2A does not depend on de novo synthesis of this protein [97, 98]. Instead, downregulation of the degradation of this receptor at the lysosomal membrane and mobilization of a luminal resident pool of LAMP-2A toward the lysosomal membrane are responsible for the increased amount of LAMP-2A present at the lysosomal membrane available to receive substrate proteins.

Vacuole Import and Degradation (*vid*)

Vacuole import and degradation (vid) pathway is a form of autophagy described in yeast that involves the transport of selective cytosolic proteins to the vacuole for their degradation in a two-step fashion [101]. So far, only two glucogenic enzymes have been identified as substrate proteins for this pathway, fructose-1,6-biphosphatase and malate dehydrogenase. Levels of both enzymes increase during starvation, but they are then targeted for degradation once a glucose source is again available [102]. Like in CMA, the *vid* substrate is recognized by chaperones that facilitate its transport in a receptor-mediated manner into small cytosolic vesicles [103]. However, in contrast to CMA, where transport is followed by degradation in the lysosomal lumen, the *vid* vesicles lack proteases, and consequently degradation of the substrate is not attained until these vesicles fuse with the vacuole [104]. Activation of *vid* usually occurs when cells in a deficient carbon source are switched to a new medium containing elevated glucose. Although some of the components that participate in *vid* are shared with other essential cellular processes and are conserved through evolution, the *vid* vesicles themselves have never been identified in organisms other than yeast and the receptor for this pathway does not seem to be evolutionarily conserved. Although *vid* recapitulates aspects of both CMA and macroautophagy, and it has been proposed that it might be the predecessor pathway of CMA in yeast (where CMA components are non-existent), there are however clear differences between these two pathways. First, the only identified *vid* substrate does not contain a CMA-targeting motif in its sequence, and consequently would not be able to interact with hsc70 for CMA degradation in mammals. In fact, the ubiquitin proteasome pathway seems the most common mechanism for degradation of this enzyme in higher organisms. In addition, activation of CMA occurs during starvation whereas *vid* is activated during refeeding. The evolutive relationship between CMA and *vid* thus requires further investigation.

The Cytosol to Vacuole Targeting Pathway (*cvt*)

The cytosol to vacuole targeting pathway (cvt) is an autophagy-like pathway exclusive to yeast by which vacuolar resident hydrolases are transported to this organelle from the cytosol [105, 106]. Despite the fact that *cvt* fits better as a biogenic pathway as opposed to a catabolic pathway, it is often included as a type of autophagy because many genes required for macroautophagy are shared by the *cvt* pathway [107, 108]. In fact, the mechanism of cargo sequestration in the cytosol and delivery to the vacuoles resemble, in many aspects, those described for macroautophagy. The same conjugation cascades described for macroautophagy are also responsible for the formation of the double membrane that sequesters *cvt* cargo, although in this case, the elongation of the membrane is negatively regulated through changes in the phosphorylation state of one of the Atgs involved in this step, resulting in a smaller sized double membrane vesicle [108, 109]. This vesicle then fuses with the vacuole, delivering the cargo into its lumen through mechanisms shared with macroautophagy. Another difference between the classic autophagosome and the double membrane *cvt* vesicle is that only enzymes of the vacuole, but not substrates, are sequestered in the latter. This selective recognition of cargo is mediated by Atg proteins exclusive for the *cvt* pathway. Delivery of enzymes to the vacuole changes from *cvt* to macroautophagy during nutrient deprivation, allowing simultaneous supply of both cargo and enzymes required for its degradation [110]. Homologues of yeast genes unique for *cvt* have not been identified in any other organism and *cvt* has only been described so far in yeast, supporting that this early close relationship between lysosomal biogenesis and its catabolic activity diverged into two completely independent pathways with evolution.

Functions of the Autophagic Pathways

Although the different types of autophagy are often described as individual entities with their own specific molecular components and regulatory characteristics, they do not function as completely independent processes. In fact, a considerable amount of cross-talk and synchronism has recently been described for some of these pathways. For example, although both macroautophagy and CMA are stress-induced pathways, they are not mutually exclusive and often work in a coordinated manner as part of the cellular response to stress. Macroautophagy represents the first line of cellular defense during starvation, in order to fulfill the cellular energetic requirements under these conditions by degrading other intracellular components. However, this random bulk degradation cannot be maintained if starvation persists beyond a certain point, as degradation of components essential for the stress response would compromise cellular survival. Thus, a gradual decrease in macroautophagy is observed during prolonged starvation (>6 h), while it is replaced by progressive upregulation of CMA [111]. The selectivity of CMA allows for the break down of non essential proteins to produce the amino acids required for cell fuelling under these conditions. Likewise, activation of both CMA and macroautophagy also occur under conditions resulting in cellular damage, such as oxidative stress, heat shock, or exposure to UV.

Overall, the functions of lysosomal degradation recapitulate those described for protein degradation in the previous section. Basal autophagy activity is essential for maintenance of cellular homeostasis by guaranteeing the continuous renewal of the proteome [1]. This continuous lysosomal turnover makes this organelle a center for cellular recycling and an additional source of essential macromolecules required for cellular anabolic processes [2].

Lysosomal degradation is also an essential component of the cellular response to stress. Although autophagic activation was first described under conditions of nutritional stress in mammals, it has become evident that this pro-survival role of autophagy as a recycler and energy provider during nutritional stress is a generalized response to low nutrient intake both in multicellular and unicellular organisms. In fact, the notion that activation of autophagy was required for yeast survival when switched to a low nitrogen source constituted the basis of a genetic screen for mutants defective in autophagy [56–58]. Autophagy has also proven to be a defense mechanism against a growing number of intra- and extracellular stressors. The unique function of autophagy in organelle turnover has been shown to be critical under conditions resulting in organelle damage or stress such as mitochondria depolarization and endoplasmic reticulum stress [79, 80]. Furthermore, the high hydrolytic capabilities of lysosomes makes them essential in the destruction of extracellular pathogens that might otherwise compromise cellular viability [112].

In addition to these general functions of autophagy in the maintenance of cellular homeostasis and the cellular response to stress, there are also individual functions unique for each of the autophagic pathways. Thus for example, macroautophagy has been shown to participate in the control of lipid cellular content due to its ability to sequester lipid droplets, the main intracellular storage of lipids [113]. Likewise, CMA has been proposed to contribute to the presentation of antigens that originate from partial cleavage of intracellular cytosolic components [114]. As our understanding of the individual characteristics of each of these lysosomal pathways grows, new cellular functions may become associated to various lysosomal compartments.

Lysosomes and Aging

Alterations in the lysosomal system with age have been described in almost all organisms and tissues. In fact, the lysosome has often been considered an "additional source of cellular damage" in aging and under extreme stress conditions because alterations in the stability of the lysosomal membrane result in the cytosolic release of potent lysosomal hydrolases and the associated unregulated destruction of intracellular components (reviewed in [115]). However, in recent years, there has been a growing interest in understanding the consequences of the functional failure of this system in aging organisms from the point of view of its critical role in maintenance of cellular homeostasis and quality control.

Protein Degradation and Aging

Alterations in protein homeostasis and intracellular accumulation of altered and damaged proteins with age are well documented [116–120]. One of the plausible explanations for the increased amount of damaged proteins in aging tissues is their higher rate of production as a result, for example, of increased production of free radicals with age. Oxidative damage often leads to protein cross-linking rendering them resistant to degradation. Post-translational modifications such as amino acid side chain or sulfydryl group oxidation, aspartyl and asparaginyl residue racemization, or asparaginyl and glutaminyl residue deamidation have been shown to increase with age [121]. In addition, accumulation of these damaged products results in part from their inefficient handling by the quality control mechanisms in aging tissues that include both protein repair mechanisms and protein degradation systems [12, 117].

Decreased protein degradation and the consequent accumulation of altered proteins and organelles are particularly detrimental in non-dividing differentiated cells, such as neurons and cardiomyocites. In these cells, the lack of division prevents the dilution of damage between the two daughter cells, and progressive aberrant protein accumulation can result in degeneration and cell death [1, 122, 123]. In fact, alterations in the proteolytic systems have been proposed to underline the pathogenesis of neurodegenerative diseases such as Alzheimer's and Parkinson's diseases, where failure to remove pathogenic proteins accelerates their accumulation in affected neurons and consequently results in cellular failure (reviewed in [123, 124]). Both the UPS and the lysosomal system have been shown to undergo major age-dependent changes. Due to the emphasis of this chapter on lysosomes, we will not describe here the effect of aging on the UPS, but readers are directed to recent reviews on this topic [22–24, 125].

Primary Changes in Lysosomes with Age

Detailed electron microscopy analysis of different tissues in old organism have revealed common morphological changes in their lysosomal system such as increased area, elongated shape, altered density and accumulation of undigested substrates in the form of an autofluorescent pigment called lipofuscin [126, 127]. Lipofuscin is made up of lipids, carbohydrates and aldehyde-cross-linked proteins that form an uncatabolizable material inside lysosomes [127]. The accumulation of lipofuscin inside lysosomes increases their susceptibility to oxidative damage, disrupts the lysosomal pH gradient and alters the membrane permeability often leading to leakage [127]. Lysosomal storage diseases (LSD) provide insight as to how the accumulation of undegraded products inside lysosomes affects both their function as well as other cellular functions depending on lysosomal activity [128]. LSDs are caused by primary deficits of a particular lysosomal hydrolase which result in accumulation of the corresponding undegraded substrate inside lysosomes. In most cases, the undigested material alters the chemical and physical properties of the lysosomal lumen and eventually causes leakage of lysosomal components into the cytosol [128, 129]. All lysosomal pathways are indirectly affected by these changes in the lysosomal compartment, as for example, proper lysosomal pH is required for autophagosome/lysosomal fusion in macroautophagy, and to preserve the stability of the luminal chaperone required for CMA [130, 131]. In addition, as reviewed in the following sections, primary changes in the different autophagic pathways have also been detected in aging organisms.

Changes in Autophagy with Age

Macroautophagy and CMA activity decrease with age, in part due to the aforementioned changes in lysosomal morphology, stability and proteolytic capability, but also because of primary defects in components of these autophagic pathways (reviewed in [21, 132]). Studies in old rodent livers have revealed alterations in the hormonal regulation of macroautophagy. Induction of autophagy during starvation in liver is attained by the combined effect of reduced levels of insulin – a negative regulator of this pathway – and increased levels of glucagon, a known activator of macroautophagy. The inability of old rodent livers to maximally activate macroautophagy results from the combined effect of lower stimulatory effect of glucagon and persisting inhibition by insulin-independent signaling through the insulin receptor even under starvation conditions [133–136]. In addition to the problems in activation, the inability of autophagosomes to efficiently fuse with lysosomes also contributes to the reduced levels of macroautophagy in old organisms [130]. Although lysosomal accumulation of lipofuscin has been proposed to contribute to poor autophagosome clearance, the reasons for this failure are still under investigation [130, 137].

A decline in CMA activity with age has also been described in different mammalian tissues. Degradation of cytosolic proteins through this pathway decreases due to lower levels of LAMP-2A in the membrane of lysosomes in old organisms [138]. This decrease in LAMP-2A, observed even before changes in CMA activity are detectable, is initially compensated for by an increase in the number of lysosomes able to perform CMA inside cells [138]. Eventually, this compensatory mechanism is insufficient and the CMA deficit becomes evident. The decrease in LAMP-2A levels does not result from transcriptional downregulation or altered splicing of this variant of the single *lamp2* gene, but instead is attributable to problems in the stability of the receptor protein once at the lysosomal membrane [139]. The organized dynamic distribution of LAMP-2A in and out of the lysosomal membrane microdomains, which is necessary to regulate its functionality and renewal, is altered as organisms age, resulting in abnormal unregulated degradation of this receptor in the lysosomal lumen [139]. The slow turnover of LAMP-2A in older organisms allows the receptor to be exposed to the harsh and potentially damaging cellular environment for longer time than normal, which increases its propensity to become damaged [139, 140].

Reduced levels of LAMP-2A seem to be a primary defect, or at least the most important factor that contributes to the decline of CMA with age, since our laboratory has recently restored normal levels of LAMP-2A in the livers of old rodents through genetic manipulation, and this intervention has proven sufficient to prevent the age-dependent decline in CMA activity with age [141]. These studies have also underlined the importance of CMA in maintenance of cellular homeostasis and as part of the response to stress, since the livers from the old animals with restored CMA activity presented lower levels of damaged intracellular proteins and higher rates of cellular survival in response to experimentally inflicted hepatotoxicity [141].

Autophagy and Longevity

The first connections between autophagic activity and longevity resulted from studies in livers of caloric restricted animals, the only intervention that has been shown to significantly slow aging and increase life-span in mammals [142]. Caloric restriction also slows down the decline in macroautophagic activity with age and constitutively activates CMA. Rather than an increase in maximal activation of macroautophagy, caloric restriction restores the hormonal responsiveness of this pathway [135, 143, 144]. This could be attained in part by decreasing the glycolytic flux which normally occurs between periods of food consumption and fasting, thereby allowing cells to shift between glycolysis and fat metabolism [121]. Caloric restriction also downregulates TOR, which should favor more efficient activation of autophagy. The initial beneficial effect of caloric restriction on autophagy in liver has been confirmed now in other organs such as heart, which could be partially responsible for the cardioprotective function of caloric restriction [145].

The first genetic evidence linking longevity and autophagy was obtained in *Caenorhabditis elegan (C. elegans)*, an experimental model in which different mutations in the insulin signaling pathway have been shown to extend life-span (Fig. 4). Thus, mutation of *daf-2* – homologue to the mammalian insulin receptor in worms – results in an estimated 300% increase in lifespan [146, 147]. The fact that genetic blockage of autophagy in the *daf-2* mutant prevented their increased life-span supports the need of properly functioning autophagic activity in order to attain full life-span extension [148]. A similar requirement for autophagy was later proven to be necessary for the increase in life-span obtained through genetic manipulations in other components of the insulin signaling pathway or by caloric restriction [149, 150]. In fact, feeding defective mutant nematodes (*eat-2*) and flies have increased levels of autophagy [151], and blockage of essential autophagy genes reduces life-span in both worms and flies. However, the fact that activation of autophagy alone is not enough to increase life-span in most worm models suggests that autophagy is necessary to extend life-span but it is not sufficient. Thus, autophagy probably works in concert with other downstream signaling pathways such as the insulin/IGF-1 pathway [152].

Surprisingly, although activation of autophagy does not seem sufficient to increase life-span in worms, it has been shown to be sufficient in flies where over-expression of the *ATG8a* autophagy gene results in a 56% increase in longevity,

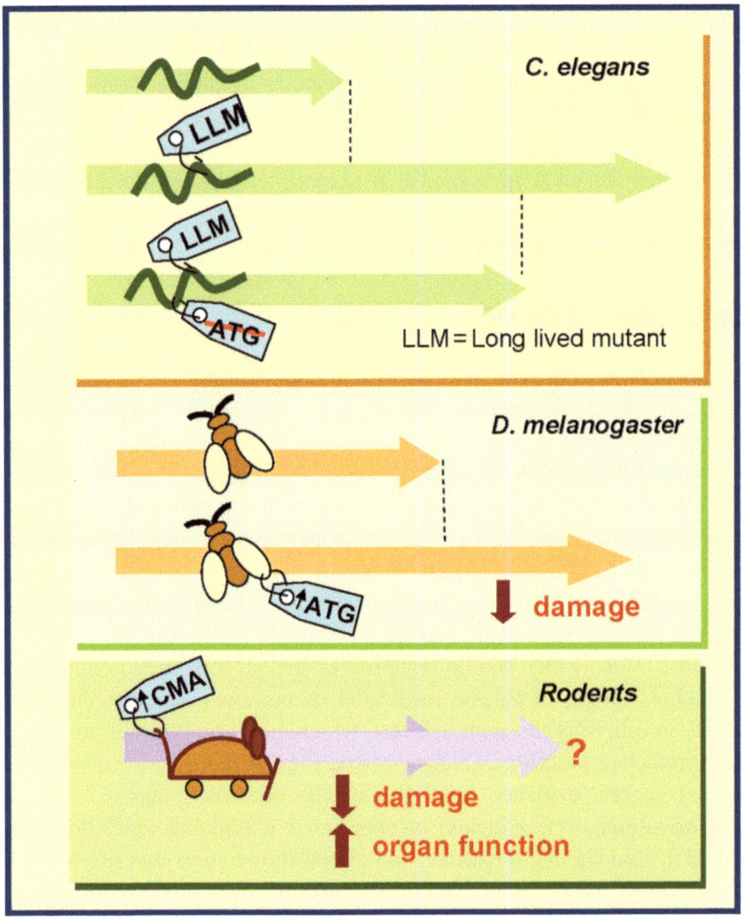

Fig. 4 Effect of changes in autophagy on longevity in different species. Studies in worms have revealed a need for macroautophagy in order to attain the maximal life-extension induced by mutations in the insulin signaling pathway or by caloric restriction. However, an increase in macroautophagy is not sufficient to increase life-span. In contrast, activation of macroautophagy (by overexpressing ATG genes) increases life-span in wild type flies. There is currently no information available on the effect of changes in macroautophagy on life-span of mammals, but restoration of normal chaperone-mediated autophagy (CMA) activity in old rodents has been shown to enhance cellular homeostasis and organ function

a greater resistance to oxidative stress, and decreased intracellular damage due to oxidizing proteins (Fig. 4) [153]. Whether this is due to specie specific differences, or to different experimental conditions will require future investigation. In addition, there has not yet been any effort to repair the macroautophagic defect in old rodents, which makes it impossible to predict whether maintenance of normal macroautophagic activity until later in life might preserve organ function, as was recently shown to be the case for CMA (Fig. 4) [140].

Lysosomal Dysfunction in Age-Related Pathologies

A detailed discussion of the multiple connections between altered lysosomal activity and disease is beyond the scope of this chapter. Instead, we briefly summarize here recent evidence supporting the contribution of lysosomal dysfunction to common age related disorders, as this provides a new conceptual context for the aggravating effect of aging in the progression of these pathologies.

The best studied examples are common neurodegenerative diseases characterized by intracellular protein aggregation of particular pathogenic proteins. In the case of Parkinson's Disease, pathogenic forms of alpha-synuclein accumulate inside affected neurons as toxic aggregates known as Lewy bodies. The inability of CMA to degrade the abnormal forms of this protein seems to contribute to their intracellular deposition [8, 154]. Furthermore, the abnormal interaction of pathogenic alpha-synucleins with LAMP-2A at the lysosomal membrane blocks CMA degradation of other cytosolic proteins, rendering the cells susceptible to stressors against which proper CMA activity is needed. Further reduction in the activity of an already impaired CMA due to the age-dependent decrease of this pathway could explain the progressive aggravation of the disease with age.

Similarly, changes in autophagy with age may also contribute to the higher incidence of cancer in aging organisms. An anti-oncogenic role has been proposed for autophagy and is supported by the fact that different tumor suppressor genes such as *PTEN, TSC1/TSC2*, and *p53*, which all inhibit TOR, function to stimulate autophagy [155, 156]. Furthermore, blockage of autophagy in certain types of cancer favors their progression by allowing rapid cell division. However, the role of autophagy in cancer progression may be more complicated and consideration should be paid to both changes in autophagy in the host as well as the tumor. Thus, upregulation of autophagy in cancer cells is beneficial for tumor survival in low nutrient conditions (such as the hypovascularized center of the tumor), and confers cancer cells resistance against anti-cancer treatments [155–157]. Although future investigation is required, it is possible that declined autophagy in the aging host cells of old individuals but not in the cancer cells could explain the higher rates of cancer in old individuals.

Despite the well characterized involvement of autophagy in the cellular immune response and defense against infection, different pathogens have evolved to not only circumvent their "autophagic death" but also use the cell's autophagic machinery to further their own pro-survival agenda [112]. For example, the bacteria *Staphylococcus aureus*, infiltrates autophagosomes and hinders their ability to fuse with lysosomes for degradation, at the same time using the nutrients sequestered in the autophagosome to enhance their replicative abilities [158]. The decline in the activity of the autophagic system with age could foster invasion and proliferation of pathogens by reducing the cellular defenses, thus contributing to the higher propensity of old organisms for infection.

Although still poorly explored, autophagic dysfunction has also been linked to some common metabolic disorders whose incidence increases dramatically with age such as diabetes. The high circulating glucose levels in diabetic patients induces persistent activation of mammalian TOR (mTOR) and consequent inhibition of macroautophagy [159]. In fact, administration of the inhibitor of mTOR, rapamycin, to patients suffering from diabetic nephropathy might represent an exciting new potential drug target [160, 161].

Concluding Remarks

Lysosomal degradation is a highly conserved intracellular pathway that serves mainly catabolic purposes. However, this degradation, or better yet degradation/recycling of intra-and extracellular components in the lysosome is not only necessary for maintenance of a positive cellular energetic balance, but also fulfils many other cellular needs. The lysosomal system contributes to the maintenance of cellular homeostasis, quality control, removal of damaged cellular components, defense against intra- and extracellular pathogens and major cellular remodeling. This multifunctional nature of the lysosomal system provides an explanation as to why alterations in lysosomes, such as those described in aging, have drastic cellular consequences and often lead to disease. On the other hand, the complex interrelation of the different lysosomal pathways is probably the reason why restoration of the activity of only one of these pathways in old organisms has a major beneficial effect in overall lysosomal function. These findings provide rational support for anti-aging interventions aimed at repairing or preventing the functional decline of this system.

References

1. Mizushima N, Levine B, Cuervo A et al. (2008). Autophagy fights disease through cellular self-digestion. *Nature* 451: 1069–1075.
2. Mizushima N and Klionsky D (2007). Protein turnover via autophagy: implications for metabolism. *Annu Rev Nutr* 27: 19–40.
3. Goldberg AL (2007). Functions of the proteasome: from protein degradation and immune surveillance to cancer therapy. *Biochem Soc Trans* 35: 12–17.
4. Lecker SH, Goldberg AL, and Mitch WE (2006). Protein degradation by the ubiquitin-proteasome pathway in normal and disease states. *J Am Soc Nephrol* 17: 1807–1819.
5. Martinez-Vicente M, Sovak G, and Cuervo AM (2005). Protein degradation and aging. *Exp Gerontol* 40: 622–633.
6. Ward W (2002). Protein degradation in the aging organism. *Prog Mol Subcell Biol* 29: 35–42.
7. Squier T (2001). Oxidative stress and protein aggregation during biological aging. *Exp Gerontol* 36: 1539–1550.
8. Cuervo AM, Stefanis L, Fredenburg R, et al. (2004). Impaired degradation of mutant alpha-synuclein by chaperone-mediated autophagy. *Science* 305: 1292–1295.
9. Goldberg AL (2003). Protein degradation and protection against misfolded or damaged proteins. *Nature* 18: 895–899.
10. Balch WE, Morimoto RI, Dillin A, et al. (2008). Adapting proteostasis for disease intervention. *Science* 319: 916–919.
11. Finkbeiner S, Cuervo AM, Morimoto RI, et al. (2006). Disease-modifying pathways in neurodegeneration. *J Neurosci* 26: 10349–10357.
12. Morimoto RI (2008). Proteotoxic stress and inducible chaperone networks in neurodegenerative disease and aging. *Genes Dev* 22: 1427–1438.
13. Kopito RR and Sitia R (2000). Aggresomes and Russell bodies. Symptoms of cellular indigestion? *EMBO Rep* 1: 225–231.

14. Ciechanover A (2006). Intracellular protein degradation: from a vague idea through the lysosome and the ubiquitin-proteasome system and onto human diseases and drug targeting. *Hematology Am Soc Hematol Educ Program* 1: 505–6.
15. Cuervo AM (2004). Autophagy: many pathways to the same end. *Mol Cell Biochem* 263: 55–72.
16. Mizushima N (2005). The pleiotropic role of autophagy: from protein metabolism to bactericide. *Cell Death Differ* 12: 1535–1541.
17. Tsukamoto S, Kuma A, Murakami M, et al. (2008). Autophagy is essential for preimplantation development of mouse embryos. *Science* 321: 117–120.
18. Fimia GM, Stoykova A, Romagnoli A, et al. (2007). Ambra1 regulates autophagy and development of the nervous system. *Nature* 447: 1121–1125.
19. Levine B and Klionsky D (2004). Development by self-digestion: molecular mechanisms and biological functions of autophagy. *Dev Cell* 6: 463–477.
20. Yun C, Stanhill A, Yang Y, et al. (2008). Proteasomal adaptation to environmental stress links resistance to proteotoxicity with longevity in Caenorhabditis elegans. *Proc Natl Acad Sci USA* 105: 7094–7099.
21. Cuervo AM (2008). Autophagy and Aging: keeping that old broom working. *Trends Genet* 24: 604–612.
22. Chondrogianni N and Gonos ES (2008). Proteasome activation as a novel antiaging strategy. *IUBMB Life* 60: 651–655.
23. Hwang JS, Chang I, and Kim S (2007). Age-associated decrease in proteasome content and activities in human dermal fibroblasts: restoration of normal level of proteasome subunits reduces aging markers in fibroblasts from elderly persons. *J Gerontol A Biol Sci Med Sci* 62: 490–499.
24. Farout L and Friguet B (2006). Proteasome function in aging and oxidative stress: implications in protein maintenance failure. *Antioxid Redox Signal* 8: 205–216.
25. Pickart CM and Cohen RE (2004). Proteasomes and their kin: proteases in the machine age. *Nat Rev Mol Cell Biol* 5: 177–187.
26. Ciechanover A (2005). Proteolysis: from the lysosome to ubiquitin and the proteasome. *Nat Rev Mol Cell Biol* 6: 79–87.
27. Ravid T and Hochstrasser M (2008). Diversity of degradation signals in the ubiquitin-proteasome system. *Nat Rev Mol Cell Biol* 9: 679–690.
28. Hanna J and Finley D (2007). A proteasome for all occasions. *FEBS Lett* 581: 2854–2861.
29. Zhou P (2005). Targeted protein degradation. *Curr Opin Chem Biol* 9: 51–55.
30. De Duve C and Wattiaux R (1966). Functions of lysosomes. [Review]. *Ann Rev Physiol* 28: 435–492.
31. Weisman LS (2003). Yeast vacuole inheritance and dynamics. *Annu Rev Genet* 37: 435–460.
32. Bowers K and Stevens TH (2005). Protein transport from the late Golgi to the vacuole in the yeast Saccharomyces cerevisiae. *Biochim Biophys Acta* 1744: 438–454.
33. Weisman LS (2006). Organelles on the move: insights from yeast vacuole inheritance. *Nat Rev Mol Cell Biol* 7: 243–252.
34. Thumm M (2000). Structure and function of the yeast vacuole and its role in autophagy. *Microsc Res Tech* 51: 563–572.
35. Pruyne D, Legesse-Miller A, Gao L, et al. (2004). Mechanisms of polarized growth and organelle segregation in yeast. *Annu Rev Cell Dev Biol* 20: 559–591.
36. Sekito T, Fujiki Y, Ohsumi Y, et al. (2008). Novel families of vacuolar amino acid transporters. *IUBMB Life* 60: 519–525.
37. Bassham DC and Raikhel NV (2000). Unique features of the plant vacuolar sorting machinery. *Curr Opin Cell Biol* 12: 491–495.
38. Bassham DC, Laporte M, Marty F, et al. (2006). Autophagy in development and stress responses of plants. *Autophagy* 2: 2–11.
39. Mo B, Tse YC, and Jiang L (2006). Plant prevacuolar/endosomal compartments. *Int Rev Cytol* 253: 95–129.

40. Muntz K (2007). Protein dynamics and proteolysis in plant vacuoles. *J Exp Bot* 58: 2391–2407.
41. Seaman MN (2008). Endosome protein sorting: motifs and machinery. *Cell Mol Life Sci* 65: 2842–2858.
42. Cullen PJ (2008). Endosomal sorting and signalling: an emerging role for sorting nexins. *Nat Rev Mol Cell Biol* 9: 574–582.
43. Geldner N and Jurgens G (2006). Endocytosis in signalling and development. *Curr Opin Plant Biol* 9: 589–594.
44. Besterman JM and Low RB (1983). Endocytosis: a review of mechanisms and plasma membrane dynamics. *Biochem J* 210: 1–13.
45. Marshall JG, Booth JW, Stambolic V, et al. (2001). Restricted accumulation of phosphatidylinositol 3-kinase products in a plasmalemmal subdomain during Fc gamma receptor-mediated phagocytosis. *J Cell Biol* 153: 1369–1380.
46. Desjardins M, Houde M, and Gagnon E (2005). Phagocytosis: the convoluted way from nutrition to adaptive immunity. *Immunol Rev* 207: 158–165.
47. Rappoport JZ (2008). Focusing on clathrin-mediated endocytosis. *Biochem J* 412: 415–423.
48. Casey CA, Lee SM, Aziz-Seible R, et al. (2008). Impaired receptor-mediated endocytosis: its role in alcohol-induced apoptosis. *J Gastroenterol Hepatol* 23(Suppl 1): S46–S49.
49. D'Hondt K, Heese-Peck A, and Riezman H (2000). Protein and lipid requirements for endocytosis. *Ann Rev Genetics* 34: 255–295.
50. Saksena S, Sun J, Chu T, et al. (2007). ESCRTing proteins in the endocytic pathway. *Trends Biochem Sci* 32: 561–573.
51. Predescu SA, Predescu DN, and Malik AB (2007). Molecular determinants of endothelial transcytosis and their role in endothelial permeability. *Am J Physiol Lung Cell Mol Physiol* 293: L823–L842.
52. Levine B (2005). Eating oneself and uninvited guests: autophagy-related pathways in cellular defense. *Cell* 120: 159–162.
53. Klionsky DJ (2005). The molecular machinery of autophagy: unanswered questions. *J Cell Sci* 118: 7–18.
54. Cuervo AM (2004). Autophagy: in sickness and in health. *Trends Cell Biol* 14: 70–77.
55. Mortimore G, Miotto G, Venerando R, et al. (1996). Autophagy. *Biochemistry* 27: 93–135.
56. Tsukada M and Ohsumi M (1993). Isolation and characterization of autophagy-defective mutants of *Saccharomyces cerevisiae*. *FEBS Lett* 333: 169–174.
57. Thumm M (1994). Isolation of autophagocytosis mutants of *Saccharomyces cerevisiae*. *FEBS Lett* 349: 275–280.
58. Harding TM, Hefner-Gravink A, Thumm M, et al. (1996). Genetic and phenotypic overlap between autophagy and the cytoplasm to vacuole protein targeting pathway. *J Biol Chem* 271: 17621–17624.
59. Mizushima N, Ohsumi Y, and Yoshimori T (2002). Autophagosome formation in mammalian cells. *Cell Struct Funct* 27: 421–429.
60. Seglen PO, Berg TO, Blankson H, et al. (1996). Structural aspects of autophagy. *Advanc Exp Med Biol* 389: 103–111.
61. Noda T, Suzuki K, and Ohsumi Y (2002). Yeast autophagosomes: de novo formation of a membrane structure. *Trends Cell Biol* 12: 231–235.
62. Yorimitsu T and Klionsky DJ (2005). Autophagy: molecular machinery for self-eating. *Cell Death Differ* 12: 1542–1552.
63. Ohsumi Y (2001). Molecular dissection of autophagy: two ubiquitin-like systems. *Nat Rev Mol Cell Biol* 2: 211–216.
64. Tassa A, Roux MP, Attaix D, and Bechet DM (2003). Class III phosphoinositide 3-kinase-Beclin1 complex mediates the amino acid-dependent regulation of autophagy in C2C12 myotubes. *Biochem J* 376: 577–586.
65. Mordier S, Deval C, Bechet D, et al. (2000). Leucine limitation induces autophagy and activation of lysosome-dependent proteolysis in C2C12 myotubes through a

mammalian target of rapamycin-independent signaling pathway. *J Biol Chem* 275: 29900–29906.
66. Kanazawa T, Taneike I, Akaishi R, et al. (2004). Amino acids and insulin control autophagic proteolysis through different signaling pathways in relation to mTOR in isolated rat hepatocytes. *J Biol Chem* 279: 8452–8459.
67. Kadowaki M and Kanazawa T (2003). Amino acids as regulators of proteolysis. *J Nutr* 133: 2052S–2056S.
68. Meijer AJ and Codogno P (2004). Regulation and role of autophagy in mammalian cells. *Int J Biochem Cell Biol* 36: 2445–2462.
69. Dunn W (1994). Autophagy and related mechanisms of lysosome-mediated protein degradation. *Trends Cell Biol* 4: 139–143.
70. Klionsky D, Cregg J, Dunn WJ, et al. (2003). A unified nomenclature for yeast autophagy-related genes. *Dev Cell* 539–545.
71. Tanida I, Tanida-Miyake E, Komatsu M, et al. (2002). Human Apg3p/Aut1p homologue is an authentic E2 enzyme for multiple substrates, GATE-16, GABARAP, and MAP-LC3, and facilitates the conjugation of hApg12p to hApg5p. *J Biol Chem* 277: 13739–13744.
72. Mortimore GE and Poso AR (1987). Intracellular protein catabolism and its control during nutrient deprivation and supply. *Ann Rev Nutri* 7: 539–564.
73. Blommaart EF, Luiken JJ, and Meijer AJ (1997). Autophagic proteolysis: control and specificity. *Histochem J* 29: 365–385.
74. Kuma A, Hatano M, Matsui M, et al. (2004). The role of autophagy during the early neonatal starvation period. *Nature* 432: 1032–1036.
75. Hara T, Nakamura K, Matsui M, et al. (2006). Suppression of basal autophagy in neural cells causes neurodegenerative disease in mice. *Nature* 441: 885–889.
76. Komatsu M, Waguri S, Chiba T, et al. (2006). Loss of autophagy in the central nervous system causes neurodegeneration in mice. *Nature* 441: 880–884.
77. Nakai A, Yamaguchi O, Takeda T, et al. (2007). The role of autophagy in cardiomyocites in the basal state and in response to hemodynamic stress. *Nat Med* 13: 619–624.
78. Komatsu M, Waguri S, Ueno T, Iwata J, Murata S, Tanida I, Ezaki J, Mizushima N, Ohsumi Y, Uchiyama Y, Kominami E, Tanaka K, and Chiba T (2005). Impairment of starvation-induced and constitutive autophagy in Atg7-deficient mice. *J Cell Biol* 169: 425–434.
79. Bernales S, McDonald KL, and Walter P (2006). Autophagy counterbalances endoplasmic reticulum expansion during the unfolded protein response. *PLoS Biol* 4: e423.
80. Lemasters J (2005). Selective mitochondrial autophagy, or mitophagy, as a targeted defense against oxidative stress, mitochondrial dysfunction, and aging. *Rejuvenation Res* 8: 3–5.
81. Mortimore GE, Lardeux BR, and Adams CE (1988). Regulation of microautophagy and basal protein turnover in rat liver. Effects of short-term starvation. *J Biol Chem* 263: 2506–2512.
82. Ahlberg J, Marzella L, and Glaumann H (1982). Uptake and degradation of proteins by isolated rat liver lysosomes. Suggestion of a microautophagic pathway of proteolysis. *Lab Invest* 47: 523–532.
83. Marzella L, Ahlberg J, and Glaumann H (1981). Autophagy, heterophagy, microautophagy and crinophagy as the means for intracellular degradation. *Virchows Archiv B Cell Pathol* 36: 219–234.
84. Mukaiyama H, Oku M, Baba M, et al. (2002). Paz2 and 13 other PAZ gene products regulate vacuolar engulfment of peroxisomes during micropexophagy. *Genes Cells* 7: 75–90.
85. Veenhuis M, Salomons FA, and Van Der Klei IJ (2000). Peroxisome biogenesis and degradation in yeast: a structure/function analysis. *Microsc Res Tech* 51: 584–600.
86. Roberts P, Moshitch-Moshkovitz S, Kvam E, et al. (2003). Piecemeal microautophagy of nucleus in Saccharomyces cerevisiae. *Mol Biol Cell* 14: 129–141.
87. Farre JC and Subramani S (2004). Peroxisome turnover by micropexophagy: an autophagy-related process. *Trends Cell Biol* 14: 515–523.

88. Dice J (1990). Peptide sequences that target cytosolic proteins for lysosomal proteolysis. *Trends Biochem Sci* 15: 305–309.
89. Chiang H, Terlecky S, Plant C, et al. (1989). A role for a 70 kDa heat shock protein in lysosomal degradation of intracellular protein. *Science* 246: 382–385.
90. Agarraberes F and Dice JF (2001). A molecular chaperone complex at the lysosomal membrane is required for protein translocation. *J Cell Sci* 114: 2491–2499.
91. Cuervo A and Dice J (1996). A receptor for the selective uptake and degradation of proteins by lysosomes. *Science* 273: 501–503.
92. Bandhyopadhyay U, Kaushik S, Vartikovsky L, et al. (2008). Dynamic organization of the receptor for chaperone-mediated autophagy at the lysosomal membrane. *Mol Cell Biol* 28: 5747–5763.
93. Agarraberes F, Terlecky S, and Dice J (1997). An intralysosomal hsp70 is required for a selective pathway of lysosomal protein degradation. *J Cell Biol* 137: 825–834.
94. Cuervo A, Dice J, and Knecht E (1997). A lysosomal population responsible for the hsc73-mediated degradation of cytosolic proteins in lysosomes. *J Biol Chem* 272: 5606–5615.
95. Dice J (2007). Chaperone-mediated autophagy. *Autophagy* 3: 295–299.
96. Massey A, Zhang C, and Cuervo A (2006). Chaperone-mediated autophagy in aging and disease. *Curr Top Dev Biol* 73: 205–235.
97. Cuervo A and Dice J (2000). Unique properties of lamp2a compared to other lamp2 isoforms. *J Cell Sci* 113: 4441–4450.
98. Cuervo A and Dice J (2000). Regulation of lamp2a levels in the lysosomal membrane. *Traffic* 1: 570–583.
99. Kaushik S, Massey AC, and Cuervo AM (2006). Lysosome membrane lipid microdomains: novel regulators of chaperone-mediated autophagy. *EMBO J* 25: 3921–3933.
100. Kiffin R, Christian C, Knecht E, et al. (2004). Activation of chaperone-mediated autophagy during oxidative stress. *Mol Biol Cell* 15: 4829–4840.
101. Chiang H, Schekman R, and Hamamoto S (1996). Selective uptake of cytosolic, peroxisomal and plasma membrane proteins into the yeast lysosome for degradation. *J Biol Chem* 271: 9934–9941.
102. Hung GC, Brown CR, Wolfe AB, et al. (2004). Degradation of the gluconeogenic enzymes fructose-1,6-bisphosphatase and malate dehydrogenase is mediated by distinct proteolytic pathways and signaling events. *J Biol Chem* 279: 49138–49150.
103. Huang PH and Chiang HL (1997). Identification of novel vesicles in the cytosol to vacuole protein degradation pathway. *J Cell Biol* 136: 803–810.
104. Brown CR, Liu J, Hung GC, et al. (2003). The Vid vesicle to vacuole trafficking event requires components of the SNARE membrane fusion machinery. *J Biol Chem* 278: 25688–25699.
105. Harding TM (1995). Isolation and characterization of yeast mutants in the cytoplasm to vacuoles protein targeting pathway. *J Cell Biol* 131: 591–602.
106. Khalfan WA and Klionsky DJ (2002). Molecular machinery required for autophagy and the cytoplasm to vacuole targeting (Cvt) pathway in S. cerevisiae. *Curr Opin Cell Biol* 14: 468–475.
107. Kim J, Dalton VM, Eggerton KP, et al. (1999). Apg7p/Cvt2p is required for the cytoplasm-to-vacuole targeting, macroautophagy, and peroxisome degradation pathways. *Mol Biol Cell* 10: 1337–1351.
108. Guan J, Stromhaug P, George M, et al. (2001). Cvt18/Gsa12 is required for cytoplasm-to-vacuole transport, pexophagy, and autophagy in Saccharomyces cerevisiae and Pichia pastoris. *Mol Biol Cell* 12: 3821–3838.
109. Huang WP and Klionsky DJ (2002). Autophagy in yeast: a review of the molecular machinery. *Cell Struct Funct* 27: 409–420.
110. Qin ZH, Wang Y, Kegel KB, et al. (2003). Autophagy regulates the processing of amino terminal huntingtin fragments. *Hum Mol Genet* 12: 3231–3244.

111. Massey AC, Kaushik S, Sovak G, et al. (2006). Consequences of the selective blockage of chaperone-mediated autophagy. *Proc Nat Acad Sci USA* 103: 5905–5910.
112. Levine B and Deretic V (2007). Unveiling the roles of autophagy in innate and adaptive immunity. *Nat Rev Immunol* 7: 767–777.
113. Singh R, Kaushik S, Wang Y, et al. (2009). Autophagy regulates lipid metabolism. *Nature*, 458: 1131–1135.
114. Zhou D, Li P, Lin Y, et al. (2005). Lamp-2a facilitates MHC class II presentation of cytoplasmic antigens. *Immunity* 22: 571–581.
115. Kiffin R, Bandyopadhyay U, and Cuervo A (2006). Oxidative stress and autophagy. *Antioxid Redox Signal* 8: 152–162.
116. Stadtman E (2001). Protein oxidation in aging and age-related diseases. *Ann NY Acad Sci* 928: 22–38.
117. Ryazanov A and Nefsky B (2002). Protein turnover plays a key role in aging. *Mech Ageing Dev* 123: 207–213.
118. Gershon H and Gershon D (1970). Detection of inactive molecules in aging organisms. *Nature* 227: 1214–1217.
119. Miquel J, Tapperl A, Dillard C, et al. (1974). Fluorescent products and lysosomal components in aging. *Drosophila Melanogaster J Gerontol* 29: 622–637.
120. Goldstein S, Stotland D, and Cordeiro R (1976). Decreased proteolysis and increased amino acid efflux in aging human fibroblasts. *Mech Ageing Dev* 5: 221–233.
121. Hipkiss AR (2006). Accumulation of altered proteins and ageing: causes and effects. *Exp Gerontol* 41: 464–473.
122. Kurz T, Terman A, and Brunk UT (2007). Autophagy, ageing and apoptosis: the role of oxidative stress and lysosomal iron. *Arch Biochem Biophys* 462: 220–230.
123. Cuervo AM (2006). Autophagy in neurons: it is not all about food. *Trends Mol Med* 12: 461–464.
124. Martinez-Vicente M and Cuervo AM (2007). Autophagy and neurodegeneration: when the cleaning crew goes on strike. *Lancet Neurol* 6: 352–361.
125. Viteri G, Carrard G, Birlouez-Aragon I, et al. (2004). Age-dependent protein modifications and declining proteasome activity in the human lens. *Arch Biochem Biophys* 427: 197–203.
126. Terman A and Brunk U (1998). Lipofuscin – Mechanisms of formation and increase with age. *APMIS* 106: 265–276.
127. Terman A and Brunk UT (2004). Lipofuscin. *Int J Biochem Cell Biol* 36: 1400–1404.
128. Neufeld EF (1991). Lysosomal storage diseases. *Annu Rev Biochem* 60: 257–280.
129. Wenger D, Coppola S, and Liu S (2002). Lysosomal storage disorders: diagnostic dilemmas and prospects for therapy. *Genet Med* 4: 412–419.
130. Terman A (1995). The effect of age on formation an elimination of autophagic vacuoles in mouse hepatocyte. *Gerontology* 41: 319–325.
131. Cuervo AM, Dice JF, and Knecht E (1997). A population of rat liver lysosomes responsible for the selective uptake and degradation of cytosolic proteins. *J Biol Chem* 272: 5606–5615.
132. Cuervo AM, Bergamini E, Brunk UT, et al. (2005). Autophagy and aging: the importance of maintaining "clean" cells. *Autophagy* 1: 131–140.
133. Bergamini E, Bombara M, Roso AD, et al. (1995). The regulation of liver protein degradation by amino acids in vivo: Effect of glutamine and leucine. *Arch Physiol Biochem* 103: 512–515.
134. Vittorini S, Paradiso C, Donati A, et al. (1999). The age-related accumulation of protein carbonyl in rat liver correlates with the age-related decline in liver proteolytic activities. *J Gerontol* 54: B318–B323.
135. Donati A, Cavallini G, Paradiso C, et al. (2001). Age-related changes in the autophagic proteolysis of rat isolated liver cells: effects of antiaging dietary restrictions. *J Gerontol* 56: B375–B383.
136. Donati A, Ventruti A, Cavallini G, et al. (2008). In vivo effect of an antilipolytic drug (3,5′-dimethylpyrazole) on autophagic proteolysis and autophagy-related gene expression in rat liver. *Biochem Biophys Res Commun* 366: 786–792.

137. Brunk U and Terman A (2002). Lipofuscin: mechanisms of age-related accumulation and influence on cell function. *Free Rad Biol Med* 33: 611–619.
138. Cuervo AM and Dice JF (2000). Age-related decline in chaperone-mediated autophagy. *J Biol Chem* 275: 31505–31513.
139. Kiffin R, Kaushik S, Zeng M, et al. (2007). Altered dynamics of the lysosomal receptor for chaperone-mediated autophagy with age. *J Cell Sci* 120: 782–791.
140. Kaushik S, Kiffin R, and Cuervo A (2007). Chaperone-mediated autophagy and aging: a novel regulatory role of lipids revealed. *Autophagy* 3: 387–389.
141. Zhang C and Cuervo AM (2008). Restoration of chaperone-mediated autophagy in aging liver improves cellular maintenance and hepatic function. *Nat Med* 14: 959–965.
142. McCay CM, Cromwell MF, and Maynard LA (1935). The effect of retarded growth upon the length of life and upon the ultimate body size. *J Nutr* 10: 63–79.
143. Cavallini G, Donati A, Gori Z, et al. (2001). The protection of rat liver autophagic proteolysis from the age-related decline co-varies with the duration of anti-ageing food restriction. *Exp Gerontol* 36: 497–506.
144. Bergamini E, Cavallini G, Donati A, et al. (2003). The anti-ageing effects of caloric restriction may involve stimulation of macroautophagy and lysosomal degradation, and can be intensified pharmacologically. *Biomed Pharmacother* 53: 203–208.
145. Wohlgemuth SE, Julian D, Akin DE, et al. (2007). Autophagy in the heart and liver during normal aging and calorie restriction. *Rejuvenation Res* 10: 281–292.
146. Johnson TE (1990). Increased life-span of age-1 mutants in Caenorhabditis elegans and lower Gompertz rate of aging. *Science* 249: 908–912.
147. Kenyon C, Chang J, Gensch E, et al. (1993). C. elegans mutant that lives twice as long as wild type. *Nature* 366: 461–464.
148. Melendez A, Talloczy Z, Seaman M, et al. (2003). Autophagy genes are essential for dauer development and life-span extension in C. elegans. *Science* 301: 1387–1391.
149. Jia K and Levine B (2007). Autophagy is required for dietary restriction-mediated life span extension in C. elegans. *Autophagy* 3: 597–599.
150. Tóth M, Sigmond T, Borsos E, et al. (2008). Longevity pathways converge on autophagy genes to regulate life span in caenorhabditis elegans. *Autophagy* 4: 330–338.
151. Morck C and Pilon M (2006). C. elegans feeding defective mutants have shorter body lengths and increased autophagy. *BMC Dev Biol* 6: 39.
152. Hansen M, Chandra A, Mitic L, et al. (2008). A role for autophagy in the extension of lifespan by dietary restriction in C. elegans. *PLoS Genet* 4: e24.
153. Simonsen A, Cumming R, Brech A, et al. (2007). Promoting basal levels of autophagy in the nervous system enhances longevity and oxidant resistance in adult Drosophila. *Autophagy* 4: 176–184.
154. Martinez-Vicente M, Talloczy Z, Kaushik S, et al. (2008). Dopamine-modified alpha-synuclein blocks chaperone-mediated autophagy. *J Clin Invest* 118: 777–788.
155. Levine B (2007). Cell biology: autophagy and cancer. *Nature* 446: 745–747.
156. Mathew R, Karantza-Wadsworth V, and White E (2007). Role of autophagy in cancer. *Nat Rev Cancer* 7: 961–967.
157. Ogier-Denis E and Codogno P (2003). Autophagy: a barrier or an adaptive response to cancer. *Biochim Biophys Acta* 1603: 113–128.
158. Schnaith A, Kashkar H, Leggio S, et al. (2006). *Staphylococcus aureas* subvert autophagy for induction of caspase-independent host cell death. *J Biol Chem* 282: 2695–2706.
159. Inoki K (2008). Role of TSC-mTOR pathway in diabetic nephropathy. *Diabetes Res Clin Pract* 82(Suppl 1): S59–S62.
160. Lloberas N, Cruzado JM, Franquesa M, et al. (2006). Mammalian target of rapamycin pathway blockade slows progression of diabetic kidney disease in rats. *J Am Soc Nephrol* 17: 1395–1404.
161. Sakaguchi M, Isono M, Isshiki K, et al. (2006). Inhibition of mTOR signaling with rapamycin attenuates renal hypertrophy in the early diabetic mice. *Biochem Biophys Res Commun* 340: 296–301.

Telomeres and Telomerase

N.M.V. Gomes, J.W. Shay, and W.E. Wright

Abstract Telomere-based replicative senescence is thought to function as a potent mechanism of tumor protection in humans. Whether this mechanism is conserved in other species is still unclear. In this review we present an inter-species critical overview of some of the available literature on the fundamental biology of telomeres and telomerase during development, regeneration, cancer and aging of living organisms during their evolutionary journey through time.

Keywords Evolution · Telomeres · Telomerase · Invertebrates · Vertebrates · Mammals · Amphibians · Fish · Birds

Introduction

Telomeres are the repetitive DNA sequences found at the ends of linear chromosomes [1, 2]. Each of the human 92 telomere ends contains thousand of repeats of the six nucleotide sequence TTAGGG and telomere-associated proteins [3–5]. During DNA replication the leading strand of linear chromosomes is synthesized as a continuous molecule that can potentially replicate all the way to the end of a linear template. The lagging strand is made as a discontinuous set of short Okazaki fragments, each requiring a new RNA primer to be laid down on the template that is then ligated to make a continuous strand. As there is no DNA beyond the end for a priming event to fill the gap between the last Okazaki fragment and the terminus, the lagging strand cannot replicate all the way to the end of a linear chromosome.

W.E. Wright (✉)
Department of Cell Biology, The University of Texas Southwestern Medical Center at Dallas, Dallas, TX 75390-9039, USA
e-mail: Woodring.Wright@UTSouthwestern.edu

N.M.V. Gomes has been co-supported by the European Union Programs POCI 2010 & FSE and by national funds from the Portuguese Ministry for Science, Technology and Superior Education (SFRH/BD/8826/2002). Also supported by the Keck Foundation.

N.S. Wolf (ed.), *The Comparative Biology of Aging*,
DOI 10.1007/978-90-481-3465-6_11, © Springer Science+Business Media B.V. 2010

This leaves a 3′ overhang that cannot be filled, and this has been called the "end replication problem" [6, 7]. The leading strands are also processed to leave a 3′overhang [8]. Since one strand cannot replicate its end, telomere shortening will occur, and once inherited by the daughter cells, the process repeats itself in subsequent divisions [7]. Human telomeres sizes range from ~15 kb at birth to sometimes less than 5 kb in chronic disease states [9].

Normal human somatic cells display a limited capacity to proliferate, a phenomenon known as the "Hayflick limit" [10]. Fetal cells divide more times in culture than those from a child, which in turn, divide more than those from an adult. Telomeres are the molecular clocks that allow for cells to keep track of their number of replications. The length of telomeres decreases both as a function of donor age in tissues and with the number of times a cell has divided in culture [8]. Replicative aging can be divided into 2 stages: Mortality stage 1 (M1 or Senescence) and Mortality stage 2 (M2 or Crisis). M1 occurs when most chromosomes still have several thousand base pairs of telomeric sequences left at their ends [11]. This stage is thought to be induced by DNA damage signals produced by one or a few particularly short telomere ends. DNA damage signaling from short telomeres, loss of the 3′ G-rich telomere single-strand overhangs and telomere position effects all have been suggested as potential inducers of M1. In the absence of cell-cycle checkpoint pathways (e.g. p53 and or p16/Rb), cells bypass M1 senescence and telomeres continue to shorten, eventually resulting in M2/crisis [11]. M2 represents the result of multiple critically short telomeres when cells are no longer able to protect the ends of chromosomes so that end-to-end fusions occur, leading to genomic instability and growth arrest or cell death. Rarely cells escape from M2 and become immortal, almost universally due to the upregulation or reactivation of the enzyme telomerase, which is able to repair and maintain the telomeres. Senescent cells (due to telomere shortening as well as other inducers of irreversible growth arrest) can be stained by senescence associated β-galactosidase, and exhibit alterations in protein expression, such as increased secreted growth factors, cytokines, extracelular matrix, and degradative enzymes [12].

Telomerase is a ribonucleoprotein cellular reverse transcriptase that uses its catalytic component (hTERT) to synthesize telomeric DNA $(TTAGGG)_n$ directly onto the ends of chromosomes. The internal RNA component (hTR or hTERC) contains the template complementary to the telomeric single-strand overhang [13, 14]. After adding six bases, the enzyme pauses while it translocates the template RNA for the synthesis of the next 3′ DNA repeat. This leads to additional rounds of replication of the 3′ end of the G-rich strand (i.e. telomerase is a processive enzyme), thus compensating for telomeric losses due to the end replication problem and perhaps other end processing events [11]. In humans, this enzyme is expressed in embryonic tissues and specific germline cells. Telomerase is found in fetal and adult testis and in female ovary but is undetectable in mature spermatozoids, oocytes, and in most normal somatic cells [11, 15, 16]. The exception are specific proliferative cells of renewal tissues (e.g. hematopoietic stem cells, activated lymphocytes, basal cells of the epidermis, proliferative endometrium, and intestinal crypt cells) [11]. Many of these stem or stem-like cells in adult humans can activate telomerase activity when stimulated to divide. Low levels of telomerase activity may be sufficient to slow but not to prevent telomere shortening. Human intestine or skin telomeres shorten as a function of age although low levels of telomerase can be found in crypt cells and basal keratinocytes. In normal somatic cells and even in stem-like cells expressing telomerase, progressive telomere shortening occurs, eventually leading to senescence [16]. Introduction of the telomerase catalytic protein component (hTERT) into normal telomerase negative cells results is restoration of telomerase activity and telomere maintenance or elongation and immortalization [17]. In some cell types in which the culture conditions are inadequate, it has been demonstrated that growth inhibitory genes can be activated due to a variety of environmental stresses in a process variously termed premature senescence, culture shock, stress-induced senescence or STASIS (STress or Aberrant Signaling Induced Senescence) [9]. In cell culture if the conditions are inadequate, hTERT alone will not immortalize cells.

There are specific proteins (shelterin) associated with human telomeres. TTAGGG is recognized directly at least by the three shelterin subunits, TRF1, TRF2, and POT1. These are interconnected by at least three additional shelterin proteins, TIN2, TPP1, and Rap1, forming a structure that enables cells to distinguish telomeres from sites of DNA damage. Without TRF2, telomeres are no longer hidden from the DNA damage surveillance and chromosome ends are inappropriately processed by the DNA repair machinery [3]. Shelterin is implicated in the formation of T-loops, first identified in human and mouse cells [18]. The telomeric overhang has been proposed to invade the double-stranded telomeric DNA forming a lariat structure, base pairing with the C-strand and displacing the G-strand (Fig. 1). T-loops are a conserved aspect of telomere structure and have been speculated to protect telomeres and regulate telomerase [3].

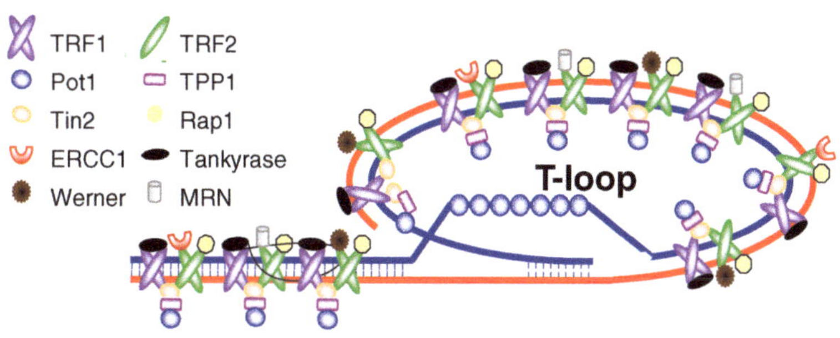

Fig. 1 The telomeric t-loop and associated protein complex

Replicative aging may have evolved as an antitumor mechanism in order to protect long-lived organisms such as humans against the early development of cancer [11]. Normal human fibroblasts essentially never immortalize in culture in part because at least three independent tumor prevention pathways (p53, p16INK4a/pRB, telomere shortening) have to be altered to allow immortal cell growth [19]. Cancer cells must acquire many mutations before they became malignant [20]. Replicative aging blocks this progression by halting cell division before many mutations are able to accumulate within a single cell. Each mutation probably requires at least 20–30 cell divisions: the cell in which an initial mutation occurs must expand to perhaps 1 million cells before there is a reasonable probability of a second mutation occurring. Since most mutations are recessive, an additional clonal expansion is required to eliminate the remaining wild-type allele (usually through loss of heterozygosity). Limiting the number of available cell divisions to less than 100 would thus prevent pre-malignant cells from dividing after accumulating only a few mutations, and thus block their progression [9]. This hypothesis is supported by the finding that ~85% of human tumors have upregulated or reactivated telomerase activity and are able to maintain their telomeres. Immortalization may occur by mutation in a gene(s) in the telomerase repression pathway [11]. The maintenance of telomeres by mechanisms other than telomerase have been observed and referred to as alternative lengthening of telomeres (ALT) [21]. ALT is an extremely rare and difficult mechanism to engage and involves DNA recombination to maintain telomeres. The ALT pathway is not observed in the most common cancers of epithelial tissues (i.e. carcinomas), but is detected in a fraction of rarer cancers (e.g. sarcomas).

Telomeres are essential to prevent chromosome ends from being recognized as double-strand breaks. In addition, telomeres regulate cellular proliferation, survival, chromosome positioning, prevent DNA recombination, and participate in proper mitotic and meiotic divisions (Table 1) [22]. As telomeres shorten during cellular aging there may be de-repression of genes near telomeres eventually leading to reactivation of other previously silenced genes. This process could occur on all or only in a subset of chromosome ends and is known as telomere position effects (TPE) [23]. Telomere dysfunction has been implicated in a variety of human age related diseases (e.g. Werner syndrome) [24]. Mutations in telomerase genes have also been linked to some pathologies such as idiopathic pulmonary fibrosis, aplastic anemia and dyskeratosis congenita [25, 26].

Table 1 Telomere functions [22, 100, 157]

- Prevent chromosome ends from being recognized as double-stand breaks.
- Regulate cellular proliferation (replicative aging/tumor prevention)
- Regulate cellular survival
- Chromosome positioning
- Prevent DNA recombination
- Role in mitotic division
- Role in meiotic division
- Telomere Position Effect (TPE)
- Participate in karyotype evolution/speciation

Evolution of Telomeres

Unicellular Organisms

Telomerase-based end maintenance is likely to be a very ancient mechanism since it is found in widely divergent species that represent many of the major eukaryote lineages (ciliates, animals, fungi, green plants). The loss of telomerase is a catastrophic event unless there is immediate (within a few generations) replacement by an alternative system.

In 1978, Elizabeth Blackburn found that the telomeres of the ciliated protozoan *Tetrahymena thermophila,* consisted of a simple sequence of the hexameric repeat of nucleotides TTGGGG [4]. Telomerase is necessary for the replication of chromosome ends in this protozoan, and telomeric elongation activity occurs massively during the macronuclear development when telomeres are formed and replicated [13]. Elongation by recombination is also seen as a backup mechanism in yeast [27, 28]. In the protozoan *Oxytricha fallax*, the telomeric sequence is similar to that of *Tetrahymena* but the terminal sequence is very short (36 bp) [29]. Gene conversion based on strand invasion and copy-choice replication has also been observed in *Tetrahymena* [30].

Easy laboratory cultivation conditions and powerful genetics have resulted in *Saccharomyces cerevisiae*, *Kluveromyces lactis* and *Schizosaccharomyces pombe* being used as crucial model organisms for telomere biology research. *Saccharomyces cerevisiae* (Sc) and *Schizosaccharomyces pombe* (Sp) are almost as different from each other as either is from vertebrates: their ancestors separated about 420–330 million years ago. The telomeric proteins of *S. pombe* are more similar to the mammalian ones [22]. In the yeast *Saccharomyces*, $(TG_{1\text{-}3})$ or $TG_{2\text{-}3}(TG_{1\text{-}6})$ telomere repeats are observed [22]. In other fungi $(TTAGGG)_n$ is observed in *Cladosporium* but more complex repeats such as $(ACACCAAGAAGTTAGACATCCGT)_n$ are found in *Candida albicans* (Table 2) [31–34]. Today's yeast telomerase enzymatic activity appears to be adapted for both TTAGGG and TG-degenerated sequences [35]. Telomeres of *Candida parapsilosis* are composed of long tandem repeats and also t-circle intermediates [36, 37]. The widespread occurrence of t-circles across eukaryote lineages suggests that t-circles (which permit telomere elongation by rolling-circles replication) may not only represent a backup if telomerase dysfunction occurs, but also may be the ancestral system for telomere maintenance [38]. Telomeres also play an important role in the nuclear architecture in some organisms. In yeast, telomeres are anchored to nuclear membranes through a protein complex [39].

In the causative agent of malaria, the intracellular protozoa *Plasmodium falciparum*, telomeres are followed by a non-coding sub-telomere region (TAS), and telomerase not only maintains telomeres, but also participates in the repair of broken chromosome ends. One of *P. falciparum's* telomere associated proteins, a homologue of the yeast Sir2, is required for the establishment of a heterochromatic structure at the telomeres, leading to silencing of sub-telomeric genes. PfSir2 associates with promoter regions of silenced genes involved in antigenic variation [40]. In kinetoplastid pathogens such as *Trypanosoma brucei*, *Trypanosoma cruzi* and *Leishmania major* subtelomeres are closely related to antigenic variation, a process which allows the clonal switch of surface antigens, enabling escape from acquired immune responses [41]. T-loops have been found in *Oxytricha fallax* and *Trypanosoma brucei.* Although trypanosome telomeres have the same size as human telomeres, their t-loops are very small (less than 1 kb in length) [27, 42].

Table 2 Telomere sequences and replicative aging during evolution [58, 59, 78]

Group/specie		Telomere sequences	Telomere-based replicative aging
Vertebrates	Mammals	TTAGGG	Probable in many Orders
	Birds		Probable in many Orders
	Reptiles		Not likely
	Amphibians		
	Fish		
Invertebrates (Urochordata)	Sea Squirts (*Ciona intestinalis* and *Ciona savignyi*)		No
Echinodermata	Purple Sea Urchin (*Strongylocentrotus purpuratus*)		
Invertebrates (Mollusca)	Wedgeshell Clam (*Donax trunculus*)		
	Bay scallop (*Argopecten irradians*)		
Invertebrates (Porifera)	Sponges	TTAGGG	
Invertebrates (Cnidaria)	Corals and jellyfish		
Invertebrates (Ctenophora)	Comb jellies		
Invertebrates (Placozoa)	*Trichoplax adhaerens*		
Invertebrates	Choanozoa		
Invertebrates	Freshwater shrimp (*Gammarus pulex*)	TTAGG	
	Lobster (*Homarus americanus*)		
Invertebrates (Insects)	Insects (except some coleoptera and Diptera)		
	Fruit Fly (*Drosophila melanogaster*)	Retrotransposons	
	Fly (*Drosophila virilis*)	Retrotransposons Satellite sequence	
	Fly (*Chironomus tentans*)	Satellite sequence	
	African malaria mosquito (*Anopheles gambiae*)	Unequal recombination/ gene convertion	
Invertebrates (Nematodes)	Nematodes	TTAGGC	
	Parascaris univalens	TTGCA and TTTGTGCGTG	
Fungi (Saccharomycotina)	Fission yest (*Schizosaccharomyces pombe*)	$G_{2\text{-}8}$ TTAC(A)	
	Baker's yeast (*Saccharomyces cerevisiae*)	$T(G)_{2\text{-}3}(TG)_{1\text{-}6}$	
	Candida albicans	ACGGATGTCTAACT-TCTTGGTGT	

Table 2 (continued)

Group/specie		Telomere sequences	Telomere-based replicative aging
	Others	Diverse complex sequences	
Fungi (Pezizomycotina)	Most	TTAGGG	
	Aspergillus oryzae	TTAGGGTCAACA	
Fungi (Basidomycotina)	*Cryptococcus neoformans*	$TTA(G)_{4-6}$	
Mold	*Dictyostelium discoideum*	$A(G)_{1-8}$	
	Physarum polycephalum	TTAGGG	
	Didymium iridis		
Plants	*Plants sp*	TTTAGGG	
Plants (Eudicots)	Common Tabacco (*Nicotiana tabacum*)	TTAGGG	
	Tomato (*Solanum lycopersicum*)	TT[T/A]GGG	
	Italian olive ash (*Strombosia pustulata*)	TTTTAGGG	
Plants	*Aloe* sp.	TTAGGG	
	Hyacinthella dalmatica		
	Othocallis siberica		
Algae	Green Alga (*Chlamydomonas reinhardtii*)	TTTTAGGG	
Ciliates (Oligohymenophorea)	*Tetrahymena thermophila*	TTGGGG	
	Paramecium sp.	TT[T/G]GGG	
Ciliates (Spirotrich)	*Euplotes* sp.	TTTTGGGG	
	Oxytricha sp.	TTTTGGGG	
Other Protists	*Plasmodium* sp.	TT[T/C]AGGG	
	Theileria annulata	TTTTAGGG	
	Cryptosporidium parvum	TTTAGG	
	Giardia lamblia	TTAGG	
	Giardia intestinalis	TAGGG	
	Leishmania major	TTAGGG	
	Trypanosoma brucei	TTAGGG	

Other ways exist to overcome terminal telomere loss and are exhibited by viruses, prokaryotes and some eukaryotes. Poxvirus has a covalently-closed hairpin at each end of its dsDNA genome. Controlled nicking of the hairpin provides the 3′OH group that is necessary for DNA replication. The linear DNA of the spirochete *Borrelia burgdorferi* displays a similar strategy. A complication of this replication strategy is the generation of circular dimers requiring a specialized conversion into monomers [27]. Retroviruses reverse transcriptase executes a complex terminal jump in order to maintain their chromosome ends and in adenoviruses the solution to the end-replication problem is provided by a terminal protein primer, which is covalently attached to the 5′ ends of its genome [27, 43]

Plants

In most plants the telomeric sequence $(TTTAGGG)_n$ is observed (Table 2) [44, 45]. Both needle and root samples of long-lived trees such as the coastal redwood (*Sequoia sempervirens*) and the bristlecone pine (*Pinus aristata*) (2,000–5,000 year lifespan) were found to have higher average telomere lengths of the longest, mean, and shortest telomeres compared with aged matched medium- and short-lived trees such as the longleaf pine (*Pinus palustris*) (100–200 years lifespan) [46]. In needle, root, and core samples, long-lived trees also display higher telomerase activity compared with both short and medium-lived trees. A direct correlation has been found between telomere length and telomerase activity and the expected lifespan of these trees. In the longest lived tree, the Great Basin bristlecone pine (*P. longaeva)* there was no evidence of overall telomere shortening or decrease in telomerase activity with age (up to 3,500 years). One living bristlecone tree "Methuselah" had estimated germination at 2838 BC [46–48].

In almost all angiosperms, telomeric DNA is composed of many repeats of the heptanucleotide TTTAGGG [49]. However, *Alliacaeae*, a group of monocots that includes the onions and *Aloe* seems to be an exception, and several alternative telomeric DNA structures have been proposed [50]. Thus in Asparagales (includes *Allium* and *Aloe*) there have been at least two switch-points in the evolution of telomeres. The first occurred with the replacement of the Arabidopsis-type telomere for a "TTAGGG vertebrate-like" sequence. A low fidelity of telomerase (with implications for telomere-binding proteins) may have favored a second switch point in the ancestor to *Allium*, leading to a still unclear mechanism [38]. It has been proposed that elongation of minisatellite repeats using recombination/replication processes initially compensated for the loss of telomerase function. In more established ALT groups, subtelomeric satellite repeats may replace the telomeric minisatellite repeat while keeping the recombination/replication mechanisms for telomere elongation in place. Retrotransposition-based mechanisms may also subsequently become established [38].

Telomeric length is variable among species, from very short telomeres in the plant model Arabidopsis (*Arabidopsis thaliana*) (2–4 kb) to the extremely long telomeres of tobacco (*Nicotiana tabacum*) (up to 150 kb) [51, 52]. Telomere length also varies within the same species [49]. Despite having much shorter telomeres than mice, telomerase null Arabidopsis generated through a T-DNA disruption of the single *At-TERT* gene can survive up to ten generations [49, 53, 54]. The last five generations of telomerase deficient mutant plants display increased cytogenetic damage and in late-generation chromosome fusions occurr in over 40% of the cells, with some cells surviving with only half of their chromosomes. Amazingly, some plants manage to flower and set seeds until the ninth generation [49]. Differences in the consequences of the massive genome damage probably reflect the greater developmental and genomic plasticity of plants. It is known, for example, that chromosomal rearrangements and ploidy changes are better tolerated in plants [54, 55]. Telomere dysfunction in plants, leading to end-to-end chromosome fusions, can have a profound effect on chromosome evolution and even speciation [38]. T loops have been found in plants. Extremely large t-loops, up to 50 kb in size, are seen in peas (*Pisum sativum*) [3, 56].

In plants, telomerase is expressed abundantly in reproductive organs and dividing tissues such as the dedifferentiated callus cells but it is expressed at low or undetectable levels in most post-mitotic vegetative organs [49]. Most cell division takes place in the apical meristem, a group of stem cells that gives rise to all tissues including germ-line cells. It is believe these cells and can undergo approximately 1,000 divisions from seed to seed and differentiate into an array of cell types that make a shoot, root, and flower [52, 57]. Therefore we can conclude that it is unlikely that plants use telomere shortening as a tumor protection mechanism [16, 57].

Metazoa

Invertebrates

Lower Metazoan

As an evolutionary bridge between fungi and higher animals, there are the Lower Metazoans, which includes the phyla Porifera (sponges), Cnidaria (corals and jellyfish), Ctenophora (comb jellies), and Placozoa (*Trichoplax adhaerens*) [34]. The "vertebrate" telomeric motif is found among all these phyla, as well as in the unicellular metazoan sister group Choanozoa [58]. Information about telomere sequences and telomerase TERT and TR/TERC sequences and structure in invertebrates is now readily available (Table 2) [59].

In Porifera, the lowest metazoan phylum, many species present with negligible senescence. These species use both sexual and vegetative forms of reproduction and have extensive regenerative capacity [60]. Most sponges grow continuously, have a long lifespan, and an extremely flexible cell lineage determination. In vivo and in vitro studies in marine demosponges *Suberites domuncula* and *Geodia cydonium* show they have elevated telomerase activity in their immortal germ/somatic tissues. After dissociation into single cell suspensions, isolated cells retain their proliferative capacity but become telomerase negative, possibly due to lack of contact/adhesion factors. After the formation of Primmorphs tissues, these cells regain their telomerase activity [61].

These simple multicellular animals provide excellent models for the study of the separation of soma and germ-cell lineages. In the sponges studied, the number of germ-cells is much reduced or null, so the levels of telomerase observed should come from elevated levels of telomerase in the somatic cells that display unlimited replication potency. Alternatively, there might be a high number of somatic stem cells capable of unlimited replication that would undergo subsequent differentiation. Although Archaeocytes in sponges are pluripotent (stem-cell like), with the potential for differentiation into all major cell types, morphological data seem to support the hypothesis that the proliferation of all major somatic cells types is the major contributor for tissue growth. Furthermore, the plasticity of sex determination and the ability of fully differentiated cells to produce gametes also favor the first hypothesis [61, 62].

In Calcarea (*Leucosolenia* sp and *Sycon* sp.) telomere sizes seem to range from below 1 kb to over 20 kb. One study has failed to detect telomerase activity in Calcarea and in the demosponge *Suberites* [58]. This is unexpected and could be due to technical reasons so further studies on telomerase activity in these species are necessary.

Reef corals can grow vegetatively for hundreds of years and the larger display lower mortality rates compared to the smaller corals. Also many species tend to behave as plants, increasing fecundity as the colonies grow larger [60]. Nonetheless, signs of senescence have been observed in reef corals, with declining growth, calcification and reproduction prior to the death of colonies in *Styrophora pistilla* [60, 63]. One class of Cnidarians, the Anthozoans (Corals) are the most basal organism reported to exhibit the $(TTAGGG)_n$ telomeric sequence. This repeat is found in DNA from the Scleractian corals: *Acropora surculosa*, *Favia pallida*, *Leptoria phrygia*, and *Goniastrea retiformis*. *Acropora surculosa* is thought to have an average telomere length of about 3.5 kb and reveals the extent of conservation of these sequences among vertebrates and invertebrates [34].

Cnidaria (*Chrysaora hysoscella*, *Cyanea lamarcki*) and Ctenophora (*Pleurobrachia pileus*) telomere sizes reportedly range from bellow 1 kb to over 20 kb. In the Cnidaria *Hydra vulgaris* sizes seem to be around 20 kb. Telomerase activity has been found in gonad extracts of Cnidaria (*Aurelia aurita*) and the ctenophore (*Pleurobrachia pileus*). However, similar studies in Cnidarians such as hydra or in Pacozoan (*Trichoplax*) do not have detectable telomerase activity [58].

Bilateria Invertebrates

Among Bilateria, the clade Onychophora, the subphylum Urochordata, and phyla Echinodermata, Platyhelminthes, most Annelida and Mollusca seem to have the "vertebrate" telomere motif $(TTAGGG)_n$ [64–71]. In Deuterostomate, which includes the phyla Chordata and Echinodermata (e.g. sea urchins), many examples of long-lived species have been found. Many sea urchins appear to live a decade or more, and in fact, mortality rates decrease with size in adults [60]. The $(TTAGGG)_n$ telomeric sequence has been found in the moderately long-lived species *S. purpuratus* [72]. The Red Sea urchin (*Strongylocentrotus franciscanus*) grows indeterminately during a lifespan that can go beyond 100 years without evidence of age-related disease or decline in reproductive potential, while other species such as *Lytechinus variegatus* are fast growing and short lived, with a maximum lifespan of 3–4 years. Telomere studies in the Red Sea urchin reveals telomerase activity in mature eggs, and also during early stages of development of *L. variegatus* and in tissues during adulthood in both species (Aristotle's lantern muscle, ampullae, esophagus, intestine, tube feet, male and female gonads). Analysis of the telomere length of *L. variegatus* reveals long telomere lengths (>20 kb) in both germ and somatic tissues. The adult tissues of *S. franciscanus* have short telomere lengths ($\approx$ 5 kb), similar to *S. purpuratus* (6 kb). Telomeres in both species seem to be maintained throughout life, with no telomere shortening occurring during the life of these species [72, 73]. It is also known that sea urchin embryo telomeres need to be maintained. The use of cationic porphyrins as telomere interfering agent decreases the rate of cell proliferation and leads to increased chromosome destabilization [74]. These results seem to indicate that neither short nor long-lived sea urchins use replicative aging as a tumor protective mechanism [73]. The number of reported cases of neoplasia in sea urchins, a very intensively studied model organism, is significantly reduced [75]. This suggests that these species have evolved other mechanisms of tumor prevention/suppression, such as efficient cellular or molecular protection against damage or free radicals and/or a good capacity of replenishment to damaged cells [73]. These species may be excellent candidates for future senescence and tumor protection mechanism studies [73].

The model Urochordate, the golden star tunicate (*Botryllus schlosseri*) is a colonial organism that propagates both asexually and sexually during the 2–5 years of colony life. Assexual budding occurs continually from the progenitor body wall and when the colony reaches a critical size sexual reproduction initiates with the production of gonads. It has been proposed that pools of stem cells assure renovation throughout the lifespan. Chromosome ends are capped by heterogeneous telomeres (6–15 kb) and it has been reported that germ and embryonic tissues contain high levels of telomerase [64]. Telomerase activity peaks in tissues containing bud rudiments, then decreases in buds that are going through organogenesis and drops to even lower levels in functional zooids, in individual organs and blood [64]. This could be due to the lack of necessity of such high levels of enzyme in long chromosome ends protecting chromosomes of a species that regenerates itself by budding weekly. It has been hypothesized that telomerase activity needs to be retained in progenitor and stem cells, is downregulated during differentiation, and is not necessary to maintain the relatively short-lived somatic tissues of *Botryllus* [64].

Among marine invertebrates, the telomeric sequence $(TTAGGG)_n$ is also found in the sea urchin (*Strongylocentrotus purpuratus*), ragworm (*Platynereis dumerilii)*, keel worm (*Pomatoceros lamarcki*), pacific oyster (*Crassostrea gigas*), bay scallop (*Argopecten irradians*), neogastropod (*Fasciolaria lignaria*), blue mussel (*Mytilus galloprovincialis*), sea cocumber (*Holothuria tubulosa*), and wedgeshell clam (*Donax truncatus*) [34, 66, 68, 69, 72, 76, 77].

Nematodes and Platyhelmintes

Another motif discovered in Metazoans is the so called nematode motif $(TTAGGC)_n$, which is found in the worms *Ascaris lumbricoides and Ascaris sum* [78]. In *Parascaris univalens* the haploid germline is contained in a large chromosome and the somatic genome is surrounded by the heterochromatin (HET) blocks that is constituted by segments of the repeats TTGCA and TTTGTGCGTG [78]. Chromatin diminution in *Ascaris* is a complex molecular process that includes site-specific chromosomal breakage, new telomere formation, and DNA degradation [50, 58]. In the free-living nematode *Caenorhabditis elegans*, the 4–9 kb of telomeric repeats $(TTAGGC)_n$ are sufficient for chromosome capping [79]. In the Trematode *Schistosoma mansoni* chromosomes are also protected from degradation by telomeres [80].

Arthropods

All of the major arthropod groups (Chelicerata-except spiders, Pycnogonida, Myriapoda, Crustacea, Hexapoda) have the $(TTAGG)_n$ telomere motif [58]. Unlike mammals that stop growing after adulthood, some invertebrates, such as the lobster (*Homarus americanus*) show continuous growth during their lifespan, with decreasing growth rates with age. Decapod crustaceans like the lobster that shows asymptotic growth and can occasionally weigh over forty pounds, are good candidates for negligible or very slow gradual senescence. Lobsters have a very long lifespan (50–100 years) and neither sex exhibits a post-reproductive phase nor molting cessation They also exhibit limb regeneration ability even at advanced ages [60, 81]. Telomere analysis reveals the sequence $(TTAGG)_n$ and telomerase expression has been found in fully differentiated tissues of all organs, with high levels detected in the hepatopancreas and heart and moderate levels in skin and muscle tissues [81]. Tumors are rare in adult lobsters and do not seem to correlate with lifespan or bodyweight [60].

With the exception of the heterogeneous Coleoptera, most insect orders can be divided into those that use the telomeric repeat $(TTAGG)_n$ (e.g. Lepidoptera) or the ones that do not (e.g. Diptera) [34, 82–85]. Telomerase activity has recently been detected in crickets, cockroaches, and species of Lepidoptera [86]. The telomerase reverse transcriptase (TERT) subunit has been identified and characterized in the domestic silkworm (*Bombix mori*) and the flour beetle (*Tribolium castaneum*) [87]. In the group of insects with the largest number of species, the beetle (order Coleoptera), the telomerase-dependent $(TTAGG)_n$ motif has been repeatedly lost (5–6 times) in different phylogenetic branches and was likely replaced with the alternative mechanisms of telomere elongation [88]. The order Diptera seems to be an exception, having evolved arrays of complex long satellite repeats at the ends of their chromosomes (e.g. *Chironomus* & *Anopheles gambiae*) [30, 58, 89]. Elongation of telomeres in the mosquito (*Anopheles*) is done through gene conversion between complex terminal satellite repeats that are present at natural telomeres [30]. One hypothesis is that Diptera may have lost the telomerase gene and was forced to use alternative mechanisms of telomere elongation [30, 90]. Telomerase independent mechanisms such as chromosome end capping with non-LTR retrotransposons are found in the fruit fly (*Drosophila melanogaster*). Chromosome end-elongation is predominantly achieved by terminal insertion of two classes of telomere-specific LINE-like retrotransposable elements, HeT-A and TART [91]. However, *Drosophila* telomeres can also be extended by gene conversion [92] and perhaps by recombination between telomeric HeT-A elements [93]. Damselflies (Zygoptera) and spiders (Araneae) have a still unknown telomere structure [71, 85].

Vertebrates

The telomere sequence $(TTAGGG)_n$ is conserved in the phylum Chordata and is thought to have arisen 400 million years ago [94]. The essential core structure of telomerase RNA seems to be preserved in vertebrates [95]. Telomere and telomerase sequences of many vertebrates are now known and available online (Table 2) [59].

Fish

Unlike mammals, several fish species can grow throughout life with high proliferative capacity displayed by all somatic cells [96]. In many species of fish, organs continue to grow throughout life and growth after the larval stage is dependent on both cellular hyperplasia and hypertrophy [97, 98]. Among Elasmobranchs, dogfish (*Squalus acanthias*) is the longest lived (70 years) but in general, reported lifespan in cartilagenous fish is much lower than 15 years in captured specimens of sharks [60]. On the other end, many reports show that eels, sturgeons, and teleosts can live 80 years or more. In teleosts the record lifespan is held by the lake sturgeon (*Acipenser fulvescens*) that can reach 152 years and the beluga sturgeon (*Huso*), with 118 years and being able to reach weights of over 3 tons [60, 99].

Telomeric $(TTAGGG)_n$ sequences are present in cartilagenous fish [100]. Telomerase activity is highly expressed in the dogfish shark [101]. Telomere bands of 3 kb are common to four species of Batoidea (*Torpedo marmorata*, *Torpedo ocellata*, *Raja asterias*, *Raja montagui*) and two species of Galeomorphii (*Mustelus asterias*, *Scyliorhinus stellaris*). In rays, intense telomeric bands varying in length from 0.5 to 2 kb, are visible [100]. Localization of telomeric sequences in the paracentromeric and/or interstitial regions is observed in chromosomes of two out of four Batoidea, the blue-spotted stingray (*Taeniura lymma*) and the electric ray (*Torpedo ocellata*). This finding supports the hypothesis that in cartilaginous fish Robertsonian fusions involving telomeres could have led to an increase in bi-armed chromosomes and a decrease of the acrocentric ones, thus playing an important role in karyotype evolution [100, 102].

Teleosts represent more than half of the forty to fifty thousand vertebrate species [60]. Teleost fishes are unique in that they exhibit different patterns of aging. The pacific salmon (*Oncorhynchus*) and eel (*Anguilla anguilla*) exhibit rapid senescence and death at first spawning, while other fish such as medaka (*Oryzias latipes*) and guppy (*Poecilia reticulata*) show gradual "mammalian-like" senescence [103]. In Cyprinidae, species with very different lifespans such as carp (*Cyprinus carpio*, which may live more than 100 years) and zebrafish (*Danio rerio*, which has a lifespan of approximately 5 years) exhibit growth characteristics that imply very slow or negligible senescence [103]. The short life, short generation time (3–5 months), and seemingly unlimited capacity to regenerate their fins in 7–10 days of zebrafish place it in a privileged spot as a vertebrate model for studying functional aging and gradual senescence [103, 104].

Telomerase activity is detected in cells and tissues of several teleost fish (e.g. fugu, zebrafish, rainbow trout, Japanese medaka, flounder) [81, 101, 103, 105].

Telomerase TR/TERCs (RNA component) from five teleost fish, *Danio rerio, Oryzias latipes, Gasterosteus aculeatus, Takifugu rubripes* and *Tetraodon nigroviridis* have been characterized [106]. The gene encoding the TERT subunit of telomerase has been isolated and cloned in pufferfish (*Fugu rubripes*) and zebrafish (*Danio rerio*) [98, 107]. In *Fugo*, the fTERT mRNA is found at low levels in several tissues such as skin, spleen, heart, brain, stomach and eye, with high expression in the gill, testis and ovary. fTERT expression is detected in an immortalized eye-derived cell line from *Fugu*. The level of expression is higher in actively dividing cells and is reduced at quiescence, suggesting cell cycle regulation of TERT [107]. In zebrafish, TERT mRNA expression and telomerase activity correlate closely and are detected in all somatic tissues, including retina and brain, with the highest activities found in gills and in the ovary, where the highly proliferative germ cells are found [98]. In trout (*Oncorhynchus mykiss*), erythrocytes have average telomeric terminal restriction fragment (TRF) lengths of 20 kb [72]. In the class Actinopterygii, telomerase activity is found in several somatic tissues of the ray-finned fish American eel (*Anguilla rostrata*). Telomeres of the intestine of this eel have mean telomere lengths of about 7 kb [101].

The different patterns of senescence reported in fish make them unique models for studying the aging process. Most marine species with their high regenerative capacities and long lifespans seem to maintain telomerase in their tissues. The lack of telomerase repression in somatic tissues suggests that they do not use telomere shortening and replicative aging as a tumor-protection mechanism. Also, many of these species may prove excellent models for studies in regeneration, stem cells, DNA repair, cancer and aging.

Amphibians

Senescence and mortality rates in the class Amphibian are not well studied but relatively long lifespans have been reported, mainly among the larger species, such as the giant salamandra (*Megalobachus japonicus*), which can live at least 55 years and the toad (*Bufo*) that can reach at least 36 years. Many other species exceed the age of 15 years. Increase in fitness with age is reported in some species such as bullfrogs (*Rana catesbeiana*) [60]. Most data on the experimental model African clawed frog (*Xenopus*), which can live at least 15 years, suggests that senescence in amphibians is negligible or very slow [60]. Telomeres in this specie range from less than 10 kb to over 50 kb, in a polymorphic pattern between individuals [108]. Mud puppy (*Necturus maculosus*) erythrocytes have been found to have huge telomere lengths (average 100 kb) [72]. A TERT gene from *Xenopus,* designated xTERT has been identified [109]. Oocytes, embryos, and tissues from adult (>1–2 years) frogs (*Xenopus laevis*) express telomerase activity. Telomerase activity is most abundant in testis, spleen, liver, and embryos [110, 111]. Telomerase activity is lower but still readily detectable in brain and muscle tissues. Furthermore, this activity does not seem to be limited to the polyploid members of the genus since telomerase activity is also found in somatic tissues of the diploid *Xenopus tropicalis* [111]. Also, when *Xenopus* telomeres from whole embryos are compared to telomeres in parent spleens, the inheritance pattern of some bands is unusual. In some crossings the telomeres of the embryo or in the male testis are shorter than the telomeres of the parents' spleen, consistent with a model for chromosome behavior that involves a significant amount of DNA rearrangement at telomeres. It is possible that length regulation of *Xenopus* telomeres is different from that reported in mammals. Telomere data in *Xenopus* is also consistent with the occurrence of some degree of meiotic rearrangement [108]. Based on the current literature we can conclude that telomerase repression during differentiation does not occur in *Xenopus*.

Reptiles

Although the telomeric sequence $(TTAGGG)_n$ has been identified in species from the orders Sauria and Serpentia [94], very few reports in the literature are available on telomere biology in the class Reptilia. Some lizards are known to have excellent tissue regeneration capacity. Telomerase activity has been observed in all tissues of a teiid lizard, the six-lined racerunner (*Cnemidophorus sexlineatus*), which is thought to have a maximum natural lifespan of about 4 years [112]. The same study found that skin fibroblasts of a juvenile blue racer (*Columber constrictor*) can undergo more than 124 population doublings (PD) with strong telomerase activity detected after 100 PD, which is suggestive of immortalization of the culture [113]. A study of fibroblast-like cells from the lizard Carolina anole (*Anolis carolinensis*) showed great cellular proliferative capacities compared to human diploid cells [114].

Turtles have been reported to live more than 100 years in captivity and have very high annual survival rates in natural conditions. Senescence has not been proven to occur in these species. Mortality does not seem to increase during aging, the reproductive capacity of females grows during their lifespan, and apart from carapace alteration from soil abrasion, no age-specific diseases are known [115]. Studies in mature breeding sea turtles (*Chelonia mydas*) have reported absence of a decline in growth rate [116]. In fibroblast cultures from young Galapagos tortoises (*Geochelene nigra*), population doublings of 100–130 have been observed [117]. A turtle species (*Pseudemys scripta*) has been found to have long average telomere lengths ($\approx$ 50 kb) [72]. However, in a study, cell culture senescence has been observed between PD 18–45 in yellow mud turtles (*Kinosternon flavescens*), which have a lifespan of 30–45 years. It has been shown that fibroblasts from hatchlings undergo about twice as many population doublings in culture as those from 25 year old mud turtles. Telomere shortening of about 30–50% was observed between hatchlings and adults, and apart from the gonad, no telomerase was found in tissues from these turtles [113]. More studies are needed to clarify if this cellular growth arrest is due to culture stress from inadequate growth conditions leading to STASIS or from telomere-based replicative aging. In the same study, cells from the long-lived snapping turtle (*Chelydra serpentina*), which is thought to be able to live greater than 100 years reportedly multiplied in culture for over 265 PD. In these snapping turtles, telomerase activity went from barely detectable at 157 PD to very strong at 191 PD. Telomerase activity was also detected in old painted turtles (*Chrysemys picta*), which can live more than 35 years, and cultured cells from this specie were still dividing well at PD 120. Telomerase activity was found in gonads of two ornate box turtles (*Terrapene ornata*) hatchlings and in other organs of one of them [113]. Telomerase activity has therefore been found in two divergent families of turtle (Chelydridae and Emydidae).

A study of the European freshwater turtles (*Emys orbicularis*) compared the mean telomeric sizes of 14 embryos and 15 adults (older than 20 years) [115]. Large sharp telomeric bands of about 20 kb were found in both embryos and adult erythrocytes. This species has a similar longevity to humans but is not known to display signs of senescence. Although it's interesting to observe that telomeric shortening did not occur in European freshwater turtles, these results should be interpreted with caution due to lack of information on telomere biology in nucleated erythrocytes and to the fact that age determination of the adult turtles was not very precise. Also, information about telomerase activity in the tissues of this specie and many other species of Chelonian (and Reptilia in general) is not available. However, we can conclude, based on the few studies available, that telomerase is often found in adult somatic tissues of reptiles and telomere based replicative senescence does not seem to occur in most of the species studied to date.

Birds

Birds (class Aves) and other homeothermic vertebrates exhibit gradual senescence with a definite lifespan [60]. Also, bird species are clearly longer lived than mammals of similar body weight (up to 3 times longer) [118]. This finding of slow senescence rates is a paradox since, compared to similar sized-mammals, birds have 2–2.5 times higher metabolic rates, higher body temperatures (3°C higher) and elevated glucose levels (2- to 4-fold). According to most biochemical theories of aging this should have led to increased tissue cellular damage and accelerated aging [118].

In Aves, rates and patterns of aging can be extremely variable among different orders. In Galliformes, including the domestic chicken (*Gallus gallus*) and quail (*Coturnix coturnix*) short lifespans and fast aging rates have been observed. Exceptionally long-lived for their body size are some raptors

(Falconiformes), hummingbirds (Apodiformes), parrots (Psittaciformes), sea birds (Charadriiformes), and songbirds (Passeriformes). Delayed maturity and low annual fecundity have been linked to slow avian senescence [118]. Some birds, such as the male zebrafinch (*Taeniopygia guttata*) have shown neuroregeneration capacity during song learning [119]. Prevention of tissue damage by ROS or glycosylation endproducts has also been reported [118].

The chicken telomerase reverse transcriptase (chTERT) component has been well characterized [120]. In chickens, telomeric DNA represents 3–4% of the genomic DNA, 10 times higher than what has been found in the human genome. During southern blot analysis of chickens three overlapping sizes of telomere arrays have been found and classified as: Class I (0.5–10 kb), Class II (10–40 kb) and Class III (40 kb to 2 Mb) [121]. These Class III telomere arrays sizes are the largest reported in all vertebrates. Class III arrays are quickly digested by Bal 31 exonuclease indicating a terminal location. These elements are highly polymorphic and map specifically to the microchromosomes, perhaps serving a protective function for these small genetic elements of 7–23 Mb. In vivo, chicken Class II telomeres seem to shorten in an age-related fashion, similarly to human telomeres. Class I bands do not exhibit age-related telomere shortening and are resistant to digestion by Bal 31 exonuclease, indicating that these arrays are located internally rather than at the end of the chromosomes [121, 122]. It has been reported that truly interstitial (non-centromeric and non-telomeric) $(TTAGGG)_n$ sites are particularly common in the chicken and primitive Palaeognathae birds [123]. Telomerase activity is high in early stage embryos and developing organs but is down-regulated during late embryogenesis or postnatally in most somatic tissues. Renewable tissues such as reproductive and immune organs retain high levels of telomerase activity even in adults (4–5 years). In general, telomerase activity in chickens correlates with the proliferative potential of the tissue. The telomere arrays of the somatic and germ tissues in the embryo display similar telomeric sizes, but telomeres in adult somatic tissues arrays are shorter, exhibiting an average decrease in size of 3.2 kb. Telomere shortening in erythrocytes was reported in a variety of avian species by comparing erythrocyte and sperm telomere length [121, 124]. Telomere shortening is detected in telomerase positive adult tissues (kidney, intestine, spleen), a pattern also reported in some human tissues [122, 125]. Primary cultures of embryonic cells have telomerase activity which, after serial culture passages, is downregulated and cells growth arrest at about 35 PD. At senescence, these cells exhibit mean telomere sizes of about 5 kb [126]. This value is similar to the one observed in human senescent cell cultures (5–6 kb). However, this growth arrest could have been driven by inadequate growth conditions leading to senescence so the critical experiment would be to immortalize these cells through ectopic telomerase expression[16]. In a study of 18 species of birds, most displayed the Class I, II and III telomeric arrays [121]. Extremely long arrays, ranging from hundreds of kilobases to 1–2 Mb (Class III) were observed in all but two raptor species, the northern goshawk (*Accipiter gentilis*) and the American bald eagle (*Haliaeetus leucocephalus*). The mean TRF length decreases with age in erythrocytes of zebra finch (*Taeniopygia guttata*), tree swallows (*Tachycineta bicolor*), Adélie penguins (*Pygoscelis adeliae*), and common terns (*Sterna hirundo*). Lifespans of these species range from 5 to 26 years. Surprisingly, in Leach's storm-petrel (*Oceanodroma leucorhoa*) erythrocytes, TRF length did not decrease but actually increased with age. This species is long-lived, with observed lifespans of 36 years [127]. Bird species with shorter lifespans lost more telomeric repeats per year than species with longer lifespans [127]. In the same long-lived species (apart from penguin that was not included) telomerase was not down-regulated at early developmental stages compared to the short-lived species. In the gonad, telomerase is detected in all species. Higher activities are observed in the Leach's storm-petrel in most tissues studied (bone marrow, intestine, brain, kidney, liver). Across these four species and all age-classes, telomerase activity is generally higher in the proliferative tissues than in the post-mitotic tissues. The short-lived zebra finch and tree swallow sharply down-regulate bone marrow telomerase before adulthood, whereas the long-lived common tern and Leach's storm-petrel express bone marrow telomerase at high levels throughout life that could produce the slower rates of erythrocyte telomere shortening observed. Post-natal telomerase activity is generally absent in the brain, kidney, skeletal muscle, and liver in all species, although higher telomerase activity is observed in the skeletal muscle, kidney and brain of hatchling common terns and Leach's storm-petrels than what is reported in chickens. Telomerase profiles in the bone marrow, gonads and intestine are elevated at all stages of life. Few cancer rate studies in long-lived bird species are available but reports tends to indicate a low incidence of cancer in wild birds, and specifically in long-lived seabirds [128, 129]. Damage susceptibility, repair abilities, shelterin proteins (which control the synthesis of telomeric DNA by telomerase) are also likely to be important in determining these telomeric shortening rates.

Telomeric $(TTAGGG)_n$ sequences are abundant in avian microchromosomes [123]. In our own unpublished studies of Japanese quail (*Coturnix coturnix japonica*) interstitial bands are preferentially localized to the 66 microchromosomes ($2n=78$). A study the chromosomal distribution of $(TTAGGG)_n$ sequences in 16 bird species representative of seven different orders, show that many species, in particular the ratites, display $(TTAGGG)_n$ hybridization signals in interstitial and centromeric regions of their macrochromosomes. The microchromosomes of most species are enriched with $(TTAGGG)_n$ sequences, displaying heterogeneous hybridization patterns, and it has been proposed that this high density of $(TTAGGG)_n$ repeats plays an important role in the exceptionally high meiotic recombination rates of avian microchromosomes [123]. However, other studies in birds claim otherwise [130].

Mammals

Telomere-based replicative senescence is thought to function as a potent mechanism of tumor protection in humans. Whether it is conserved in other species of the class Mammalia remains unclear. Available data in laboratory rodents indicates that laboratory rodents do not use telomere based replicative aging as a mechanism to limit cell proliferation [19, 131]. Hepatocyte telomeres of rat (*Rattus norvegicus*) and mouse (*Mus musculus*) have been found to have very long average telomere lengths (50 and 40 kb respectively) [72]. Other studies in inbred laboratory rodents have revealed extremely long telomeres (25–150 Kb) indicating that telomere shortening as a tumor protection mechanism does not occur in these inbred laboratory strains of rodents. In situ hybridization reveal that all the telomeres are long, with no telomeres sufficiently short to induce replicative senescence [16, 19, 131, 132]. Shortening observed in brain telomeres cannot be due to replicative aging due to lack of cellular proliferation in this organ and oxidative damage has been proposed as a contributor to this event [133]. Oxidative damage has also been implicated in the post-mitotic telomere shortening observed in the heart of mouse and rat [16, 134]. Most murine tissues display telomerase activity and those that do not are likely to reflect a quiescent nature rather than lack of telomerase competency. In two mouse strains telomerase activity was reported in adult testes, ovary, breast, colon and liver, but absent in skin, brain, heart, stomach and muscle [16, 132, 135]. Although mTERT protein is only found in telomerase positive tissues, the finding of mTERT mRNA in all tissues (including telomerase negative ones), suggests telomerase competency in these tissues [136]. Lack of expression of telomerase might be due to mechanisms of alternate splicing triggered by quiescence [16, 136].

Many current publications still reveal confusion between replicative aging and stasis/senescence [137–139]. Studies of the telomerase negative $mTR^{-/-}$ mouse show that the growth arrest observed after 10–15 doublings in mice is not due to telomere shortening and does not restrict tumor development [140]. Blasco showed that $mTR^{-/-}$ mouse cells escaped from this growth arrest barrier as frequently as wild-type mice and went on to divide for at least 200–300 PD. Senescence in mouse culture occurs as part of a stress response due to inadequate growth conditions similar to reports in some types of human cells. Human skin keratinocytes grown in defined media suffered from p16/RB mediated growth arrest but this could be prevented growing cells on appropriate feeder layers [141]. There are now several reports showing that rodent cells have an indefinite replication capacity given proper growth conditions [142, 143]. MEFs (mouse embryo fibroblasts) from mice defective in DNA repair factors such as Ku80, ATM (mutated in Ataxia Telangiectasis) or BrCA2 (mutated in some breast cancers) growth arrest after only 3–4 PD and exhibit high levels of p53 and $p21^{Cip1}$ [19, 144–146]. Since these cells divide adequately in vivo, the premature growth arrest observed in vitro reflects an induction of DNA damage upon putting these cells under conventional culture conditions, in which ambient oxygen is a major contributor to damage [147]. Furthermore, the Rb pathway is not involved in cellular growth arrest in mice and abrogation of ARF/p53 is sufficient to escape this cell growth barrier [19, 148].

If we take into account the stochastic nature of mutations, their number will be the result of the product of both time and pool size. Multiplying the weight and lifespan of humans versus mice, humans need to be about 100,000 times more resistant to the formation of tumors than rodents. The normal frequency of tumor formation observed by Blasco in the telomerase knockout mouse $mTR^{-/-}$ suggests that escaping replicative aging by telomerase activation is not needed for murine tumorigenesis and other mechanisms of tumor protection such as DNA repair, cell cycle checkpoints or immune surveillance are sufficient for tumor protection during the short lifespan observed in these small sized rodents [16, 19, 140]. There are, however, examples of wild mice with "human sized' telomeres such as Algerian mouse (*Mus spretus*) [135].

There are a few reports about telomere biology in a limited number of domestic species. Sheep dermal and lung fibroblasts have a finite lifespan in culture, after which the cells enter replicative senescence. Terminal restriction fragment lengths from sheep tissues reveal "human-like" telomere lengths (9–23 kb). Telomerase activity is found in the testis but suppressed in somatic tissues. Similarly to humans, senescent sheep skin fibroblasts have increased levels of p53 and $p21^{WAF1}$ compared to young cells [149]. Pigs (*Sus scrofa*) also seem to display replicative aging [150]. In blood and other tissues obtained from 2 domestic shorthair cats (*Felis catus*) mean TRF values ranged from 4.7–26.3 kb and there is significant telomeric attrition with increasing age of cats. Telomerase activity was not detected in a wide range of normal tissues [151]. In another study of lymphocytes and granulocyte of cats, average telomere lengths analyzed by fluorescence in situ hybridization and flow cytometry (Flow FISH) are reported as being 5- to 10-fold longer than in humans. However, much higher telomeric shortening rates both in vivo as in vitro are observed (500 bp/PD in T cells), suggesting that this shortening might not be caused by the end replication problem but by other mechanisms [152]. Heterogeneity in telomere lengths is observed in a variety of somatic tissues of several dog (*Canis lupus familiaris*) breeds. Mean TRFs ranged between 12 and 23 kb. Telomerase activity was low or absent in normal somatic tissues and was present in testis and tumor tissues. Soft tissue sarcomas were identified with mean TRFs of 22.2 and 18.2 kb [153] Telomere lengths in peripheral blood samples from donkey (*Equus asinus*) ranging from 2 to 30 years of age were 7–21 kb and showed a statistically significant inverse correlation between telomere lengths and donor age. In horse (*Equus equus*), fibroblasts cultured to senescence displayed telomeric loss. No telomerase activity was observed in primary cell cultures. Similarly, no telomerase activity could be detected in normal equine tissues or equine benign tumor samples of the sarcoid or papilloma type [154].

During the last years we have been investigating the role of telomeres and replicative aging in warm-blooded animals. We previously reported that almost all primates examined are similar to humans in exhibiting replicative senescence [155]. In skin fibroblasts from the "New World" primates [spider monkey (*Ateles geoffroyi*) and squirrel monkey (*Saimiri sciureus*)] and the "Old World" primates [rhesus monkey (*Macaca mulatta*), orangutan (*Pongo pygmaeus*), and pigmy chimpanzee (*Pan paniscus*)] telomere shortening limits the replicative capacity of anthropoid fibroblasts and the expression of human telomerase produced telomere elongation and the extension of their in vitro lifespan. In contrast to the rigorous control of replicative aging by telomere shortening conserved among anthropoid primates, in the prosimian ring-tailed lemur (*Lemur catta*) this control seemed to be less effective. Lemur cells have both long and short telomeres and telomere shortening did not provide an absolute barrier to immortalization. Following a transient growth arrest a subset of lemur cells showing a reduced number of chromosomes overgrew the cultures without activation of telomerase [155]. The small Asian barking deer, the Indian Muntjac (*Muntiacus muntjak*) is an ideal model to study telomere biology since it has the fewest number of diploid chromosomes of all mammals with only six chromosomes (1, 2, 3) in the female and seven in the male (1, 2, 3 + X) [156]. In our studies we found that Indian muntjac skin fibroblast reached senescence at PD 89 and could be immortalized with human TERT expression [157]. Near senescence, ends became telomere signal-free and chromosome abnormalities increased dramatically. Interstitial telomere sequences coincided with fragile sites, suggesting that these remnants of chromosome fusion events might contribute to genome instability. This species is a good candidate as a telomere-based replicative senescence model for human cells [157]. Intrachromosomal TTAGGG sequence sites are known to be fragile "hot spots" prone to breakage and recombination in the Armenian hamster (*Cricetulus migratorius*) and Chinese hamster (*Cricetulus griseus)* [158–160] and are thought to be involved in the process of karyotype evolution during speciation due to Robertsonian fusions [100, 161, 162].

In an ongoing unpublished study, we have analyzed fibroblasts from skin and other organs from 55 animals representing most orders of the mammalian radiation. We addressed the question of whether there was a relationship between senescence in cell culture, cellular telomerase expression, telomere size, telomere shortening rates, ability of hTERT to immortalize versus longevity or the respective damage susceptibility and repair abilities in different species. Our results to date show that the telomere-based tumor protection mechanism has deep roots in the mammalian evolutionary tree. However, there is widespread presence of animals with long "mouse sized" telomeres indicating that there are likely trade-offs between repressing telomerase/having short telomeres to count cell divisions/tumor protection or maintaining telomerase activity and having very long telomeres. Species from the orders Cetacea, Artiodactyla, Perissodactyla, Hyracoidea, Proboscidea, and Xenartha seem to exhibit telomere-based cellular replicative aging. One of the species we found to display replicative aging was the bowhead whale (*Balaena mysticetus*). The presence of nineteenth century stone harpoon points and changes in aspartic acid levels in 48 eye lenses indicates at least one bowhead whale lived approximately 211 years (between 177 and 240 years) [163]. The oldest known ages for other whales are 100 years for a blue whale and 114 for a fin whale (based on counting of waxy laminates on the inner ear plug). The challenges of living in Arctic waters may nurture slow growth and long life [163, 164].

However, we also find that most species from important orders such as Rodentia, Chiroptera, Eulotypia, and Macroscelidea, do not seem to exhibit replicative aging, with their cells maintaining good telomerase activity and having very long telomeres. In another study, telomerase activity was detected in several tissues of 15 rodent species, and long telomere lengths (>30 kb) were observed in most species. Telomerase activity is detected in some somatic tissues of all of these species, with the lowest values observed in beaver and capybara, which, together with guinea-pig and deer mouse displayed shorter "human-like" telomeres [84]. We have previously reported that Lagomorpha cells, although telomerase negative, do not growth arrest in culture due to their extremely long telomeric arrays. In the North American pika (*Ochotona princes*), endogenous telomerase activity is present. These data suggest it is unlikely that lagomorphs use telomere shortening and replicative senescence as a tumor protective mechanism [165]. Results are less clear in Xenarthra and Carnivora where we observe more heterogeneity. Moreover, non-placental mammalian orders such as Marsupials and Monotremata show evidence of alternative mechanisms of telomere maintenance including the presence of restriction enzyme recognition sites intercalated between the telomeric $(TTAGGG)_n$ sequences. Marsupials are particularly interesting species, not only due to their placement at the very base of the mammalian evolutionary tree, but also for their low metabolic rates (70–80% of similar sized eutherians) [166]. According to the rate of living theory they should be longer lived, but in fact, they are short-lived for their size [166].

Our studies also indicate that species using replicative aging tend to have longer lifespans and higher adult body weights. Results within the order Rodentia suggest that telomerase activity coevolves inversely with body mass, not lifespan, with longer lived rodents displaying lower telomerase activities [84]. However, our results with cultured cells and comparing different orders show this correlation within rodents does not extend to all orders, since animals such as the rock hyrax (*Procavia Capensis*, Hyracoidea) with small body size seem to use replicative aging and repress telomerase while others, such as the tiger (*Panthera tigris corbetti*, Carnivora) have very long telomeres and do not repress telomerase.

Some studies claiming that the replicative potential of fibroblasts positively correlates with body mass or longevity [138, 139] have to be re-evaluated given the current understanding of the consequences of inadequate growth conditions. The studies of Lorenzini included the early growth arrest in culture of fibroblasts from smaller, shorter lived species such as rodents (half of the species) and carnivores as an example of telomere-based replicative aging. In fact, studies show that fibroblasts from many of these species, given adequate media and more physiological (2% O_2) growth conditions, can grow over 100 PD [167, 168].

Overall telomere length tends to be conserved within evolutionary blocks (e.g. the bulk of rodents and nearby species have very long telomeres, although individual species such as the deer mouse can have short telomeres). The presence of several large clades of species having long telomeres interspersed with large clades having short telomeres suggests that the switch between these telomeric strategies has happened several times, reinforcing the concept that there must be advantages/tradeoffs between each pattern of telomere biology. One working hypothesis is that if a long-lived animal that used replicative aging as a tumor-protection mechanism occupied a short-lived niche, it would then be investing excess resources in tumor protection. Since it would already have adequate DNA-repair/immune surveillance/etc. mechanisms to prevent tumors during its short lifespan without the additional barrier of replicative aging, it might abandon replicative aging if there was a compensatory advantage. One such advantage might come from the ability to reduce levels of oxidative protection. Telomeres are very sensitive to oxidative damage, both because triplet Gs are a preferential target for free radicals [169] and there are triplet Gs within every TTAGGG repeat, and because the proteins/structures that hide the end of the chromosome from being recognized as a double strand break also reduce the efficiency with which oxidative damage is recognized and repaired [170]. Having very long telomeres would permit one to lose large pieces due to oxidative damage without denuding the telomere, and not repressing telomerase would permit the repair and elongation of excessively shortened telomeres. Thus abandoning small telomeres that shortened in a well-regulated fashion to count cell divisions to serve as a tumor protection mechanism in favor of very large telomeres and not repressing telomerase would have the potential advantage of permitting a reduction in the energy invested in oxidative damage protection. We have observed a good correlation between having long telomeres and telomerase activity and the rapid appearance of culture stasis, indicating that in general species using the long-telomere strategy are sensitive to the stresses of the tissue culture environment. Preliminary studies in members of several orders of mammals suggest that a much greater correlation exists between resistance to some inducers of oxidative stress and the telomere strategy of the group than between resistance and lifespan [147, 171–174]. Species that use replicative aging seem to have better cellular protection/repair mechanisms to some types of stress than species that are telomerase positive that do not use this tumor protection mechanism.

This ongoing study has been providing insights into the role of mammalian telomeres as tumor protectors, novel ALT mechanisms, telomere regulatory strategies and the role of replicative senescence in human aging. The results of these experiments should help to clarify the biological importance and evolutionary flexibility of telomere-based replicative aging.

Animal Cloning

An initial report comparing telomere lengths of sheep derived by natural mating and nuclear transfer suggested that somatic telomeres decrease in length with age, and that Dolly, derived by the transfer of a 6 year-old adult somatic nucleus, exhibited diminished telomere lengths [175]. This was proposed to limit the utility of cloning for replacement of cells and tissue for human transplantation. However, the reported difference was well within the normal TRF variation range. Given the activation of telomerase at the blastocyst stage, reprogramming of the adult nucleus is likely to involve reactivation of telomerase and resetting of the telomeres to normal levels [16]. In fact, subsequent studies have shown that aged adult fibroblasts were suitable as nuclear donors [176]. In cloned calves derived from senescent donors, somatic cell nuclear transfer prolonged the replicative lifespan of senescent cells and telomeres were extended beyond those of newborn (2 weeks old) and age-matched control animals [73]. Moreover, telomerase activity has been found in the blastocysts of post-clonal embryos, independently of the age of the nuclear donor [16, 177]. The ability of nuclear transfer to restore somatic cells to a phenotypically young state has important implications for agriculture and medicine [73].

Conclusion

Telomerase plays a vital role in chromosomal maintenance and stability in unicellular and multicellular organisms. In invertebrates, fish, amphibian, and reptiles persistent telomerase activity in somatic tissues also allows the maintenance of the incredible regenerative potentials of these species. The lack of telomerase repression in poikylotherms suggests that these animals do not use replicative aging, and that replicative aging may have evolved to provide an additional barrier to tumor protection only under the additional mutational load that occurs in eutherians. In birds and many mammals, the efficient tissue repression of telomerase suggests that they might use replicative aging as a tumor protection mechanism, similar to humans, while other mammals appear to have adopted another telomere strategy that has abandoned replicative aging. The link between replicative senescence and aging is more controversial but it has been established in some age-related human diseases [26]. Also, the role of telomeres and telomerase regulation in embryonic and adult stem cells has placed telomerase "back in the game" of this exploding field of stem cell biology [26]. The understanding of telomere biology has already led to the development of several telomerase inhibitor drugs that are in advanced clinical trials and can soon be part of human chemotherapy cocktails [178]. Telomerase activators which can potently lead to increased tissue regeneration are already commercialized in the United States [179]. The recent addition of more species to the genome and protein databases, will allow an emergence of more in depth studies on the role of the shelterin proteins in telomeric regulation during development and aging in many multicellular organisms.

Acknowledgments We thank Agnel Sfeir for Fig. 1.

References

1. McClintock B (1941). The stability of broken ends of chromo somes in Zea mays. *Genetics* 26: 234–282.
2. Muller HJ (1938). The remaking of chromosomes. *The Collecting Net* 13: 181–198.
3. DeLange T (2005). Shelterin: the protein complex that shapes and safeguards human telomeres. *Genes Dev* 19: 2100–2110.
4. Blackburn EH and Gall JG (1978). A tandemly repeated sequence at the termini of the extrachromosomal ribosomal RNA genes in *Tetrahymena*. *J Mol Biol* 120: 33–53.
5. Moyzis RK, Buckingham JM, Cram LS, Dani M, Deaven LL, Jones MD, Meyne J, Ratliff RL, and Wu JR (1988). A highly conserved repetitive DNA sequence, $(TTAGGG)_n$, present at the telomeres of human chromosomes. *Proc Natl Acad Sci U S A* 85: 6622–6626.
6. Watson JD (1972). Origin of concatameric T4 DNA. *Nature* 239: 197–201.
7. Olovnikov AM (1973). A theory of marginotomy. The incomplete copying of template margin in enzymic synthesis of polynucleotides and biological significance of the phenomenon. *J Theor Biol* 41: 181–190.
8. Shay JW and Wright WE (2000). Hayflick, his limit, and cellular ageing. *Nature Rev Mol Cell Biol* 1: 72–76.
9. Shay JW and Wright WE (2004). Senescence and immortalization: role of telomeres and telomerase. *Carcinogenesis* 26: 867–874.

10. Hayflick L and Moorhead PS (1961). The serial cultivation of human diploid cell strains. *Exp Cell Res* 25: 585–621.
11. Shay JW and Wright WE (2001). Forward: aging and cancer: are telomeres and telomerase the connection? In *Telomerase, Aging and Disease*. (MP Mattson, ed.). Baltimore, MD: Elsevier, p. 231.
12. Krtolica A, Parrinello S, Lockett S, Desprez P-Y, and Campisi J (2001). Senescent fibroblasts promote epitelial cell growth and tumorigenesis: a link between cancer and aging. *Proc Natl Acad Sci U S A* 98: 12072–12077.
13. Greider CW and Blackburn EH (1985). Identification of a specific telomere terminal transferase activity in *Tetrahymena* extracts. *Cell* 43: 405–413.
14. Morin GB (1989). The human telomere terminal transferase enzyme is a ribonucleoprotein that synthesizes TTAGGG repeats. *Cell* 59: 521–529.
15. Kim NW, Piatyszek MA, Prowse KR, Harley CB, West MD, Ho PL, Coviello GM, Wright WE, Weinrich SL, and Shay JW (1994). Specific association of human telomerase activity with immortal cells and cancer. *Science* 266: 2011–2015.
16. Forsyth NR, Wright WE, and Shay JW (2002). Telomerase and differentiation in multicellular organisms: turn it off, turn it on, and turn it off again. *Differentiation* 69: 188–197.
17. Bodnar AG, Ouellete M, Frolkis M, Kolt SE, Chiu C-P, Morin GB, Harley CB, Shay JW, Lichtsteiner S, and Wright WE (1998). Extension of lifespan by introduction of telomerase into normal human cells. *Science* 279: 349–352.
18. Griffith JD, Comeau L, Rosenfield S, Stansel RM, Bianchi A, Moss H, and de Lange T (1999). Mammalian telomeres end in a large duplex loop. *Cell* 97: 503–514.
19. Wright WE and Shay JW (2000). Telomere dynamics in cancer progression and prevention: fundamental differences in human and mouse telomere biology. *Nat Med* 6: 849–851.
20. Shay JW and Roninson IB (2004). Hallmarks of senescence in carcinogenesis and cancer therapy. *Oncogene* 23: 2919–2933.
21. Bryan TM, Englezou A, Dalla-Pozza L, Dunham MA, and Reddel RR (1997). Evidence for an alternative mechanism for maintaining telomere length in human tumors and tumor-derived cell lines. *Nat Med* 3: 1271–1274.
22. Teixeira MT, Arneric M, Sperisen P, and Lingner J (2004). Telomere length homeostasis is achieved via a switch between telomerase- extendible and -nonextendible states. *Cell* 117: 323–335.
23. Baur JA, Zou Y, Shay JW, and Wright WE (2001). Telomere position effect in human cells. *Science* 292: 2075–2077.
24. Crabbe L, Jauch A, Naeger CM, Holt-Grez H, and Karlseder J (2007). Telomere dysfunction as a cause of genomic instability in Werner syndrome. *Proc Natl Acad Sci U S A* 104: 2205–2210.
25. Armanios MY, Chen JJ, Cogan JD, Alder JK, Ingersoll RG, Markin C, Lawson WE, Xie M, Vulto I, Phillips JA, 3rd, Lansdorp PM, Greider CW, and Loyd JE (2007). Telomerase mutations in families with idiopathic pulmonary fibrosis. *N Eng J Med* 356: 1370–1372.
26. Blasco MA (2007). Telomere length, stem cells and aging. *Nat Chem Biol* 3: 640–649.
27. DeLange T (2004). T-loops and the origin of telomeres. *Nat Rev Mol Cell Biol* 5: 323–329.
28. Lundblad V and Blackburn EH (1993). An alternative pathway for yeast telomere maintenance rescues est1 – senescence. *Cell* 73: 347–360.
29. Pluta AF, Kaine BP, and Spear BB (1982). The terminal organization of macromolecular DNA in *Oxytricha fallax*. *Nucleic Acids Res* 10: 8145–8154.
30. Walter MF, Bozorgnia L, Mahesshwari A, and Biessmann H (2001). The rate of terminal nucleotide loss from a telomere of the mosquito *Anopheles gambiae*. *Insect Mol Biol* 10: 105–110.
31. Coleman MJ, McHale MT, Arnau J, Watson A, and Oliver RP (1993). Cloning and characterization of telomeric DNA from *Cladosporium fulvum*. *Gene* 132: 67–73.
32. Shampay J, Szostak JW, and Backburn EH (1984). DNA sequences of telomeres maintained in yeast. *Nature* 310: 154–157.

33. McEachern MJ and Hicks JB (1993). Unusually large telomeric repeats in the yeast *Candida albicans*. *Mol Cell Biol* 13: 551–560.
34. Wood JG and Sinclair DA (2002). TPE or not TPE? It's no longer a question. *Trends Pharmacol Sci* 23: 1–4.
35. Forstemann K, Zaug AJ, Cech TR, and Lingner J (2003). Yeast telomerase is specialized for C/A-rich RNA templates. *Nucleic Acids Res* 31: 1646–1655.
36. Tomaska L, McEachern MJ, and Nosek J (2004). Alternatives to telomerase: keeping linear chromosomes via telomeric circles. *FEBS Lett* 567: 142–146.
37. Nosek J, Rycovska A, Makhov AM, Griffith JD, and Tomaska L (2005). Amplification of telomeric arrays via rolling-circle mechanism. *J Biol Chem* 280: 10840–10845.
38. Fajkus J, Sykorova E, and Leitch AR (2005). Telomeres in evolution and evolution of telomeres. *Chromosome Res* 13: 469–479.
39. Galy V, Olivo-Marin J-C, Scherthan H, Doye V, Rascalou N, and Nehrbass U (2000). Nuclear pore complexes in the organization of silent telomeric chromatin. *Nature* 403: 108–112.
40. Figueiredo L and Scherf A (2005). Plasmodium telomeres and telomerase: the usual actor in an unusual scenario. *Chromosome Res* 13: 517–524.
41. Horn D and Barry D (2005). The central roles of telomeres and subtelomeres in antigenic variation in African trypanosomes. *Chromosome Res* 13: 525–533.
42. Munoz-Jordan JL, Cross GA, de Lange T, and Griffith JD (2001). T-loops at trypanosome telomeres. *EMBO J* 20: 579–588.
43. Murphy WJ, Eizirik E, O'Brien SJ, Madsen O, Scally M, Douady CJ, Teeling E, Ryder OA, Stanhope MJ, de Jong WW, and Springer MS (2001). Resolution of the early placental mammal radiation using Bayesian phylogenetics. *Science* 294: 2348–2351.
44. Cox AV, Bennett ST, Parokonny AS, Kenton A, Callimassia MA, and Bennett MD (1993). Comparison of plant telomere locations using a PCR-generated synthetic probe. *Ann Bot* 72: 239–247.
45. Fuchs J, Brandes A, and Schubert I (1995). Telomere sequence localization and karyotype evolution in higher plants. *Plant Syst Evol* 196: 227–241.
46. Flanary BE and Kletetschka G (2005). Analysis of telomere length and telomerase activity in tree species of various life-spans, and with age in the bristlecone pine *Pinus longaeva*. *Biogerontology* 6: 101–111.
47. Flanary BE and Kletetschka G (2006). Analysis of telomere length and telomerase activity in tree species of various lifespans, and with age in the bristlecone pine *Pinus longaeva*. *Rejuvenation Res* 9: 61–63.
48. Schulman E (1958). Bristlecone pine, oldest known living thing. *Natl Geogr* 113: 354–372.
49. McKnight TD, Riha K, and Shippen DE (2002). Telomeres, telomerase, and stability of the plant genome. *Plant Mol Biol* 48: 331–337.
50. Muller F, Wicky C, Spicher A, and Tobler H (1991). New telomere formation after developmentally regulated chromosomal breakage during the process of chromatin diminution in *Ascaris lumbricoides*. *Cell* 67: 815–822.
51. Richards EJ and Ausubel FM (1888). Isolation of a higher eukaryotic telomere from *Arabidopsis thaliana*. *Cell* 53: 127–136.
52. Fajkus J, Kovarik A, Kralovics R, and Bezdek M (1995). Organization of telomeric and subtelomeric chromatin in the higher plant *Nicotiana tabacum*. *Mol Gen Genet* 247: 633–638.
53. Riha K, McKnight TD, Griffing LR, and Shippen DE (2001). Living with genome instability: plant responses to telomerase dysfunction. *Science* 291: 1797–1800.
54. Fitzgerald PH and Morris CM (1984). Telomeric association of chromosomes in B-cell lymphoid leukemia. *Hum Genet* 67: 385–390.
55. Walbot V (1996). Sources and consequences of phenotypic and genotypic plasticity in flowering plants. *Trends Plant Sci* 1: 27–32.

56. Fouche N, Cesare AJ, Willcox S, Ozgur S, Compton SA, and Griffith JD (2006). The basic domain of TRF2 directs binding to DNA junctions irrespective of the presence of TTAGGG repeats. *J Biol Chem* 281: 37486–37495.
57. Ogilvie M, Yu X, Nicolas-Metral V, Pulido SM, Liu C, Ruegg UT, and Noguchi CT (2000). Erythropoietin stimulates proliferation and interferes with differentiation of myoblasts. *J Biol Chem* 275: 39754–39761.
58. Traut W, Szczepanowski M, Vitkova M, Opitz C, Marec F, and Zrzavy J (2007). The telomere repeat motif of basal Metazoa. *Chromosome Res* 15: 371–382.
59. Podlevsky JD, Bley CJ, Omana RV, Qi X, and Chen J, (2007). The telomerase database, *Nucleic Acids Res*. D339–D343 http://telomerase.asu.edu/
60. Finch CE (1990). *Longevity, Senescence, and the Genome*. Chicago, IL: University of Chicago Press, p. 922.
61. Koziol C, Borojevic R, Steffen R, and Muller WEG (1998). Sponges (Porifera) model systems to study the shift from immortal to senescent somatic cells: the telomerase activity in somatic cells. *Mech Ageing Dev* 100: 107–120.
62. Muller WEG and Muller IM (2003). The hypothetical ancestral animal. The Urmetazoa: telomerase activity in sponges (Porifera). *J Serb Chem Soc* 68: 257–268.
63. Rinkevich B and Loya Y (1986). Senescence and dying signals in a reef-building coral. *Experientia* 42: 320–322.
64. Laird DJ and Weissman L (2004). Telomerase maintained in self-renewing tissues during serial regeneration of the urochordate *Bothryllus schosserii*. *Dev Biol* 273: 185–194.
65. Castro LFC and Holand P (2002). Fluorescent in situ hybridisation to amphioxus chromosomes. *Zool Sci* 19: 1349–1353.
66. Plohl M, Prats E, Martinez-Lage A, Gonzalez-Tizon A, Mendez J, and Cornudella L (2002). Telomeric localization of the vertebrate type hexamer repeat $(TTAGGG)_n$ in the wedgeshell clam *Donax trunculus* and other marine invertebrate genomes. *J Biol Chem* 277: 19839–19846.
67. Joffe BI, Solovei I, and Macgregor HC (1998). Ends of chromosomes in *Polycelis tenuis* (Platyhelminthes) have telomeric epeat TTAGGG. *Chromosome Res* 4: 323–324.
68. Jha AN, Dominguez I, Balajee AS, Hutchinson TH, Dixon DR, and Natarajan AT (1995). Localization of a vertebrate telomeric sequence in the chromosomes of two marine worms (phylum Annelida: class polychaeta). *Chromosome Res* 3: 507–508.
69. Vitturi R, Colomba MS, Gianguzza P, and Pirrone AM (2000). Chromosomal location of ribosomal DNA (rDNA), $(GATA)_n$ and $(TTAGGG)_n$ telomeric repeats in the neogastropod *Fasciolaria lignaria* (Mollusca: Prosobranchia). *Genetica* 108: 253–257.
70. Wang YP and Guo XM (2001). Chromosomal mapping of the vertebrate telomeric sequence $(TTAGGG)_n$ in four bivalve molluscs by fluorescence in situ hybridization. *J Shellfish Res* 20: 1187–1190.
71. Vitkova M, Kral J, Traut W, Zrzavy J, and Marec F (2005). The evolutionary origin of insect telomeric repeats, $(TTAGG)_n$. *Chromosome Res* 13: 145–156.
72. Lejnine S, Makarov VL, and Langmore JP (1995). Conserved nucleoprotein structurear the ends of vertebrate and invertebrate chromosomes. *Proc Natl Acad Sci U S A* 94: 2393–2397.
73. Lanza RP, Cibelli JB, Blackwell C, Cristofalo VJ, Francis MK, Baerlocher GM, Mak J, Schertzer M, Chavez EA, Sawyer N, Lansdorp PM, and West MD (2000). Extension of cell life-span and telomere length in animals cloned from senescent somatic cells. *Science* 288: 665–669.
74. Izbicka E, Nishioka D, Marcell V, Raymond E, Davidson KK, Lawrence RA, Lawrence RA, Wheelhouse RT, Hurley L, Wu RS, and Von Hoff D (1999). Telomere-interactive agents affect proliferation rates and induce chromosomal destabilization in sea urchin embryos. *Anticancer Drug Des* 4: 355–365.
75. www.pathology-registry.org

76. Guo X and Allen SJ (1997). Fluorescence in situ hybridization of vertebrate telomere sequence to chromosome ends of the pacific oyster, *Crassostrea gigas thunberg*. *J Shellfish Res* 16: 87–89.
77. Estabrooks SL (1999). The telomeres of the bay scallop, *Argopecten irradians* (Lamark). *J Shellfish Res* 18: 401–404.
78. Niedermaier J and Moritz KB (2000). Organization and dynamics of satellite and telomere DNAs in *Ascaris*: implications for formation and programmed breakdown of compound chromosomes. *Chromosoma* 109: 439–452.
79. Wicky C, Villeneuve AM, Lauper N, Codoureyy L, Tobler H, and Muller F (1996). Telomeric repeats $(TTAGGC)_n$ are sufficient for chromosome capping function in Caenorhabditis elegans. *Proc Natl Acad Sci U S A* 93: 8983–8988.
80. Hirai H and LoVerde PT (1996). Identification of the telomeres on *Schistosoma mansoni* chromosomes by FISH. *J Parasitol* 82: 511–512.
81. Klapper W, Heidorn K, Kuhne K, Parwaresch R, and Krupp G (1998). Telomerase activity in 'immortal' fish. *FEBS Lett* 434: 409–412.
82. Okazaki S, Tsuchida K, Mackawa H, and Fugiwara H (1993). Identification of a pentanucleotide telomere sequence $(TTAGG)_n$ in the silkworm Bombyx mori and in other insects. *Mol Cell Biol* 13: 1424–1432.
83. Meyne JH and Imai HT (1995). FISH analysis of the telomere sequences of bulldog ants (Myrmecia: Formicidae). *Chromosoma* 104: 14–18.
84. Seluanov A, Chen Z, Hine C, Sasahara TH, Ribeiro AA, Catania KC, Presgraves DC, and Gorbunova V (2007). Telomerase activity coevolves with body mass not lifespan. *Aging Cell* 6: 45–52.
85. Frydrychova R, Grossmann P, Trubac P, Vitkova M, and Marec F (2004). Phylogenetic distribution of TTAGG telomeric repeats in insects. *Genome* 47: 163–178.
86. Sasaki T and Fujiwara H (2000). Detection and distribution patterns of telomerase activity in insects. *Eur J Biochem* 267: 3025–3031.
87. Osanai M, Kojima KK, Futahashi R, Yaguchi S, and Fujiwara H (2006). Identification and characterization of the telomerase reverse transcriptase of *Bombyx mori* (silkworm) and *Tribolium castaneum* (flour beetle). *Gene* 376: 281–289.
88. Frydrychova R and Marec F (2002). Repeated losses of TTAGG telomere repeats in evolution of beetles (Coleoptera). *Genetica* 115: 179–187.
89. Nagaraju K, Casciola-Rosen L, Rosen A, Thompson C, Loeffler L, Parker T, Danning C, Rochon PJ, Gillespie J, and Plotz P (2000). The inhibition of apoptosis in myositis and in normal muscle cells. *J Immunol* 164: 5459–5465.
90. Biessmann H and Mason JM (1997). Telomere maintenance without telomerase. *Chromosoma* 106: 63–69.
91. Mason JM and Biessmann H (1995). The unusual telomeres of *Drosophila*. *Trends Genet* 11: 58–62.
92. Mikhailovsky S, Belenkaya T, and Georgiev P (1999). Broken chromosomal ends can be elongated by conversion in *Drosophila melanogaster*. *Chromosoma* 108: 114–120.
93. Kahn T, Savitsky M, and Georgiev P (2000). Attachment of HeTA sequences to chromosomal termini in *Drosophila* may occur by different mechanisms. *Mol Cell Biol* 20: 7634–7642.
94. Meyne J, Ratliff RL, and Moyzis RK (1989). Conservation of the Human Telomere Sequence $(TTAGGG)_n$ among Vertebrates. *Proc Natl Acad Sci U S A* 86: 7049–7053.
95. Chen J-L, Blasco MA, and Greider CW (2000). Secondary structure of vertebrate telomerase RNA. *Cell* 100: 503–514.
96. Patbaik B, Mahapatro N, and Jena B (1994). Ageing in fishes. *Gerontology* 40: 113–132.
97. Mommsen TP (2001). Paradigms of growth in fish. *Comp Biochem Physiol B Biochem Mol Biol* 129: 207–219.
98. Lau BW-M, Wong AO-L, Tsao GS-W, So K-F, and Yip HK-F (2008). Molecular cloning and characterization of the zebrafish (Danio rerio) telomerase catalytic subunit (telomerase reverse transcriptase, TERT). *J Mol Neurosci* 34: 63–75.

99. Tsepkin YA and Sokolov LI (1971). The maximum size and age of some sturgeons. *J Ichthyol* 11: 444–446.
100. Rocco L, Costagliola D, and Stingo V (2001). $(TTAGG)_n$ telomeric sequence in selachian chromosomes. *Heredity* 87: 583–588.
101. McChesney PA, Aisner DL, Frank BC, Wright WE, and Shay JW (2000). Telomere dynamics in cells with introduced telomerase: a rapid assay for telomerase activity on telomeres. *Mol Cell Biol Res Commun* 3: 312–318.
102. Rocco L, Morescalchi MA, Costagliola D, and Stingo V (2002). Karyotype and genome characterization in four cartilaginous fishes. *Gene* 295: 289–298.
103. Kishi S, Uchiyama J, Baughman AM, Tadateru G, Lin MC, and Stephanie BT (2003). The zebrafish as a vertebrate model of funcional aging and very gradual senescence. *Exp Gerontol* 38: 777–786.
104. Pierce AJ, Johnson RD, Thompson LH, and Jasin M (1999). XRCC3 promotes homology-directed repair of DNA damage in mammalian cells. *Genes Dev* 13: 2633–2638.
105. Bradford GB, Williams B, Rossi R, and Bertoncello I (1997). Quiescence, cycling, and turnover in the primitive hematopoietic stem cell compartment. *Exp Hematol* 25: 445–453.
106. Xie M, Mosig A, Qi X, Li Y, Stadler PF, and Chen J (2008). Structure and function of the smallest vertebrate telomerase RNA from teleost fish. *J Biol Chem* 283: 2049–2059.
107. Jeyapalan JN, Varley H, Foxon JL, Pollock RE, Jeffreys AJ, Henson JD, Reddel RR, and Royle NJ (2005). Activation of the ALT pathway for telomere maintenance can affect other sequences in the human genome. *Hum Mol Genet* 14: 1785–1794.
108. Bassham S, Beam A, and Shampay J (1998). Telomere variation in Xenopus laevis. *Mol Cell Biol* 18: 269–275.
109. Kuramoto M, Ohsumi K, Kishimoto T, and Ishikawa F (2001). Identification and analyses of the Xenopus TERT gene that encodes the catalytic subunit of telomerase. *Gene* 277: 101–110.
110. Mantell LL and Greider CW (1994). Telomerase activity in germline and embryonic cells of Xenopus. *EMBO J* 13: 3211–3217.
111. Bousman S, Schneider G, and Shampay J (2003). Telomerase activity is widespread in adult somatic tissues of xenopus. *J Exp Zool (Mol Dev Evol)* 295B: 82–86.
112. Christiansen JL, Henderson ER, Budke B, Lynch M, Lu Q, and Johnson J (2001). A final report of studies of the Hayflick limit in reptiles, a test of potential immortality. In *Proceedings of the Iowa Space Grant Consortium.*
113. Christiansen J, Johnson J, Henderson ER, Budke B, and Lynch M (2001). The relationship between telomeres, telomerase, reptilian lifespan, and reptilian tissue regeneration. In *Proceedings of the Iowa Space Grant Consortium.*
114. Simpson SB and Rauch DM (1989). Cells from the lizard anolis do not exhibit senescence. *Gerontologist*, Special Issue, 29: 284A (Abstract).
115. Girondot M and Garcia J, *Senescence and longevity in turtles: what telomeres tell us*, in *Proceedings of the 9th Ordinary General Meeting of the Societas Europaea Herpetologica In Current Studies in Herpetology:*, C.M.a. R. Guyetant, Editor. 1999: Le Bourget du Lac, France, p. 133–137.
116. Carr A and Goodman D (1970). Ecologic implications of size and growth in Chelonia. *Copeia* 4: 783–786.
117. Goldstein S (1974). Growth of cultured cells from the galapagos tortoise. *Exp Cell Res* 83: 279–302.
118. Ogburn CE, Carlberg K, Ottinger MA, Holmes DJ, Martin GM, and Austad SN (2001). Exceptional cellular resistance to oxidative damage in long-lived birds requires active gene expression. *J Gerontol A Biol Sci Med Sci* 56: B468–B474.
119. Scharff C, Kirn JR, Grossman M, Macklis J, and Nottebohm F (2000). Targeted neuronal death affects neuronal replacement and vocal behavior in adult songbirds. *Neuron* 25: 481–492.
120. Delany ME and Daniels LM (2004). The chicken telomerase reverse transcriptase (chTERT): molecular and cytogenetic characterization with a comparative analysis. *Gene* 339: 61–69.

121. Delany ME, Krupkin AB, and Miller MM (2000). Organization of telomere sequences in birds: evidence for arrays of extreme length and for in vivo shortening. *Cytogenet Cell Genet* 90: 139–145.
122. Delany ME, Daniels LM, Swanberg SE, and Taylor HA (2003). Telomeres in the chicken: genome stability and chromosome ends. *Poult Sci* 82: 917–926.
123. Nanda I, Schrama D, Feichtinger W, Haaf T, Schartl M, and Schmid M (2002). Distribution of telomeric $(TTAGGG)_n$ sequences in avian chromosomes. *Chromosoma* 111: 215–227.
124. Taylor G, Lehrer MS, Jensen PJ, Sun TT, and Lavker RM (2000). Involvement of follicular stem cells in forming not only the follicle but also the epidermis. *Cell* 102: 451–461.
125. Hiyama E, Kiyama K, Tatsumoto N, Kodama T, Shay J, and Yokoyama T (1996). Telomerase activity in human intestine. *Int J Oncol* 9: 453–458.
126. Venkatesan RN and Price C (1998). Telomerase expression in chickens: constitutive activity in somatic tissues and down-regulation in culture. *Proc Natl Acad Sci U S A* 95: 14763–14768.
127. Haussmann MF, Winkler DW, O'Reilly KM, Huntington CE, Nisbet ICT, and Vleck CM (2003). Telomeres shorten more slowly in long-lived birds and mammals than in short-lived ones. *Proc R Soc Lond B* 270: 1387–1392.
128. Siegfried M (1983). Neoplasms identified in free-living birds. *Avian Dis* 27: 86–99.
129. Haussmann MF, Winkler DW, Huntington CE, Nisbet ICT, and Vleck CM (2007). Telomerase activity is maintained throughout the lifespan of long-lived birds. *Exp Gerontol* 42: 610–618.
130. Galkina S, Lukina N, Zakharova K, and Rodionov AV (2005). Interstitial $(TTAGGG)_n$ sequences are not hot spots of recombination in the chicken lampbrush macrochromosomes 1-3. *Chromosome Res* 13: 551–557.
131. Sherr CJ and DePinho RA (2000). Cellular senescence: mitotic clock or culture shock? *Cell* 102: 407–410.
132. Prowse KR and Greider CW (1995). Developmental and tissue-specific regulation of mouse telomerase and telomere length. *Proc Natl Acad Sci U S A* 92: 4818–4822.
133. von Zglinicki T, Serra V, Lorenz M, Saretzki G, Lenzen-Grossimlighaus R, Gessner R, Risch A, and Steinhagen-Thiessen E (2000). Short telomeres in patients with vascular dementia: an indicator of low antioxidative capacity and a possible risk factor? *Lab Invest* 80: 1739–1747.
134. Oh H, Taffet GE, Youker KA, Entman ML, Overbeek PA, Michael LH, and Schneider MD (2001). Telomerase reverse transcriptase promotes cardiac muscle cell proliferation, hypertrophy and survival. *Proc Natl Acad Sci U S A* 98: 10308–10313.
135. Coviello-McLaughlin GM and Prowse KR (1997). Telomere length regulation during postnatal development and ageing in Mus spretus. *Nucleic Acids Res* 25: 3051–3058.
136. Martin-Rivera L, Herrera E, Albar JP, and Blasco MA (1998). Expression of mouse telomerase catalytic subunit in embryos and adult tissues. *Proc Natl Acad Sci U S A* 95: 10471–10476.
137. Carman TA, Afshari CA, and Barrett JC (1998). Cellular senescence in telomerase-expressing syrian hamsterembryo cells. *Exp Cell Res* 24: 33–42.
138. Rohme D (1981). Evidence for a relationship between longevity of mammalian species and lifespans of normal fibroblasts in vitro and erytrocytes in vivo. *Proc Natl Acad Sci U S A* 78: 5009–5013.
139. Lorenzini A, Tresini M, Austad SN, and Cristofalo VJ (2005). Cellular replicative capacity correlates primarily with species body mass not longevity. *Mech Ageing Dev* 126: 1130–1133.
140. Blasco MA, Lee HW, Hande MP, Samper E, Lansdorp PM, DePinho RA, and Greider CW (1997). Telomere shortening and tumor formation by mouse cells lacking telomerase RNA. *Cell* 91: 25–34.
141. Ramirez RD, Morales CP, Herbert BS, Rohde JM, Passons C, Shay JW, and Wright WE (2001). Putative telomere-independent mechanisms of replicative aging reflect inadequate growth conditions. *Genes Dev* 15: 398–403.

142. Mathon NF, Malcolm DS, Harrisingh MC, Cheng L, and Lloyd AC (2001). Lack of replicative senescence in normal rodent glia. *Science* 291: 872–875.
143. Loo DT, Fuquay JI, Rawson CL, and Barnes DW (1987). Extended culture of mouse embryo cells without senescence: inhibition by serum. *Science* 236: 200–202.
144. Nussenzweig A, Chen C, da Costa Soares V, Sanchez M, Sokol K, Nussenzweig MC, and Li GC (1996). Requirement for Ku80 in growth and immunoglobulin V(D)J recombination. *Nature* 382: 551–555.
145. Barlow C, Hirotsune S, Paylor R, Liyanage M, Eckhaus M, Collins F, Shiloh Y, Crawley JN, Ried T, Tagle D, and Wynshaw-Boris A (1996). Atm-deficient mice: a paradigm of ataxia telangiectasia. *Cell* 86: 159–171.
146. Connor F, Bertwistle D, Mee PJ, Ross GM, Swift S, Grigorieva E, Tybulewicz VL, and Ashworth A (1997). Tumorigenesis and a DNA repair defect in mice with a truncating Brca2 mutation. *Nat Genet* 17: 423–430.
147. Parrinello S, Samper E, Krtolica A, Goldstein J, Melov S, and Campisi J (2003). Oxygen sensitivity severely limits the replicative lifespan of murine fibroblasts. *Nat Cell Biol* 5: 741–747.
148. Zalvide J and DeCaprio JA (1995). Role of pRb-related proteins in simian virus 40 large-T-antigen- mediated transformation. *Mol Cell Biol* 15: 5800–5810.
149. Davisa T, Skinnera JW, Faragherb RGA, Jonesa CJ, and Kipling D (2005). Replicative senescence in sheep fibroblasts is a p53 dependent process. *Exp Gerontol* 40: 17–26.
150. Pathak S, Multani AS, McConkey DJ, Imam AS, and Amoss MS (2000). Spontaneous regression of cutaneous melanoma in sinclair swine is associated with defective telomerase activity and extensive telomere erosion. *Int J Oncol* 17: 1219–1224.
151. McKevitt TP, Nasir L, Wallis CV, and Argyle DJ (2003). A cohort study of telomere and telomerase biology in cats. *Am J Vet Res* 64: 1496–1499.
152. Brummendorf TH, Mak J, Sabo KM, Baerlocher GM, Dietz K, Abkowitz JL, and Lansdorp PM (2002). Longitudinal studies of telomere length in feline blood cells. *Exp Hematol* 30: 1147–1152.
153. Nasir L, Devliny P, Mckevitt T, Ruttemanz G, and Argyle DJ (2001). Telomere lengths and telomerase activity in dog tissues: a potential model system to study human telomere and telomerase biology. *Neoplasia* 3: 351–359.
154. Argyle DJ, Ellsmore V, Gault EA, Munro AF, and Nasir L (2003). Equine telomeres and telomerase in cellular immortalization and ageing. *Mech Ageing Dev* 124: 759–764.
155. Steinert S, White DM, Zou Y, Shay JW, and Wright WE (2002). Telomere biology and cellular aging in nonhuman primate cells. *Exp Cell Res* 272: 146–152.
156. Wurster DH and Benirschke K (1970). Indian muntjac, Muntiacus munjak: a deer with a low diploid chromosome number. *Science* 168: 1364–1366.
157. Zou Y, Yi X, Wright WE, and Shay JW (2002). Human telomerase can immortalize Indian muntjac cells. *Exp Cell Res* 281: 63–76.
158. Baxter SM, Greizerstein MB, Kushlan DM, and Ashley GW (1993). Conformational properties of DNA hairpins with TTT and TTTT loops. *Biochemistry* 32: 8702–8711.
159. Day JP, Limoli CL, and Morgan WF (1998). Recombination involving interstitial telomere repeat-like sequences promotes chromosomal instability in Chinese hamster cells. *Carcinogenesis* 19: 259–265.
160. Zou Y, Yi X, Wright W, and Shay J (2002). Human telomerase can immortalize Indian muntjac cells. *Exp Cell Res* 281: 63–76.
161. Meyne J, Baker RJ, Hobart HH, Hsu TC, Ryder OA, Ward OG, Wiley JE, Wurster-Hill DH, Yates TL, and Moyzis RK (1990). Distribution of non-telomeric sites of the $(TTAGGG)_n$ telomeric sequence in vertebrate chromosomes. *Chromosoma* 99: 3–10.
162. Rocco L, Costagliola D, and Stingo V (2001). $(TTAGGG)_n$ telomeric sequence in selachian chromosomes. *Heredity* 87: 583–588.
163. George JC, et al. (1999). Age and growth estimates of bowhead whales (Balaena mysticetus) via aspartic acid racemization. *Can. J. Zool.* 77:571–580.

164. Rozell N (2001). Bowhead whales may be the world's oldest mammals. In *Alaska Science Forum*.
165. Forsyth NR, Elder FF, Shay JW, and Wright WE (2005). Lagomorphs (rabbits, pikas and hares) do not use telomere-directed replicative aging in vitro. *Mech Ageing Dev* 126: 685–691.
166. Austad SN (1997). *Why We Age: What Science Is Discovering About the Body's Journey Through Life*. New York: John Wiley & Sons Inc., p. 244.
167. Guyton AC and Hall JE (1966). Transport of oxygen and carbon dioxide in the blood and bodily fluids. In *Textbook of Medical Physiology*. Philadelphia, PA: Saunders, pp. 513–523.
168. Wright WE and Shay JW (2006). Inexpensive low-oxygen incubators. *Nat Protoc* 1: 2088–2090.
169. Oikawa S and Kawanishi S (1999). Site-specific DNA damage at GGG sequence by oxidative stress may accelerate telomere shortening. *FEBS Lett* 453: 365–368.
170. von Zglinicki T (2002). Oxidative stress shortens telomeres. *Trends Biochem Sci* 27: 339–344.
171. Lorenz M, Saretzki G, Sitte N, Metzkow S, and von Zglinicki T (2001). BJ fibroblasts display high antioxidant capacity and slow telomere shortening independent of hTERT transfection. *Free Radic Biol Med* 31: 824–831.
172. Masutomi K, Yu EY, Khurts S, Ben-Porath I, Currier JL, Metz GB, Brooks MW, Kaneko S, Murakami S, DeCaprio JA, Weinberg RA, Stewart SA, and Hahn WC (2003). Telomerase maintains telomere structure in normal human cells. *Cell* 114: 241–253.
173. Harper JM, Salmon AB, Leiser SF, Galecki AT, and Miller RA (2007). Skin-derived fibroblasts from long-lived species are resistant to some, but not all, lethal stresses and to the mitochondrial inhibitor rotenone. *Aging Cell* 6: 1–13.
174. Kapahi P, Boulton ME, and Kirkwood TB (1999). Positive correlation between mammalian life span and cellular resistance to stress. *Free Radic Biol Med* 26: 495–500.
175. Shiels PG, Kind AJ, Campbell KHS, Wilmut I, Waddington D, Colman A, and Schnieke AE (1999). Analysis of telomere length in dolly, a sheep derived by nuclear transfer. *Cloning* 1: 119–125.
176. Kubota C, Yamakuchi H, Todoroki J, Mizoshita K, Tabara N, Barber M, and Yang X (2000). Six cloned calves produced from adult fiboblast cells after long term culture. *Proc Natl Acad Sci U S A* 97: 990–995.
177. Betts D, Bordignon V, Hill J, Winger Q, Westhusin M, Smith L, and King W (2001). Reprograming of telomerase activity and rebuilding of telomere length in cloned cattle. *Proc Natl Acad Sci U S A* 98: 1077–1082.
178. Dikmen ZG, Gellert GC, Jackson S, Gryaznov S, Tressle R, Dogan P, Wright WE, and Shay JW (2005). In vivo inhibition of lung cancer by GRN163L: a novel human telomerase inhibitor. *Cancer Res* 65: 7866–7873.
179. http://www.tasciences.com/index.html

Cardiac Aging

Dao-Fu Dai, Robert J. Wessells, Rolf Bodmer, and Peter S. Rabinovitch

Abstract Cardiovascular diseases are the leading causes of death in the western world. The fact that cardiovascular mortality and morbidity rates increase exponentially in the elderly suggests that age per se is a major risk factor for cardiovascular diseases. Data from the Framingham Heart study and the Baltimore Longitudinal Study on Aging showed an age-dependent increase in left ventricular hypertrophy, diastolic dysfunction, atrial fibrillation, and decline in exercise capacity. Experimental evidence shows that cardiac aging in the mouse closely recapitulates that found in humans. The evolutionary conservation of intrinsic cardiac aging is demonstrated by studies in *Drosophila melanogaster*, and this model offers unique genetic insights into cardiac aging. In this chapter we summarize the biology of cardiac aging in humans, rodents, flies, dogs and primates. Murine and *Drosophila* models of cardiac aging have been valuable to elucidate the molecular mechanisms of cardiac aging and increase vulnerability in the aged heart. This chapter highlights the mechanistic role of mitochondrial oxidative stress, insulin-IGF1, the renin-angiotensin system and adrenergic signaling, as well as the aging of cardiac stem cells. The mechanism of progression to heart failure in aged hearts and the effect of dietary restriction are also discussed. As the number of elderly persons is predicted to double in the next 25 years and the prevalence of age-related cardiovascular disabilities continues to increase, there is an urgent need to understand the biology of the aging heart, the mechanisms for age-mediated cardiac vulnerability and to use these insights to develop strategies to ameliorate myocardial dysfunction in the elderly.

Keywords Aging · Cardiac · Hypertrophy · Diastolic function · Heart failure · Arrhythmia · Stem cell · Mitochondria · ROS · Angiotensin · Adrenergic · Insulin-like growth factor · Dietary restriction

---

P.S. Rabinovitch (✉)
Department of Pathology, University of Washington, Seattle, WA, USA
e-mail: PeterR@medicine.washington.edu

N.S. Wolf (ed.), *The Comparative Biology of Aging*,
DOI 10.1007/978-90-481-3465-6_12,

Introduction

Age is a major risk factor for cardiovascular disease, at least in part because it prolongs exposure to hypertension, diabetes, hypercholesterolemia, smoking and other cardiovascular risk factors. However, intrinsic cardiac aging, the slowly progressive structural changes and functional declines with age, also makes the heart more susceptible to stress and contributes to increased cardiovascular mortality and morbidity in elderly humans. Intrinsic cardiac aging is also evident in rodents, dogs and flies, even though the risk factors common in humans are generally absent in these species. Thus, these model organisms can be very useful for study of the pathophysiology and genetics of intrinsic cardiac aging.

Cardiac Aging in Humans

The importance of understanding cardiac aging is indicated by the high prevalence of cardiovascular diseases in the human geriatric population. The American Heart Association Statistics Committee Report showed that the elderly (>65 y/o) account for greater than 80% of patients with ischemic heart disease, more than 75% of patients with congestive heart failure and greater than 70% of patients with atrial fibrillation, which is known to be a major risk factor for thromboembolic stroke (AHA Statistics Committee report, 2007 update [1]). Furthermore, cardiovascular diseases are the leading cause of human death, as shown by an exponential increase in mortality rate due to coronary heart disease, cardiomyopathy and heart failure in the US elderly population (NHLBI mortality and morbidity chart book [2]). This is illustrated in Fig. 1.

Echocardiography in healthy populations from the Framingham Heart Study and Baltimore Longitudinal Study on Aging (BLSA) showed an age-dependent increase in the prevalence of left ventricular hypertrophy, a decline in diastolic function, and relatively preserved systolic function at rest but a decline in exercise capacity, as well as an increase in the prevalence of atrial fibrillation (reviewed in Lakatta and Levy, 2003 [3–5]). These cross-sectional studies of subjects without hypertension or clinically apparent cardiovascular disease indicate that left ventricular (LV) wall thickness, as measured by echocardiography, increases progressively with age in both sexes (Fig. 2a). LV filling in early diastole is progressively compromised in age, presumably due to fibrosis and decreased elasticity of the ventricle, coupled with reduced rates of calcium reuptake in myocardial cells, which further delays relaxation. LV filling is maintained in aging largely by an increased contribution from atrial contraction, however, this itself contributes to atrial hypertrophy and an increased risk of atrial fibrillation. The Doppler measurement of mitral inflow E/A ratio, the ratio of early to late diastolic LV filling, declines dramatically with age (Fig. 2b and c); the decline in this parameter is interpreted clinically as an evidence of diastolic dysfunction. Diastolic heart failure, defined as symptoms of heart failure in the setting of diminished diastolic function but preserved systolic function, is

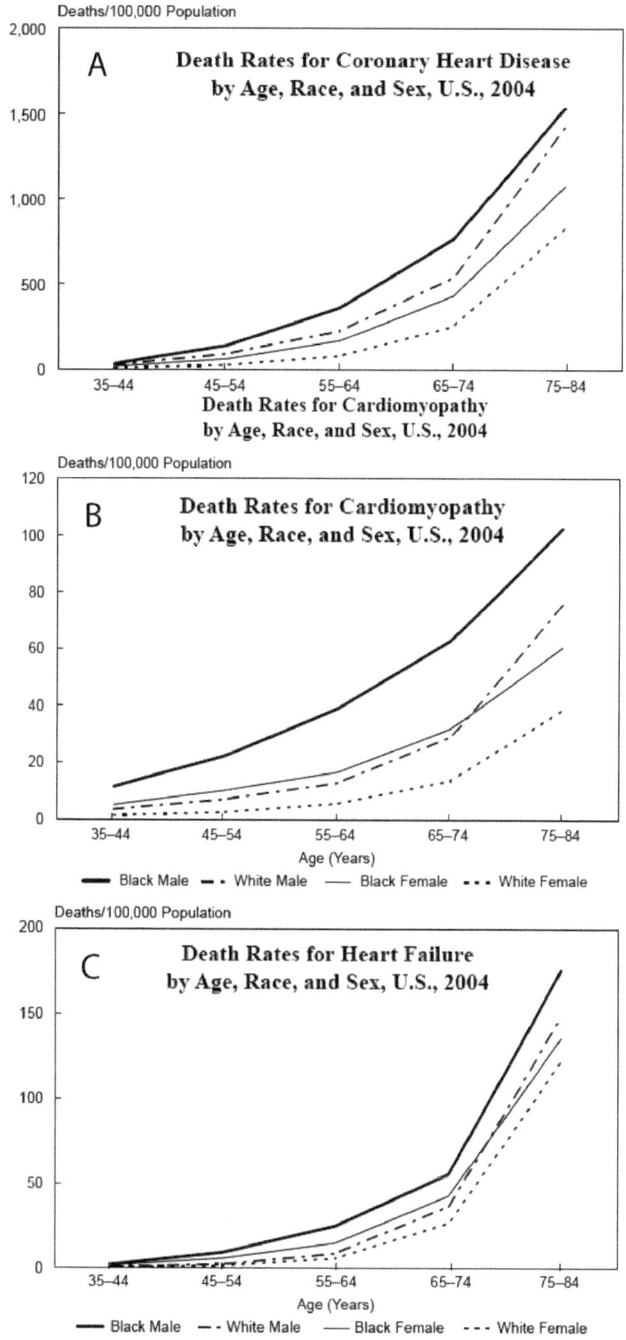

Fig. 1 Exponential increases in death rates for coronary hart disease (**a**), cardiomyopathy (**b**) and heart failure (**c**) in the US population. Data from the 2007 NHLBI Morbidity and Mortality Chart Book [2], reproduced by permission

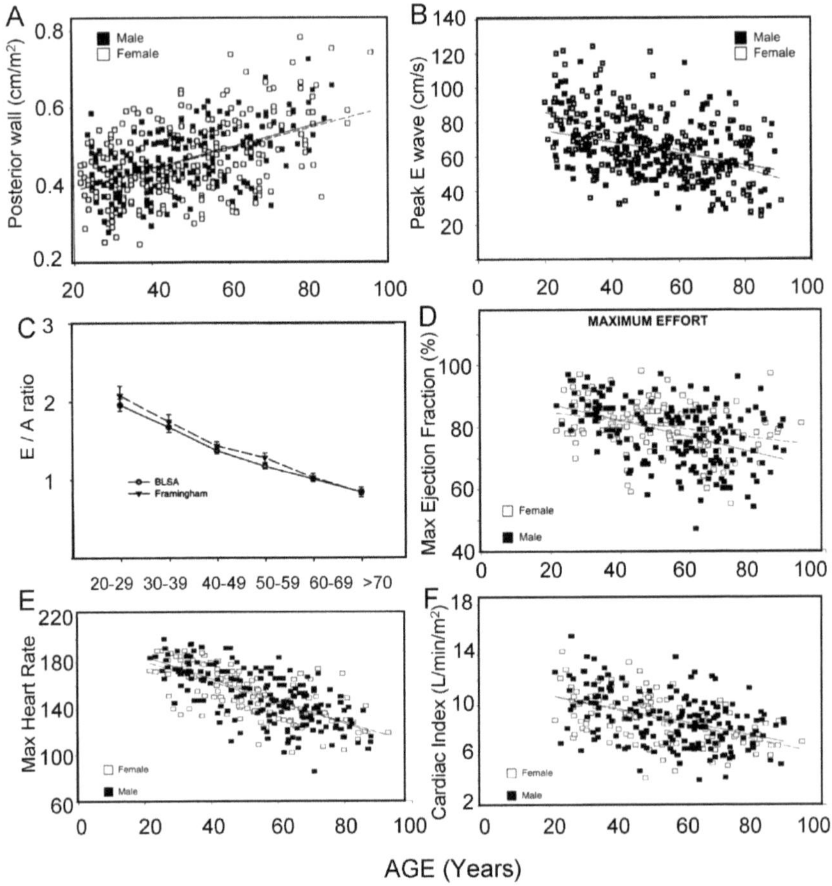

Fig. 2 Physical and functional cardiac changes with age. Left ventricular hypertrophy increases with age (**a**). A decline in diastolic function is shown by the age-associated reduction in early diastolic left ventricular filling (**b**), with an increased contribution to filling by atrial contraction (not shown), resulting in a decline in the early (ventricular relaxation) to late (atrial) filling ratio (**c**), as seen in healthy participants in both the Baltimore Longitudinal Study on Aging (BLSA) and in the Framingham Study. Left ventricular ejection fraction after maximal exercise (**d**), maximum exercise heart rate (**e**) and normalized cardiac output (**f**) are reduced in older persons. Reproduced with permission from Lakatta and Levy (2003)

pervasive in older individuals and markedly increases the risk of mortality [6]. Greater than half of individuals over the age of 75 with validated congestive heart failure had diastolic dysfunction and in many individuals this was clinically unrecognized and untreated. Diastolic dysfunction is also a major contributor to exercise intolerance in the elderly population. While the resting heart rate in the supine position does not change with age, the maximal heart rate under exhaustive exercise declines dramatically with age (Fig. 2e) [7]. The decline in maximal heart rate is the largest contribution to the reduction in maximum cardiac index (cardiac index = cardiac output normalized to body surface area) with age in healthy individuals (Fig. 2f), though the modest reduction of maximal ejection fraction with age might also contribute to reduced cardiac index and exercise capacity in the elderly (Fig. 2d and f).

These changes in physiological aging result in a severe compromise in the cardiac reserve capacity and lower the threshold for symptoms and signs of heart failure [8]. Coupled with increased exposure to several cardiovascular risk factors, this makes the aged heart much more susceptible to stresses and disease-related challenges, thus contributing to increased heart failure and cardiovascular mortality in the elderly. This is illustrated by the responses to mild myocardial ischemia or tachycardia, which may be asymptomatic in younger individual, but can precipitate the symptoms of heart failure in the elderly. Similarly, the development of atrial fibrillation, together with the consequent tachycardia, further reduces the already compromised diastolic filling and can cause acute exacerbation of heart failure symptoms.

Murine Model of Cardiac Aging

The availability of genetically modified mice and the relatively short mouse lifespan have made this a premier model of mammalian aging for gerontologic studies. Diabetes and hypertension, which are highly prevalent in the human elderly population, have been shown to accelerate cardiovascular senescence in rodents and humans [9–13], and the changes seen in diabetic or hypertensive cardiomyopathies might obscure the intrinsic cardiac aging changes. However, commonly used strains of laboratory mice do not have the same age-related cardiovascular risk factors as humans, such as elevated blood pressure or adverse blood glucose and lipid profiles [14, 15]; therefore, the cardiac changes seen in aged mice are likely to be intrinsic to cardiac aging. The histopathologic changes in old mouse hearts include subendocardial interstitial fibrosis, hyaline cytoplasmic change, vacuolization of cytoplasm, variable myocyte fiber size, hypercellularity, collapse of sarcomeres, mineralization, and arteriolosclerosis, which was designated as age-associated cardiomyopathy [16]. Morphometric analysis shows cardiomyocytes hypertrophy (increased myocardial fiber size), increased cardiomyocytes apoptosis [17] and increased deposition of collagen and amyloid [14]. Interestingly, fibrosis in aged mice was more commonly observed in the ventricular endocardium, which might due to exposure to higher wall stress in the endocardial layers.

However, the functional decline of the aging murine heart is most readily quantified using echocardiography. Echocardiography performed on a mouse longevity cohort in our lab [14] revealed that there were significant age-dependent linear trends for several cardiac parameters (Fig. 3; $p < 0.05$ for all). Left ventricular mass index (LVMI, Fig. 3a) was 76% higher in the oldest group compared to the young adult group. Left atrial dimension was significantly increased by 35% with age (Fig. 3b). Systolic function measured by fractional shortening showed a 12% decline from young adult to the oldest group (Fig. 3c). Tissue Doppler Imaging revealed an age-dependent decline in Ea/Aa, from 1.69 ± 0.3 in young adult to 0.95 ± 0.4 in the oldest (Fig. 3d). Ea/Aa by tissue Doppler imaging is the ratio of myocardial velocities of early and late ventricular filling during diastole; early filling normally predominates, but can be reduced by fibrosis or slow calcium reuptake, in which case, atrial-dependent late filling predominates. The prevalence of diastolic dysfunction, defined in humans as Ea/Aa <1 [18], was dramatically increased to 55% in the oldest age group (Fig. 3e). The myocardial performance index (calculated as the ratio of the sum of isovolemic contraction and relaxation time to LV ejection time) was significantly increased (worsened) with age [19] (Fig. 3f), consistent with the age-related declines in systolic and diastolic function. An increase in MPI indicates that a greater fraction of systole is spent to cope with the pressure changes during isovolemic phases, and has been shown to reflect both LV systolic and diastolic dysfunction [20]. These abnormalities mimic closely the age-related echocardiographic changes in human cardiac aging previously reported in healthy human populations [5]. This indicates that the mouse heart becomes hypertrophic with age and that while the systolic function defined by FS only declines slightly with age, the decline in diastolic function is more prominent. Furthermore, there is a significant decline in general myocardial performance, shown by worsening (increase) in myocardial performance index (MPI). The above changes have also been observed in aging C3H x Bl/6 and BALB/c x Bl/6 F1 mice (unpublished results).

Figure 3 also shows echocardiographic results obtained from transgenic mice that overexpress catalase that is targeted to mitochodria using the ornithine transcarbamylase mitochondrial targeting sequence (mCAT). These mice have ~20% extension of mean and maximal lifespan, while in contrast, mice overexpressing wild type human catalase (naturally delivered to peroxisomes, pCAT) or catalase

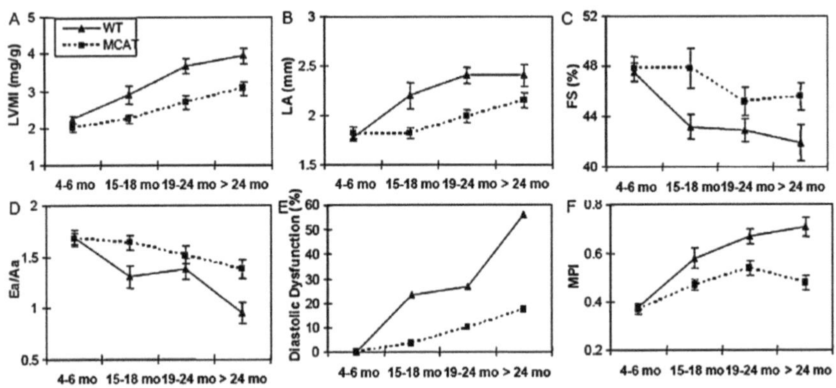

Fig. 3 Echocardiographic parameters of cardiac aging in WT and mCAT C57Bl/6 mice in the four age ranges shown (~20 mice per group/genotype, N=170 total). Left ventricular mass index (LVMI) (**a**), left atrial dimension (**b**), fractional shorting of the left ventricle (**c**), Ea/Aa (see text) (**d**), the proportion of mice with diastolic dysfunction, defined as Ea/Aa<1, (**d**) and the myocardial performance (Tei) index (**f**) were quantitated. The linear trend between genotypes is significant, p<0.01 in all cases, except fractional shortening, p<0.05

with a nuclear localizing signal (nCAT) had, at most, 0–5% extension of mean lifespan, with no effect on maximal lifespan [54]. As shown in Fig. 3, for each of the echocardiographic parameters, the mCAT genotype was protective for the rate of decline in aging cardiac function. This was most significant for diastolic dysfunction and the MPI. This is consistent with the decrease in cardiac histopathology seen in cross-sectional analysis of older mCAT mice [54] or at end of life [16]. The role of mitochondria and mt ROS in cardiac aging is discussed further below.

Molecular Mechanisms of Cardiac Aging (Data from Mice and Humans)

Mechanism of Age-Dependent LV Hypertrophy and Diastolic Dysfunction

The molecular signaling of cardiac hypertrophy is complex (see review by Heinecke et al., 2006 [21]). Age-related cardiac hypertrophy is associated with activation of the calcineurin-NFAT pathway, which has been shown to play a key role in pathological hypertrophy [21]. Calcineurin activity was increased by approximately 4-fold in the aged mouse heart [14]. This phosphatase activates the transcription factor NFAT. Activated NFAT translocates into nucleus where it interacts with the transcription factor (GATA4) to initiate transcription of hypertrophic fetal genes, such as atrial natriuretic peptides and brain natriuretic peptides (ANP and BNP). Consistently, age-related cardiac hypertrophy is also associated with GATA4 phosphorylation at Ser105, which has been reported to enhance its activity of DNA binding and transcription activation [22]. In contrast, our study showed that ERK1/2 signaling in compensated hypertrophy is not significantly changed with old age.

Changes in the function or expression of several proteins involved in excitation-contraction coupling have been observed in cardiac aging. In aged rodent hearts, downregulation of SERCA protein levels [23], concomitant with compensatory increase in the levels of Na^+/Ca^{2+} exchanger have been documented [24]. Furthermore, oxidative damage to particular cysteine thiols has been shown to impair SERCA2 activity stimulated by NO [25]. In aged mice we and others found that chronic reduction of SERCA protein level/function could lead to prolongation of Ca^{2+} decay rate, reduction in SR Ca^{2+}-load and hence smaller amplitude of Ca^{2+} transients [14, 26, 27]. Indeed, the decline in SERCA2 protein level has been shown as a major contributor to the age-dependent diastolic dysfunction [14]. It is suggested that the aged heart utilizes the compensatory increase in the L-type Ca^{2+} currents [28] and the significant prolongation of action potential duration to preserve SR loading and to keep the amplitude of intracellular Ca^{2+}-transients and contractions in old cardiomyocytes [29]. The other factors contributing to diastolic dysfunction in the aged mice include increased myocardial stiffness related to cardiac hypertrophy and fibrosis [14].

Decreased Cardiac Functional Reserve in Aging Mice and Humans

Impairment in the myocardial response to β-adrenergic stimulation and an increase in circulating catecholamines have been widely demonstrated in the aging mice and humans. This is manifested as reduced maximal heart rate and LV contractility during vigorous exercise, despite preserved heart rate and contractility at rest. The most likely cause includes chronic cathecholamine stimulation that reduces β-adrenergic responsivity, which further increases production of cathecholamine. The mechanisms underlying reduced adrenergic sensitivity include downregulation of β-receptors and the defect in the G-protein-coupled-receptor-adenylyl cyclase-cAMP-PKA signaling cascades. Previous studies documented age-related decreases in both β1 and β2-adrenergic subtype densities and a reduction in membrane adenylate cyclase activity, as well as increased the inhibitory Gαi activity and decreased levels of cAMP [30, 31]. Furthermore, upregulation of opioid peptide receptor-signaling with age also significantly dampens β-adrenergic response [32, 33]. Cardiac muscarinic receptor density and function is diminished with age and might contribute to the decrease in baroreflex activity observed in aged human subjects [33]. These age-associated changes in receptor density, function or coupling to downstream signaling mediators might significantly impair the adaptive response of the old heart to multiple stresses. Exercise training and beta-blocker treatment has been shown to ameliorate age-dependent impairment of β-adrenergic receptor signaling and enhance cardiac responsiveness to adrenergic stimulation. This resensitization effect might be mediated through reduction of G-protein receptor kinase-2 protein, a negative regulator of β-receptors-G-protein signaling [34].

Compared to the young adult myocardium, the senescent myocardium is more sensitive to ischemia and hemodynamic stress [35, 36], suggesting that several protective mechanisms in the young adult myocardium are impaired by aging. For example, several studies in rodents and humans have shown that the endogenous cardioprotective mechanism incited by repetitive ischemia (known as ischemic preconditioning) is impaired in the aged myocardium (reviewed by Juhaszova et al., 2004 [37]).The mechanism underlying this impairment include diminished Hsp 70 expression [38], reduced NO bioavailability [39, 40], damage to mitochondria that increases their susceptibility to stress (such as ischemia) and diminished PKC translocation, which is required for the protective effect of ischemic preconditioning [41, 42]. In summary, aging induces major changes in structure, function and molecular signaling in the heart which may reduce adaptative capacity to stress as well as the functional reserve, and thus increases the risk of heart failure.

Aging of Cardiac Stem/Progenitor Cells

Emerging evidences have shown that the adult heart is capable of regeneration by the resident stem and progenitor cells [43, 44]. These cells include c-kit$^+$ cells, Sca-1$^+$/c-kit$^-$ cells and side population cells (reviewed by Reinecke et al. [45]).

Beltrami et al. clonogenic and multipotent, capable of self-renewing and differentiating into myocardial, endothelial and smooth muscle lineages [44]. Oh et al. discovered another group of cardiac resident stem cells expressing stem cell antigen-1 (Sca-1) but not c-kit. These Sca-1^+/c-kit^- cells isolated from adult murine hearts were shown to differentiate into cardiomyocytes in the infarct border zone [46]. Induction of Sca-1^+/c-kit^- cells with oxytocin resulted in spontaneous beating activity in a small fraction of these cells, which also express several cardiac markers [46, 47]. Transplantation of these cells has been shown to improve cardiac function in rodent models of experimental myocardial infarction. Although these cells have been shown to play a critical role in continuous turnover of the cardiomyocytes in adult hearts, however, they fail to prevent the progression of cardiovascular diseases. A possible explanation includes limited capacity of these cells to repair or regenerate myocardium in the presence of continuous damage resulted from chronic stress, such as pressure overload (e.g. hypertension) and ischemia (e.g. coronary heart disease). In addition, aging of cardiac stem cells may impair their regenerative capacity in aged heart, either by senescence intrinsic to the stem cells or by an extrinsic hostile microenvironment associated with advanced age.

Experimental data in rodents has shown that cardiac stem cells in older animals had a higher rate of apoptosis and shorter telomeres [48]. Furthermore, these cell populations are sensitive to ROS mediated damage in diabetic cardiomyopathy, as shown by the presence of telomere shortening, expression of senescence markers p53 and p16INK4a as well as increases in apoptosis. All of the above changes were attenuated by the ablation of p66Shc [49]. Anversa et al. reported that c-kit^+ cardiac stem cells underwent apoptosis and increased expression of senescence marker p16INK4a in the heart of patients with cardiovascular diseases [48]. A recent study showed that cardiac progenitor cell aging is mediated by attenuation of the IGF1/IGF1 receptor and hepatocyte growth factor/c-Met systems, which results in cellular senescence, growth arrest, and apoptosis [50]. Furthermore, a recent study by Bergmann et al. measuring the ^{14}C labeling (a retrospective birth dating method) in human hearts showed that the turnover or renewal rate of cardiomyocytes in young adult was around 1% annualy, and this was significantly reduced to 0.45% in the hearts of the elderly [51]. Thus, the decline in number and regenerative capacity of cardiac stem cells might explain part of the increased susceptibility to heart failure in the elderly.

Role of Mitochondria and mt ROS in Cardiac Aging

The free radical theory of aging postulates that the production of intracellular reactive oxygen species (ROS) is the major determinant of lifespan [52]. An age-dependent functional decline of cells and organ systems, as well as associated degenerative diseases, could be attributed to deleterious effects of ROS on various cell and organ components. ROS are generated in multiple compartments and by multiple enzymes within the cell, such as NADPH oxidase at plasma membrane, lipid oxidation within peroxisomes, oxidative phosphorylation within mitochondria, as well as various cyclooxygenases and xanthine oxidase in the cytoplasm. Although all of these sources contribute to the overall oxidative burden, the majority of ROS are produced during oxidative phosphorylation and ATP generation within the mitochondria. This has led to the mitochondrial variant of the free radical theory of aging (reviewed by Balaban et al. 2005 [53]), which proposes that mitochondrial ROS attack mitochondrial constituents, causing mitochondrial DNA damage and mitochondrial dysfunction, causing further damage to mitochondria, further production of ROS and eventual functional declines of cellular and organ function that lead progressively to death [53]. As a vital organ rich in mitochondria and high in oxygen utilization, the heart is especially prone to oxidative damage. As discussed above, direct evidence of this was demonstrated in mice overexpressing catalase targeted to the mitochondria (mCAT), which had 18% prolongation of lifespan [54] and are better protected from age-dependent cardiac hypertrophy and diastolic dysfunction (Fig. 3), in parallel with less oxidative damage to mitochondria [14, 54]. Another line of evidence indicating the role of mitochondria in aging was demonstrated by mice with homozygous mutations of mitochondrial polymerase gamma, which impair the proofreading capacity and thus induced substantial increase in mt-DNA point mutations and deletions [55, 56]. These mice had shortened lifespan and the phenotype of accelerated aging, including kyphosis, alopecia, anemia, osteoporosis and cardiac enlargement [55]. The accumulation of mitochondrial DNA mutations have been shown to increase apoptotic rate, though whether this effect was mediated through mt-ROS is still controversial [56]. Nevertheless, it has been shown that accumulation of mt-DNA deletions drives the premature aging phenotypes in these mice [57].

Furthermore, mice with a targeted mutation of the $p66^{Shc}$ gene display prolonged lifespan, reduced production of ROS and increased resistance to ROS-mediated apoptosis [58]. The $p66^{Shc}$ acts as a mitochondrial redox enzyme that shuffles the electron flow to produce H_2O_2 [59]. A recent study from the same group showed that $p66^{Shc}$ was phosphorylated by PKC-beta together with prolyl isomerase Pin-1, then the phosphorylated $p66^{Shc}$ accumulated within mitochondria to activate mitochondrial Ca^{2+} response, and subsequently induce apoptosis [60]. Disruption of $p66^{Shc}$ prevents Ang II induced LV hypertrophy and cardiomyocytes apoptosis as well as reduces oxidative damage in cardiac progenitor cells and myocytes in diabetic mouse model, as discussed above.

Neurohormonal Regulation: The Role of Insulin/IGF1, the Renin Angiotensin System (RAS) and Adrenergic Signaling in Cardiac Aging

Renin Angiotensin Aldosterone System

Angiotensin II (Ang II), a key molecule in hypertension, directly induces cardiomyocyte hypertrophy and apoptosis, increases cardiac fibrosis and impairs cardiomyocyte relaxation [61], all of which were found in cardiac aging [14]. Indeed, cardiac Ang II concentrations has been shown to increase significantly in aged rodent hearts [14, 62], probably related to increased tissue level of angiotensin II converting enzyme (ACE) [3]. The mechanism of increased ACE in the aged heart is not well understood. Long-term inhibition of Ang II has been shown to reduce age-dependent cardiac pathology and prolongs rat survival [63]. A recent study demonstrated that disruption of Angiotensin receptor type I prolongs mouse survival [64]. Furthermore, Angiotensin II exposure induces mitochondrial ROS and subsequently caused mitochondrial damage, increased mitochondrial autophagy and signaling of mitochondrial biogenesis (Dai et al., ms. in preparation). As discussed above, mitochondrial ROS has been implicated in cardiac aging and several cardiovascular diseases.

Adrenergic Signaling

It is well known that chronic enhancement of β-adrenergic signaling by stimulation of the sympathetic nervous system is deleterious to the heart. Activation of the sympathetic nervous system increased cardiac metabolic demand secondary to increase in heart rate, contractility, afterload (blood pressure) and wall stress. In humans, several clinical trials have shown that inhibition of β-adrenergic signaling by beta-blockers provide survival benefit in patients with heart failure. Adenylate cyclase is a key enzyme producing c-AMP as a secondary messenger downstream to β-adrenergic signaling and adenylate cyclase type 5 (AC5) is the major form in the heart. Disruption of AC5 has been shown to protect against chronic pressure overload-induced cardiac hypertrophy, apoptosis and failure by chronic cathecholamine stimulation or aortic banding [65, 66]. AC-5 knock-out mice were also shown to protect against cardiac aging, which include age-dependent cardiac hypertrophy, systolic dysfunction, apoptosis and fibrosis [17]. Furthermore, these mice were shown to have prolonged lifespan which might be mediated through upregulation of Raf-1/pMEK/pERK pathway, which confers protection against stress including oxidative stress [17].

Insulin/IGF1 Signaling

Insulin/IGF-1 signaling has been implicated in the regulation of lifespan in vertebrate and invertebrate animal models (see Chapter ["Hormonal Influences on Aging and Lifespan"]). Both the Ames Dwarf mice (deficiency in growth hormone and prolactin) and mice with genetic mutation of the IGF-1 receptor are known to have prolonged lifespans [67]. Deficiency in insulin/IGF-1 signaling was shown to improve cardiac performance at advanced age in *Drosophila* as well as to attenuate age-associated cardiomyocytes dysfunction in mice [27, 68] (see below). In contrast, the circulating level of IGF-1 is significantly reduced with advanced age [69, 70] Vasan et al. reported that low serum IGF-1 levels in patients in the Framingham Heart Study were associated with increased risk of heart failure in the elderly without a prior history of myocardial infarction [70]. Furthermore, low levels of GH and

IGF-1 have been correlated with systolic dysfunction in heart failure patients and GH replacement therapy has been shown to reduce heart failure symptoms and attenuate cardiac remodeling in both patients [71, 72] and experimental rat model of heart failure [73–75]. Moreover, GH replacement, which increased IGF-1 signaling, was shown to attenuate age-associated diastolic dysfunction and increased cardiac angiotensin II in senescent rats [62]. Thus, in the face of these seemingly contradictory observations, the role of insulin/IGF1 signaling on cardiac aging remains controversial.

Mechanisms of Progression from Cardiac Hypertrophy to Heart Failure in the Old Age

As the result of chronic stress, the aged myocardium remodels by a complex of events that includes myocyte growth or hypertrophy, re-expression of fetal gene program and remodeling of extracellular matrix. These predispose to the development of heart failure in the aged hearts. Several molecular mechanisms have been proposed for the transition of hypertrophy to failure, as discussed below [76].

Increased Cardiomyocyte Death

Cardiomyocyte death has been reported in ischemic and dilated cardiomyopathy, hypertensive cardiomyopathy and aging [77]. Compelling evidence has shown that cell death contributes to a decline in pumping and induces ventricular remodeling, which may result in symptomatic heart failure [78]. In the failing heart, signaling pathways for cell death overwhelm those that promote cell survival and result in a decrease in the number of cardiomyocytes, which may occur through necrosis, apoptosis or autophagy. Augmentation of Ca^{2+} entry through the L-type Ca^{2+} channel triggers the opening of mitochondrial permeability transition pores by activating cyclophilin D, thereby inducing necrosis [79]. Cardiomyocyte apoptosis can be initiated by multiple ROS-mediated pathways, including Ang II, sympathetic stimulation and cytokines. It has been shown that an increased rate of cardiomyocyte apoptosis contributes to the phenotype of heart failure [78, 80].

Autophagic cell death can be induced by starvation and is characterized by recycling of proteins within organelles. Increased levels of autophagy have been documented in human heart failure [81]. Furthermore, a recent study using mice overexpressing cardiac specific diphtheria toxin receptors showed that diphtheria toxin-induced autophagic cardiomyocyte death induced heart failure [82]. In contrast, mice with cardiac specific disruption of Atg5 (autophagy-deficient hearts) were found to have cardiac hypertrophy, left ventricular dilatation and contractile dysfunction, accompanied by increased levels of ubiquitination [83]. Zhu et al., showed that pressure overload induced by aortic banding greatly increased cardiac autophagy and thereby induced heart failure [84]. Heterozygous disruption of the gene coding for beclin 1, a protein required for early autophagosome formation, decreased cardiomyocyte autophagy and diminished pathological remodeling induced by severe pressure overload. Conversely, beclin 1 overexpression increased autophagic activity and accentuated pathological remodeling [84]. Taken together, these studies indicate that constitutive autophagy is normally a homeostatic mechanism in the heart that is required to maintain cardiac structure and function, and that upregulation of autophagy in failing hearts can be a maladaptive response to hemodynamic stress, such as from pressure overload.

Extracellular Matrix Remodeling

Myocardial remodeling is part of the compensatory and pathologic response of the heart to various injury stimuli. Alterations in the composition and structure of the extracellular matrix (ECM) can make an important contribution to changes in the cardiac size, structure, and function and might contribute to heart failure [85–88]. ECM remodeling is an active and continuous process in which the ultrastructure of ECM is altered. This process involves the degradation of collagens by matrix metalloproteinases (MMPs) and the synthesis of ECM by myofibroblasts through TGF-β dependent signaling [89]. Adverse ECM remodeling (fibrosis) increases cardiac stiffness and reduces cardiac compliance as well as disrupts the coordination of myocardial excitation-contraction coupling (reviewed by Berk [90] and Spinale [91]). Indeed, several studies in animal models of heart failure revealed that inhibition of adverse ECM remodeling (e.g. by MMP inhibitors) attenuates cardiac dysfunction [85–89].

Alteration of Calcium Handling Proteins

Ca^{2+} is the key regulator of excitation-contraction coupling. This coupling is initiated by an entry of small amount of Ca^{2+} through the LTCC during depolarization (action potential), which triggers a larger scale Ca^{2+} release from sarcoplasmic reticulum Ca^{2+} storage ($SRCa^{2+}$) through the ryanodine receptor. The increase in cytoplasmic Ca^{2+} binds and activates Troponin C within the myofilaments and induces myocyte contraction. Relaxation is initiated by reuptake of cytoplasmic Ca^{2+} into the SR through phospholamban -regulated Calcium-ATPase (SERCA2a) and subsequent trans-sarcolemmal Ca^{2+} removal through the sodium calcium exchanger (NCX). In failing hearts, Ca^{2+} reuptake into the SR is impaired, and consequently SR Ca^{2+} storage is decreased [92, 93]. The decline in SERCA2 function (including that of aging [14]) may be explained by the decline in SERCA2 protein concentration, oxidative modification that impairs SERCA2 protein function, reduced levels of PLN phosphorylation, and the depletion of SR Ca^{2+} through leaky RyR channels [94]. Genetic manipulation of mice that increased SERCA2 activity, such as overexpression of SERCA2 [95] or disruption of phospholamban, has been reported to attenuate heart failure in experimental animals [96, 97].

Hypoxic Response and Angiogenesis

Pressure overload to the heart induces cardiac hypertrophy, in parallel with increased myocardial oxygen demand as well as decreased coronary perfusion pressure, resulting from compression of the coronary microvasculature. This supply-demand mismatch is believed to induce a relative ischemia in the hypertrophic heart, which eventually causes energetic failure. Previous studies in mice with Akt-induced cardiac hypertrophy showed a concomitant increase in angiogenic growth factors such as vascular endothelial growth factor (VEGF) and angiopeitin 2 during the hypertrophic phase, and that inhibition of VEGF signaling resulted in relative ischemia and accelerated transition to heart failure [98, 99]. Moreover, the cardiac transcription factor GATA4, which is upregulated during hypertrophy, has been shown to stimulate angiogenesis to help maintain the balance between growth of hypertrophic muscle and new capillaries [100].

Mitochondrial Dysfunction and Abnormalities in Energetics

Proliferation of mitochondria as a powerhouse does not keep pace with the increasing energy demand of the heart during hypertrophy [101]. This might have important contributions to cardiac failure. Studies on human hearts using ^{31}P NMR spectroscopy indicated that the ATP content of failing hearts is generally 20–30% lower than that of normal hearts [102]. Furthermore, phosphocreatine, an important short-term reserve energy source that maintains a high phosphorylation potential to cope with acute increases in energy demand (e.g. exercise), significantly declined by up to 60% in the elderly heart failure patients [103]. The magnitude of this reduction is related to the severity of heart failure [104] and has been shown to predict mortality in patients with dilated cardiomyopathy [105].

Although all of the above mechanisms contribute to the development of heart failure, the precise interplay of mechanisms regulating the transition of cardiac hypertrophy to failure remains unknown.

Beneficial Effects of DR on Cardiac Function in Aging

Dietary restriction has been shown to extend longevity, reduce the onset of chronic diseases and impact morbidity and mortality in numerous animal models. This relationship is so well established that the capacity of a physiological parameter to respond to DR is often used as evidence of its being an intrinsic characteristic of the biology of aging. Literature from both rhesus monkey studies and short term dietary restriction in humans provides strong evidence that DR reduces risk factors associated with heart disease: resting heart rate and blood pressure are decreased, insulin sensitivity is enhanced, lipid profiles are improved, and inflammatory processes that likely contribute to atherosclerosis are reduced (reviewed by Mattson and Wan [106]). A recent study suggests that in humans undertaking DR for a mean of 6.5 years there is lower blood pressure, lower systemic oxidative stress, and improved diastolic function [107]. This finding of enhanced diastolic function has been reproduced in individuals maintained on one-year DR; similar effects were seen whether weight loss (approximately 12%) was induced by DR or exercise [108].

Surprisingly, there have been relatively few direct studies of the effects of DR on cardiac function in non-obese rodent models. Taffett et al. found that DR of mice had a large positive effect on age-related impaired diastolic function [109]. In another recent study, the Dahl salt-sensitive rat, which develops gradual, hypertension-associated diastolic dysfunction, was compared in DR vs. AL animals. Moderate calorie restriction markedly attenuated changes in heart weight, left ventricular mass, and wall thickness in these rats and echocardiography demonstrated that DR reduced cardiac diastolic dysfunction in this model [110].

Drosophila: An Invertebrate Model of Cardiac Senescence

In recent years, the development of new techniques for analysis of *Drosophila* cardiac function has allowed the fruit fly genetic system to be used to study age-related functional changes in cardiac tissue. The following section will discuss the array of functional changes that occur during normal aging in the fruit fly heart, then highlight some of the genetic components that have been shown to regulate such changes in flies.

Normal Aging of the Drosophila *Heart*

High-speed video imaging of either intact or semi-intact preparations of *Drosophila* hearts has revealed several parameters of cardiac function that reproducibly change in an age-dependent manner.

Heart Rate

The average heart rate of intact adult fruit flies under anesthesia in the first week after eclosion at 25°C is approximately 3 Hz [68, 111]. This correlates well with results from both semi-intact and isolated hearts [112, 113]. This rate declines in a linear fashion with age in both genders and multiple genetic backgrounds, with flies at 5 weeks of age beating at 2.4 Hz [68]. Results generated by using the ultrasound-like OCT imaging technique on wakeful flies, by contrast, did not show an appreciable age-related decline in heart rate, although flies were only observed to 30 days of age, or less than half the mean lifespan [114]. Although the heartbeat in flies is myogenic [115, 116], flies exhibit a multi-directional heartbeat, with reversal of flow direction in adults controlled by neuronal input [117]. Additional flexibility in heart rate is conferred by the ability of the heart to accelerate or decelerate in response to a variety of hormones and neurotransmitters [143].

Rhythmicity

In addition to an increase with age in the average period length per heartbeat, a dramatic increase in the variability of the heartbeat length has also been observed

[118]. Using a novel analysis program of beating hearts in semi-dissected abdomens recorded with a high-speed camera, the standard deviation of heart period length has also been used to derive an arrhythmia index (AI). This index increases several-fold and fairly linearly with age [118].

Fiber Structure

The cardiac tube in young flies exhibits a transverse or spiral shaped myofibril array that is clearly visible through antibody staining with sarcomeric markers such as alpha-actinin [112, 114, 119–121]. As flies age, sarcomeric staining reveals increasing levels of disorganization in the myofibrillar arrays within the cardiomyocytes. By 5 weeks of age, there are signs of misalignment that may contribute to impairment of age-dependent cardiac functionality [112].

Stress Resistance

Resistance to external electrical pacing stress has been used both to identify genetic mutations affecting cardiac performance throughout life [112, 118, 122] and as a marker for age-related functional decline [68, 123]. The percentage of flies that respond to a stereotyped 30-second pacing protocol by entering either fibrillation or arrest is reported as "failure rate" [111]. The failure rate of multiple wild-type genetic backgrounds, as well as wild-caught isogenic lines [122] has been seen to increase steadily during the first 5 weeks of age in both genders [68] at which point it reaches a maximum prior to the mean lifespan of these backgrounds (Fig. 4).

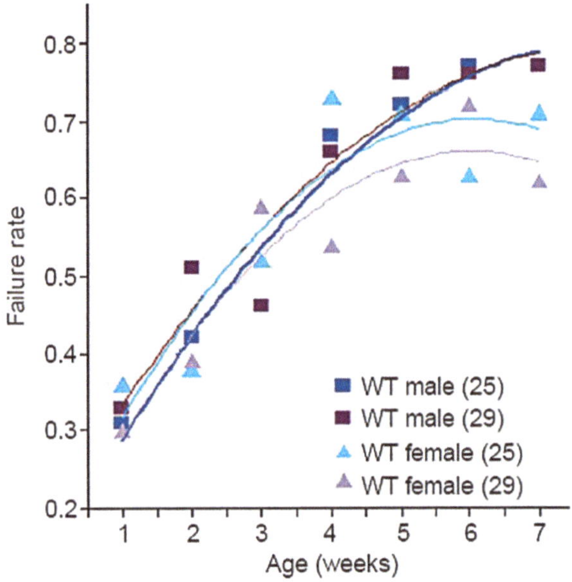

Fig. 4 Heart failure in *Drosophila* as a function of age after external electrical pacing from outbred wild-type offspring (WT; *yw* x Canton S). Experiments were done at 25°C and at 29°C for 7 weeks. Test temperature alone had no effect on failure rate. Pacing-induced failure rate was age-dependent for both genders at both temperatures ($\chi^2 > 40$, $P < 0.0001$)

This suite of measurement techniques provides an opportunity to use the power of *Drosophila* genetics to uncover conserved genetic factors that influence the aging physiology of cardiac tissue. The next section will briefly discuss some of the factors that have been identified by these means thus far.

Genetic Regulation

A combination of unbiased genetic screens and candidate-gene approaches has identified several overlapping mechanisms that contribute to age-related functional decline in the *Drosophila* heart, including alterations in ion channel expression, contractile protein localization and arrangement, and accumulation of reactive oxygen species. Upstream signaling pathways have also been identified that may act by regulating any or all of these downstream phenomena.

Ion Channels

The gradual increase in arrhythmia with age suggests a dysregulation in ion flux through the cardiomyocytes. Indeed, multiple potassium channel-encoding genes have been identified as a direct contributor both to increasing arrhythmia and decreasing cardiac stress tolerance with age [118] (Ocorr and R.B., unpublished).

Flies carrying a mutation in the *KCNQ* gene, for example, exhibit early declines in stress tolerance and early increases in arrhythmia index. Such flies exhibit a delayed repolarization and delayed myocardial relaxation. These changes lead to a high incidence of spontaneous fibrillations, an accelerated increase in the arrhythmia index with age and an extraordinary sensitivity to external electrical pacing already in young flies [118]. These phenotypes resemble those seen in human Torsades des Pointes, which has also been correlated with functional alterations in the human ortholog of *Drosophila KCNQ* [124].

Importantly, the expression level of *KCNQ* RNA decreases with age in wild-type flies, consistent with the idea that changes in the function of this channel are linked with age-related loss of rhythmicity [118]. This hypothesis is strongly supported by cardiac-specific overexpression of a wildtype *KCNQ* transgene in old flies, which significantly reverses the age-dependent increase in arrhythmias. Thus, it is likely that ion channel dysfunction is in part the cause for the age-related decline in cardiac functionality.

Contractile Proteins

The observation that regularity of microfibrillar arrays within aging cardiomyocytes exhibit increased gaps and disorganization with age [112] suggests that the activity and localization of contractile proteins and subcellular structures may be an important contributing factor to age-related functional decline. Flies carrying mutations or heart-specific knockdowns in genes encoding several different contractile or other structural proteins have now been examined.

Flies with reduced expression of the highly-conserved *Dystrophin* gene show early, progressive deterioration of myofibril alignment that resembles an accelerated version of changes seen in aging wild-type flies. Additionally, and perhaps in consequence, these flies also show age-dependent functional abnormalities [112]. Interestingly, *Dystrophin* mutant hearts have a faster rate than wildtype at old ages, perhaps as a compensatory mechanism to counteract their impaired contractility. Consistent with this idea is the observation that these flies exhibit a significantly dilated heart phenotype that results in a reduction in fractional shortening, similar to that seen in vertebrates with dilated cardiomyopathy [112]. In normal flies, however, reduced fractional shortening has not consistently been observed during cardiac aging [112, 121, 125]. Interestingly, the *Dystrophin* mutant heart phenotype can be rescued by overexpressing a short C-terminal isoform of mammalian *dystrophin* [112, 126].

Mutations that perturb myosin motor function in *Drosophila* cardiac tissue also generate progressive phenotypes that resemble those seen during normal aging in several ways [121], including increased incidence of spontaneous fibrillations and arrhythmias, as well as a prolonged heart period. Cardiac abnormalities due to mutations in myosin, as in *dystrophin* (see above), are also accentuated with age. Thus, dysfunction of contractile machinery also leads to an accelerated functional decline in heart function, similar to disturbances in electrical properties of cardiac myocytes, discussed above.

ROS-Scavenging Proteins

Accumulation of reactive oxygen species has been proposed to be an important mechanism in functional senescence [52, 127]. In a fly model for increased ROS accumulation, the *Sod2* null mutant fly, several aspects of cardiac performance decline rapidly and prematurely in a way that seems to mimic normal aging. In particular, such flies have a dramatic increase in spontaneous fibrillations and arrhythmias, while exhibiting an increased heart period and decreased relaxation velocity [125].

Nutrient-Sensing Signaling Pathways

Gradual deterioration or dysregulation of homeostasis in ion channels, contractile machinery, or oxidative stress levels are all attractive candidates for factors contributing to the functional senescence of cardiac tissue in flies as in vertebrates. However, the observation that single-gene mutations can extend both cardiac functionality and lifespan [113, 128–130] suggests that global regulatory systems must be in place to coordinate such changes.

Two closely-related and interconnected signaling pathways have been shown to control the rate of cardiac functional senescence in a tissue-autonomous fashion, the insulin signaling pathway [68] and the TOR kinase signaling pathway [123].

Long-lived flies mutant for either insulin or Tor signaling components have a corresponding delay in their age-related decline in cardiac stress resistance. Moreover, heart-specific expression of molecules reducing the level of signaling through either pathway blocks the age-related decrease in cardiac stress resistance. Conversely, heart-specific upregulation of such signaling molecules creates flies that exhibit cardiac stress resistance characteristic of old flies even at young ages [68, 123] (R.J.W and R.B, unpublished).

Thus, these signaling pathways are attractive candidates for overarching regulatory cassettes that connect environmental conditions, such as nutrition, to the control of functional aging parameters, including those of the heart.

Exercise

A recently developed system for providing *Drosophila* with modular exercise-training via enforced, reiterated negative geotaxis has made it possible to utilize flies as a model for exercise-induced changes in myocardial function. Providing young flies with exercise-training has been demonstrated to delay age-related changes in relaxation velocity and cardiac stress response in multiple genetic backgrounds [125] (unpublished data). Both candidate gene and unbiased approaches are underway to identify genetic factors necessary to modulate the response of cardiac function to exercise-training in *Drosophila*.

Future work will continue to utilize the *Drosophila* model system to identify genetic factors that play important roles in mediating cardiac functional senescence. In particular, the fly system will be useful for combining genetics with changes in environmental conditions such as diet or exercise levels to identify genetic components necessary for the heart to respond to global environmental change across ages.

Cardiac Aging in Dogs

Canine models have been valuable to investigate several cardiovascular diseases ranging from cardiac arrhythmia, ischemia reperfusion injury (a model of myocardial infarction), cardiac failure to genetic cardiomyopathy such as in Duchene muscular dystrophy [131–133]. The dog model is particularly useful for the electrophysiological study of cardiac arrhythmia because the distribution of Purkinje fibers (part of the electrical conduction system) in dogs resembles that in humans and the cardiac activation sequence of the dog mimics quite closely to that in humans [134]. Age-related changes of the canine heart include myocardial hypertrophy, increased cardiac stiffness, prolonged action potential duration, and a decline in cardiovascular responsiveness to β-adrenergic stimulation. These changes induced progressive loss of cardiac reserve and adaptability, and hence increased susceptibility to heart diseases in the aged dogs [135]. Eichelberg et al. reported that cardiovascular diseases are among the most common causes of death in the dogs, accounting for 16.3% of all death [136]. Cardiovascular diseases with age-dependent increases in prevalence include chronic degenerative valvular disease, cardiac hypertrophy, dilated cardiomyopathy, amyloidosis, lipofuscinosis and sick sinus syndrome [137].

Among these, chronic degenerative valvular disease is the most prevalent, reaching about 75% for dogs over 16 years old [138]. The mitral valve is most frequently involved, causing significant mitral regurgitation that leads to chronic left-sided volume overload and finally results in congestive heart failure.

Cardiac Aging in Non-Human Primates

Although rodents remain the most widely used animal model for aging research, models that are phylogenetically close to humans are useful for the study of complex physiology. Research using nonhuman primates provides a valuable tool to investigate aging process which closely recapitulates human aging and allows the evaluation of potential anti-aging interventions before human clinical trials. Data from longitudinal study of aging in rhesus monkeys (*Macaca mulatta*) conducted by the National Institute of Aging revealed that under normal diets rhesus monkeys develop several aging-related cardiac pathologies, including aortic and mitral valves degenerative calcifications, loss or degeneration of myocardial fibers with hypertrophy of remaining cardiomyocytes, lipofuscin accumulation and variable degree of myocarditis, multifocal interstitial fibrosis, myocardial infarction and congestive heart failure [139–142]. Atherosclerotic plaques were observed only in rhesus monkeys fed high fat diets.

Summary

As summarized in Table 1, parameters of cardiac aging are highly conserved among divergent species. This illustrates that mechanisms of cardiac aging are evolutionary conserved and a fundamental feature of physiological aging.

Table 1 Summary of evolutionary conservation

	Human	Dog	Mouse	*Drosophila*
Age associated failure	X	x	x	x
Age-associated conduction defects	X	x	x	x
Protected by CR	X	?	x	?
Affected by growth hormone/IGF-1	x	?	x	x
Mitochondrial changes	x	?	x	x
Affected by antioxidant capacity	?	?	x	x

References

1. Rosamond WFK, Friday G et al. (2007). Heart disease and stroke statistics – 2007 update: a report from the American Heart Association Statistics Committee and Stroke Statistics Subcommittee. *Circulation* 115: e69–e171.

2. *Morbidity and Mortality: 2007 Chart Book on Cardiovascular, Lung and Blood Diseases* (2007). Bethesda, MD: National Heart, Lung, and Blood Institute, National Institutes of Health.
3. Lakatta EG (2003). Arterial and cardiac aging: major shareholders in cardiovascular disease enterprises: Part III: cellular and molecular clues to heart and arterial aging. *Circulation* 107: 490–497.
4. Lakatta EG and Levy D (2003). Arterial and cardiac aging: major shareholders in cardiovascular disease enterprises: Part II: the aging heart in health: links to heart disease. *Circulation* 107: 346–354.
5. Lakatta EG and Levy D (2003). Arterial and cardiac aging: major shareholders in cardiovascular disease enterprises: Part I: aging arteries: a "set up" for vascular disease. *Circulation* 107: 139–146.
6. Bursi F, Weston SA, Redfield MM, Jacobsen SJ, Pakhomov S, Nkomo VT, Meverden RA, and Roger VL (2006). Systolic and diastolic heart failure in the community. *J Am Med Assoc* 296: 2209–2216.
7. Fleg JL, O'Connor F, Gerstenblith G, Becker LC, Clulow J, Schulman SP, and Lakatta EG (1995). Impact of age on the cardiovascular response to dynamic upright exercise in healthy men and women. *J Appl Physiol* 78: 890–900.
8. Correia LC, Lakatta EG, O'Connor FC, Becker LC, Clulow J, Townsend S, Gerstenblith G, and Fleg JL (2002). Attenuated cardiovascular reserve during prolonged submaximal cycle exercise in healthy older subjects. *J Am Coll Cardiol* 40: 1290–1297.
9. Kosugi R, Shioi T, Watanabe-Maeda K, Yoshida Y, Takahashi K, Machida Y, and Izumi T (2006). Angiotensin II receptor antagonist attenuates expression of aging markers in diabetic mouse heart. *Circ J* 70: 482–488.
10. Brodsky SV, Gealekman O, Chen J, Zhang F, Togashi N, Crabtree M, Gross SS, Nasjletti A, and Goligorsky MS (2004). Prevention and reversal of premature endothelial cell senescence and vasculopathy in obesity-induced diabetes by ebselen. *Circ Res* 94: 377–384.
11. Hayashi T, Matsui-Hirai H, Miyazaki-Akita A, Fukatsu A, Funami J, Ding QF, Kamalanathan S, Hattori Y, Ignarro LJ, and Iguchi A (2006). Endothelial cellular senescence is inhibited by nitric oxide: implications in atherosclerosis associated with menopause and diabetes. *Proc Natl Acad Sci U S A* 103: 17018–17023.
12. Tsirpanlis G (2008). Cellular senescence, cardiovascular risk, and CKD: a review of established and hypothetical interconnections. *Am J Kidney Dis* 51: 131–144.
13. Orimo M, Minamino T, Miyauchi H, Tateno K, Okada S, Moriya J, and Komuro I (2009). Protective role of SIRT1 in diabetic vascular dysfunction. *Arterioscler Thromb Vasc Biol* 29: 889–894.
14. Dai DF, Santana LF, Vermulst M, Tomazela DM, Emonds MJ, Maccoss MJ, Gollahon K, Martin GM, Loeb LA, Ladiges WC, and Rabinovitch PS (2009). Overexpression of catalase targeted to mitochondria attenuates murine cardiac aging. *Circulation* 119: 2789–2797.
15. Zheng F, Plati AR, Potier M, Schulman Y, Berho M, Banerjee A, Leclercq B, Zisman A, Striker LJ, and Striker GE (2003). Resistance to glomerulosclerosis in B6 mice disappears after menopause. *Am J Pathol* 162: 1339–1348.
16. Treuting PM, Linford NJ, Knoblaugh SE, Emond MJ, Morton JF, Martin GM, Rabinovitch PS, and Ladiges WC (2008). Reduction of age-associated pathology in old mice by overexpression of catalase in mitochondria. *J Gerontol A Biol Sci Med Sci* 63: 813–822.
17. Yan L, Vatner DE, O'Connor JP, Ivessa A, Ge H, Chen W, Hirotani S, Ishikawa Y, Sadoshima J, and Vatner SF (2007). Type 5 adenylyl cyclase disruption increases longevity and protects against stress. *Cell* 130: 247–258.
18. Khouri SJ, Maly GT, Suh DD, and Walsh TE (2004). A practical approach to the echocardiographic evaluation of diastolic function. *J Am Soc Echocardiogr* 17: 290–297.
19. Barger JL, Kayo T, Vann JM, Arias EB, Wang J, Hacker TA, Wang Y, Raederstorff D, Morrow JD, Leeuwenburgh C, Allison DB, Saupe KW, Cartee GD, Weindruch R, and Prolla TA (2008). A low dose of dietary resveratrol partially mimics caloric restriction and retards aging parameters in mice. *PLoS ONE* 3: e2264.

20. Tei C, Nishimura RA, Seward JB, and Tajik AJ (1997). Noninvasive Doppler-derived myocardial performance index: correlation with simultaneous measurements of cardiac catheterization measurements. *J Am Soc Echocardiogr* 10: 169–178.
21. Heineke J and Molkentin JD (2006). Regulation of cardiac hypertrophy by intracellular signalling pathways. *Nat Rev Mol Cell Biol* 7: 589–600.
22. Liang Q, Wiese RJ, Bueno OF, Dai YS, Markham BE, and Molkentin JD (2001). The transcription factor GATA4 is activated by extracellular signal-regulated kinase 1- and 2-mediated phosphorylation of serine 105 in cardiomyocytes. *Mol Cell Biol* 21: 7460–7469.
23. Xu A and Narayanan N (1998). Effects of aging on sarcoplasmic reticulum Ca^{2+}-cycling proteins and their phosphorylation in rat myocardium. *Am J Physiol* 275: H2087–H2094.
24. Koban MU, Moorman AF, Holtz J, Yacoub MH, and Boheler KR (1998). Expressional analysis of the cardiac Na-Ca exchanger in rat development and senescence. *Cardiovasc Res* 37: 405–423.
25. Adachi T, Weisbrod RM, Pimentel DR, Ying J, Sharov VS, Schoneich C, and Cohen RA (2004). S-Glutathiolation by peroxynitrite activates SERCA during arterial relaxation by nitric oxide. *Nat Med* 10: 1200–1207.
26. Li SY, Du M, Dolence EK, Fang CX, Mayer GE, Ceylan-Isik AF, LaCour KH, Yang X, Wilbert CJ, Sreejayan N, and Ren J (2005). Aging induces cardiac diastolic dysfunction, oxidative stress, accumulation of advanced glycation endproducts and protein modification. *Aging Cell* 4: 57–64.
27. Li Q, Ceylan-Isik AF, Li J, and Ren J (2008). Deficiency of insulin-like growth factor 1 reduces sensitivity to aging-associated cardiomyocyte dysfunction. *Rejuvenation Res* 11: 725–733.
28. Josephson IR, Guia A, Stern MD, and Lakatta EG (2002). Alterations in properties of L-type Ca channels in aging rat heart. *J Mol Cell Cardiol* 34: 297–308.
29. Janczewski AM, Spurgeon HA, and Lakatta EG (2002). Action potential prolongation in cardiac myocytes of old rats is an adaptation to sustain youthful intracellular Ca2+ regulation. *J Mol Cell Cardiol* 34: 641–648.
30. Xiao RP, Tomhave ED, Wang DJ, Ji X, Boluyt MO, Cheng H, Lakatta EG, and Koch WJ (1998). Age-associated reductions in cardiac beta1- and beta2-adrenergic responses without changes in inhibitory G proteins or receptor kinases. *J Clin Invest* 101: 1273–1282.
31. Kilts JD, Akazawa T, Richardson MD, and Kwatra MM (2002). Age increases cardiac Galpha(i2) expression, resulting in enhanced coupling to G protein-coupled receptors. *J Biol Chem* 277: 31257–31262.
32. Pepe S, Xiao RP, Hohl C, Altschuld R, and Lakatta EG (1997). ‘Cross talk’ between opioid peptide and adrenergic receptor signaling in isolated rat heart. *Circulation* 95: 2122–2129.
33. Brodde OE, Konschak U, Becker K, Ruter F, Poller U, Jakubetz J, Radke J, and Zerkowski HR (1998). Cardiac muscarinic receptors decrease with age. In vitro and in vivo studies. *J Clin Invest* 101: 471–478.
34. Leosco D, Rengo G, Iaccarino G, Filippelli A, Lymperopoulos A, Zincarelli C, Fortunato F, Golino L, Marchese M, Esposito G, Rapacciuolo A, Rinaldi B, Ferrara N, Koch WJ, and Rengo F (2007). Exercise training and beta-blocker treatment ameliorate age-dependent impairment of beta-adrenergic receptor signaling and enhance cardiac responsiveness to adrenergic stimulation. *Am J Physiol Heart Circ Physiol* 293: H1596–H1603.
35. Tani M, Suganuma Y, Hasegawa H, Shinmura K, Ebihara Y, Hayashi Y, Guo X, and Takayama M (1997). Decrease in ischemic tolerance with aging in isolated perfused Fischer 344 rat hearts: relation to increases in intracellular Na^{+} after ischemia. *J Mol Cell Cardiol* 29: 3081–3089.
36. Isoyama S and Nitta-Komatsubara Y (2002). Acute and chronic adaptation to hemodynamic overload and ischemia in the aged heart. *Heart Fail Rev* 7: 63–69.
37. Juhaszova M, Rabuel C, Zorov DB, Lakatta EG, and Sollott SJ (2005). Protection in the aged heart: preventing the heart-break of old age?. *Cardiovasc Res* 66: 233–244.

38. Nitta Y, Abe K, Aoki M, Ohno I, and Isoyama S (1994). Diminished heat shock protein 70 mRNA induction in aged rat hearts after ischemia. *Am J Physiol* 267: H1795–H1803.
39. Bolli R (2001). Cardioprotective function of inducible nitric oxide synthase and role of nitric oxide in myocardial ischemia and preconditioning: an overview of a decade of research. *J Mol Cell Cardiol* 33: 1897–1918.
40. Chou TC, Yen MH, Li CY, and Ding YA (1998). Alterations of nitric oxide synthase expression with aging and hypertension in rats. *Hypertension* 31: 643–648.
41. Korzick DH, Holiman DA, Boluyt MO, Laughlin MH, and Lakatta EG (2001). Diminished alpha1-adrenergic-mediated contraction and translocation of PKC in senescent rat heart. *Am J Physiol Heart Circ Physiol* 281: H581–H589.
42. Tani M, Honma Y, Hasegawa H, and Tamaki K (2001). Direct activation of mitochondrial K(ATP) channels mimics preconditioning but protein kinase C activation is less effective in middle-aged rat hearts. *Cardiovasc Res* 49: 56–68.
43. Hsieh PC, Segers VF, Davis ME, MacGillivray C, Gannon J, Molkentin JD, Robbins J, and Lee RT (2007). Evidence from a genetic fate-mapping study that stem cells refresh adult mammalian cardiomyocytes after injury. *Nat Med* 13: 970–974.
44. Beltrami AP, Barlucchi L, Torella D, Baker M, Limana F, Chimenti S, Kasahara H, Rota M, Musso E, Urbanek K, Leri A, Kajstura J, Nadal-Ginard B, and Anversa P (2003). Adult cardiac stem cells are multipotent and support myocardial regeneration. *Cell* 114: 763–776.
45. Reinecke H, Minami E, Zhu WZ, and Laflamme MA (2008). Cardiogenic differentiation and transdifferentiation of progenitor cells. *Circ Res* 103: 1058–1071.
46. Oh H, Bradfute SB, Gallardo TD, Nakamura T, Gaussin V, Mishina Y, Pocius J, Michael LH, Behringer RR, Garry DJ, Entman ML, and Schneider MD (2003). Cardiac progenitor cells from adult myocardium: homing, differentiation, and fusion after infarction. *Proc Natl Acad Sci U S A* 100: 12313–12318.
47. Matsuura K, Wada H, Nagai T, Iijima Y, Minamino T, Sano M, Akazawa H, Molkentin JD, Kasanuki H, and Komuro I (2004). Cardiomyocytes fuse with surrounding noncardiomyocytes and reenter the cell cycle. *J Cell Biol* 167: 351–363.
48. Anversa P, Rota M, Urbanek K, Hosoda T, Sonnenblick EH, Leri A, Kajstura J, and Bolli R (2005). Myocardial aging – a stem cell problem. *Basic Res Cardiol* 100: 482–493.
49. Rota M, LeCapitaine N, Hosoda T, Boni A, De Angelis A, Padin-Iruegas ME, Esposito G, Vitale S, Urbanek K, Casarsa C, Giorgio M, Luscher TF, Pelicci PG, Anversa P, Leri A, and Kajstura J (2006). Diabetes promotes cardiac stem cell aging and heart failure, which are prevented by deletion of the p66shc gene. *Circ Res* 99: 42–52.
50. Gonzalez A, Rota M, Nurzynska D, Misao Y, Tillmanns J, Ojaimi C, Padin-Iruegas ME, Muller P, Esposito G, Bearzi C, Vitale S, Dawn B, Sanganalmath SK, Baker M, Hintze TH, Bolli R, Urbanek K, Hosoda T, Anversa P, Kajstura J, and Leri A (2008). Activation of cardiac progenitor cells reverses the failing heart senescent phenotype and prolongs lifespan. *Circ Res* 102: 597–606.
51. Bergmann O, Bhardwaj RD, Bernard S, Zdunek S, Barnabe-Heider F, Walsh S, Zupicich J, Alkass K, Buchholz BA, Druid H, Jovinge S, and Frisen J (2009). Evidence for cardiomyocyte renewal in humans. *Science* 324: 98–102.
52. Harman D (1956). Aging: a theory based on free radical and radiation chemistry. *J Gerontol* 11: 298–300.
53. Balaban RS, Nemoto S, and Finkel T (2005). Mitochondria, oxidants, and aging. *Cell* 120: 483–495.
54. Schriner SE, Linford NJ, Martin GM, Treuting P, Ogburn CE, Emond M, Coskun PE, Ladiges W, Wolf N, Van Remmen H, Wallace DC, and Rabinovitch PS (2005). Extension of murine life span by overexpression of catalase targeted to mitochondria. *Science* 308: 1909–1911.
55. Trifunovic A, Wredenberg A, Falkenberg M, Spelbrink JN, Rovio AT, Bruder CE, Bohlooly YM, Gidlof S, Oldfors A, Wibom R, Tornell J, Jacobs HT, and Larsson NG (2004). Premature ageing in mice expressing defective mitochondrial DNA polymerase. *Nature* 429: 417–423.

56. Kujoth GC, Hiona A, Pugh TD, Someya S, Panzer K, Wohlgemuth SE, Hofer T, Seo AY, Sullivan R, Jobling WA, Morrow JD, Van Remmen H, Sedivy JM, Yamasoba T, Tanokura M, Weindruch R, Leeuwenburgh C, and Prolla TA (2005). Mitochondrial DNA mutations, oxidative stress, and apoptosis in mammalian aging. *Science* 309: 481–484.
57. Vermulst M, Wanagat J, Kujoth GC, Bielas JH, Rabinovitch PS, Prolla TA, and Loeb LA (2008). DNA deletions and clonal mutations drive premature aging in mitochondrial mutator mice. *Nat Genet* 40: 392–394.
58. Migliaccio E, Giorgio M, Mele S, Pelicci G, Reboldi P, Pandolfi PP, Lanfrancone L, and Pelicci PG (1999). The p66shc adaptor protein controls oxidative stress response and life span in mammals. *Nature* 402: 309–313.
59. Orsini F, Migliaccio E, Moroni M, Contursi C, Raker VA, Piccini D, Martin-Padura I, Pelliccia G, Trinei M, Bono M, Puri C, Tacchetti C, Ferrini M, Mannucci R, Nicoletti I, Lanfrancone L, Giorgio M, and Pelicci PG (2004). The life span determinant p66Shc localizes to mitochondria where it associates with mitochondrial heat shock protein 70 and regulates trans-membrane potential. *J Biol Chem* 279: 25689–25695.
60. Pinton P, Rimessi A, Marchi S, Orsini F, Migliaccio E, Giorgio M, Contursi C, Minucci S, Mantovani F, Wieckowski MR, Del Sal G, Pelicci PG, and Rizzuto R (2007). Protein kinase C beta and prolyl isomerase 1 regulate mitochondrial effects of the life-span determinant p66Shc. *Science* 315: 659–663.
61. Domenighetti AA, Wang Q, Egger M, Richards SM, Pedrazzini T, and Delbridge LM (2005). Angiotensin II-mediated phenotypic cardiomyocyte remodeling leads to age-dependent cardiac dysfunction and failure. *Hypertension* 46: 426–432.
62. Groban L, Pailes NA, Bennett CD, Carter CS, Chappell MC, Kitzman DW, and Sonntag WE (2006). Growth hormone replacement attenuates diastolic dysfunction and cardiac angiotensin II expression in senescent rats. *J Gerontol A Biol Sci Med Sci* 61: 28–35.
63. Basso N, Cini R, Pietrelli A, Ferder L, Terragno NA, and Inserra F (2007). Protective effect of long-term angiotensin II inhibition. *Am J Physiol Heart Circ Physiol* 293: H1351–H1358.
64. Benigni A, Corna D, Zoja C, Sonzogni A, Latini R, Salio M, Conti S, Rottoli D, Longaretti L, Cassis P, Morigi M, Coffman TM, and Remuzzi G (2009). Disruption of the Ang II type 1 receptor promotes longevity in mice. *J Clin Invest* 119: 524–530.
65. Okumura S, Vatner DE, Kurotani R, Bai Y, Gao S, Yuan Z, Iwatsubo K, Ulucan C, Kawabe J, Ghosh K, Vatner SF, and Ishikawa Y (2007). Disruption of type 5 adenylyl cyclase enhances desensitization of cyclic adenosine monophosphate signal and increases Akt signal with chronic catecholamine stress. *Circulation* 116: 1776–1783.
66. Okumura S, Takagi G, Kawabe J, Yang G, Lee MC, Hong C, Liu J, Vatner DE, Sadoshima J, Vatner SF, and Ishikawa Y (2003). Disruption of type 5 adenylyl cyclase gene preserves cardiac function against pressure overload. *Proc Natl Acad Sci U S A* 100: 9986–9990.
67. Holzenberger M, Dupont J, Ducos B, Leneuve P, Geloen A, Even PC, Cervera P, and Le Bouc Y (2003). IGF-1 receptor regulates lifespan and resistance to oxidative stress in mice. *Nature* 421: 182–187.
68. Wessells RJ, Fitzgerald E, Cypser JR, Tatar M, and Bodmer R (2004). Insulin regulation of heart function in aging fruit flies. *Nat Genet* 36: 1275–1281.
69. Corpas E, Harman SM, and Blackman MR (1993). Human growth hormone and human aging. *Endocr Rev* 14: 20–39.
70. Vasan RS, Sullivan LM, D'Agostino RB, Roubenoff R, Harris T, Sawyer DB, Levy D, and Wilson PW (2003). Serum insulin-like growth factor I and risk for heart failure in elderly individuals without a previous myocardial infarction: the Framingham Heart Study. *Ann Intern Med* 139: 642–648.
71. Giustina A, Lorusso R, Borghetti V, Bugari G, Misitano V, and Alfieri O (1996). Impaired spontaneous growth hormone secretion in severe dialated cardiomyopathy. *Am Heart J* 131: 620–622.
72. Broglio F, Fubini A, Morello M, Arvat E, Aimaretti G, Gianotti L, Boghen MF, Deghenghi R, Mangiardi L, and Ghigo E (1999). Activity of GH/IGF-I axis in patients with dilated cardiomyopathy. *Clin Endocrinol (Oxf)* 50: 417–430.

73. Yang R, Bunting S, Gillett N, Clark R, and Jin H (1995). Growth hormone improves cardiac performance in experimental heart failure. *Circulation* 92: 262–267.
74. Tajima M, Weinberg EO, Bartunek J, Jin H, Yang R, Paoni NF, and Lorell BH (1999). Treatment with growth hormone enhances contractile reserve and intracellular calcium transients in myocytes from rats with postinfarction heart failure. *Circulation* 99: 127–134.
75. Cittadini A, Grossman JD, Napoli R, Katz SE, Stromer H, Smith RJ, Clark R, Morgan JP, and Douglas PS (1997). Growth hormone attenuates early left ventricular remodeling and improves cardiac function in rats with large myocardial infarction. *J Am Coll Cardiol* 29: 1109–1116.
76. Oka T and Komuro I (2008). Molecular mechanisms underlying the transition of cardiac hypertrophy to heart failure. *Circulation J* Supp.A: A13–A16.
77. Nadal-Ginard B, Kajstura J, Leri A, and Anversa P (2003). Myocyte death, growth, and regeneration in cardiac hypertrophy and failure. *Circ Res* 92: 139–150.
78. Foo RS, Mani K, and Kitsis RN (2005). Death begets failure in the heart. *J Clin Invest* 115: 565–571.
79. Nakayama H, Chen X, Baines CP, Klevitsky R, Zhang X, Zhang H, Jaleel N, Chua BH, Hewett TE, Robbins J, Houser SR, and Molkentin JD (2007). Ca^{2+}- and mitochondrial-dependent cardiomyocyte necrosis as a primary mediator of heart failure. *J Clin Invest* 117: 2431–2444.
80. Wencker D, Chandra M, Nguyen K, Miao W, Garantziotis S, Factor SM, Shirani J, Armstrong RC, and Kitsis RN (2003). A mechanistic role for cardiac myocyte apoptosis in heart failure. *J Clin Invest* 111: 1497–1504.
81. Kostin S, Pool L, Elsasser A, Hein S, Drexler HC, Arnon E, Hayakawa Y, Zimmermann R, Bauer E, Klovekorn WP, and Schaper J (2003). Myocytes die by multiple mechanisms in failing human hearts. *Circ Res* 92: 715–724.
82. Akazawa H, Komazaki S, Shimomura H, Terasaki F, Zou Y, Takano H, Nagai T, and Komuro I (2004). Diphtheria toxin-induced autophagic cardiomyocyte death plays a pathogenic role in mouse model of heart failure. *J Biol Chem* 279: 41095–41103.
83. Nakai A, Yamaguchi O, Takeda T, Higuchi Y, Hikoso S, Taniike M, Omiya S, Mizote I, Matsumura Y, Asahi M, Nishida K, Hori M, Mizushima N, and Otsu K (2007). The role of autophagy in cardiomyocytes in the basal state and in response to hemodynamic stress. *Nat Med* 13: 619–624.
84. Zhu H, Tannous P, Johnstone JL, Kong Y, Shelton JM, Richardson JA, Le V, Levine B, Rothermel BA, and Hill JA (2007). Cardiac autophagy is a maladaptive response to hemodynamic stress. *J Clin Invest* 117: 1782–1793.
85. Ducharme A, Frantz S, Aikawa M, Rabkin E, Lindsey M, Rohde LE, Schoen FJ, Kelly RA, Werb Z, Libby P, and Lee RT (2000). Targeted deletion of matrix metalloproteinase-9 attenuates left ventricular enlargement and collagen accumulation after experimental myocardial infarction. *J Clin Invest* 106: 55–62.
86. Lindsey ML, Gannon J, Aikawa M, Schoen FJ, Rabkin E, Lopresti-Morrow L, Crawford J, Black S, Libby P, Mitchell PG, and Lee RT (2002). Selective matrix metalloproteinase inhibition reduces left ventricular remodeling but does not inhibit angiogenesis after myocardial infarction. *Circulation* 105: 753–758.
87. Rohde LE, Ducharme A, Arroyo LH, Aikawa M, Sukhova GH, Lopez-Anaya A, McClure KF, Mitchell PG, Libby P, and Lee RT (1999). Matrix metalloproteinase inhibition attenuates early left ventricular enlargement after experimental myocardial infarction in mice. *Circulation* 99: 3063–3070.
88. Lindsey ML, Escobar GP, Dobrucki LW, Goshorn DK, Bouges S, Mingoia JT, McClister DM, Jr., Su H, Gannon J, MacGillivray C, Lee RT, Sinusas AJ, and Spinale FG (2006). Matrix metalloproteinase-9 gene deletion facilitates angiogenesis after myocardial infarction. *Am J Physiol Heart Circ Physiol* 290: H232–H239.
89. Lee RT (2001). Matrix metalloproteinase inhibition and the prevention of heart failure. *Trends Cardiovasc Med* 11: 202–205.

90. Berk BC, Fujiwara K, and Lehoux S (2007). ECM remodeling in hypertensive heart disease. *J Clin Invest* 117: 568–575.
91. Spinale FG (2007). Myocardial matrix remodeling and the matrix metalloproteinases: influence on cardiac form and function. *Physiol Rev* 87: 1285–1342.
92. O'Rourke B, Kass DA, Tomaselli GF, Kaab S, Tunin R, and Marban E (1999). Mechanisms of altered excitation-contraction coupling in canine tachycardia-induced heart failure. I: experimental studies. *Circ Res* 84: 562–570.
93. Hobai IA and O'Rourke B (2001). Decreased sarcoplasmic reticulum calcium content is responsible for defective excitation-contraction coupling in canine heart failure. *Circulation* 103: 1577–1584.
94. Bers DM (2006). Altered cardiac myocyte Ca regulation in heart failure. *Physiology (Bethesda)* 21: 380–387.
95. Miyamoto MI, del Monte F, Schmidt U, DiSalvo TS, Kang ZB, Matsui T, Guerrero JL, Gwathmey JK, Rosenzweig A, and Hajjar RJ (2000). Adenoviral gene transfer of SERCA2a improves left-ventricular function in aortic-banded rats in transition to heart failure. *Proc Natl Acad Sci U S A* 97: 793–798.
96. Minamisawa S, Hoshijima M, Chu G, Ward CA, Frank K, Gu Y, Martone ME, Wang Y, Ross J, Jr., Kranias EG, Giles WR, and Chien KR (1999). Chronic phospholamban-sarcoplasmic reticulum calcium ATPase interaction is the critical calcium cycling defect in dilated cardiomyopathy. *Cell* 99: 313–322.
97. Hoshijima M, Ikeda Y, Iwanaga Y, Minamisawa S, Date MO, Gu Y, Iwatate M, Li M, Wang L, Wilson JM, Wang Y, Ross J, Jr., and Chien KR (2002). Chronic suppression of heart-failure progression by a pseudophosphorylated mutant of phospholamban via in vivo cardiac rAAV gene delivery. *Nat Med* 8: 864–871.
98. Shiojima I, Sato K, Izumiya Y, Schiekofer S, Ito M, Liao R, Colucci WS, and Walsh K (2005). Disruption of coordinated cardiac hypertrophy and angiogenesis contributes to the transition to heart failure. *J Clin Invest* 115: 2108–2118.
99. Izumiya Y, Shiojima I, Sato K, Sawyer DB, Colucci WS, and Walsh K (2006). Vascular endothelial growth factor blockade promotes the transition from compensatory cardiac hypertrophy to failure in response to pressure overload. *Hypertension* 47: 887–893.
100. Heineke J, Auger-Messier M, Xu J, Oka T, Sargent MA, York A, Klevitsky R, Vaikunth S, Duncan SA, Aronow BJ, Robbins J, Crombleholme TM, and Molkentin JD (2007). Cardiomyocyte GATA4 functions as a stress-responsive regulator of angiogenesis in the murine heart. *J Clin Invest* 117: 3198–3210.
101. Rabinowitz M and Zak R (1975). Mitochondria and cardiac hypertrophy. *Circ Res* 36: 367–376.
102. Beer M, Seyfarth T, Sandstede J, Landschutz W, Lipke C, Kostler H, von Kienlin M, Harre K, Hahn D, and Neubauer S (2002). Absolute concentrations of high-energy phosphate metabolites in normal, hypertrophied, and failing human myocardium measured noninvasively with (31)P-SLOOP magnetic resonance spectroscopy. *J Am Coll Cardiol* 40: 1267–1274.
103. Weiss RG, Gerstenblith G, and Bottomley PA (2005). ATP flux through creatine kinase in the normal, stressed, and failing human heart. *Proc Natl Acad Sci U S A* 102: 808–813.
104. Nakae I, Mitsunami K, Omura T, Yabe T, Tsutamoto T, Matsuo S, Takahashi M, Morikawa S, Inubushi T, Nakamura Y, Kinoshita M, and Horie M (2003). Proton magnetic resonance spectroscopy can detect creatine depletion associated with the progression of heart failure in cardiomyopathy. *J Am Coll Cardiol* 42: 1587–1593.
105. Neubauer S, Horn M, Cramer M, Harre K, Newell JB, Peters W, Pabst T, Ertl G, Hahn D, Ingwall JS, and Kochsiek K (1997). Myocardial phosphocreatine-to-ATP ratio is a predictor of mortality in patients with dilated cardiomyopathy. *Circulation* 96: 2190–2196.
106. Mattson MP and Wan R (2005). Beneficial effects of intermittent fasting and caloric restriction on the cardiovascular and cerebrovascular systems. *J Nutr Biochem* 16: 129–137.
107. Meyer TE, Kovacs SJ, Ehsani AA, Klein S, Holloszy JO, and Fontana L (2006). Long-term caloric restriction ameliorates the decline in diastolic function in humans. *J Am Coll Cardiol* 47: 398–402.

108. Riordan MM, Weiss EP, Meyer TE, Ehsani AA, Racette SB, Villareal DT, Fontana L, Holloszy JO, and Kovacs SJ (2008). The effects of caloric restriction- and exercise-induced weight loss on left ventricular diastolic function. *Am J Physiol Heart Circ Physiol* 294: H1174–H1182.
109. Taffet GE, Pham TT, and Hartley CJ (1997). The age-associated alterations in late diastolic function in mice are improved by caloric restriction. *J Gerontol A Biol Sci Med Sci* 52: B285–B290.
110. Seymour EM, Parikh RV, Singer AA, and Bolling SF (2006). Moderate calorie restriction improves cardiac remodeling and diastolic dysfunction in the Dahl-SS rat. *J Mol Cell Cardiol* 41: 661–668.
111. Wessells RJ and Bodmer R (2004). Screening assays for heart function mutants in *Drosophila*. *Biotechniques* 37: 58–60.
112. Taghli-Lamallem O, Akasaka T, Hogg G, Nudel U, Yaffe D, Chamberlain JS, Ocorr K, and Bodmer R (2008). Dystrophin deficiency in *Drosophila* reduces lifespan and causes a dilated cardiomyopathy phenotype. *Aging Cell* 7: 237–249.
113. Ocorr K, Akasaka T, and Bodmer R (2007). Age-related cardiac disease model of *Drosophila*. *Mech Ageing Dev* 128: 112–116.
114. Wolf MJ, Amrein H, Izatt JA, Choma MA, Reedy MC, and Rockman HA (2006). *Drosophila* as a model for the identification of genes causing adult human heart disease. *Proc Natl Acad Sci U S A* 103: 1394–1399.
115. Dowse H, Ringo J, Power J, Johnson E, Kinney K, and White L (1995). A congenital heart defect in *Drosophila* caused by an action-potential mutation. *J Neurogenet* 10: 153–168.
116. Johnson E, Ringo J, and Dowse H (1997). Modulation of *Drosophila* heartbeat by neurotransmitters. *J Comp Physiol [B]* 167: 89–97.
117. Dulcis D and Levine RB (2003). Innervation of the heart of the adult fruit fly, *Drosophila melanogaster*. *J Comp Neurol* 465: 560–578.
118. Ocorr K, Reeves NL, Wessells RJ, Fink M, Chen HS, Akasaka T, Yasuda S, Metzger JM, Giles W, Posakony JW, and Bodmer R (2007). KCNQ potassium channel mutations cause cardiac arrhythmias in *Drosophila* that mimic the effects of aging. *Proc Natl Acad Sci U S A* 104: 3943–3948.
119. Curtis NJ, Ringo JM, and Dowse HB (1999). Morphology of the pupal heart, adult heart, and associated tissues in the fruit fly, *Drosophila melanogaster*. *J Morphol.* 240: 225–235.
120. Molina MR and Cripps RM (2001). Ostia, the inflow tracts of the *Drosophila* heart, develop from a genetically distinct subset of cardial cells. *Mech Dev* 109: 51–59.
121. Cammarato A, Dambacher CM, Knowles AF, Kronert WA, Bodmer R, Ocorr K, and Bernstein SI (2008). Myosin transducer mutations differentially affect motor function, myofibril structure, and the performance of skeletal and cardiac muscles. *Mol Biol Cell* 19: 553–562.
122. Ocorr K, Perrin L, Lim HY, Qian L, Wu X, and Bodmer R (2007). Genetic control of heart function and aging in *Drosophila*. *Trends Cardiovasc Med* 17: 177–182.
123. Luong N, Davies CR, Wessells RJ, Graham SM, King MT, Veech R, Bodmer R, and Oldham SM (2006). Activated FOXO-mediated insulin resistance is blocked by reduction of TOR activity. *Cell Metab* 4: 133–142.
124. Robbins J (2001). KCNQ potassium channels: physiology, pathophysiology, and pharmacology. *Pharmacol Ther* 90: 1–19.
125. Piazza N, Hayes M, Martin I, Duttaroy A, Grotewiel M, and Wessells R (2009). Multiple measures of functionality exhibit progressive decline in a parallel, stochastic fashion in *Drosophila* Sod2 null mutants. *Biogerontology* 10: 637–648.
126. Judge LM, Haraguchiln M, and Chamberlain JS (2006). Dissecting the signaling and mechanical functions of the dystrophin-glycoprotein complex. *J Cell Sci* 119: 1537–1546.
127. Morrow G and Tanguay RM (2008). Mitochondria and ageing in *Drosophila*. *Biotechnol J* 3: 728–739.

128. Kenyon C (2005). The plasticity of aging: insights from long-lived mutants. *Cell* 120: 449–460.
129. Kim SK (2007). Common aging pathways in worms, flies, mice and humans. *J Exp Biol* 210: 1607–1612.
130. Wessells RJ and Bodmer R (2007). Age-related cardiac deterioration: insights from *Drosophila*. *Front Biosci* 12: 39–48.
131. Hamlin RL (2007). Animal models of ventricular arrhythmias. *Pharmacol Ther* 113: 276–295.
132. Koh GY, Soonpaa MH, Klug MG, Pride HP, Cooper BJ, Zipes DP, and Field LJ (1995). Stable fetal cardiomyocyte grafts in the hearts of dystrophic mice and dogs. *J Clin Invest* 96: 2034–2042.
133. Jugdutt BI (2002). The dog model of left ventricular remodeling after myocardial infarction. *J Card Fail* 8: S472–S475.
134. Hamlin RL and Smith CR (1960). Anatomical and physiologic basis for interpretation of the electrocardiogram. *Am J Vet Res* 21: 701–718.
135. Guglielmini C (2003). Cardiovascular diseases in the ageing dog: diagnostic and therapeutic problems. *Vet Res Commun* 27(Suppl 1): 555–560.
136. Eichelberg H and Seine R (1996). [Life expectancy and cause of death in dogs. I. The situation in mixed breeds and various dog breeds]. *Berl Munch Tierarztl Wochenschr* 109: 292–303.
137. Van Vleet JF (2001). Age-related non-neoplastic lesions of the heart. In *Patholbiology of the Ageing Dog*. (U Mohr et al., ed.). Ames: ISU Press, pp. 101–107.
138. Kwart C and Haggstrom J (2000). Acquired valvular heart disease. In *Textbook of Veterinary Internal Medicine*. (SJ Ettinger, EC Feldman, ed.). Philadelphia: W.B. Saunders.
139. Lane MA, Ingram DK, and Roth GS (1999). Calorie restriction in nonhuman primates: effects on diabetes and cardiovascular disease risk. *Toxicol Sci* 52: 41–48.
140. Lane MA, Mattison J, Ingram DK, and Roth GS (2002). Caloric restriction and aging in primates: relevance to humans and possible CR mimetics. *Microsc Res Tech* 59: 335–338.
141. Mattison JA, Lane MA, Roth GS, and Ingram DK (2003). Calorie restriction in rhesus monkeys. *Exp Gerontol* 38: 35–46.
142. Roth GS, Mattison JA, Ottinger MA, Chachich ME, Lane MA, and Ingram DK (2004). Aging in rhesus monkeys: relevance to human health interventions. *Science* 305: 1423–1426.
143. Bodmer R, Wessells RJ, Johnson EC, Dowse H (2005). Heart development and function. In: Comprehensive Molecular Insect Science (volumes 1–7). (eds. Gilbert LI, Iatrou K, and Gill S.), Elsevier, vol. 2, p. 199–250.

Comparative Skeletal Muscle Aging

David J. Marcinek, Jonathan Wanagat, and Jason J. Villarin

Abstract The decline in skeletal muscle function is characteristic of aging organisms across species ranging from *C. elegans* to humans. This is a multifaceted process involving changes in both the quantity (sarcopenia) and quality of skeletal muscle with age. This chapter describes age-related changes in several aspects of muscle function, including energy metabolism, muscle contraction, stem cell and injury repair and response to cellular stress. We pay particular attention to the effect of aging on different muscle fiber types. These fiber type differences, which are often overlooked, may explain some of the apparent contradictory results in the literature.

Keywords Skeletal muscle · Atrophy · Sarcopenia · Mitochondrial · Dysfunction · Strength

Introduction

The decline in skeletal muscle mass and function is a well-recognized characteristic of aging animals. In fact, all animals examined demonstrate a significant loss of skeletal muscle function with age, including some combination of metabolic defects, reduced efficiency, poor contractile function, reduced recovery from injury, and muscle atrophy (i.e. sarcopenia). Skeletal muscle is responsible for all voluntary movement in vertebrates and constitutes approximately 40% of the body mass of humans and most other vertebrates. The main functions of muscle are to generate force and power; both of these functions decline with age. The ubiquity of this effect is demonstrated by the significant declines in muscle performance in several species (see Table 1): humans, rats, mice, horses, *Drosophila*, and *C. elegans*. In humans this is manifest in the gradual, but steady decline of the ability to perform everyday activities. Even among highly trained competitive athletes performance

D.J. Marcinek (✉)
Department of Radiology, University of Washington, Seattle, WA 98195, USA
e-mail: dmarc@u.washington.edu

N.S. Wolf (ed.), *The Comparative Biology of Aging*,
DOI 10.1007/978-90-481-3465-6_13,

Table 1 Summary of contributions to skeletal muscle dysfunction with age in various animal models

	Fiber loss	Fiber atrophy	Reduced oxidative capacity	Mitochondrial dysfunction	Contractile dysfunction	Apoptosis	Impaired satellite cell function	Loss of motorneurons	Improvement with exercise
Human	[27]	[27]	[62]	[80]	[44]		[174, 175]	[110, 216]	[196, 197, 201]
Non-human primate	[145]	[152]		[145]					
Horse			[68, 94]						[68, 217]
Rat	[16]	[10]	[66]	[26]	[43, 47]	[218, 219]	[182, 220]	[110]	[204, 205]
Mouse	[221]	[222]		[81]	[38, 49]	[223]	[167, 170]	[224]	[202]
Drosophila			[95]	[95]	[225]	[20]			
C. elegans	[13]	[213]							

steadily declines after the third or fourth decade of life [1]. This suggests the phenomenon is an inherent part of aging, although reduced physical activity exacerbates the age-related decline.

The decline in skeletal muscle function has been said to be "the gateway into frailty" and loss of independence in the elderly. The loss of independence is a function of both the inability to perform activities of daily living and co-morbidities associated with the impaired skeletal muscle function. For example, reduced sustained power output results in reduced endurance that will limit the range of walking and declines in peak force and power output can limit the ability to get out of bed or lift groceries and contribute to increased risk of falls in the elderly [2]. In addition, the large volume and high energy demand of muscle in the body makes it an important contributor to the systemic energy balance of the body. For example, muscle is an important site uptake for glucose and ATP use in the body and reduced muscle mass and metabolism or weakness induced inactivity are all significant risk factors for the onset of obesity and insulin resistance and the associated complications, including cardiovascular problems [2]. As a result, the healthcare costs associated with loss of muscle function in the elderly are significant. The US healthcare costs attributable to sarcopenia and related comorbidities in the year 2000 was $18 billion [3]. An aging US population means the costs to society are going to increase in the coming years in both economic and quality of life measures. Thus the decline in muscle function is a significant and immediate healthcare concern.

Sarcopenia

Muscle Atrophy in Humans

Humans muscle mass declines about 10–15% per decade between 50 and 70 years of age, followed by an accelerated decline of up to another 30% from 70 to 80 [4]. This cumulative loss of muscle results in a dramatic decline between young adults and the elderly. For example, in vivo imaging indicates that there is a 25% reduction in human quadriceps cross sectional area between young and old subjects [5], while [6] reported a 20% change in cross-sectional area between 65 and 80 years.

The loss of muscle mass is a function of both a loss of muscle fibers and a reduced cross sectional area of the fibers. Accurate analysis of muscle fiber number is difficult and prone to error. The most thorough methods involve counting fibers in the entire cross section of a muscle, however this approach can be impractical. Other methods are to digest the muscle and count the dissociated fibers or estimate fiber number from the mean fiber cross sectional area and the muscle cross section [7]. Due to the laborious nature of these analyses, there are relatively few quantitative analyses of changes in muscle fiber number with age. An analysis of vastus lateralis (quadriceps) muscle from human males from 20 to 80 years old found a loss of muscle fibers that begins between the ages of 30 and 40, progressing to a 40% drop in fiber number between 40 and 80 years of age [8]. Distinguishing whether the decrease in total muscle cross sectional area is due mainly to loss of fibers or to atrophy of the same number of fibers is important as it guides intervention strategies. If the primary pathology is loss of fiber number, then maintaining cell number is most important. However, if individual fiber atrophy is the greater pathology, then driving fiber hypertrophy becomes of greater importance. Available evidence suggests both conditions occur in aging muscle.

Atrophy in Animal Models

Age-related loss of muscle mass and cross-sectional area is also found in rodent models of sarcopenia [9]. Rats experience significant declines in muscle weight in the quadriceps, gastrocnemius, and plantaris muscles between adult (9–10 months) and old age (28–30 months) [10]. Declines in mouse muscle mass of 10–25% are reported for a variety of hindlimb muscles between adult (8–14 months) and old age (27 months) [11, 12]. The quantification of the loss of muscle mass and cross-sectional area is more difficult in invertebrate systems and therefore there is little information regarding the change in muscle mass or cross sectional area with age. However, the loss of muscle and deterioration of muscle structure with age extends to invertebrate systems in which it has been studied. The locomotory dysfunction in Caenorhabditis elegans noted earlier is correlated with a disruption of muscle morphology and loss of myofilaments and muscle nuclei in the body wall [13, 14].

Fiber Loss

Muscle fiber loss also plays a significant role in rodents, although the extent of fiber loss in rodent models appears to be less than observed in human tissue. Early studies concluded that the reduction in cross sectional area in multiple muscles of the rat hindlimb almost entirely explained the muscle atrophy with age [10]. However, studies in which fiber number was actually counted indicate a 5–10% loss of muscle fibers between adult and 27-month old in soleus, extensor digitorum longus (edl) [15], and gastrocnemius [16] between adult and 24- to 27-month old rats. Another study found that fiber number remained constant through middle-age in rats, but then fell by approximately 25% in the rectus femoris, vastus lateralis, and soleus muscle of 36-month old rats [17]. This pattern of fiber loss in the oldest ages is in contrast to that reported for human muscle in which there is a progressive loss of muscle fibers throughout middle and old-age.

Apoptosis

A role for fiber loss in sarcopenia in rodent models is supported by molecular and biochemical evidence of markers of apoptosis in aging muscle in multiple taxa. Increases in DNA fragmentation indicative of apoptosis have been observed in aging rat [18, 19] and *Drosophila* [20] skeletal muscle and load induced susceptibility to apoptosis is also greater in aged quail muscle [21]. Both mitochondrially mediated

[18, 21–23] and mitochondria-independent [24, 25] pathways are responsible for the increase in markers of apoptosis with age. Despite the evidence for the presence of apoptotic markers and nuclei in aged skeletal muscle, it is not clear what role apoptosis plays in sarcopenia. One reason for this is that skeletal muscle fibers are multinucleated, so the effect of apoptotic nuclei on the survival of the fiber is more complicated than for mononucleated cells. The localized atrophy and fiber breakage leading to fiber loss demonstrated by the Aiken laboratory [26] provides a potential mechanism as to how loss of a small number of nuclei in a muscle fiber could lead to fiber loss (see below for more discussion).

Preferential Loss of Type II Fibers

There is a preferential loss of type-II fibers in the muscles of elderly humans [27, 28]. The fiber loss in the human male quadriceps is associated with a significant decrease in relative mean fiber area of type II (fast-twitch) fibers [29] [27], which results in an increase in the type I fiber contribution to the muscle. Although there is considerable variation in the exact numbers, there is general agreement that there is a shift in human muscle toward a greater relative slow-twitch fiber area with age. The anatomical effect is accompanied by a shift in the relative expression levels of myosin heavy chain isoforms (increased MHC I/MHCII) with age [30]. This preferential loss of the type-II fiber area and shift toward the type-I has even been observed in sprint trained aging athletes [30], where the explosive muscle contractions would be expected to stimulate fast motor units and preserve type II muscle fibers with age.

Data for rodent models are also consistent with that for human muscles in demonstrating greater effects on type II than type I muscle fibers. One advantage to working with rodent models of muscle aging is that there is a greater segregation of fiber type between muscles than is typically found in human muscles. Accordingly, the different effects of aging on muscle fiber type can be examined by looking at different muscles in mice and rats. The standard model for a type I muscle is the soleus, which is composed of 80–90% type I muscle fibers, while most other muscles of the mouse and rat hindlimb comprise mostly type II fibers [31]. Studies typically find that the muscles composed of primarily type I fibers (slow-twitch) lose less muscle mass [10, 11, 17, 24, 32–34], and fewer fibers with age [17] than muscles with a greater percentage of type II fibers. These results suggest that there is something intrinsic to fast-twitch muscles that makes them more susceptible to age-related atrophy. Possible mechanisms are greater exposure to oxidative stress, as has been demonstrated in rodent muscle [35], or less frequent stimulation compared to type I fibers [36].

Force Production

In skeletal muscle force is produced by the interaction of the myosin and actin components of the cross-bridges. A larger cross-sectional area of muscle contains more cross-bridges and, therefore, produces more force than smaller muscles.

Thus the decline in muscle cross-sectional area with age results in a reduced force production by the muscle. The reduced cross-sectional area in aging muscle has been found to account for approximately 2/3 of the drop in maximal force production in vivo. Studies in the human quadriceps have shown that a 20–25% drop in cross-sectional area with age is associated with 35–40% decline in strength [5, 37]. Both rat and mouse models demonstrate similar reductions in peak force. In mice maximal specific force (force per cross sectional area) developed by isolated soleus and edl muscles declines 23 and 35%, respectively, between 5 and 28 months of age [38], indicating that the differences between the young and old muscles are not solely due to changes in muscle size. This loss of force contributes to the impaired locomotory (voluntary movement) activity that occurs with age in humans, rodents, and invertebrate model systems [39, 40].

Changes in Contractile Apparatus

The effects of aging on muscle are also evident in tissue and cellular morphology. The reduction in muscle quantity is accompanied by a decline in muscle quality, giving rise to the functional frailty described in the preceding sections. Histological evaluation of aged muscle shows characteristic changes including increased heterogeneity of fiber size, increased infiltration of connective tissue, and changes in capillary morphometry. This change in relative proportion of contractile fibers begins to explain why the force per unit cross sectional area is reduced in aged muscle. Furthermore, the volume of muscle tissue still occupied by fibers may contain contractile elements with reduced function. This is supported by contractile performance characteristics that are below the predicted value even when including the true fiber volume in the estimate or when evaluating single isolated fibers [41].

Accumulation of damage to muscle proteins and the progression of sarcopenia have functional consequences to muscle contractile performance (reviewed by [42]). When comparing single soleus muscle fibers from adult (12 months) versus old (36 months) rats, there is a 36% decrease in cross-sectional area, a 30% decrease in specific maximal force, and 36% decrease in shortening velocity [43]. Similar results have been reported for mouse [12, 38] and human muscles [44], as well. In vitro motility assays, used to isolate the function of the acto-myosin complex, have demonstrated an age-related decrease in the sliding velocity of actin on myosin isolated from muscles from mice, rats, and humans [45–47]. However, this decline only accounts for a fraction of the decrease in shortening velocity. The remaining deficit in shortening velocity and peak force generation are most likely due to modifications of ATPase activities of myosin [46, 47] and Ca^{2+} regulation [48]. Delbono has demonstrated significant changes in excitation contraction coupling in aging skeletal muscle, including reduced Ca^{2+} release from sarcoplasmic reticulum [49], increased dependence on extracellular calcium [50], decreased cross-bridge sensitivity to Ca^{2+} [51]. Thus aged muscle dysfunction is clearly present at the molecular and single fiber levels of organization in addition to the compromised function seen in intact tissue and the whole organism.

Changes in Muscle Innervation

Aging muscle dysfunction is not limited to intrinsic muscle modifications. An important component of muscle dysfunction is the change in muscle innervation properties. There is a decrease of function starting at the level of the alpha-motoneurons that activate the muscles [52]. Motor units (MU) are composed of a motoneuron and the innervated muscle fibers. These functional increments of muscle activation are of variable size as defined by the number of muscle fibers innervated by the single neuron. MUs are generally composed of similar fiber types and are of different sizes. This allows a given muscle to be activated in different ways such that it performs contractile tasks ranging from fine movements to recruitment of all available strength. During aging the MU's are restructured in a way that affects muscle function [53]. First, there are fewer motor units. This means the muscle is activated in larger increments producing a loss of fine motor coordination. Second, there is an elongation of the interspike interval of discharge [54] producing a slower tonic frequency of muscle activation. Since muscle fiber type is largely determined by chronic innervation frequency [36], this change in neuronal activity may contribute to muscle fiber type composition. That is, a shift in myosin isoform due to the rate of age-associated denervation outpacing the process of reinnervation [55]. The loss of motor units in rodents is predominantly from the type-2 (fast twitch) pool. This leaves some muscle fibers denervated, allowing them to be reincorporated into slow motor units due to sprouting by the slow motor unit neurons. Fibers that are not re-innervated undergo atrophy and are lost from the muscle. A reduced ability of senescent motor neurons to reinnervate muscle may exacerbate the effect of denervation on fiber loss [56, 57]. The net effect is a decrease in the total number of fast motor units and an increase in the size of the slow motor units.

Muscle Oxidative Metabolism

Muscle work is dependent on an adequate supply of ATP to meet increased demand by myosin ATPases for cross-bridge cycling and ion pumping, particularly for Ca^{2+} cycling, during muscle contraction. ATP demand for short term bursts of muscle activity are met by the breakdown of Phosphocreatine (PCr) through creatine kinase and glycolysis, but ATP demand for sustained muscle contraction (more than a few seconds) must be met by oxidative phosphorylation in the mitochondria. In vivo mitochondrial ATP production can increase over 20-fold in mouse (Marcinek, unpublished data) and over 100-fold in human [58] skeletal muscle to meet this increased demand. Thus reductions in mitochondrial function will limit the ATP supply resulting in increased fatiguability and reduced endurance performance.

A reduction in exercise capacity and oxidative ATP supply is a hallmark of aging muscle. This decline is apparent at the whole body level as well as tissue level measurements. Elderly humans exercising at VO_2max consistently demonstrate reduced oxygen uptake and power output of the exercising muscle [59–64]. This decline in oxidative capacity is also found in other models of aging muscle. Both mice and rats also demonstrate reductions in VO_2max and sustained power output with age [65, 66]. Horses were also found to undergo significant declines in VO_2max and cardiovascular function with age [67, 68]. Changes in cardiovascular function [69, 70] and muscle microcirculation [71, 72] certainly contribute to the reduced whole-body and tissue level VO_2max of muscle with age, but it is clear that alterations in the metabolic capacities of the muscle themselves are a key determinant in the decline with age. This was demonstrated by a nice set of experiments by Hepple et al. [66], where they found decreases in peak force, force at VO_2max, and VO_2max of 30% with age in the plantar flexion muscles of the rat hindlimb (gastrocnemius, plantaris, soleus) in situ despite matching muscle perfusion and O_2 delivery in the different age groups.

Mitochondrial Content vs. Dysfunction

The capacity for mitochondrial ATP synthesis of a tissue is the product of the structural and functional aspects of mitochondria. The structural component consists of the mitochondrial content (i.e. mitochondrial volume density, mitochondrial enzyme concentrations) of the cell. The functional component (i.e. how well the parts work) is reflected in the capacity for ATP production or O_2 consumption per unit mitochondria. Therefore, reduced mitochondrial capacity could be due to reduced mitochondrial content (fewer mitochondria) or mitochondrial dysfunction (reduced function of existing mitochondria) [73]. This is an important distinction because these components may be caused by independent mechanisms and lead to different outcomes. Reduced mitochondrial content may be the result of reduced physical activity and be easily reversed with exercise [74]. Mitochondrial dysfunction, on the other hand, may be indicative of a pathological condition associated with age or disease and has been linked to oxidative damage to mitochondrial DNA (mtDNA), proteins, and lipids in mitochondria isolated from aged tissues [75–78]. In addition to the obvious effects on muscle work, oxidative damage to electron transport chain components has also been found to be linked to increased ROS production [79] and uncoupling of oxidative phosphorylation to reduced in vivo resting ATP levels in both human [80] and mouse skeletal muscle [81]. Both reduced ATP and increased oxidative stress are triggers for the induction of apoptotic and/or necrotic cell death [82, 83]. Therefore, mitochondrial dysfunction can lead not only to impairment of ATP production, but also may sensitize the cell to induction of cell death pathways [84–86].

Reduced Mitochondrial Content

Reduced mitochondrial content [62, 87, 88] or lower activities of oxidative enzymes [87, 89, 90] are associated with declines in ATP production and/or O_2 consumption in vivo with age [62, 63]. Reduced maximal oxygen consumption and ATP production has also been found with age in isolated muscles of aged rodents [66, 91]. The decline in oxidative capacity is supported by decreased activities of oxidative enzymes in rodents [92, 93] and horses [94] with age. Decreases in maximal O_2 consumption and cytochrome c oxidase activity have also been reported to decline with age in the flight muscles of *Drosophila* [95–97]. However, the decline was not observed for other components of the electron transport chain. This is consistent with data in mammalian models demonstrating a dysregulation of the multiple components of mitochondria with age [98], potentially due mitochondrial mutations that impair the production of mitochondrial encoded proteins [26].

Reduced mitochondrial content with age is supported by analysis of age-related changes in the transcriptome. Genes encoding subunits of the electron transport chain are downregulated in aged human skeletal muscle [99, 100]. Interestingly, this pattern of decreased electron transport chain (ETC) subunit expression in also found in aged human brain and kidney tissues [99]. ETC genes are also downregulated in aged tissues across disparate species, such as *Drosophila* [99, 101] and mouse kidney [99], but not *C. elegans* [99, 102]. Downregulation of genes involved in mitochondrial energy metabolism has also been reported for mouse skeletal muscle [103]. The consistent decline in mitochondrially associated genes in skeletal muscle is particularly interesting given the apparent enhancement of the more oxidative type I fiber type in aging skeletal muscles, which would be expected in to increase the relative mitochondrial content in aged muscle.

Despite the in vivo, tissue, biochemical, and molecular evidence supporting a decline in mitochondrial content, other studies have reported no effects of age on in vivo mitochondrial capacity [104–106] or function [107, 108]. These authors conclude that reduced activity and not aging per se is responsible for the decline in mitochondrial function in skeletal muscle. However, the consistent declines in human athletic performance [109] and energy metabolism with age, even among master athletes [110] in both power [111] and endurance [112–114] sports supports the hypothesis that the loss of function is an intrinsic property of aging muscle.

A potential explanation for the some of the inconsistent results reported in the literature is variation between individual muscles in the effects of age on mitochondrial function. Kent-Braun's work demonstrated no effect of age on energetics and exercise performance in the tibialis anterior, which is a predominantly slow-twitch muscle in humans (75% type I) [115], while the vastus lateralis [31, 115] and the gastrocnemius [31, 115], two commonly studied muscles in humans and rodents that demonstrate a decline in mitochondrial function with age are primarily type II. The fiber type differences in mitochondrial function with age are consistent with preferential loss and atrophy of type II fibers discussed above. Inter-muscle variation in the effects of age on mitochondrial function are supported by numerous studies demonstrating different effects of age on oxidative enzyme activities and expression between muscles in human [116], and rodent models [10, 33], although these differences do not necessarily breakdown along type I/type II fiber dichotomy. One confounding factor in the interpretation of measurements of oxidative enzyme activities from age muscle homogenates is the shift toward a more type I phenotype with age. Since type I fibers typically have greater mitochondrial content, the preferential atrophy and loss of type II would be expected to lead to an increase in mitochondrial enzymes per gram muscle that may mask more subtle changes with age.

Mitochondrial Dysfunction

Mitochondrial dysfunction, defined as a reduced capacity per unit mitochondria, may be a better marker of pathogical effects of age on energy metabolism, because it relates to the functional quality of the existing pool of mitochondria. In one study where both content and dysfunction were assessed, both were found to contribute to the decline in ATP production with age. Conley et al. [62] combined in vivo measures of ATP production in human vastus lateralis muscle with analysis of the mitochondrial volume density of the muscle to quantify the contributions of both reduced content and mitochondrial dysfunction to the loss of mitochondrial capacity with age. Their results indicate that the loss of mitochondrial content accounted for only about 50% of the total loss of oxidative capacity of muscle. The other 50% was due to a reduced ATP production per mitochondrial content. These in vivo results are supported by work in isolated mitochondria from human [87] and rodent models [91] demonstrating that a reduced capacity for ATP production in mitochondrial isolated from aged skeletal muscle.

Mechanisms of Dysfunction

Mitochondria produce ATP by coupling the generation of the proton gradient by the electron transport chain (ETC) to ATP synthesis through the inner membrane F_1F_2 ATP synthase. This process is called oxidative phosphorylation because the oxidation of substrates by the electron transport chain is linked to phosphorylation of ADP by the ATP synthase. Protons are pumped from the matrix to the inner membrane space of mitochondria by the flux of electrons through the complexes of the ETC. This electron flux culminates in the reduction of O_2 to water at complex IV (cytochrome oxidase) of the ETC. The proton pumping establishes a proton motive force creating a membrane potential and pH gradient across the inner mitochondrial membrane that is used to drive the synthesis of ATP. Therefore damage to ETC proteins or associated lipids will limit mitochondrial ATP production and may result in increased ROS production by the mitochondria and increased proton leak [117–119] all of which would contribute to a loss of resting ATP levels and potentially induce cell death.

Reduced Coupling with Age

We have recently demonstrated significant mitochondrial uncoupling (lower P/O values) in vivo in aged skeletal muscles of both human hand muscle, first digitorum interosseous (FDI) [80], and mouse hindlimb [81]. Using a combination of magnetic resonance and optical spectroscopy to directly measure ATP use and O_2 uptake in vivo we found that aged skeletal muscles in mouse leg produce on average 40–50% fewer ATP per mitochondrial O_2 uptake (reduced P/O) than young muscle. This mitochondrial uncoupling is consistent with the reduced in vivo maximal mitochondrial ATP production (ATPmax) in vivo found in the quadriceps from elderly humans [62]. The reduced ATPmax will contribute to exercise intolerance in the elderly. In addition, the reduced mitochondrial membrane potential is expected to lead to lower ATP concentrations in the muscle [120]. In fact, in both aged mouse and human muscles reduced mitochondrial uncoupling was associated with a disruption of resting energetics and lower resting ATP [80, 81]. This loss of ATP, as well as the reduced membrane potential itself can sensitize the cell to the initiation of apoptosis [121–123] that may contribute to fiber loss and sarcopenia.

We have found that mitochondrial uncoupling with age is also fiber-type dependent in humans. The predominantly slow-twitch tibialis anterior did not change significantly with age [80], while the FDI, which is has greater type II contribution [115], experienced a greater than 50% loss of coupling in vivo and resting ATP levels. These data are supported by other in vivo investigations of human muscles demonstrating a lack of uncoupling with age in vivo in the human soleus [124] (89% type I [115]), but significant mitochondrial dysfunction in the aged human quadriceps [62] that is consistent with mitochondrial uncoupling.

Uncoupling Paradox

Reduced mitochondrial coupling in aging tissues presents an interesting paradox. In addition to the negative consequences of mitochondrial uncoupling cited above, reduced coupling has been demonstrated to reduce the generation of ROS by the mitochondria [125–127] and provide protection against age-related disease [128, 129]. Overexpression of uncoupling protein 1 (UCP1) in mouse skeletal muscle was found to reduce artherosclerosis and lymphoma, reverse diabetes and hypertension [129] and increase insulin sensitivity [128]. Treating mice with low-level 2,4-dinitrophenol (DNP) in the drinking water was also shown to improve health and lifespan [130]. This mild uncoupling in the skeletal muscle of healthy adults is most likely physiologically regulated through UCP and/or the adenine nucleotide translocator (ANT). In an interesting study, Speakman et al. [131] found that isolated mitochondria from the longest-lived quartile of a mouse population had significantly greater proton leak levels than those from the shortest-lived quartile at the same ages. They also demonstrated that this difference was UCP and ANT dependent.

The in vivo data from human skeletal muscle support a protective effect of mild uncoupling. In healthy adults the tibialis anterior, which did not demonstrate mitochondrial dysfunction and loss of ATP with age, was more uncoupled in healthy adults than the flexor digitorum interosseous, which did undergo mitochondrial dysfunction with age [80]. This suggests that the mild uncoupling in the adult tibialis anterior was compensated for by the young skeletal muscle and protected this tissue from age-related mitochondrial dysfunction. These results support the hypothesis that mild uncoupling can be protective against cell pathology, possible by reducing ROS production [127]. The difference between pathological and protective effects of uncoupling may be the ability of the cell to induce mitochondrial biogenesis to offset the associated energetic stress. Mitochondrial uncoupling was recently shown to induce mitochondrial biogenesis in cell culture [132]. The increased mitochondrial function restored resting ATP levels following acute uncoupling. An impaired ability to induce mitochondrial biogenesis in response to energetic stress in aged skeletal muscle [133] may prevent the cell from compensating for uncoupling and lead to loss of ATP and tissue degeneration.

Focal Electron Transport Chain Defects

Global age-related changes in mitochondrial function receive the greatest share of attention. However, another bioenergetic feature of skeletal muscle aging is the occurrence of focal electron transport chain (ETC) defects within individual skeletal muscle fibers. This section of the chapter will review the emerging characteristics of these abnormalities and recent work that suggests their biological impact and close association with mtDNA deletion mutations. The issue of the abundance of these abnormalities will be addressed followed by a brief discussion of interventions that have been shown to impact occurrence of these defects.

In the late 1980s Mueller-Hoecker demonstrated cells lacking cytochrome oxidase (COX, complex IV) activity in hearts of older individuals. After the age of 50, COX-negative cardiomyocytes were found in all hearts examined, but only in some hearts less than 50 years old and none in hearts of children <7 year old [134]. Similar age-associated ETC abnormal muscle fibers were identified in skeletal muscle, even in the absence of muscular disease [135]. These abnormal fibers contain clusters of abnormal mitochondria and excessive lipid droplets which Olson and colleagues [136] termed "ragged-red" fibers (RRF). Additional studies in humans demonstrated age-related increases in RRFs in extraocular eye muscles and diaphragm in additional to limb and cardiac muscle [135, 137]. In these studies, the histochemical ETC phenotype most closely associated with focal ETC defects was a loss of COX activity accompanied by a normal or higher level of succinate dehydrogenase activity (SDH, complex II).

mtDNA Mutations in Focal ETC Defect

The RRFs and ETC abnormalities observed in aging muscle have been linked to mitochondrial DNA (mtDNA) mutations [138–140]. mtDNA point and deletion mutations increase with age in homogenate studies from a variety of tissues including skeletal muscle (reviewed in [141]) and have their highest frequencies in post-mitotic tissues with the greatest energy demands such as brain, heart and skeletal muscle (reviewed in [142]).

Early studies using in situ hybridization techniques [139, 140] linking the histochemical phenotype of RRFs to mtDNA mutations, could not identify the precise mutation event leading to the RRF. More recent studies have used laser capture microdissection to isolate individual skeletal muscle fibers for DNA extraction and PCR analyses [143]. These methods facilitated the identification of breakpoint sequences, flanking repeats, if present, and even quantitation of wild-type and mutant mtDNA levels as will be discussed later in this section.

Shared mtDNA Mutation Characteristics

The mtDNA mutations present in age-associated focal ETC defects of skeletal muscle have been well characterized. In over 150 RRFs studied in rats (69 fibers), monkeys (39 fibers) and humans (48 fibers), all have revealed the presence of mtDNA deletion mutation. Similar numbers of ETC normal fibers have been microdissected and interrogated for deletion mutations and none have been found. The size of deletion mutations ranges from a few hundred base pairs to 12 kb. Generally, the deletion breakpoints fall between the two origins of replication, thereby removing genes from the major arc. Occasionally the deletion event will remove the light strand origin of replication, but this is relatively rare and suggests accessory replication origins.

Regardless of the species examined, the deletion events in RRFs are nearly always unique mutation events. Of the 39 fibers analyzed in old rhesus monkeys, only 3 RRFs contained the same deletion event, while in 48 RRFs from older humans, one "common" deletion event occurred in five fibers, another in two fibers and a third in two fibers [144, 145]. Interestingly, while the DNA sequences flanking the deletion breakpoint often involve direct repeats in monkeys and human, in the rat, they do not [146].

Elegant work done recently to apply quantitative PCR to the study of mtDNA deletion mutations in RRFs revealed a number of other characteristics [147]. As demonstrated previously, the deletion mutations are clonal within the RRF portion of the fiber and when the mutation level exceeds 90%, there is a loss of COX activity and the concomitant increase in SDH activity. This study also demonstrated that full-length mitochondrial genomes are present in the RRF segment at a similar absolute number as in the ETC normal portions of the fiber. This suggests an advantage to the mutant genome, rather than a preferential loss or turnover of the wild-type genome.

Biological Impact of Focal ETC Defects

The biological impact of age-associated RRFs is best demonstrated in histological studies, which demonstrate an association between the histochemical abnormality and pathological changes in the affected fibers. Throughout the species examined, focal ETC defects are closely linked to decreases in fiber cross sectional area

[26, 147, 148]. Fiber atrophy may occur with or without fiber splitting and, in fact, may completely sever a fiber along its length. In rat studies, the degree of atrophy is positively correlated with the length of the affected region of the fiber. This suggests that in smaller defect regions, wild type mitochondria or mitochondrial metabolites are able to diffuse in from flanking normal fiber segments. Nuclear number is also decreased or absent within the atrophic regions, which may be related to triggering apoptotic phenomena in a multinucleate cell (Wanagat, unpublished observations). Numerous studies have examined the gene expression effects of RRFs, but these have only been examined in mitochondrial myopathies, not in age-associated RRFs [149, 150]. Gene expression changes in age-associated focal ETC defects may suggest points of intervention to lessen or prevent their pathological consequences.

Abundance of Focal ETC Defects and Sarcopenia

The causal role of focal ETC defects in sarcopenia is unclear and this uncertainty derives primarily from concerns about the abundance of these defects. The defects are three dimensional objects within a three dimensional tissue, therefore, quantification from a single section is inherently inaccurate as it depends on morphometric characteristics of the defects, in this case, length. Morphometric analyses that utilize the volume density of the defects are able to account for this and provide more accurate measurements that can also be extrapolated to entire tissues. For example, an early report in rat skeletal muscle reported one to two RRFs per muscle section in vastus lateralis muscle [151]. Volume density measurements in 38-month old rat rectus femoris demonstrated about 1 RRF in every cubic millimeter of tissue, which when extrapolated to the volume of the entire muscle predicts over 1,000 RRFs. This would represent over 15% of the fibers in an old rat rectus femoris. Volume density measurements in old monkeys yielded an estimate of 60% [152]. In humans, subjects were examined across many ages and showed a steady increase in the predicted percentage of affected fibers, from 6% at 49 years to 22% at 67 years and 31% at 92 years [153]. Another important point regarding estimates of RRF abundance is the fact that these are steady-state levels because new RRFs are being created as older ones are being lost from the population. Therefore, the total number of affected fibers throughout the lifespan is necessarily even greater than the above estimates. In addition to an age and species effect, the abundance of RRFs differs between muscle groups. Examination of individual muscles shows that muscles resistant to sarcopenia (e.g., soleus, adductor longus) demonstrate far fewer RRFs than muscles that are prone to atrophy [154]. The resistance to atrophy is often attributed to a predominance of type I fibers which is supported by a study of hybrid rat rectus femoris where RRFs were more likely to occur in type II fibers [26].

The abundance of focal ETC defects in skeletal muscle has been argued to be too low and too late to explain age-related fiber loss and sarcopenia. In mitochondrial myopathies, these abnormalities are more abundant, but those diseases progress within a few decades. However, as noted above RRFs begin to accumulate in middle age, which is consistent with the early onset of muscle atrophy and dysfunction. It would seem that the lower percentages found in otherwise healthy aging muscle would be consistent with the gradual loss of muscle that spans 40–50 years in humans. The data reviewed above indicates that RRFs are one of multiple potential mechanisms leading to sarcopenia in aging muscle. Future tests of their causal role should focus on the prevention of age-associated ETC defects and the resulting impact on sarcopenia.

Interventions that Alter ETC Defect Abundance

A number of interventions, primarily transgenic animal models, have been shown to modulate the abundance of ETC defects. Most of these models recapitulate the various mitochondrial myopathies and include mutations in polgamma, Twinkle, and the ANT [155–158]. In general, these models demonstrate disruption of ETC activities, but as with all models in decreased lifespan, the pathological effects may be more related to disruption of a required biological pathway, than recapitulation of an aging phenotype.

Only one intervention has been demonstrated to reduce the number of focal skeletal muscle ETC defects – caloric restriction (CR) [144]. In this study, early onset CR preserved muscle mass with age and caused a substantial decrease in the number of focal ETC defects. Interestingly, while the number of defects was lower, the length of the involved muscle fiber in each RRF was unchanged between CR and *ad libidum* fed animals. This suggests that CR acts on the inciting event in the formation of RRFs, rather than their progression or resulting biological impact.

Future Directions in Study of ETC Defects

Many other interventions that have been shown to lengthen lifespan and have beneficial effects on sarcopenia have not been studied for their impact on focal skeletal muscle defects. These include exercise, antioxidant therapies that have been shown to decrease mitochondrial mutations [159], and small molecule compounds that may also protect the mitochondrial genome from mutations. Focal ETC defects have not been reported in the skeletal muscle of model organisms such as worms or flies, but as more is understood about the specific mechanisms behind the formation, progression, accumulation and biological impact of these defects, these models will be useful in further dissection of these pathways.

Muscle Injury

Cycles of injury and repair are a normal part of skeletal muscle biology. Contraction induced injury is caused by lengthening (eccentric) contractions [160], such as that of the quadriceps muscle when walking downstairs or sitting in a chair. This type of injury can be the result of the accumulation of small injury due to repetitive motion or larger single events such as a fall. The injury is the result of stretching the sarcomere while it is contracted and leads to disruption of sarcomeric structure and myofiber integrity. Data from both human and rodents indicates that aged muscles are more susceptible to contraction-induced damage than younger muscles when exposed to identical loads [161–165]. In one of the few studies to extend beyond human and rodent models Sui and Alway [21] demonstrated that aged quail muscle was also more susceptible to overload-induced damage. The effect of increased damage is compounded by slower or incomplete recovery from injury that can result in reduced muscle mass and force production following lengthening contractions in aged muscle [163, 165, 166]. Muscle injury is the result of disruption of myofiber integrity and an increase in apoptosis of muscle nuclei leading to the loss of nuclei and muscle atrophy. In young muscle damaged myofibers are repaired and nuclei replaced by the myogenic differentiation and incorporation of muscle satellite cells into the damaged myofiber. In old muscle the ability of the satellite cells to differentiate into myogenic cells and repair damaged muscle is impaired [167–169].

Muscle Satellite Cells and Injury Repair

Muscle satellite cells are the stem cells that are located adjacent to the myofiber under the basal lamina. Their role in the muscle is to differentiate and fuse with existing myofibers during muscle hypertrophy and muscle repair or, in the case of extreme muscle damage, to form new myofibers. Reduced myogenic potential of the muscle satellite cell pool is a key aspect of the reduced ability of aged muscle to recover from injury. Reduced number of satellite cells have been reported for mouse EDL, soleus, and tibialis anterior [167, 170–172] and rat tibialis anterior [173] and human vastus lateralis [174, 175]. In the mouse the pattern of loss varied between muscles, with most of the reduction occurring earlier during the adult years in the soleus than in the EDL [170]. In contrast, other authors have found no loss of satellite cells with age in the rodent and human muscles [162, 176].

Despite the controversy regarding the reduction in the size of the satellite cell pool with age, it is clear that the myogenic potential of this population declines age. This decline includes an attenuated response to activation factors [177–179], decreased proliferative capacity [180, 181], and increased susceptibility to apoptosis [21, 182]. There is growing evidence that reduced myogenic potential is affected by the environment of the satellite cells. Morphological data indicate that the satellite cells cluster around the microvasculature [169] and neuromuscular junction [183–185]. Both the muscle microvasculature and innervation are remodeled with age, thus changing the physical environment of the cells. The chemical environment also changes. Endothelial cells in the microvasculature have been shown to secrete less VEGF [186] and eNOS [187], both of which effect satellite cell activation [188]. In addition to direct cell interaction, circulating factors may also affect the ability of muscle satellite cells to differentiate and repair damaged muscle.

In a series of elegant experiments Conboy et al. demonstrated that switching the circulation from young and old mice partially restores the myogenic potential of aged muscle satellite cell pool [189, 190]. This result is consistent with previous experiments in cell culture showing that exposing satellite cells from aged muscle to serum from young animals increases their ability to differentiate [191]. Reductions in circulating insulin-like growth factors [179, 192] or increased TGF-beta [193] are potential explanations for the different effects of the young and old systemic environment on muscle satellite cell function. Despite the uncertainties regarding mechanism, it is clear that there is reduced function of the muscle satellite cell pool with age and that this leads to a reduced ability to repair and regenerate muscle.

Exercise and Muscle Aging

Slowing Muscle Degeneration

The two most effective interventions to retard or reverse the decline of skeletal muscle function are caloric restriction and exercise training. Caloric restriction has been demonstrated to slow muscle degeneration in many different species including, mice, rats, non-human primates, and humans (Table 1). The effects of caloric restriction on aging are addressed in depth in another chapter. This section will focus on how exercise training can improve muscle function in aged individuals.

Chronic exercise retards the progression of age-related skeletal muscle dysfunction. Lifelong treadmill exercise in mice preserves gastrocnemius muscle fiber cross-sectional area and reduces oxidative damage in middle-aged animals [194]. In humans the study of master athletes has been a useful tool to examine the effects of long-term training on aging skeletal muscle. Master athletes consistently show greater strength and power compared to their age-matched sedentary controls [111, 114]. However, in master athletes the shift toward the aged phenotype in skeletal muscle, including the loss of muscle oxidative capacity, sarcopenia, and a shift toward a more type I phenotype, even in sprint-trained elderly athletes [30], is only slowed and not prevented [111, 114]. This decline even in highly trained athletes indicates that, despite the adaptable nature of skeletal muscle into old age, a decline in function is an intrinsic part of the aging process.

Reversal of Dysfunction

Exercise training has also proven to be the most effective intervention to reverse some of the age-related declines in skeletal muscle function. The effects of exercise training on improving skeletal muscle function are well established in healthy adults. Endurance training increases oxidative capacity through a combination of increased mitochondrial content, increased capillarity, and improved cardiovascular function, while resistance training leads to increases in muscle mass, strength, and power. Results are similar in the elderly humans, where endurance and resistance training have been shown to increase muscle oxidative capacity [195, 196] and strength [2] more than 20%, respectively. Interestingly, Jubrias et al. [196] found that both endurance and resistance training led to increases in oxidative capacity, while only resistance training resulted in an increase in muscle cross-sectional area. This dual improvement in metabolic and mechanical properties with resistance training is supported by a recent gene expression study in which resistance training led to the reversal of aging-associated changes in gene expression, including those affecting mitochondrial function [197]. This suggests that the effects of training extend to both muscle quantity and quality. In aged muscle increases in strength with resistance training frequently exceed that expected by increases in muscle cross-sectional area alone [2]. Such changes could be due to improved motor unit activation [198, 199], reversal of age-related changes in lipid encroachment into muscle, changes in connective tissue [200], or improved contractile function [201]. This improvement in muscle quality with exercise training may extend to mitochondrial function as well. In the study by Jubrias et al. [196] cited above, the significant improvements in muscle oxidative capacity (maximal mitochondrial ATP production) following endurance training were not associated with increases in muscle mitochondrial volume. This indicates that there was an increase in ATP produced per mitochondria, which could reflect improved mitochondrial coupling, following exercise. Thus, age-related declines in both mechanical and metabolic properties appear to be reversible with exercise training in human skeletal muscle.

In non-human models of aging endurance training is the most common exercise intervention. This is probably because of the strong interest in mitochondrial function and aging and because there are fewer well-accepted approaches for resistance training in animal models. Several studies in mice, rats, and horses have demonstrated increased mitochondrial content, increased oxidative enzymes, and/or oxidative capacity following exercise training [68, 202–206], although negative effects of exercise on muscle oxidative enzyme expression have been noted in old mice [202]. In addition to improvement in metabolic capacity and mitochondrial function, endurance training may also be effective at reducing fiber loss in aging muscle. The age-associated increase in cleaved caspases and DNA fragmentation in rat skeletal muscle was reversed with treadmill training (4–12 weeks) [19, 23], suggesting that short-term endurance training may be an effective tool at mitigating the loss of muscle mass with age, even in the absence of hypertrophy.

There are currently no well-established invertebrate models to study the effects of activity and exercise on muscle function and sarcopenia. One study did find that lifelong increased muscular activity in *C. elegans* led to increased rates of muscular degeneration with age. This study genetically modified the rate of pharyngeal pumping, the feeding behavior in *C. elegans*, and found the increased activity of the pharyngeal muscles throughout life led to greater and more rapid degeneration [207]. This result is in contrast to mammalian models where increased activity provides a protective effect against muscle degeneration with age. However, it is not clear if these results from the *C. elegans* model are directly relevant to mammalian systems because *C. elegans* as an adult lacks muscle satellite cells that are key to muscle repair after injury in vertebrate systems.

Reduced Response to Cellular Stress

Recent reports suggest changes in the muscle cell signaling response to stress between young and old muscle. An example of this is regulation of AMP-activated protein kinase (AMPK). AMPK is a key regulator of muscle function because it appears to play a role both in regulating energy metabolism and protein synthesis [208, 209]. A key component of the role of AMPK in regulating energy metabolism is its control over the transcription and activity PGC1a, which in turn regulates mitochondrial biogenesis [132, 209]. A study by Reznick et al. demonstrated that the AMPK-PGC1a signaling pathway is not activated in response to energetic stress induced by treatment with both AICAR (aminoimidazole carboxamide ribonucleotide) and βGPA (beta-guanidinopropionic acid) in aged rat EDL (fast-twitch) resulting in no change in mitochondrial content [133]. In adult muscle the same treatments led to activation of this pathway and significant increases in mitochondrial content and markers of oxidative metabolism.

Interestingly, the effects of resistance training or muscle overload on AMPK activation in aged fast and slow-twitch muscle are different. Muscle overload in the rat plantaris and soleus muscle induced by ablating the gastrocnemius led to an increase in phosphorylated AMPK in the fast-twitch plantaris, but no change in the slow-twitch soleus [210, 211]. The increase in p-AMPK in the aged plantaris was associated with decreased hypertrophic response to the overload treatment, while there was no difference in hypertrophy between the soleus muscles from young and old animals. A potential mechanism for the involvement of AMPK in limiting muscle hypertrophy is its reported inhibitory effect on protein synthesis [209–211], presumeably as part of an energy conserving response. This is supported by a study by Drummond et al. [212] reporting lower rates of protein synthesis and increased phosphorylation of AMPK in aged human vastus lateralis muscle following an acute bout of resistance exercise. This effect of age on AMPK signaling provides one example of a potential mechanistic link between changes in cellular adaptive responses with age and the decline in skeletal muscle function.

Summary

The decline of skeletal muscle with age is a multifaceted process with many contributing mechanisms. Although sarcopenia has received the most attention, it is clear that there are qualitative changes in muscle function, such as a decline in specific force production and mitochondrial uncoupling, in addition to muscle atrophy with age. Both the qualitative and quantitative changes in muscle function are similar between human and rodent models. This allows investigation of the biochemical and molecular nature of the changes in rodents to complement less invasive studies of function and exercise interventions in humans. Although invertebrate models of muscle degeneration have not been exploited to a great extent (see [13, 207, 213, 214] for exceptions), their ease of genetic manipulation and short lifespans make them attractive models to better understand the molecular mechanisms leading to skeletal muscle dysfunction with age.

Fiber type differences in the decline of muscle function provide yet another comparative approach to understanding the underlying mechanisms and potential interventions involved. Slow-twitch muscles are more resistant to many of the age-related changes in both muscle quality [38, 41, 80, 215] and quantity [11, 27, 154] with age than type II muscle fibers, even when compared within the same individuals. These differences provide an excellent opportunity to examine the effect of cellular environment on the ability of skeletal muscle to adapt to and resist the pathological effects of aging.

Acknowledgements The authors would like to thank Kevin Conley and Martin Kushmerick for discussions related to skeletal muscle aging. The authors are supported NIH grants AG028455, AG022385, AG032873, AG029052 and the Nathan Shock Center of Excellence in the Basic Biology of Aging at the University of Washington and an Ellison Medical Foundation New Scholar Award in Aging.

References

1. Wright VJ and Perricelli BC (2008 Mar). Age-related rates of decline in performance among elite senior athletes. *Am J Sports Med* 36(3): 443–450.
2. Hunter GR, McCarthy JP, and Bamman MM (2004). Effects of resistance training on older adults. *Sports Med* 34(5): 329–348.
3. Janssen I, Shepard DS, Katzmarzyk PT, and Roubenoff R (2004 Jan). The healthcare costs of sarcopenia in the United States. *J Am Geriatr Soc* 52(1): 80–85.
4. Danneskiold-Samsoe B, Kofod V, Munter J, Grimby G, Schnohr P, and Jensen G (1984). Muscle strength and functional capacity in 78-81-year-old men and women. *Eur J Appl Physiol Occup Physiol* 52(3): 310–314.
5. Overend TJ, Cunningham DA, Paterson DH, and Lefcoe MS (1992 Nov). Thigh composition in young and elderly men determined by computed tomography. *Clin Physiol* 12(6): 629–640.
6. Jubrias SA, Odderson IR, Esselman PC, and Conley KE (1997 Jul). Decline in isokinetic force with age: muscle cross-sectional area and specific force. *Pflugers Arch* 434(3): 246–253.
7. Lexell J, Henriksson-Larsen K, Winblad B, and Sjostrom M (1983 Oct). Distribution of different fiber types in human skeletal muscles: effects of aging studied in whole muscle cross sections. *Muscle Nerve* 6(8): 588–595.
8. Lexell J, Taylor CC, and Sjostrom M (1988 Apr). What is the cause of the ageing atrophy? Total number, size and proportion of different fiber types studied in whole vastus lateralis muscle from 15- to 83-year-old men. *J Neurol Sci* 84(2–3): 275–294.
9. Brown M and Hasser EM (1996 Mar). Complexity of age-related change in skeletal muscle. *J Gerontol A Biol Sci Med Sci* 51(2): B117–B123.
10. Holloszy JO, Chien M, Cartee GD, and Young JC (1991). Skeletal muscle atrophy in old rats: differential changes in the three fiber types. *Mech Ageing Dev* 60: 199–213.
11. Muller FL, Song W, Liu Y, Chaudhuri A, Pieke-Dahl S, Strong R et al (2006 Jun 1). Absence of CuZn superoxide dismutase leads to elevated oxidative stress and acceleration of age-dependent skeletal muscle atrophy. *Free Radic Biol Med* 40(11): 1993–2004.
12. Brooks SV and Faulkner JA (1988 Oct). Contractile properties of skeletal muscles from young, adult and aged mice. *J Physiol* 404: 71–82.

13. Herndon LA, Schmeissner PJ, Dudaronek JM, Brown PA, Listner KM, Sakano Y et al (2002 Oct 24). Stochastic and genetic factors influence tissue-specific decline in ageing *C. elegans*. *Nature* 419(6909): 808–814.
14. Cao Z, Wu Y, Curry K, Wu Z, Christen Y, and Luo Y (2007 Dec). Ginkgo biloba extract EGb 761 and Wisconsin Ginseng delay sarcopenia in Caenorhabditis elegans. *J Gerontol A Biol Sci Med Sci* 62(12): 1337–1345.
15. Daw CK, Starnes JW, and White TP (1988 Jun). Muscle atrophy and hypoplasia with aging: impact of training and food restriction. *J Appl Physiol* 64(6): 2428–2432.
16. Larkin LM, Kuzon WM, and Halter JB (2003 May). Effects of age and nerve-repair grafts on reinnervation and fiber type distribution of rat medial gastrocnemius muscles. *Mech Ageing Dev* 124(5): 653–661.
17. McKiernan SH, Bua E, McGorray J, and Aiken J (2004 Mar). Early-onset calorie restriction conserves fiber number in aging rat skeletal muscle. *Faseb J* 18(3): 580–581.
18. Chung L and Ng YC (2006 Jan). Age-related alterations in expression of apoptosis regulatory proteins and heat shock proteins in rat skeletal muscle. *Biochim Biophys Acta* 1762(1): 103–109.
19. Song W, Kwak HB, and Lawler JM (2006 Mar–Apr). Exercise training attenuates age-induced changes in apoptotic signaling in rat skeletal muscle. *Antioxid Redox Signal* 8(3–4): 517–528.
20. Zheng J, Edelman SW, Tharmarajah G, Walker DW, Pletcher SD, and Seroude L (2005 Aug 23). Differential patterns of apoptosis in response to aging in *Drosophila*. *Proc Natl Acad Sci U S A* 102(34): 12083–12088.
21. Siu PM and Alway SE (2005 Nov). Age-related apoptotic responses to stretch-induced hypertrophy in quail slow-tonic skeletal muscle. *Am J Physiol Cell Physiol* 289(5): C1105–C1113.
22. Dirks AJ and Leeuwenburgh C (2004 Jan 1). Aging and lifelong calorie restriction result in adaptations of skeletal muscle apoptosis repressor, apoptosis-inducing factor, X-linked inhibitor of apoptosis, caspase-3, and caspase-12. *Free Radic Biol Med* 36(1): 27–39.
23. Marzetti E, Lawler JM, Hiona A, Manini T, Seo AY, and Leeuwenburgh C (2008 Jan 15). Modulation of age-induced apoptotic signaling and cellular remodeling by exercise and calorie restriction in skeletal muscle. *Free Radic Biol Med* 44(2): 160–168.
24. Phillips T and Leeuwenburgh C (2005 Apr). Muscle fiber specific apoptosis and TNF-alpha signaling in sarcopenia are attenuated by life-long calorie restriction. *Faseb J* 19(6): 668–670.
25. Dirks AJ and Leeuwenburgh C (2006 Aug). Tumor necrosis factor alpha signaling in skeletal muscle: effects of age and caloric restriction. *J Nutr Biochem* 17(8): 501–508.
26. Wanagat J, Cao Z, Pathare P, and Aiken JM (2001). Mitochondrial DNA deletion mutations colocalize with segmental electron transport system abnormalities, muscle fiber atrophy, fiber splitting, and oxidative damage in sarcopenia. *Faseb J* 15(2): 322–332.
27. Lexell J (1995 Nov). Human aging, muscle mass, and fiber type composition. *J Gerontol A Biol Sci Med Sci* 50 Spec No: 11–16.
28. Lee WS, Cheung WH, Qin L, Tang N, and Leung KS (2006 Sep). Age-associated decrease of type IIA/B human skeletal muscle fibers. *Clin Orthop Relat Res* 450: 231–237.
29. Lexell J and Downham D (1992 Feb). What is the effect of ageing on type 2 muscle fibres? *J Neurol Sci* 107(2): 250–251.
30. Korhonen MT, Cristea A, Alen M, Hakkinen K, Sipila S, Mero A et al (2006 Sep). Aging, muscle fiber type, and contractile function in sprint-trained athletes. *J Appl Physiol* 101(3): 906–917.
31. Armstrong RB and Phelps RO (1984 Nov). Muscle fiber type composition of the rat hindlimb. *Am J Anat* 171(3): 259–272.
32. Pistilli EE, Jackson JR, and Alway SE (2006 Dec). Death receptor-associated pro-apoptotic signaling in aged skeletal muscle. *Apoptosis* 11(12): 2115–2126.
33. Lyons CN, Mathieu-Costello O, and Moyes CD (2006 Jan). Regulation of skeletal muscle mitochondrial content during aging. *J Gerontol A Biol Sci Med Sci* 61(1): 3–13.

34. Rice KM and Blough ER (2006 Aug). Sarcopenia-related apoptosis is regulated differently in fast- and slow-twitch muscles of the aging F344/N x BN rat model. *Mech Ageing Dev* 127(8): 670–679.
35. Anderson EJ and Neufer PD (2006 Mar). Type II skeletal myofibers possess unique properties that potentiate mitochondrial H(2)O(2) generation. *Am J Physiol Cell Physiol* 290(3): C844–C851.
36. Pette D and Staron RS (2000). Myosin isoforms, muscle fiber types, and transitions. *Microsc Res Tech* 50: 500–509.
37. Vandervoort AA and McComas AJ (1986 Jul). Contractile changes in opposing muscles of the human ankle joint with aging. *J Appl Physiol* 61(1): 361–367.
38. Lynch GS, Hinkle RT, Chamberlain JS, Brooks SV, and Faulkner JA (2001). Force and power output of fast and slow skeletal muscles from mdx mice 6–28 months old. *J Physiol* 535 (Pt 2): 591–600.
39. Bolanowski MA, Russell RL, and Jacobson LA (1981 Mar). Quantitative measures of aging in the nematode Caenorhabditis elegans. I. Population and longitudinal studies of two behavioral parameters. *Mech Ageing Dev* 15(3): 279–295.
40. Ingram DK (2000 Sep). Age-related decline in physical activity: generalization to nonhumans. *Med Sci Sports Exerc* 32(9): 1623–1629.
41. Gonzalez E, Messi ML, and Delbono O (2000 Dec 1). The specific force of single intact extensor digitorum longus and soleus mouse muscle fibers declines with aging. *J Membr Biol* 178(3): 175–183.
42. Prochniewicz E, Thompson LV, and Thomas DD (2007 Oct). Age-related decline in actomyosin structure and function. *Exp Gerontol* 42(10): 931–938.
43. Thompson LV and Brown M (1999 Mar). Age-related changes in contractile properties of single skeletal fibers from the soleus muscle. *J Appl Physiol* 86(3): 881–886.
44. Yu F, Hedstrom M, Cristea A, Dalen N, and Larsson L (2007 Jul). Effects of ageing and gender on contractile properties in human skeletal muscle and single fibres. *Acta Physiol (Oxf)* 190(3): 229–241.
45. D'Antona G, Pellegrino MA, Adami R, Rossi R, Carlizzi CN, Canepari M et al (2003 Oct 15). The effect of ageing and immobilization on structure and function of human skeletal muscle fibres. *J Physiol* 552(Pt 2): 499–511.
46. Hook P, Sriramoju V, and Larsson L (2001 Apr). Effects of aging on actin sliding speed on myosin from single skeletal muscle cells of mice, rats, and humans. *Am J Physiol Cell Physiol* 280(4): C782–C788.
47. Hook P, Li X, Sleep J, Hughes S, and Larsson L (1999 Oct 15). In vitro motility speed of slow myosin extracted from single soleus fibres from young and old rats. *J Physiol* 520 (Pt 2): 463–471.
48. Fugere NA, Ferrington DA, and Thompson LV (2006 Aug). Protein nitration with aging in the rat semimembranosus and soleus muscles. *J Gerontol A Biol Sci Med Sci* 61(8): 806–812.
49. Jimenez-Moreno R, Wang ZM, Gerring RC, and Delbono O (2008 Apr 15). Sarcoplasmic reticulum Ca^{2+} release declines in muscle fibers from aging mice. *Biophys J* 94(8): 3178–3188.
50. Payne AM, Zheng Z, Gonzalez E, Wang ZM, Messi ML, and Delbono O (2004 Oct 1). External Ca(2+)-dependent excitation–contraction coupling in a population of ageing mouse skeletal muscle fibres. *J Physiol* 560(Pt 1): 137–155.
51. Brooks SV and Faulkner JA (1994 Aug). Isometric, shortening, and lengthening contractions of muscle fiber segments from adult and old mice. *Am J Physiol* 267(2 Pt 1): C507–C513.
52. Lexell J (1997 May). Evidence for nervous system degeneration with advancing age. *J Nutr* 127(5 Suppl): 1011S–1013S.
53. Gordon T, Hegedus J, and Tam SL (2004 Mar). Adaptive and maladaptive motor axonal sprouting in aging and motoneuron disease. *Neurol Res* 26(2): 174–185.
54. Nelson RM, Soderberg GL, and Urbscheit NL (1984 Jan). Alteration of motor-unit discharge characteristics in aged humans. *Physical Therapy* 64(1): 29–34.

55. Snow LM, McLoon LK, and Thompson LV (2005 Sep). Adult and developmental myosin heavy chain isoforms in soleus muscle of aging Fischer Brown Norway rat. *Anat Rec A Discov Mol Cell Evol Biol* 286(1): 866–873.
56. Wang YJ, Zhou CJ, Shi Q, Smith N, and Li TF (2007 May). Aging delays the regeneration process following sciatic nerve injury in rats. *J Neurotrauma* 24(5): 885–894.
57. Edstrom E, Altun M, Bergman E, Johnson H, Kullberg S, Ramirez-Leon V et al (2007 Sep 10). Factors contributing to neuromuscular impairment and sarcopenia during aging. *Physiol Behav* 92(1–2): 129–135.
58. Jubrias SA and Conley KE (2001). *Aging Effects on Muscle Properties and Human Performance. Functional Neurobiology of Aging*. New York: Academic Press, pp. 661–673.
59. Buskirk ER and Hodgson JL (1987). Age and aerobic power: the rate of change in men and women. *Fed Proc* 46: 1824–1829.
60. Coggan AR, Abduljalil AM, Swanson SC, Earle MS, Farris JW, Mendenhall LA et al (1993). Muscle metabolism during exercise in young and older untrained and endurance-trained men. *J Appl Physiol* 75(5): 2125–2133.
61. Conley KE, Esselman PC, Jubrias SA, Cress ME, Inglin B, Mogadam C et al (2000). Ageing, muscle properties and maximal O_2 uptake rate in humans. *J Physiol* 526(1): 211–217.
62. Conley KE, Jubrias SA, and Esselman PC (2000). Oxidative capacity and ageing in human muscle. *J Physiol* 526(1): 203–210.
63. McCully K, Fielding K, Evans RA, Leigh WJ, Jr. JS, and Posner J (1993). Relationships between in vivo and in vitro measurements of metabolism in young and old human calf muscles. *J Appl Physiol* 75(2): 813–819.
64. Proctor DN and Joyner MJ (1997). Skeletal muscle mass and the reduction of VO_2max in trained older subjects. *J Appl Physiol* 82: 1411–1415.
65. Schefer V and Talan MI (1996 May–Jun). Oxygen consumption in adult and AGED C57BL/6 J mice during acute treadmill exercise of different intensity. *Exp Gerontol* 31(3): 387–392.
66. Hepple RT, Hagen JL, Krause DJ, and Jackson CC (2003 Mar). Aerobic power declines with aging in rat skeletal muscles perfused at matched convective O_2 delivery. *J Appl Physiol* 94(2): 744–751.
67. McKeever KH and Malinowski K (1997 Dec). Exercise capacity in young and old mares. *Am J Vet Res* 58(12): 1468–1472.
68. Betros CL, McKeever KH, Kearns CF, and Malinowski K (2002). Effects of ageing and training on maximal heart rate and VO_2max. *Equine Vet J* Suppl. (Sep.) 34: 100–105.
69. Proctor DN, Shen PH, Dietz NM, Eickhoff TJ, Lawler LA, Ebersold EJ et al (1998 Jul). Reduced leg blood flow during dynamic exercise in older endurance-trained men. *J Appl Physiol* 85(1): 68–75.
70. Betik AC and Hepple RT (2008 Feb). Determinants of VO_2max decline with aging: an integrated perspective. *Appl Physiol Nutr Metab* 33(1): 130–140.
71. Payne GW (2006 Jun). Effect of inflammation on the aging microcirculation: impact on skeletal muscle blood flow control. *Microcirculation* 13(4): 343–352.
72. Payne GW and Bearden SE (2006 Jun). The microcirculation of skeletal muscle in aging. *Microcirculation* 13(4): 275–277.
73. Marcinek DJ (2004 Dec). Mitochondrial dysfunction measured in vivo. *Acta Physiol Scand* 182(4): 343–352.
74. Irrcher I, Adhihetty PJ, Joseph AM, Ljubicic V, and Hood DA (2003). Regulation of mitochondrial biogenesis in muscle by endurance exercise. *Sports Med* 33(11): 783–793.
75. Floyd RA, West M, and Hensley K (2001). Oxidative biochemical markers; clues to understanding aging in long- lived species. *Exp Gerontol* 36(4–6): 619–640.
76. Ji LL, Leeuwenburgh C, Leichtweis S, Gore M, Fiebig R, Hollander J et al (1998 Nov). Oxidative stress and aging. Role of exercise and its influences on antioxidant systems. *Ann N Y Acad Sci* 20(854): 102–117.
77. Sastre J, Pallardo FV, and Vina J (2003 Jul 1). The role of mitochondrial oxidative stress in aging. *Free Radic Biol Med* 35(1): 1–8.

78. Van Remmen H and Richardson A (2001). Oxidative damage to mitochondria and aging. *Exp Gerontol* 36(7): 957–968.
79. Hoppeler H and Lindstedt SL (2006 Apr). In health and in a normoxic environment, VO_2max is/is not limited primarily by cardiac output and locomotor muscle blood flow. *J Appl Physiol* 100(4): 1415–1416.
80. Amara CE, Shankland EG, Jubrias SA, Marcinek DJ, Kushmerick MJ, and Conley KE (2007 Jan 16). Mild mitochondrial uncoupling impacts cellular aging in human muscles in vivo. *Proc Natl Acad Sci U S A* 104(3): 1057–1062.
81. Marcinek DJ, Schenkman KA, Ciesielski WA, Lee D, and Conley KE (2005 Dec 1). Reduced mitochondrial coupling in vivo alters cellular energetics in aged mouse skeletal muscle. *J Physiol* 569(Pt 2): 467–473.
82. Richter C, Schweizer M, Cossarizza A, and Franceschi C (1996 Jan 8). Control of apoptosis by the cellular ATP level. *FEBS Lett* 378(2): 107–110.
83. Skulachev VP (2006 Mar 9). Bioenergetic aspects of apoptosis, necrosis and mitoptosis. *Apoptosis* 11(4): 473–485.
84. Halestrap AP, Doran E, Gillespie JP, and O'Toole A (2000 Feb). Mitochondria and cell death. *Biochem Soc Trans* 28(2): 170–177.
85. Kroemer G, Dallaporta B, and Resche-Rigon M (1998). The mitochondrial death/life regulator in apoptosis and necrosis. *Annu Rev Physiol* 60: 619–642.
86. Lemasters JJ, Qian T, Bradham CA, Brenner DA, Cascio WE, Trost LC et al (1999 Aug). Mitochondrial dysfunction in the pathogenesis of necrotic and apoptotic cell death. *J Bioenerg Biomembr* 31(4): 305–319.
87. Short KR, Bigelow ML, Kahl J, Singh R, Coenen-Schimke J, Raghavakaimal S et al (2005 Apr 12). Decline in skeletal muscle mitochondrial function with aging in humans. *Proc Natl Acad Sci U S A* 102(15): 5618–5623.
88. Barazzoni R, Short KR, and Nair KS (2000). Effects of aging on mitochondrial DNA copy number and cytochrome c oxidase gene expression in rat skeletal muscle, liver, and heart. *J Biol Chem* 275(5): 3343–3347.
89. Trounce I, Byrne E, and Marzuki S (1989 Mar 25). Decline in skeletal muscle mitochondrial respiratory chain function: possible factor in ageing. *Lancet* 1(8639): 637–639.
90. Boffoli D, Scacco SC, Vergari R, Solarino G, Santacroce G, and Papa S (1994 Apr 12). Decline with age of the respiratory chain activity in human skeletal muscle. *Biochim Biophys Acta* 1226(1): 73–82.
91. Drew B, Phaneuf S, Dirks A, Selman C, Gredilla R, Lezza A et al (2003 Feb). Effects of aging and caloric restriction on mitochondrial energy production in gastrocnemius muscle and heart. *Am J Physiol Regul Integr Comp Physiol* 284(2): R474–R480.
92. Desai VG, Weindruch R, Hart RW, and Feuers RJ (1996). Influences of age and dietary restriction on gastrocnemius electron transport system activities in mice. *Arch Biochem Biophys* 333(1): 145–151.
93. Hagen JL, Krause DJ, Baker DJ, Fu MH, Tarnopolsky MA, and Hepple RT (2004 Nov). Skeletal muscle aging in F344BN F1-hybrid rats: I. Mitochondrial dysfunction contributes to the age-associated reduction in VO_2max. *J Gerontol A Biol Sci Med Sci* 59(11): 1099–1110.
94. Kim JS, Hinchcliff KW, Yamaguchi M, Beard LA, Markert CD, and Devor ST (2005 May). Age-related changes in metabolic properties of equine skeletal muscle associated with muscle plasticity. *Vet J* 169(3): 397–403.
95. Ferguson M, Mockett RJ, Shen Y, Orr WC, and Sohal RS (2005 Sep 1). Age-associated decline in mitochondrial respiration and electron transport in *Drosophila melanogaster*. *Biochem J* 390(Pt 2): 501–511.
96. Schwarze SR, Weindruch R, and Aiken JM (1998 Oct). Oxidative stress and aging reduce COX I RNA and cytochrome oxidase activity in *Drosophila*. *Free Radic Biol Med* 25(6): 740–747.
97. Schwarze SR, Weindruch R, and Aiken JM (1998 May). Decreased mitochondrial RNA levels without accumulation of mitochondrial DNA deletions in aging *Drosophila melanogaster*. *Mutat Res* 382(3–4): 99–107.

98. Kwong LK and Sohal RS (2000 Jan 1). Age-related changes in activities of mitochondrial electron transport complexes in various tissues of the mouse. *Arch Biochem Biophys* 373(1): 16–22.
99. Zahn JM, Sonu R, Vogel H, Crane E, Mazan-Mamczarz K, Rabkin R et al (2006 Jul). Transcriptional profiling of aging in human muscle reveals a common aging signature. *PLoS Genet* 2(7): e115.
100. Welle S, Brooks AI, Delehanty JM, Needler N, and Thornton CA (2003 Jul 7). Gene expression profile of aging in human muscle. *Physiol Genomics* 14(2): 149–159.
101. Pletcher SD, Macdonald SJ, Marguerie R, Certa U, Stearns SC, Goldstein DB et al (2002 Apr 30). Genome-wide transcript profiles in aging and calorically restricted *Drosophila melanogaster*. *Curr Biol* 12(9): 712–723.
102. Lund J, Tedesco P, Duke K, Wang J, Kim SK, and Johnson TE (2002 Sep 17). Transcriptional profile of aging in *C. elegans*. *Curr Biol* 12(18): 1566–1573.
103. Lee C-K, Klopp RG, Weindruch R, and Prolla TA (1999). Gene expression profile of aging and its retardation by caloric restriction. *Science* 285: 1390–1393.
104. Kent-Braun JA and Ng AV (2000 Sep). Skeletal muscle oxidative capacity in young and older women and men. *J Appl Physiol* 89(3): 1072–1078.
105. Brierly EJ, Johnson MA, Bowman A, Ford GA, Subhan F, Reed JW et al (1997). Mitochondrial function in muscle from elderly athletes. *Ann Neurol* 41(1): 114–116.
106. Brierley E, Johnson M, James O, and Turnbull D (1996 April 1). Effects of physical activity and age on mitochondrial function. *QJM* 89(4): 251–258.
107. Rasmussen UF, Krustrup P, Kjaer M, and Rasmussen HN (2003 Aug). Experimental evidence against the mitochondrial theory of aging A study of isolated human skeletal muscle mitochondria. *Exp Gerontol* 38(8): 877–886.
108. Rasmussen UF, Krustrup P, Kjaer M, and Rasmussen HN (2003 May). Human skeletal muscle mitochondrial metabolism in youth and senescence: no signs of functional changes in ATP formation and mitochondrial oxidative capacity. *Pflugers Arch* 446(2): 270–278.
109. Rogers MA, Hagberg JM, Martin WH, 3rd, Ehsani AA, and Holloszy JO (1990 May). Decline in VO_2max with aging in master athletes and sedentary men. *J Appl Physiol* 68(5): 2195–2199.
110. Faulkner JA, Larkin LM, Claflin DR, and Brooks SV (2007 Nov). Age-related changes in the structure and function of skeletal muscles. *Clin Exp Pharmacol Physiol* 34(11): 1091–1096.
111. Ojanen T, Rauhala T, and Hakkinen K (2007 Feb). Strength and power profiles of the lower and upper extremities in master throwers at different ages. *J Strength Cond Res* 21(1): 216–222.
112. Jokl P, Sethi PM, and Cooper AJ (2004 Aug). Master's performance in the New York City Marathon 1983–1999. *Br J Sports Med* 38(4): 408–412.
113. Lazarus NR and Harridge SD (2007 Oct). Inherent ageing in humans: the case for studying master athletes. *Scand J Med Sci Sports* 17(5): 461–463.
114. Tarpenning KM, Hamilton-Wessler M, Wiswell RA, and Hawkins SA (2004 Jan). Endurance training delays age of decline in leg strength and muscle morphology. *Med Sci Sports Exerc* 36(1): 74–78.
115. Johnson MA, Polgar J, Weightman D, and Appleton D (1973 Jan). Data on the distribution of fibre types in thirty-six human muscles. An autopsy study. *J Neurol Sci* 18(1): 111–129.
116. Houmard JA, Weidner ML, Gavigan KE, Tyndall GL, Hickey MS, and Alshami A (1998 Oct). Fiber type and citrate synthase activity in the human gastrocnemius and vastus lateralis with aging. *J Appl Physiol* 85(4): 1337–1341.
117. Bejma J and Ji LL (1999 Jul). Aging and acute exercise enhance free radical generation in rat skeletal muscle. *J Appl Physiol* 87(1): 465–470.
118. Lopez-Torres M, Gredilla R, Sanz A, and Barja G (2002). Influence of aging and long-term caloric restriction on oxygen radical generation and the oxidative DNA damage in rat liver mitochondria. *Free Radic Biol Med* 32(9): 882–889.

119. Moghaddas S, Hoppel CL, and Lesnefsky EJ (2003). Aging defect at the Q(O) site of complex III augments oxyradical production in rat heart interfibrillar mitochondria. *Arch Biochem Biophys* 414(1): 59–66.
120. Nicholls DG (2004 Feb). Mitochondrial membrane potential and aging. *Aging Cell* 3(1): 35–40.
121. Heerdt BG, Houston MA, Wilson AJ, and Augenlicht LH (2003 Oct 1). The intrinsic mitochondrial membrane potential (Deltapsim) is associated with steady-state mitochondrial activity and the extent to which colonic epithelial cells undergo butyrate-mediated growth arrest and apoptosis. *Cancer Res* 63(19): 6311–6319.
122. Stoetzer OJ, Pogrebniak A, Pelka-Fleischer R, Hasmann M, Hiddemann W, and Nuessler V (2002 Feb 1). Modulation of apoptosis by mitochondrial uncouplers: apoptosis-delaying features despite intrinsic cytotoxicity. *Biochem Pharmacol* 63(3): 471–483.
123. Goodell S and Cortopassi G (1998). Analysis of oxygen consumption and mitochondrial permeability with age in mice. *Mech Ageing Dev* 101: 245–256.
124. Petersen KF, Befroy D, Dufour S, Dziura J, Ariyan C, Rothman DL et al (2003 May 16). Mitochondrial dysfunction in the elderly: possible role in insulin resistance. *Science* 300(5622): 1140–1142.
125. Korshunov SS, Skulachev VP, and Starkov AA (1997 Oct 13). High protonic potential actuates a mechanism of production of reactive oxygen species in mitochondria. *FEBS Lett* 416(1): 15–18.
126. Skulachev VP (1998). Uncoupling: new approaches to an old problem of bioenergetics. *Biochimica et Biophysica Acta* 1363: 100–124.
127. Brand MD (2000). Uncoupling to survive? The role of mitochondrial inefficiency in ageing. *Exp Gerontol* 35: 811–820.
128. Neschen S, Katterle Y, Richter J, Augustin R, Scherneck S, Mirhashemi F et al (2008 Mar 18). Uncoupling protein 1 expression in murine skeletal muscle increases AMPK activation, glucose turnover and insulin sensitivity in vivo. *Physiol Genomics* 33(3): 333–340.
129. Gates AC, Bernal-Mizrachi C, Chinault SL, Feng C, Schneider JG, Coleman T et al (2007 Dec). Respiratory uncoupling in skeletal muscle delays death and diminishes age-related disease. *Cell Metab* 6(6): 497–505.
130. Caldeira Da Silva C, Cerqueira F, Barbosa L, Medeiros M, and Kowaltowski A (2008). Mild mitochondrial uncoupling in mice affects energy metabolism, redox balance, and longevity. *Aging Cell* 7(4): 552–560.
131. Speakman JR, Talbot DA, Selman C, Snart S, McLaren JS, Redman P et al (2004 Jun). Uncoupled and surviving: individual mice with high metabolism have greater mitochondrial uncoupling and live longer. *Aging Cell* 3(3): 87–95.
132. Rohas LM, St-Pierre J, Uldry M, Jager S, Handschin C, and Spiegelman BM (2007 May 8). A fundamental system of cellular energy homeostasis regulated by PGC-1alpha. *Proc Natl Acad Sci U S A* 104(19): 7933–7938.
133. Reznick RM, Zong H, Li J, Morino K, Moore IK, Yu HJ et al (2007 Feb). Aging-associated reductions in AMP-activated protein kinase activity and mitochondrial biogenesis. *Cell Metab* 5(2): 151–156.
134. Muller-Hocker J, Droste M, Kadenbach B, Pongratz D, and Hubner G (1989 Jul). Fatal mitochondrial myopathy with cytochrome-c-oxidase deficiency and subunit-restricted reduction of enzyme protein in two siblings: an autopsy-immunocytochemical study. *Human Pathol* 20(7): 666–672.
135. Muller-Hocker J (1990 Dec). Cytochrome c oxidase deficient fibres in the limb muscle and diaphragm of man without muscular disease: an age-related alteration. *J Neurol Sci* 100 (1–2): 14–21.
136. Olson W, Engel WK, Walsh GO, and Einaugler R (1972 Mar). Oculocraniosomatic neuromuscular disease with “ragged-red” fibers. *Arch Neurol* 26(3): 193–211.
137. Muller-Hocker J, Schneiderbanger K, Stefani FH, and Kadenbach B (1992 Sep). Progressive loss of cytochrome c oxidase in the human extraocular muscles in ageing – a cytochemical-immunohistochemical study. *Mutat Res* 275(3–6): 115–124.

138. Liu VW, Zhang C, and Nagley P (1998 Mar 1). Mutations in mitochondrial DNA accumulate differentially in three different human tissues during ageing. *Nucleic Acids Res* 26(5): 1268–1275.
139. Muller-Hocker J, Seibel P, Schneiderbanger K, and Kadenbach B (1993). Different in situ hybridization patterns of mitochondrial DNA in cytochrome c oxidase-deficient extraocular muscle fibres in the elderly. *Virchows Arch A Pathol Anat Histopathol* 422(1): 7–15.
140. Lee CM, Chung SS, Kaczkowski JM, Weindruch R, and Aiken JM (1993 Nov). Multiple mitochondrial DNA deletions associated with age in skeletal muscle of rhesus monkeys. *J Gerontol* 48(6): B201–B205.
141. Cortopassi GA and Wong A (1999 Feb 9). Mitochondria in organismal aging and degeneration. *Biochimica et biophysica acta* 1410(2): 183–193.
142. Wanagat J, Lopez M, and Aiken JM (2001). Alterations of the mitochondrial genome. In *Handbook of The Biology of Aging*. (Masoro E, Austad S, eds., 5th ed). San Diego: Academic Press, pp. 114–131.
143. Wanagat J, Wolff MR, and Aiken JM (2002 Jan). Age-associated changes in function, structure and mitochondrial genetic and enzymatic abnormalities in the Fischer 344 x Brown Norway F(1) hybrid rat heart. *J Mol Cell Cardiol* 34(1): 17–28.
144. Bua EA, McKiernan SH, and Aiken JM (2004 Jan 20). Calorie restriction limits the generation but not the progression of mitochondrial abnormalities in aging skeletal muscle. *Faseb J* 18(3): 582–584.
145. Gokey NG, Cao Z, Pak JW, Lee D, McKiernan SH, McKenzie D et al (2004 Oct). Molecular analyses of mtDNA deletion mutations in microdissected skeletal muscle fibers from aged rhesus monkeys. *Aging Cell* 3(5): 319–326.
146. Cao Z, Wanagat J, McKiernan SH, and Aiken JM (2001 Nov 1). Mitochondrial DNA deletion mutations are concomitant with ragged red regions of individual, aged muscle fibers: analysis by laser-capture microdissection. *Nucleic Acids Res* 29(21): 4502–4508.
147. Herbst A, Pak JW, McKenzie D, Bua E, Bassiouni M, and Aiken JM (2007 Mar). Accumulation of mitochondrial DNA deletion mutations in aged muscle fibers: evidence for a causal role in muscle fiber loss. *J Gerontol A Biol Sci Med Sci* 62(3): 235–245.
148. Bua E, Johnson J, Herbst A, Delong B, McKenzie D, Salamat S et al (2006 Sep). Mitochondrial DNA-deletion mutations accumulate intracellularly to detrimental levels in aged human skeletal muscle fibers. *Am J Hum Genet* 79(3): 469–480.
149. Prigione A and Cortopassi G (2007 Oct). Mitochondrial DNA deletions induce the adenosine monophosphate-activated protein kinase energy stress pathway and result in decreased secretion of some proteins. *Aging Cell* 6(5): 619–630.
150. Prigione A and Cortopassi G (2007 Jul–Aug). Mitochondrial DNA deletions and chloramphenicol treatment stimulate the autophagic transcript ATG12. *Autophagy* 3(4): 377–380.
151. Aspnes LE, Lee CM, Weindruch R, Chung SS, Roecker EB, and Aiken JM (1997 Jun). Caloric restriction reduces fiber loss and mitochondrial abnormalities in aged rat muscle. *Faseb J* 11(7): 573–581.
152. Lopez ME, Van Zeeland NL, Dahl DB, Weindruch R, and Aiken JM (2000 Jul 20). Cellular phenotypes of age-associated skeletal muscle mitochondrial abnormalities in rhesus monkeys. *Mutat Res* 452(1): 123–138.
153. Bua E, Johnson J, Herbst A, Delong B, McKenzie D, Salamat S et al (2006 Sep). Mitochondrial DNA deletion mutations accumulate intracellularly to detrimental levels in aged human skeletal muscle fibers. *Am J Hum Genet* 79(3): 469–480.
154. Bua EA, McKiernan SH, Wanagat J, McKenzie D, and Aiken JM (2002 Jun). Mitochondrial abnormalities are more frequent in muscles undergoing sarcopenia. *J Appl Physiol* 92(6): 2617–2624.
155. Tyynismaa H, Mjosund KP, Wanrooij S, Lappalainen I, Ylikallio E, Jalanko A et al (2005 Dec 6). Mutant mitochondrial helicase Twinkle causes multiple mtDNA deletions and a late-onset mitochondrial disease in mice. *Proc Natl Acad Sci U S A* 102(49): 17687–17692.

156. Kujoth GC, Hiona A, Pugh TD, Someya S, Panzer K, Wohlgemuth SE et al (2005 Jul 15). Mitochondrial DNA mutations, oxidative stress, and apoptosis in mammalian aging. *Science* 309(5733): 481–484.
157. Trifunovic A, Wredenberg A, Falkenberg M, Spelbrink JN, Rovio AT, Bruder CE et al (2004 May 27). Premature ageing in mice expressing defective mitochondrial DNA polymerase. *Nature* 429(6990): 417–423.
158. Esposito LA, Melov S, Panov A, Cottrell BA, and Wallace DC (1999 Apr 27). Mitochondrial disease in mouse results in increased oxidative stress. *Proc Natl Acad Sci U S A* 96(9): 4820–4825.
159. Schriner SE, Linford NJ, Martin GM, Treuting P, Ogburn CE, Emond M et al (2005 May 5). Extension of murine lifespan by overexpression of catalase targeted to mitochondria. *Science* 308(5730):1909–1911.
160. Faulkner JA, Brooks SV, and Zerba E (1995 Nov). Muscle atrophy and weakness with aging: contraction-induced injury as an underlying mechanism. *J Gerontol A Biol Sci Med Sci* 50 Spec No: 124–129.
161. Manfredi TG, Fielding RA, O'Reilly KP, Meredith CN, Lee HY, and Evans WJ (1991 Sep). Plasma creatine kinase activity and exercise-induced muscle damage in older men. *Med Sci Sports Exerc* 23(9): 1028–1034.
162. Roth SM, Martel GF, Ivey FM, Lemmer JT, Metter EJ, Hurley BF et al (2000 Dec 1). Skeletal muscle satellite cell populations in healthy young and older men and women. *Anat Rec* 260(4): 351–358.
163. McBride T (2000 Jun 20). Increased depolarization, prolonged recovery and reduced adaptation of the resting membrane potential in aged rat skeletal muscles following eccentric contractions. *Mech Ageing Dev* 115(3): 127–138.
164. Rader EP and Faulkner JA (2006 Sep). Effect of aging on the recovery following contraction-induced injury in muscles of female mice. *J Appl Physiol* 101(3): 887–892.
165. Rader EP and Faulkner JA (2006 Feb). Recovery from contraction-induced injury is impaired in weight-bearing muscles of old male mice. *J Appl Physiol* 100(2): 656–661.
166. McBride TA, Gorin FA, and Carlsen RC (1995 Sep 15). Prolonged recovery and reduced adaptation in aged rat muscle following eccentric exercise. *Mech Ageing Dev* 83(3): 185–200.
167. Brack AS, Bildsoe H, and Hughes SM (2005 Oct 15). Evidence that satellite cell decrement contributes to preferential decline in nuclear number from large fibres during murine age-related muscle atrophy. *J Cell Sci* 118(Pt 20): 4813–4821.
168. Carlson ME and Conboy IM (2007 Jun). Loss of stem cell regenerative capacity within aged niches. *Aging Cell* 6(3): 371–382.
169. Gopinath SD and Rando TA (2008 Aug). Stem cell review series: aging of the skeletal muscle stem cell niche. *Aging Cell* 7(4): 590–598.
170. Shefer G, Van de Mark DP, Richardson JB, and Yablonka-Reuveni Z (2006 Jun 1). Satellite-cell pool size does matter: defining the myogenic potency of aging skeletal muscle. *Dev Biol* 294(1): 50–66.
171. Snow MH (1977 Dec 19). The effects of aging on satellite cells in skeletal muscles of mice and rats. *Cell Tissue Res* 185(3): 399–408.
172. Gibson MC and Schultz E (1983 Oct). Age-related differences in absolute numbers of skeletal muscle satellite cells. *Muscle Nerve* 6(8): 574–580.
173. Dedkov EI, Borisov AB, Wernig A, and Carlson BM (2003 Jul). Aging of skeletal muscle does not affect the response of satellite cells to denervation. *J Histochem Cytochem* 51(7): 853–863.
174. Renault V, Rolland E, Thornell LE, Mouly V, and Butler-Browne G (2002 Dec). Distribution of satellite cells in the human vastus lateralis muscle during aging. *Exp Gerontol* 37(12): 1513–1514.
175. Sajko S, Kubinova L, Cvetko E, Kreft M, Wernig A, and Erzen I (2004 Feb). Frequency of M-cadherin-stained satellite cells declines in human muscles during aging. *J Histochem Cytochem* 52(2): 179–185.

176. Nnodim JO (2000 Jan 3). Satellite cell numbers in senile rat levator ani muscle. *Mech Ageing Dev* 112(2): 99–111.
177. Grounds MD (1998 Nov). Age-associated changes in the response of skeletal muscle cells to exercise and regeneration. *Ann N Y Acad Sci* 20(854): 78–91.
178. Harridge SD (2003 Feb). Ageing and local growth factors in muscle. *Scand J Med Sci Sports* 13(1): 34–39.
179. Goldspink G (2004 Jun). Age-related muscle loss and progressive dysfunction in mechanosensitive growth factor signaling. *Ann N Y Acad Sci* 1019: 294–298.
180. Jejurikar SS and Kuzon WM, Jr. (2003 Dec). Satellite cell depletion in degenerative skeletal muscle. *Apoptosis* 8(6): 573–578.
181. Chakravarthy MV, Abraha TW, Schwartz RJ, Fiorotto ML, and Booth FW (2000 Nov 17). Insulin-like growth factor-I extends in vitro replicative life span of skeletal muscle satellite cells by enhancing G1/S cell cycle progression via the activation of phosphatidylinositol 3′-kinase/Akt signaling pathway. *J Biol Chem* 275(46): 35942–35952.
182. Jejurikar SS, Henkelman EA, Cederna PS, Marcelo CL, Urbanchek MG, and Kuzon WM, Jr. (2006 Sep). Aging increases the susceptibility of skeletal muscle derived satellite cells to apoptosis. *Exp Gerontol* 41(9): 828–836.
183. Wokke JH, Van den Oord CJ, Leppink GJ, and Jennekens FG (1989 Feb). Perisynaptic satellite cells in human external intercostal muscle: a quantitative and qualitative study. *Anat Rec* 223(2): 174–180.
184. Zammit PS, Partridge TA, and Yablonka-Reuveni Z (2006 Nov). The skeletal muscle satellite cell: the stem cell that came in from the cold. *J Histochem Cytochem* 54(11): 1177–1191.
185. Kelly AM (1978 Apr). Perisynaptic satellite cells in the developing and mature rat soleus muscle. *Anat Rec* 190(4): 891–903.
186. Wagatsuma A (2006 Jan). Effect of aging on expression of angiogenesis-related factors in mouse skeletal muscle. *Exp Gerontol* 41(1): 49–54.
187. Brandes RP, Fleming I, and Busse R (2005 May 1). Endothelial aging. *Cardiovasc Res* 66(2): 286–294.
188. Anderson JE (2000 May). A role for nitric oxide in muscle repair: nitric oxide-mediated activation of muscle satellite cells. *Mol Biol Cell* 11(5): 1859–1874.
189. Conboy IM, Conboy MJ, Wagers AJ, Girma ER, Weissman IL, and Rando TA (2005 Feb 17). Rejuvenation of aged progenitor cells by exposure to a young systemic environment. *Nature* 433(7027): 760–764.
190. Conboy IM and Rando TA (2005 Mar). Aging, stem cells and tissue regeneration: lessons from muscle. *Cell Cycle* 4(3): 407–410.
191. Mezzogiorno A, Coletta M, Zani BM, Cossu G, and Molinaro M (1993 Aug 1). Paracrine stimulation of senescent satellite cell proliferation by factors released by muscle or myotubes from young mice. *Mech Ageing Dev* 70(1–2): 35–44.
192. Reeves I, Abribat T, Laramee P, Jasmin G, and Brazeau P (2000 Apr). Age-related serum levels of insulin-like growth factor-I, -II and IGF-binding protein-3 following myocardial infarction. *Growth Horm IGF Res* 10(2): 78–84.
193. Beggs ML, Nagarajan R, Taylor-Jones JM, Nolen G, Macnicol M, and Peterson CA (2004 Dec). Alterations in the TGFbeta signaling pathway in myogenic progenitors with age. *Aging Cell* 3(6): 353–361.
194. Rosa EF, Silva AC, Ihara SS, Mora OA, Aboulafia J, and Nouailhetas VL (2005 Oct). Habitual exercise program protects murine intestinal, skeletal, and cardiac muscles against aging. *J Appl Physiol* 99(4): 1569–1575.
195. Kohrt WM, Malley MT, Coggan AR, Spina RJ, Ogawa T, Ehsani AA et al (1991 Nov). Effects of gender, age, and fitness level on response of VO_2max to training in 60-71 yr olds. *J Appl Physiol* 71(5): 2004–2011.
196. Jubrias SA, Esselman PC, Price LB, Cress ME, and Conley KE (2001). Large energetic adaptations of elderly muscle to resistance and endurance training. *J Appl Physiol* 90: 1663–1670.

197. Melov S, Tarnopolsky MA, Beckman K, Felkey K, and Hubbard A (2007). Resistance exercise reverses aging in human skeletal muscle. *PLoS ONE* 2(5): e465.
198. Hakkinen K, Pakarinen A, Kraemer WJ, Hakkinen A, Valkeinen H, and Alen M (2001 Aug). Selective muscle hypertrophy, changes in EMG and force, and serum hormones during strength training in older women. *J Appl Physiol* 91(2): 569–580.
199. Hakkinen K, Alen M, and Komi PV (1985 Dec). Changes in isometric force- and relaxation-time, electromyographic and muscle fibre characteristics of human skeletal muscle during strength training and detraining. *Acta Physiol Scand* 125(4): 573–585.
200. Mackey AL, Heinemeier KM, Koskinen SO, and Kjaer M (2008). Dynamic adaptation of tendon and muscle connective tissue to mechanical loading. *Connect Tissue Res* 49(3): 165–168.
201. Suetta C, Andersen JL, Dalgas U, Berget J, Koskinen S, Aagaard P et al (2008 Jul). Resistance training induces qualitative changes in muscle morphology, muscle architecture, and muscle function in elderly postoperative patients. *J Appl Physiol* 105(1): 180–186.
202. Steinhagen-Thiessen E, Reznick AZ, and Hilz H (1981). Positive and negative adaptation of muscle enzymes in aging mice subjected to physical exercise. *Mech Ageing Dev* 16(4): 363–369.
203. Ji LL, Wu E, and Thomas DP (1991). Effect of exercise training on antioxidant and metabolic functions in senescent rat skeletal muscle. *Gerontology* 37(6): 317–325.
204. Rossiter HB, Howlett RA, Holcombe HH, Entin PL, Wagner HE, and Wagner PD (2005 Jun 15). Age is no barrier to muscle structural, biochemical and angiogenic adaptations to training up to 24 months in female rats. *J Physiol* 565(Pt 3): 993–1005.
205. Betik AC, Baker DJ, Krause DJ, McConkey MJ, and Hepple RT (2008 Jul). Exercise training in late middle-aged male Fischer 344 x Brown Norway F1-hybrid rats improves skeletal muscle aerobic function. *Exp Physiol* 93(7): 863–871.
206. German E and Hoffman-Goetz L (1986 Jul). The effect of cold acclimation and exercise training on cold tolerance in aged C57BL/6 J mice. *J Gerontol* 41(4): 453–459.
207. Chow DK, Glenn CF, Johnston JL, Goldberg IG, and Wolkow CA (2006 Mar). Sarcopenia in the Caenorhabditis elegans pharynx correlates with muscle contraction rate over lifespan. *Exp Gerontol* 41(3): 252–260.
208. Towler MC and Hardie DG (2007 Feb 16). AMP-activated protein kinase in metabolic control and insulin signaling. *Circ Res* 100(3): 328–341.
209. Atherton PJ, Babraj J, Smith K, Singh J, Rennie MJ, and Wackerhage H (2005 May). Selective activation of AMPK-PGC-1alpha or PKB-TSC2-mTOR signaling can explain specific adaptive responses to endurance or resistance training-like electrical muscle stimulation. *Faseb J* 19(7): 786–788.
210. Thomson DM and Gordon SE (2005 Feb). Diminished overload-induced hypertrophy in aged fast-twitch skeletal muscle is associated with AMPK hyperphosphorylation. *J Appl Physiol* 98(2): 557–564.
211. Thomson DM and Gordon SE (2006 Jul 1). Impaired overload-induced muscle growth is associated with diminished translational signalling in aged rat fast-twitch skeletal muscle. *J Physiol* 574(Pt 1): 291–305.
212. Drummond MJ, Dreyer HC, Pennings B, Fry CS, Dhanani S, Dillon EL et al (2008 May). Skeletal muscle protein anabolic response to resistance exercise and essential amino acids is delayed with aging. *J Appl Physiol* 104(5): 1452–1461.
213. Johnston J, Iser WB, Chow DK, Goldberg IG, and Wolkow CA (2008). Quantitative image analysis reveals distinct structural transitions during aging in Caenorhabditis elegans tissues. *PLoS ONE* 3(7): e2821.
214. Wolkow CA (2006 Oct). Identifying factors that promote functional aging in Caenorhabditis elegans. *Exp Gerontol* 41(10): 1001–1006.
215. Conley KE, Amara CE, Jubrias SA, and Marcinek DJ (2007 Mar). Mitochondrial function, fibre types and ageing: new insights from human muscle in vivo. *Exp Physiol* 92(2): 333–339.

216. Lexell J (1993 Mar). Ageing and human muscle: observations from Sweden. *Can J Appl Physiol* 18(1): 2–18.
217. Kim JS, Hinchcliff KW, Yamaguchi M, Beard LA, Markert CD, and Devor ST (2005 Jan). Exercise training increases oxidative capacity and attenuates exercise-induced ultrastructural damage in skeletal muscle of aged horses. *J Appl Physiol* 98(1): 334–342.
218. Chung L and Ng YC (2005 Sep 1). Age-related alterations in expression of apoptosis regulatory proteins and heat shock proteins in rat skeletal muscle. *Biochim Biophys Acta* 1762(1):103–109.
219. Dirks AJ and Leeuwenburgh C (2005). The role of apoptosis in age-related skeletal muscle atrophy. *Sports Med* 35(6): 473–483.
220. Dedkov EI, Borisov AB, and Carlson BM (2003 Nov). Dynamics of postdenervation atrophy of young and old skeletal muscles: differential responses of fiber types and muscle types. *J Gerontol A Biol Sci Med Sci* 58(11): 984–991.
221. Lee CM, Aspnes LE, Chung SS, Weindruch R, and Aiken JM (1998 Nov). Influences of caloric restriction on age-associated skeletal muscle fiber characteristics and mitochondrial changes in rats and mice. *Ann N Y Acad Sci* 20(854): 182–191.
222. Siriett V, Platt L, Salerno MS, Ling N, Kambadur R, and Sharma M (2006 Dec). Prolonged absence of myostatin reduces sarcopenia. *J Cell Physiol* 209(3): 866–873.
223. Braga M, Sinha Hikim AP, Datta S, Ferrini MG, Brown D, Kovacheva EL et al (2008 Jun). Involvement of oxidative stress and caspase 2-mediated intrinsic pathway signaling in age-related increase in muscle cell apoptosis in mice. *Apoptosis* 13(6): 822–832.
224. Caccia MR, Harris JB, and Johnson MA (1979 May–Jun). Morphology and physiology of skeletal muscle in aging rodents. *Muscle Nerve* 2(3): 202–212.
225. Miller MS, Lekkas P, Braddock JM, Farman GP, Ballif BA, Irving TC et al (2008 Sep). Aging enhances indirect flight muscle fiber performance yet decreases flight ability in *Drosophila*. *Biophys J* 95(5): 2391–2401.

Aging of the Nervous System

Catherine A. Wolkow, Sige Zou, and Mark P. Mattson

Abstract The aging nervous system encompasses two related research areas. One is the effect of aging itself on nervous system function, also referred to as "normal" aging. A second area of consideration is that of neurodegenerative diseases with aging-associated onset, which are not representative of normal nervous system aging. This chapter examines both aspects of the aging nervous system, with emphasis on comparative studies that have revealed important insights in these areas. Both normal and pathological nervous system aging involve elevated oxidative stress, perturbed energy metabolism and the accumulation of protein aggregates. Changes in pathways for cell replacement, regeneration and repair are also important factors that are altered in the aging nervous system. The major neurodegenerative diseases associated with aging are Alzheimer's disease, Parkinson's disease, Huntington's disease and ALS. Clinical, cellular and molecular features of each disease are described. For many of these diseases, underlying genetic and environmental causes have been identified. Environmental factors capable of modifying nervous system aging and neurodegenerative disease susceptibility are also examined. Overall, many factors impact the nervous system during aging in humans as well as other species. Recent studies have shown that some are protective (exercise, dietary energy restriction and cognitive stimulation) while others (diabetes, depression and dietary factors) enhance nervous system decline at the end of life. Comparative approaches have identified those changes that represent evolutionarily conserved aspects of nervous system aging.

Keywords Brain · Neuron · Aging · Neurodegenerative · Disease · Oxidative stress · Invertebrate · Canine · Rodent

---

C.A. Wolkow (✉)
Laboratory of Neurosciences, National Institute on Aging Intramural Research Program, Baltimore, MD, USA
e-mail: wolkowca@grc.nia.nih.gov; mattsonm@grc.nia.nih.gov

N.S. Wolf (ed.), *The Comparative Biology of Aging*,
DOI 10.1007/978-90-481-3465-6_14, © Springer Science+Business Media B.V. 2010

Introduction and Overview

Nervous system aging can be considered as two distinct, but interrelated, areas of investigation. One is the study of aging-intrinsic effects on nervous system structure and function, or "normal aging". This area of investigation falls into the general classification of "biogerontology", the study of the biology of aging. The second area is the study of aging-related nervous system diseases, which show age-related onset and are primarily neurodegenerative. The age-related onset of neurodegenerative disease may reflect changes due to "normal aging" that enhance susceptibility to these diseases. Comparative approaches have contributed greatly to the understanding of both "normal" aging and neurodegenerative diseases (Table 1). For these reasons, both facets of the aging nervous system are considered together in this chapter.

The nervous system of a mature adult relies on the proper functioning of many other body systems. A decrement in any of these systems may cause a loss in brain function that can result in cognitive or motor impairment. First is the nervous system, itself, including the brain, spinal cord and peripheral nerves. The cells that comprise these tissues include the neurons, which carry out the work of signal detection and propagation that is the central job of the CNS. Normal aging is not associated with any significant decline in neuron number or density [1]. Structural stability during aging has also been observed in the nervous system of *C. elegans* nematodes [2]. However, molecular changes consistent with a decline in synaptic connectivity have been reported to correlate with aging-related declines in spatial learning in rat brains [3]. Neurons rely on support from other nervous system cells. Oligodendrocytes and Schwann cells are one type of neuronal support cell whose primary function is to provide myelin that insulates neuronal axons,

Table 1 Summary of animal models used to study neuronal aging and neurodegenerative disease

	Age-related nervous system decline	Alzheimer's disease	Parkinson's disease	ALS	Huntington's disease (polyglutamine repeat diseases)
C. elegans	[209, 220]	[221]	[180]		[183, 182]
Drosophila	[222]	[223, 165, 224]	[223, 224]		[182]
Mouse	[225, 226]	[162]	[227]	[228, 229]	[230, 231, 182, 232]
Rat			[233]	[234]	[71, 235]
Dog	[236, 158, 237]	[158, 238, 159, 161]		[239]	
Equine			[240]		
Non-human primate	[225, 241, 160]	[160]	[242]		[71]
Yeast		[243]			[244]
Molluscs	[245]				

allowing fast axonal signaling in the human nervous system. Astrocytes provide neurons with nutrients and microglia offer immune protection. There do not appear to be stereotypical changes in these cell populations during normal aging [4]. The relative structural stability of the aged brain is consistent with clinical observations of stable mental and cognitive functioning throughout old age in the absence of disease. In contrast, neurodegenerative diseases cause extensive neuron loss in patterns characteristic for particular diseases.

The nervous system is critically dependent on the cardiovascular system to provide nutrients and remove waste. The vasculature is dramatically affected by normal aging, which is associated with a progressive loss of vascular elasticity and consequent increase in stiffness. In the brain of disease-free, aged rodents and humans, vascular changes are observed as a decline in arterioles of the brain. Interestingly, cerebral capillaries are not observed to be consistently reduced in all studies, although there appears to be a reduction in the ability of the vasculature to support growth of new capillaries in response to increased brain function, a process referred to as microvascular plasticity [5].

Conditions such as hypertension, metabolic disease and diabetes severely impair vascular function, by increasing vascular stiffness, which has secondary effects on brain function. Thus, in human patients, cognitive impairment in the absence of neurodegenerative disease may be a secondary outcome of vascular disease. Two major effects of compromised vasculature on brain function are white matter lesions (WML) and stroke. White matter lesions are apparent in CT and MRI scans of aged brains and are associated with mild, but subclinical, cognitive decline [6]. The etiology of WML is unclear, but their occurrence may reflect atherosclerotic blockage of small blood vessels in the brain and consequent cell death in the vicinity of the blockage. Aging is also associated with increased stroke incidence and severity. Stroke is also a consequence of blood vessel blockade, and may reflect a more severe form of the events that induce WML.

Nervous System Intrinsic Changes Associated with Aging

During aging there are progressive decrements in the functional performance of the nervous system as indicated by slowed reaction times, sensory and motor deficits, disturbances of circadian rhythms and a decline in cognitive performance [7, 8]. However, the rate of nervous system functional decline varies considerably among individuals, as does the particular regions of the nervous system affected. Because the probability of developing a neurodegenerative disorder increases dramatically with advancing age, it is likely that at least some of the cellular and molecular changes that occur in the nervous system during normal aging are also fundamental to disease processes. In this section we review some of the prominent molecular, biochemical and cellular changes that occur in the aging nervous system, and how the changes might result in dysfunction of neural circuits and may predispose to neurodegenerative disorders such as Alzheimer's and Parkinson's diseases (Fig. 1).

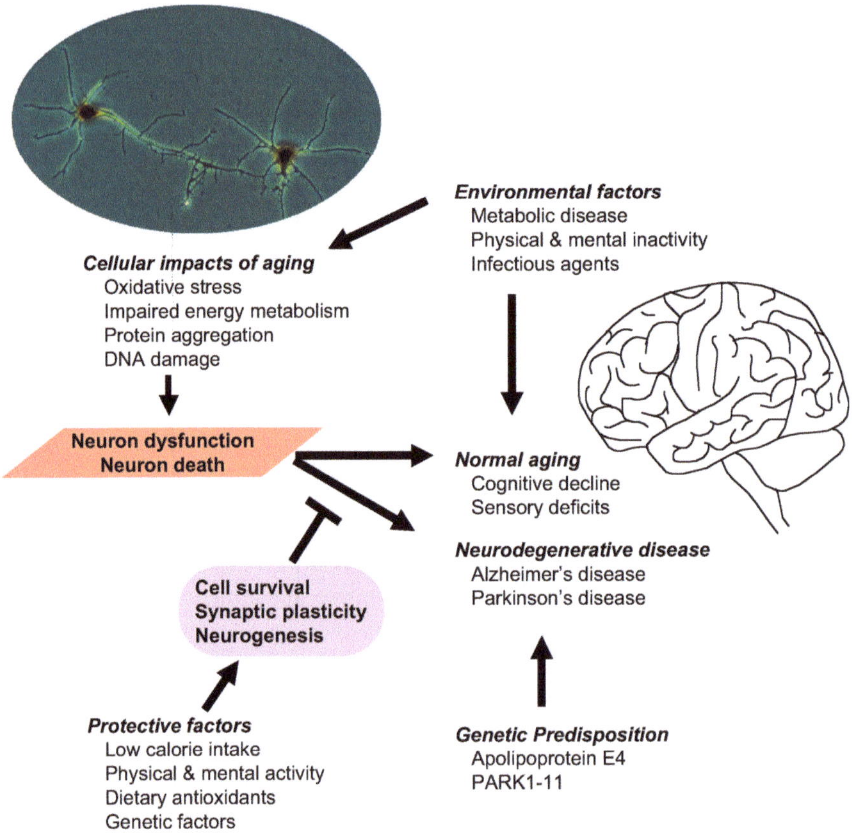

Fig. 1 Genetic and environmental factors influence nervous system aging and development of aging-related neurodegenerative diseases. Cellular changes associated with aging result in neuron dysfunction and death. Neuron dysfunction leads to aging-related changes in nervous system function, such as cognitive declines and sensory deficits. Environmental factors are tightly related to cellular aging and nervous system decline. Neurodegenerative diseases also result from neuron dysfunction and neuron loss during aging. Development of neurodegenerative disease is modulated by both environmental factors and genetic predisposition. Finally, protective environmental and genetic factors can delay aging-related declines and protect against neurodegenerative disease. These may function by promoting cellular survival, increasing synaptic plasticity and/or stimulating neurogenesis

Elevated Oxidative Stress

There is no doubt that progressive damage to cellular components by oxygen free radicals occurs during aging in most, if not all, organisms [9, 10]. Cumulative oxidative damage may result from the normal production of free radicals during energy production in mitochondria, in combination with age-related decrements in antioxidant defense systems. While there is a general increase in oxidative stress during aging, there is variability among cells, tissues and organs within an individual.

Moreover, the rate of increase in oxidative stress during a lifetime can be modified by dietary factors, with dietary energy intake being a prime example – low caloric intake reduces oxidative stress and lengthens life, while high energy intake accelerates oxidative damage and shortens lifespan [11–13]. Neurons are postmitotic cells, meaning that they cannot divide to replace damaged or deteriorated cells. For this reason, accumulated damage from oxidative stress may be a major factor in normal aging of the nervous system and, also, for some neurodegenerative diseases.

Reactive Oxygen Species: Sources and Sinks

A prominent source of ROS in all eukaryotic cells is mitochondria, which produce superoxide anion radical as a by-product of oxidative phosphorylation [14, 15]. Accordingly, there is a direct relationship between the production of energy (ATP) and superoxide. Superoxide dismutases (SOD) located within the mitochondria (Mn-SOD) and cytoplasm (Cu/Zn-SOD) convert superoxide radicals to hydrogen peroxide [16]. Hydrogen peroxide is not a free radical, but can be converted to the highly reactive hydroxyl radical in a reaction induced by Fe^{2+} and Cu^{+}. Hydroxyl radical can damage proteins and, by attacking double bonds in unsaturated fatty acids, is a very potent inducer of membrane lipid peroxidation [17]. Another type of free radical is the gas nitric oxide (NO), which is produced in a reaction catalyzed by NO synthase, an enzyme that is activated by the calcium-binding protein calmodulin [18]. NO is known to interact with superoxide to generate peroxynitrite which can cause nitration of proteins and membrane lipid peroxidation [19]. Additional ROS are generated by oxidases which produce superoxide.

In order to prevent the production and damaging actions of ROS, cells express various antioxidant enzymes. Superoxide is detoxified by the sequential activities of SODs which convert superoxide to hydrogen peroxide, and glutathione peroxidases and catalases which convert hydrogen peroxide to water [16]. Additional enzymes with important antioxidant functions include peroxiredoxins, ferroxidases (e.g., ceruloplasmin) and oxidoreductases [20]. These enzymes are typically associated with mitochondrial membranes, but are also present in the plasma membrane where they may play pivotal roles in protecting cells against aging and disease [20]. Cells also produce or sequester numerous low molecular weight molecules that serve antioxidant functions. The tripeptide glutathione (glutamate-cysteine-glycine) is a major antioxidant that functions to scavenge ROS [21]. The cysteine thiol in glutathione reduces oxidized thiols in proteins, and glutathione can reduce disulfides nonenzymatically. Many well-known low molecular weight antioxidants are obtained in the diet and/or are synthesized in cells including vitamin E (α- and γ-tocopherols), vitamin C (ascorbate) and coenzyme Q10 (ubiquinone) [22, 20].

Molecular Damage Caused by ROS

Damage to all three major classes of molecules in cells (proteins, lipids and nucleic acids) accrue during aging. Oxidative modifications to proteins include carbonylation, nitration caused by interactions with nitric oxide, and covalent binding of lipid peroxidation products such as 4-hydroxynonenal [17, 23, 24]. Proteins are nitrated on tyrosine residues, and are modified by 4-hydroxynonenal on cysteine, lysine and histidine residues. Because these amino acids are often critical for the normal function of the proteins, their oxidative modification can impair those functions. For example, phosphorylation of tyrosine residues is an important mechanism to regulate the activities of many enzymes, and nitration of those tyrosine residues may preclude their phosphorylation [25, 26]. Lipid peroxidation products can also modify proteins. Modification by 4-hydroxynonenal has been shown to alter the function of a range of proteins including ion-motive ATPases (sodium and calcium pump proteins), glucose and glutamate transporters, GTP-binding proteins and cytoskeletal proteins [27–31]. The attack by hydroxyl radical and peroxynitrite on the double bonds of unsaturated lipids generates lipid hydroperoxides resulting in the autocatalytic process of lipid peroxidation. Lipid peroxidation directly damages membranes, which may compromise their functions in the maintenance of ion homeostasis and signal transduction and, as described above, also generates toxic lipid peroxidation products including 4-hydroxynonenal, acrolein and ceramides [32–34]. During normal aging, oxidative damage to DNA is increased, and DNA repair is decreased. These changes are also exacerbated in many age-related diseases including cardiovascular disease, diabetes and neurodegenerative disorders [35–37]. Several different types of oxidative damage to DNA bases have been identified including formation of 7,8-dihydro-8-oxo-2′-deoxyguanosine [38]. Cells normally repair damaged DNA efficiently via the activities of different DNA repair enzymes [39]. However, such repair mechanisms may be compromised during aging [40]. Oxidative damage to DNA and RNA is increased in brain cells during aging, and has been proposed to contribute to age-related cognitive deficits [41, 42].

Consequences of Oxidative Damage for Nervous System Function During Aging

Oxidative damage to proteins, lipids and nucleic acids in neurons and glial cells can impair synaptic transmission, and can result in degeneration of synapses and ultimately to the death of neurons. Synapses are particularly susceptible to oxidative damage because of their high levels of ion flux, metabolic demands and associated ROS production [43]. Synaptic proteins that are targets for oxidative modification during normal aging, and/or in experimental models of age-related neurodegenerative disorders, include ion-motive ATPases, glutamate and glucose transporters, GTP-binding proteins and calcium channels, among others [44]. ROS-mediated structural damage to synaptic membranes and cytoskeletal/scaffold proteins may contribute to age-related degeneration of synapses [45–49]. Oxidative stress can also trigger neuronal apoptosis, which is implicated in Alzheimer's and Parkinson's diseases, and may also affect some neurons during normal aging [50, 51].

In addition to the association of oxidative molecular damage to synapses and neurons with aging, studies of animal models have demonstrated that increased oxidative stress can cause functional deficits similar to those observed during normal aging and/or in neurodegenerative disorders. For example, infusion of 4-hydroxynonenal into the basal forebrain of rats results in damage to cholinergic neurons and deficits in visuospatial memory [52]. The latter study further showed that infusion of Fe^{2+} into the basal forebrain caused the accumulation of 4-hydroxynonenal-modified proteins, damage to cholinergic neurons and memory deficits. Hyperoxia induced oxidative stress in the cerebral cortex of rats and, to a much greater extent, in synaptosomes isolated from the rats [53], indicating that synapses are more vulnerable to oxidative stress than other parts of the neuron or glial cells. Treatment of rats with vitamin E resulted in lower levels of hyperoxia-induced oxidative stress (accumulation of lipid hydroperoxides and protein carbonyls) in pre-synaptic membranes [54]. Of relevance to age-related cognitive deficits and Alzheimer's disease are data showing that amyloid β-peptide (Aβ) impairs hippocampal synaptic transmission by a mechanism involving induction of synaptic membrane-associated oxidative stress [45, 55]. Indeed, treatment with docosohexanoic acid protected the dendrites of neurons against damage in a mouse model of Alzheimer's disease [56]. Additional evidence that oxidative stress plays a major role in age-related impairment in synaptic function comes from studies showing that caloric restriction, which reduces oxidative stress in the brain, ameliorates age-related functional deficits and enhances synaptic plasticity in rats and mice [57–60].

Energy Metabolism

Nerve cells utilize more energy than do most other types of cells, and so are very susceptible to dysfunction and death under conditions of reduced energy availability as occurs during cardiac arrest or a stroke, for example. Considerable evidence also suggests that neurons may suffer from compromised energy metabolism during normal aging [61, 62]. It was reported 30 years ago that the activities of 3-hydroxybutyrate dehydrogenase, 3-oxo acid CoA transferase, acetoacetyl CoA thiolase and NAD-isocitrate dehydrogenase in the cerebral cortex of rats are reduced during aging [63]. Mitochondrial succinate dehydrogenase activity decreases in cerebellar Purkinje cells during aging in rats [64] and hippocampal mitochondria of aged rats exhibit lower levels of the rate of state 3 respiration with NAD-dependent substrates and lower activities of mitochondrial complexes I and IV compared to young rats [65]. Caloric restriction, which extends lifespan and delays cognitive decline during aging, also attenuates age-related declines in enzymes involved in energy metabolism, while reducing mitochondria-associated oxidative stress [66, 47, 65]. Direct evidence that age-related changes in mitochondrial energy metabolism and ROS production are pivotal determinants of aging come from studies of *Drosophila melanogaster* fruitflies in which overexpression of Mn-SOD lengthened lifespan, and resulted in a pattern of gene expression opposite to that of normal aging [67]. The latter study also showed that the pattern of gene expression caused by MnSOD overexpression was similar to that in long-lived

C. elegans insulin-like signaling mutants, elucidating a key role for mitochondria in the genetic control of lifespan. These data suggest the possibility that regulation of mitochondrial energy and ROS metabolism in neurons may be particularly important for nervous system aging.

Neurons are particularly vulnerable to impaired mitochondrial energy production and increased mitochondrial oxidative stress. Indeed, exposure of rodents, monkeys and humans to mitochondrial toxins can cause pathological changes in the brain and associated behavioral deficits similar to those in patients with age-related neurodegenerative disorders. For example, experimental exposure of mice and monkeys (and accidental exposure of humans) to MPTP results in selective damage to dopaminergic neurons and Parkinson's disease-like motor symptoms [68]. Upon entering the brain, MPTP is converted into MPP^+ by the activity of monoamine oxidase in astrocytes; the MPP^+ is then selectively transported into dopaminergic neurons where it inhibits mitochondrial complex I. The relevance of the MPTP model to the most common late-onset form of sporadic Parkinson's disease is suggested by the ability of dietary energy restriction to protect dopaminergic neurons against MPTP and improve functional outcome in mouse and monkey models [69, 70]. Increasing evidence suggests that dietary energy intake, and energy expenditure with exercise, can have major effects on aging of the nervous system and its vulnerability to age-related disease (see section "Environmental Factors and Their Effect on the Aging Nervous System" below). A different mitochondrial toxin that targets succinate dehydrogenase, 3-nitropropionic acid, selectively damages medium spiny neurons in the striatum and causes motor deficits similar to those seen in humans with Huntington's disease [71]. Dietary energy restriction protects striatal neurons from being damaged and killed by 3-nitropropionic acid [72]. Further evidence that neurons are particularly vulnerable to mitochondrial dysfunction comes from a study showing that mice lacking mitochondrial Mn-SOD exhibit selective brain and spinal cord degeneration [73].

While most attention has focused on mitochondrial changes as being key to age-related deficits in energy metabolism and increases in oxidative stress, several other energy- and redox-regulating systems are also involved in aging of the nervous system. The plasma membrane redox system (PMRS) consists of a series of enzymes and cofactors (including NADH-ascorbate free radical reductase, NADH-quinone oxidoreductase 1, NADH-ferrocyanide reductase, NADH-coenzyme Q10 reductase, and NADH-cytochrome c reductase), and alpha-tocopherol and coenzyme Q10. During normal brain aging there is a decline in activities of PMRS enzyme activities that is associated with increased levels of plasma membrane lipid peroxidation and accumulation of protein carbonyls and nitrotyrosine [74]. Dietary energy restriction attenuates the age-related decreases in levels of PMRS enzyme activities and reduces markers of oxidative stress. Alterations in glycolytic pathways in brain cells have also been documented in studies of aging. Levels of pyruvate, malate and creatine phosphate diminished in the brains of rats with advancing age, as did levels of glucose-6-phosphate, fructose-1, 6-diphosphate, and lactate [75]. The latter findings suggest that glucose and energy metabolism may diminish in brain cells during normal aging.

Protein Homeostasis and Aging

Cellular function requires the production, modification and regulation of many diverse proteins. In neurons, there are special requirements for translocating proteins to distant sites on dendrites and axons. Furthermore, cells depend on catabolic processes to remove damaged proteins and recycle their components. Two major pathways for protein degradation are the ubiquitin/proteasome pathway (UPP) and the lysosomal/autophagic pathway. Several lines of evidence suggest that aged cells have deficits in these protein turnover pathways. Furthermore, neuronal cells appear to be extraordinarily sensitive to these particular deficits, suggested by the preponderance of impaired protein degradation phenotypes associated with neurodegenerative diseases.

The Ubiquitin/Proteasome Pathway in Brain Aging

A number of cytoplasmic, endoplasmic reticulum and nuclear proteins are subject to regulated degradation by the ubiquitin/proteasome pathway (UPP) [76]. A protein targeted for UPP-mediated degradation is first modified by the covalent addition of a poly-ubiquitin chain. Ubiquitin (Ub) is a 76-amino acid polypeptide that is conjugated onto a lysine residue of the target protein in a series of sequential transfers from an E1 Ub-activating enzyme to the E2 Ub-conjugating enzyme that, in turn, transfers Ub to the target protein in a reaction catalyzed by the E3 Ub ligase. Target protein specificity is mediated by the E3 ligase. Eukaryotic genomes encode numerous E3 Ub ligases, presumably with distinct target specificities. Finally, E4 Ub factors polymerize additional Ub moieties onto the substrate protein, targeting it for proteasomal destruction. The proteasome is a 26S multisubunit complex that is composed of two large components, a 19S regulatory particle and a 20S core particle. The 19S regulatory particle recognizes ubiquitinated substrates for degradation. The 20S core particle contains proteolytic activities that degrade the target protein as it is denatured by chaperones intrinsic to the core particle. Proteasomal degradation is an ATP-dependent process and relies on cellular energy production processes.

Disruptions in proteasomal function are stressful to cells and can lead to cell death by apoptosis, which has rendered the proteasome as a potential cancer chemotherapy target [77]. There is also evidence that proteasomal function is disrupted in normal aging and this may contribute to aging-related declines in some tissues. In particular, treatment with proteasomal inhibitors shortened lifespan in *Drosophila* fruitflies and induced some changes indicative of premature aging [78]. An age-associated decline of proteasomal function during normal aging could, in turn, compromise other cellular processes and increase the likelihood of developing disease [76].

Autophagic Processes and Aging

Under conditions where the UPP pathway is unable to keep pace with cellular needs for protein degradation, alternative protein turnover pathways may be activated. One such pathway is lysosome-mediated degradation, which is normally responsible for turnover of non-UPP substrates, which include membrane proteins and organelles, such as mitochondria. In lysosome-mediated degradation, substrates are targeted to lysosomes by cellular vesicle trafficking pathways. Under nutrient or environmental stress, lysosome-mediated degradation is enhanced by activation of autophagy, which efficiently targets whole organelles into double-membraned vesicles destined for lysosomal fusion and degradation.

The lysosomal and autophagy pathways are both disrupted during normal aging. In many cell types, there is an accumulation of autofluorescent material, termed lipofuscin, which is believed to reflect substrate accumulation in dysfunctional lysosomes. Lipofuscin accumulation itself may compromise cell functions and impair survival [79]. Autophagy declines during normal aging in most organisms [80]. Autophagy is likely to be very important for promoting healthy aging in a number of settings. For example, long lifespan in *C. elegans* mutants with defective insulin-like signaling is completely dependent on an active autophagy pathway, suggesting that autophagic processes have some relationship to extreme longevity in these mutant animals [81]. Long-lived insulin-defective mutants are also resistant to the accumulation of protein aggregates like those implicated as causal agents in neurodegenerative diseases, and this activity may also be related to an upregulation of autophagy in these animals [82]. Furthermore, dietary restriction can induce autophagy. In *C. elegans*, longevity from DR is associated with and dependent on upregulation of autophagy [83, 84]. Thus, elevation of autophagy may be one approach to protecting cells from aging-intrinsic declines.

Protein Translation Rates as a Factor in Aging

An alternative to increasing protein degradation rates to alleviate burdens of cellular aging is to reduce the rate of protein synthesis. Theoretically, a reduced rate of protein synthesis could protect cells from aging-related declines in degradation pathways by lowering substrate protein levels. This theory has been tested in several model organisms, although it remains to be explicitly tested in mammals. In *C. elegans*, mutations that reduce the rate of protein translation can extend lifespan, consistent with this theory [85, 86]. It has been suggested that reduced protein translation rates may extend lifespan by lessening the burden on the cellular degradative machinery [87]. Further study of protein-translation-deficient mutants in these systems has indicated that the protein translation deficits may extend lifespan in a mechanism that overlaps, at least partially, with calorie restriction. Thus, one possibility is that calorie restriction in mammals also has a similar effect on protein translation and may also reduce the demands on the protein degradation machinery over time. Further work remains to elucidate specific mechanisms through which protein translation affects longevity in these organisms, and to determine the effects in mammals.

Pathways for Cell Replacement

Until recently it had been thought that nerve cells are not replaced during adult life. While this may be true for many invertebrates, including *C. elegans*, it has been clearly demonstrated that there are populations of self-renewing neural stem cells present in some regions of the nervous system in adult poikilotherms and mammals [88, 89]. In mammals there are neural stem cells located in the subventricular zone of the cerebral cortex and the dentate gyrus of the hippocampus that are capable of differentiating into neurons and glial cells. In the hippocampus, newly generated neurons grow axons and dendrites that form functional synapses with appropriate target and input cells [90–92]. Hippocampal neurogenesis is decreased during normal aging [93], and is increased in response to "anti-aging" environmental factors including dietary energy restriction [94], exercise [95] and environmental enrichment [96]. Studies of animal models suggest that neurogenesis may be decreased in Alzheimer's disease [97]. Transplantation of neural stem cells into relevant brain regions can at least partially restore function in animal models of stroke [98] and age-related neurodegenerative disorders including Parkinson's disease and Alzheimer's disease [99, 100]. The cellular and molecular mechanisms that may contribute to reduced cell replacement in the nervous system during aging and in neurodegenerative disorders are being elucidated. Factors include increased levels of oxidative stress and decreased levels of neurotrophic factors [97, 101].

Aging's Effects on Regeneration and Repair

Peripheral nerve injury, such as a crush injury, can sever neuromuscular and sensory signaling systems. Repair processes are initiated after nerve injury and are dependent on assistance from macrophages and neurolemmocytes, or Schwann cells. Wallerian degeneration is initiated to degrade the distal trunk of the injured nerve. Axon regeneration ensues at the proximal trunk, followed by target innervation. Finally, the myelin sheath around the regenerated axon is repaired. Axon elongation and target renervation are dependent on matrix guidance cues and on trophic factors. Exogenous addition of trophic factors can permit regeneration and target renervation by severed sensory axons leading to the spinal cord, to which renervation is normally blocked [102].

Descriptive studies have demonstrated that peripheral nerve regeneration after crush injury proceeds more slowly in aged rodents than in young animals. One study comparing sciatic nerve regeneration in 6- and 24-month old mice reported fewer regenerating axons in the older animals [103]. In addition, the regenerating axons that were present in the older animals were smaller in size and less well myelinated than in the younger animals. This report also presented evidence that Wallerian degeneration was delayed in older animals. Age-related delays in Wallerian degeneration were also observed in a separate study of regeneration after facial nerve crush. In this study, regrowth and remyelination were slower in 15-month old rats than in 3-month old rats [104]. The mechanisms contributing to the aging-related deficits in Wallerian degeneration have not been identified. It has been suggested to reflect age-associated deficits in macrophage function, which is required to engulf the decaying axon stumps [105]. Alternatively, aging could lead to alterations in the guidance cues needed to recruit macrophages and neurolemmocytes to the site of injury or stimulating the initiation of repair.

Intrinsic Changes in the Nervous System Associated with Aging

Sensory Loss

Decline of diverse neuronal functions with aging has been well documented in humans. Aging is associated with gradual functional loss in the sensory systems, such as olfaction [106], hearing and vision [107, 108], which has profound impacts on daily activities of the elderly. Population studies indicate that more than 60% of people over 80 display olfactory impairment and more than 50% of people over 75 show hearing loss. Characteristics of age-associated olfactory impairment include defects in thresholds to detect odors, recognition and discrimination of odors and perception of odor intensity [106]. Age-related hearing loss (ARHL) is associated with increased thresholds to low-frequency sound (e.g. 250–251 kHz), but decreased thresholds to the high frequency sound (e.g. 8 kHz) [109]. A common form of ARHL is called presbycusis, in which a patient has difficulty understanding speech due to inability distinguishing high-frequency tones [107]. Investigation of the sensory circuits has demonstrated that age-associated sensory loss can be due to defects in peripheral and/or central pathways. Epidemiological studies have implicated a number of environmental and genetic factors as risk factors for sensory loss. For hearing loss, noise is an obvious environmental risk factor. Quantitative trait loci (QTL) mapping studies in human and mice have identified several genes and modifiers potentially involved in ARHL [110–112]. One possible candidate involved in ARHL is ATP2B2, a plasma membrane ATPase type 2-Ca^{2+} transporter pump, located in the hair cells of the cochlea, suggesting a critical role for Ca^{2+} signaling in age-related changes of the auditory system [109, 113]. Analyses of knock-out mice with deletion of Gpx1 or Sod1 indicate that ARHL is accelerated in these mice, suggesting that oxidative stress also plays an important role in ARHL [109].

Cognitive Decline

Cognitive function gradually declines with increasing age although the impairment varies among individuals with different ages of onset [114]. As a part of cognitive function, memory is a complex process involving the peripheral and central nervous systems. In a memory process, information is received by the sensory system, transmitted to the central nervous system, and processed and stored in the brain to form memory [115]. Despite its complexity, memory can be classified into several different types based on the difference in information acquisition, retention and recall. Not all types of memory are equally affected by aging. It appears that aging primarily affects memory for recent events while having little effect long on short term memory [116, 117]. Structural studies indicate that the age-related memory impairment (AMI) appears not to result from gross morphological change since general brain structures and neuron numbers remain relatively unchanged during aging [114]. AMI can be attributed to age-related alterations of Ca^{2+} homeostasis, cyclic AMP (cAMP) level and synapse number in the neurons. One group of neurons implicated in AMI is central cholinergic neurons primarily located in the pontine reticular formation and basal forebrain [118]. A number of rodent studies have demonstrated age-related changes of calcium buffering systems and Ca^{2+} signaling, which may cause dysfunction of brain cholinergic neurons [119]. AMI has been genetically investigated in *D. melanogaster*, a genetically tractable invertebrate system. These studies have shown that AMI is delayed in flies with mutations in *amnesia*, which encodes putative neuropeptides regulating cAMP levels, and *DCO*, which encodes a cAMP dependent protein kinase [120, 121]. This suggests that the role of cAMP signaling in AMI is an evolutionarily conserved feature of aging.

Regulating Lifespan by the Nervous System – Insights from Genetics

It has been well-documented that various nervous system functions gradually decline with age, as described in previous sections. It is also possible that neurons play an active role in aging by modulating lifespan and aging processes. It is not completely clear whether this is the case in humans, although patients with neurodegenerative diseases have relatively higher mortality rates compared to the age-matched general population [122]. Studies using model organisms have provided strong support for an active role of neurons in regulating lifespan. A number of genetic pathways have been implicated in modulating lifespan in rodents, as well as two invertebrate species, *C. elegans* nematodes and *D. melanogaster* fruitflies. These include evolutionarily conserved insulin/insulin-like signaling Jun N-terminal kinase (JNK) signaling, and Sirtuin pathways [123–126]. These data provide a foundation to investigate functions of the nervous system in modulating human lifespan.

In *C. elegans*, the gene, *daf-2*, encodes an insulin receptor-like protein, which can function in a variety of cell types, including neurons, to coordinate aging of all cells throughout the body [127, 128]. Mutations that disrupt DAF-2/insulin-like signaling significantly extend adult lifespan [129, 130]. In *D. melanogaster*, several insulin-like peptides (Dilp), including Dilp2, are released from a small number of neurons in the brain [131, 132]. Genetic ablation of the Dilp2 neurons in the brain by neuron-specific induction of apoptosis results in increased lifespan compared to the control with intact Dilp neurons [133]. In addition, overexpression of dominant negative forms of p53, a key apoptosis gene, in neurons by a pan-neuronal driver extends lifespan partly through a pathway related to dietary restriction which has been shown to be able to increase lifespan in almost all the species tested so far [134, 135]. Modifications of the JNK signaling pathway in neurons only have been demonstrated to be sufficient to extend lifespan [136, 137]. Defects in the sensory and olfactory neurons have also been shown to extend *C. elegans* and *Drosophila* lifespan [138, 139]. It has recently been shown in *Drosophila* that modulation of the olfactory sensory system can extend lifespan in flies probably through DR-related pathway [139]. Finally, wildtype *C. elegans* treated with the anticonvulsant medication, ethosuccimide, live longer than untreated animals and this effect appears to be due to altered neuronal function [140]. Neuronal functions in modulating lifespan are not limited to invertebrates. Transgenic mice with deletion of IRS-2 selectively in the brain have significantly longer lifespans than wild type mice of the same genetic background [141]. These results suggest that modulation of lifespan through the nervous system is an evolutionarily conserved feature.

It is still largely unknown how neurons might modulate lifespan. One possible mechanism is that structures and/or functions of neurons fall apart ahead of other important tissues in aging and thus delaying aging of neurons would extend lifespan of the whole organism. This hypothesis might explain why the longevity effect of the anti-apoptotic gene, p53, in fly neurons. However, this may be only part of the mechanism. In *C. elegans*, little morphological degeneration has been detected in neurons in aging while there is significant age-related degeneration in muscle [2]. In *D. melanogaster*, brains have significantly lower apoptosis than the thoracic muscle during aging [142]. Recent results indicate that p53 can modulate the insulin-like signaling pathway [135]. This suggests that neurons are not necessarily a more vulnerable tissue. Alternatively, it is possible that signals from neurons or the neuroendocrine system are critical in modulating lifespan. As described above, genetic ablation experiments in flies suggest that Dilp released from the brain probably mediates reduction of the insulin-like signaling in the whole body, which in turn extends lifespan. This neuroendocrine hypothesis may similarly explain why neuron-specific modulation of genes involved in insulin-like signaling can extend lifespan in *C. elegans* and mice, although it is not clear what factors in neurons are critical. Nevertheless, further investigation of aging in the nervous system will provide key information for our understanding of aging in general.

Alzheimer's Disease

Approximately 5 million Americans are currently afflicted with Alzheimer's disease (AD), and this number will increase sharply as baby-boomers reach the age zone where the risk of AD increases (beyond age 65) [143]. There are currently no effective treatments for AD, but based on increasing knowledge of the cellular and molecular alterations that underlie the disease process, and information concerning risk factors, there is reason to be optimistic that this disease can be prevented and will be treatable soon.

Clinical, Histological and Molecular Pathology of Alzheimer's Disease

Typically the first clinical signs of AD are subtle decrements in short-term memory [144]. For example, during a conversation, someone in the early stages of AD may repeat what they had said earlier in the conversation because they did not remember that they had already said it. Short-term memory declines progressively over a period of years until it fails completely. The patients also develop emotional disturbances, disrupted circadian rhythms and decrements in function of the autonomic nervous system. In the last few years and months of the disease AD patients require round-the-clock care. The most common proximate cause of death is aspiration pneumonia. Brain imaging studies reveal progressive reductions in the size of the hippocampus and associated cortical structures, and reductions in energy metabolism in the same brain regions. At autopsy, gyri in the temporal and frontal lobes are markedly shrunken. Examination of brain sections reveals extensive loss of synapses and neurons in the affected brain regions which is associated with extensive accumulation of amyloid β-peptide (Aβ) in diffuse deposits and senile plaques [101]. In addition, there are abnormal fibrillar aggregates of hyperphosphorylated tau, a microtubule-associated protein, in many neurons – the so-called neurofibrillary tangles.

Genetic and Environmental Factors

Although the vast majority of cases of AD occur very late in life and in individuals who have no clear family history of the disease, AD is inherited in an autosomal dominant manner in some families [145]. The genomes of affected individuals in some of these families harbor mutations in the β-amyloid precursor protein (APP), while others have mutations in presenilin-1 or presenilin-2. Mutations in all three genes result in increased proteolytic cleavage of APP to generate Aβ. The APP mutations affect cleavage of APP by enzymes that cleave at either the N-terminus (β-secretase/BACE) or C-terminus (γ-secretase) of the Aβ sequence. Presenilins are integral membrane proteins and the proteolytic component of the γ-secretase enzyme complex; presenilin mutations result in increased production of the long (42 amino acid) form of Aβ. In addition, there is evidence that presenilin-1 normally functions in the regulation of cellular calcium homeostasis in neurons, and that disease-causing mutations disrupt this normal function of presenilin-1 [146]. Aβ has been shown to impair synaptic function and render neurons vulnerable to death (excitotoxicity and apoptosis) by a mechanism involving membrane-associated oxidative stress in neurons contacted by Aβ; oligomers of Aβ may be particularly toxic [101]. In addition to disease-causing mutations, genetic factors may also affect the risk of developing AD. Indeed, individuals with an E4 allele of apolipoprotein E are at increased risk of AD.

Findings from epidemiological, clinical and animal model studies suggest that several different environmental factors can affect one's risk for AD [101, 147]. Individuals who exercise regularly are at reduced risk for AD [148], and exercise suppresses Aβ accumulation and cognitive deficits in mouse models of AD [149]. Similarly, those who engage in intellectually challenging activities are at reduced risk of AD [150]. Dietary energy intake may also affect the risk of AD; high energy intake increases risk, while low energy diets reduce the risk [151]. Consistent with the latter epidemiological data, it has been shown that caloric restriction suppresses Aβ and tau pathology, and improves cognitive function in mouse models of AD [59, 152, 153]. Interestingly, alternate day fasting ameliorated cognitive deficits, without preventing Aβ accumulation in a mouse AD model [59], suggesting that some dietary factors can influence the disease process downstream of Aβ. Finally, various chemicals present in dietary plants and animals may influence the risk of AD [154]. Examples of potentially beneficial chemicals include phytochemicals such as curcumin and resveratrol, and omega-3 fatty acids present in relatively high amounts in fish [155]. Examples of potentially detrimental dietary factors include saturated fats and excitotoxins [156, 157].

Animal Models of Alzheimer's Disease

Both dogs and monkeys display an increase in endogenous Aβ accumulation during aging, making these species useful for studying the relationship between Aβ plaques and cognitive impairment [158–161]. Dogs are particularly useful in this regard for their compatibility with humans and large behavioral repertoire [159]. In dogs, Aβ levels have been correlated with some types of cognitive impairment [159, 161]. However, there are two notable differences between human AD and canine Aβ plaques. One is the absence of neurofibrillary tangles from aged dog brain and the second is that, in dogs, Aβ is only observed in the form of diffuse plaques, but not the cored plaques apparent in the human disease

The most widely used animal models of AD are mice that express mutant forms of APP and/or presenilin-1 which cause inherited disease in humans [162]. Transgenic APP mutant mice typically exhibit age-related accumulation of Aβ in the cerebral cortex, hippocampus and other brain regions, and associated cognitive deficits. Presenilin-1 mutant mice do not exhibit Aβ pathology or cognitive deficits, but their neurons are more vulnerable to dysfunction and degeneration under conditions of metabolic and oxidative stress [146]. However, the presence of mutant presenilin-1 results in a marked exacerbation of Aβ pathology in APP mutant mice. APP and presenilin-1 mutant mice do not exhibit tau (neurofibrillary tangle-like) pathology. Mice expressing APP and presenilin-1 mutations, in combination with a tau mutation that causes frontotemporal lobe dementia in humans, develop Aβ and tau pathologies in the hippocampus and cortex, synaptic dysfunction and learning and memory impairment as they age [59, 163]. Invertebrate models relevant to AD have also provided additional insight into the molecular basis of this disease. For example, transgenic *C. elegans* that overexpress Aβ exhibit cytotoxic changes that can be ameliorated by heat-shock protein chaperones [164, 165].

Parkinson's Disease (PD)

Symptoms of PD

PD is a movement and neurodegenerative disorder. Epidemiologic studies indicate that the incidence of PD is approximately five to 20 new cases per 100,000 and the prevalence ranges from 1 to several hundred per 100,000 people depending on population surveyed [166]. These rates make PD the second most common neurodegenerative disorder after Alzheimer's disease. Clinically, PD is characterized by defects in motor functions including resting tremor, bradykinesia, rigidity and postural instability and with good response to treatment with L-dopamine or other dopamine agonists [167]. The defects in PD are primarily due to gradual loss of pigmented and dopaminergic neurons in the substantial nigra pass compacta (SNpc), which in turn result in dysfunction in sensory and motor fields. Postmortem examination of PD patients indicates that eosinophilic inclusions are generally present in either the perikaryon or neuronal process in the substantia nigra. These inclusions are called Lewy bodies or Lewy neutitis respectively, which are the hallmarks of the PD [167]. Immunohistochemical studies have shown that the inclusions contain α-synuclein, ubiquitin and various other proteins [168–170]. How these inclusions are involved in causing PD is still a matter of debate.

Risk Factors for PD

Case-controlled population studies indicate that most PD cases are sporadic and approximately 10% of PD cases appear linked to family history of disease [171]. Studies of risk factors for PD have implicated a number of environmental factors [166]. Age is probably the most influential risk factor for PD, since disease symptoms do not manifest until after age of 50 for most patients. Juvenile PD is rare and most often associated with familial cases. In addition, epidemiological studies have indicated higher incidence of PD correlated with certain addictive drugs, pesticide exposure and the use of well water. These findings suggest that certain chemicals may contribute to the development of PD. One such chemical is 1-methy-4-pheny-1,2,3,6-tetrahydropyridine (MPTP), which potently induces Parkinson syndromes in humans and animal models [68]. MPTP induces loss of dopaminergic neurons partly through its conversion to toxic MPP^+ in the brain. MPP^+ impairs generation of ATP in the mitochondria and increases production of ROS in the neurons. Interestingly, several environmental factors inversely associated with PD have been identified as smoking, caffeine intake, and certain nonsteroidal anti-inflammatory drugs [172–174]. However, additional studies are required to definitely determine the effects of these factors in lowering PD incidence.

Genetics of PD

Our understanding of molecular mechanisms of PD largely comes from studies of rare familial PD cases. Genetic linkage analyses of various kindred have revealed a few genetic loci implicated in PD, which are named PARK1 to PARK11 [175]. Sequencing analyses have identified several genes in these loci associated with PD. The first PD-linked gene to be identified was α-synuclein (PARK1) [176]. Although mutations of α-synuclein are rare in familial and sporadic PD cases, α-synuclein has been found to be a major component of the Lewy bodies, which strongly supports the association of this protein with PD [169]. Subsequent studies thus far have identified five more PD-associated genes. These include PARKIN (PARK2, a ubiquitin E3 ligase), Ubiquitin carboxy-terminal hydrolase L-1 (UCHL-1, PARK5), PINK1 (PARK6), DJ-1 (PARK7) and Leucine-rich repeat kinase 2 (LRRK2, PARK8) [175]. Frequency of mutations of these genes varies significantly among familial PD cases, ranging from rare mutations of DJ-1 (approximately 1% in the recessive forms of PD cases) to relatively frequent mutations of LRRK2 (approximately 5% of PD cases with positive family history) [177].

Model Systems for Mechanistic Studies of PD

Identification of PD-associated genes provides a foundation to investigate molecular mechanisms of PD using model organisms. Orthologs of all PD-associated genes, except α-synuclein, have been identified in evolutionarily diverse species, ranging from *D. melanogaster* and *C. elegans* to rodents and monkeys [178–180]. α-synuclein appears to be absent from the genomes of invertebrates and lower organisms. The functions of PD-associated genes have been investigated by introducing wild type and mutant versions of the human genes into genetically tractable species, such as worms, flies or mice. For example, wild type and mutant human α-synuclein genes have been introduced to the genomes of mice, flies, and worms [178–180]. Several, but not all, of the features of α-synuclein-associated PD pathology were recapitulated in the model organisms. Overexpression of human α-synuclein has been shown to induce gradual loss of a subset of dopaminergic neurons in mice and flies [179]. An alternative approach is to study the functions of the endogenous loci in these species using mutation generation. Although the phenotypes of transgenic models are often dramatically different, a common feature of most of the transgenic models is that neurologic phenotypes and neuropathology manifest gradually with increasing age, supporting the observation in humans that aging is a risk factor in development of PD [181]. Functional studies of these animal models for PD have demonstrated that PD-associated genes influence neuronal functions and survival through inducing abnormal protein aggregation, mitochondrial dysfunction and increased sensitivity to oxidative stress [175]. However, experimental evidence pinpointing aspects of aging that influence PD occurrence and progression is lacking. In particular, it is not known how dopaminergic neurons are selectively targeted for PD during aging. Nevertheless, the genetic models of PD will be valuable to advance our understanding of this disease.

Huntington's Disease

Huntington's disease (HD) is a hereditary neurodegenerative disease that is transmitted in a dominant autosomal fashion. HD is somewhat rare with incidence of 1 in 10,000–1,000,000, depending on ethnicity. Although HD is inherited, the symptoms are progressive and begin at about middle age. Neurodegeneration in HD occurs primarily in the striatum, which controls movement and some cognitive functions. The major HD symptom is jerky, uncoordinated body movement. The disease is also associated with deficits in cognitive function [182].

The Role of Polyglutamine Repeat Expansions in Huntington's Disease

HD is caused by mutations that dramatically increase the number of glutamine residues within a large polyglutamine repeat region in the endogenous huntingtin gene. Expansion of this polyglutamine repeat, from ≤35 amino acids in asymptomatic individuals to 40–55 in affected patients, is also associated with aggregation of the mutant huntingtin protein [182]. Disease severity and age of onset are correlated with increasing polyglutamine repeat length, with symptoms arising in middle age for most patients. Thus, the theme of aberrant protein aggregation in age-onset neurodegenerative disease, as seen for AD and PD, resurfaces in HD.

Animal Models Establishing the Role of Polyglutamine Repeat Expansion in HD

In contrast to the relatively complicated spectrum of environmental and genetic factors implicated in the causation of AD and PD, the causal role for huntingtin polyglutamine repeat expansion in HD has been relatively clearly established. This significant advance has been made possible by several elegant genetic studies in *C. elegans* and mice. Researchers have shown in both species that introduction of transgenes carrying long polyglutamine repeats can induce symptoms related to those of human HD [183, 182]. Most significantly, transgenes carrying only the polyglutamine repeats, but not the remaining portions of the huntingtin protein, cause symptoms in a length-dependent manner. These particular experimental systems will be important for identifying genetic modifiers and therapeutic strategies that might be applicable for a spectrum of these aggregate-associated neurodegenerative diseases.

ALS

Amyotrophic lateral sclerosis (ALS), a disorder characterized by progressive neuromuscular dysfunction, paralysis and death, affects more than 30,000 Americans [184]. The risk of ALS increases with advancing age, but the average age of onset for ALS (55–65 years old) is younger than for AD and PD.

Clinical and Pathological Features of ALS

ALS patients first develop subtle neuromuscular symptoms, including weakness and fatigue localized to specific (usually distal) muscle groups [185]. The symptoms worsen and "spread" to adjacent muscle groups. Typically the disease begins in muscles of the lower and/or upper extremities and then progress to muscles of the torso and face. These symptoms are the result of the progressive degeneration of lower motor neurons in the spinal cord. Later in the disease process, cranial nerve motor neurons and upper motor neurons may be affected. Patients succumb to the disease when neurons required for breathing degenerate. Analysis of spinal cords from ALS patients have demonstrated death of motor neurons which is preceded by decreased acetylcholine production and the accumulation of abnormal aggregates of neurofilament proteins in their cell bodies and axons [186]. In addition, intracellular inclusions of Cu/Zn-SOD are prominent in the motor neuron cell bodies. Increased oxidative damage (lipid peroxidation and protein oxidation) and perturbed membrane lipid metabolism are associated with the neurodegenerative process in ALS [32, 187]. Excitotoxicity and perturbed calcium homeostasis are also implicated in the death of motor neurons in ALS [188, 189].

Genetic Mutations Cause Some Cases of ALS

Approximately 10% of all cases of ALS are inherited in an autosomal dominant manner, and mutations in Cu/Zn-SOD are responsible for many such familial ALS cases [190]. Transgenic mice that overexpress mutant Cu/Zn-SOD exhibit progressive motor neuron degeneration, paralysis and death. These mice provide a valuable animal model of the human disease [191]. Mutations in Cu/Zn-SOD do not impair the dismutase activity of the enzyme, but may, instead, result in a gain of an adverse property of the protein. Two possible consequences of the ALS Cu/Zn-SOD mutations are the generation of reactive oxygen species and interaction and sequestration of neuroprotective proteins such as heat-shock proteins [192]. Interestingly, dietary energy restriction has no beneficial effect and, indeed, exacerbates the disease process, in a mouse model of ALS [193]. Conversely, high-energy diets are beneficial for ALS mice [194].

Sporadic ALS

There is increasing evidence that environmental factors may contribute to the most common sporadic forms of ALS. Perhaps the strongest evidence comes from studies of populations native to islands near Guam where there was a very high incidence of a neurodegenerative disorder called ALS-Parkinson-dementia syndrome [195]. Epidemiologic analysis pointed to an environmental cause of this disease, and studies of the diet of the natives revealed that they consume large amounts of a seed called the cycad either by directly eating breads made from the seed, or by eating flying bats which feed on the cycad [196]. In support of this, the incidence of this ALS syndrome has decreased as the diet of the islanders has changed and, importantly, the disease risk of natives who moved to different countries was reduced. Whether environmental toxins are involved in cases of ALS in other regions of the world is unclear.

Environmental Factors and Their Effect on the Aging Nervous System

There is convincing evidence that the rate of aging of the nervous system, and vulnerability to neurodegenerative disorders and injury, can be modified by diet, lifestyle and exposure to environmental stressors. In general, factors that affect one's risk for age-related diseases in other organ systems also affect one's risk for neurodegenerative disorders. In this section we describe evidence obtained from epidemiological and experimental research in this area.

Energy Intake

Overeating is a major risk factor for diabetes, cardiovascular disease and stroke, but may also increase the risk of AD [151] and PD [197]. Diabetes, which is typically associated with excessive calorie intake, is now considered a risk factor for cognitive impairment and AD [198]. Diabetes impairs hippocampal synaptic plasticity and neurogenesis, and causes cognitive deficits in rat and mouse models [199, 200]. In rodents, dietary restriction attenuates age-related cognitive decline [58] and can stimulate neurogenesis [94]. Dietary energy restriction is also neuroprotective, suppresses the molecular disease process and improves functional outcome in animal models of AD [59], PD [69, 70], Huntington's disease [201] and stroke [202]. Reduced energy intake may protect the brain by decreasing oxidative stress and/or by inducing a mild adaptive stress response resulting in increased production of neurotrophic factors [203]. Consistent with findings from human and rodent studies, experiments in invertebrate models of aging support a beneficial effect of relatively low energy intake on nervous system aging and lifespan [204].

Exercise

Epidemiological studies have provided evidence that individuals who exercise regularly reduce their risk for AD and PD [205, 148]. Studies of animals have shown several benefits of exercise that may counteract aging processes including increased vascularization and neurogenesis, and enhanced synaptic plasticity and learning and memory [206, 95]. Exercise may exert its beneficial effects on the nervous system by stimulating the production of growth factors such as brain-derived neurotrophic factor (BDNF) and vascular endothelial growth factor (VEGF).

Stress

Chronic uncontrollable stress is detrimental to the nervous system, and stressful life events may trigger the onset of depression and AD [207]. Chronic stress results in hyperactivation of the hypothalamic-pituitary-adrenal neuroendocrine system, and the consequent increase in levels of circulating glucocorticoids (cortisol in humans and corticosterone in rats and mice) can impair synaptic function and neurogenesis, and may render neurons vulnerable to oxidative stress and excitotoxicity [208]. Studies of worms and flies have shown that genetic factors that extend lifespan also increase the resistance of the organism to environmental stressors, such as thermal and oxidative stress, and exposure to toxins [209, 210]. Further evidence that stress resistance is critical for preserving nervous system function during aging are data showing that dietary energy restriction activates adaptive stress response pathways in neurons thereby protecting them against the detrimental effects of aging and disease processes [203].

Social Interaction

Living a lonely life is detrimental for health in general, and health of the nervous system in particular. Among humans, those who do not marry have a lower life expectancy than those who do marry, and those who maintain regular social interactions with friends and relatives are more likely to live longer [211, 212]. Those who engage in social activities have a lower risk for AD than those with few social interactions [213]. In laboratory animals, social isolation results in reduced neurogenesis, hippocampal atrophy and impaired cognitive function [214, 215]. Social isolation has also been reported to accelerate the disease process in a mouse model of AD [216]. The mechanisms by which social isolation adversely affects the brain appear to be similar to those operative in depression including reduced levels of norepinephrine, serotonin and BDNF [217, 218]. The importance of social interactions for successful aging appears to be highly conserved in evolution. For example, studies of honeybees have shown that foragers that are forced to revert to hive-tasks show recovery of immunity with age [219]. This recovery is associated with an increase in circulating levels of vitellogenin, a zinc binding glycolipoprotein that has been implicated in the regulation of honeybee immune integrity.

References

1. Rapp PR and Gallagher M (1996). Preserved neuron number in the hippocampus of aged rats with spatial learning deficits. *Proc Natl Acad Sci U S A* 93: 9926–9930.
2. Herndon LA, Schmeissner PJ, Dudaronek JM, Brown PA, Listner KM, Sakano Y, Paupard MC, Hall DH, and Driscoll M (2002). Stochastic and genetic factors influence tissue-specific decline in ageing *C. elegans*. *Nature* 419: 808–814.
3. Smith TD, Adams MM, Gallagher M, Morrison JH, and Rapp PR (2000). Circuit-specific alterations in hippocampal synaptophysin immunoreactivity predict spatial learning impairment in aged rats. *J Neurosci* 20: 6587–6593.
4. Pelvig DP, Pakkenberg H, Stark AK, and Pakkenberg B (2008). Neocortical glial cell numbers in human brains. *Neurobiol Aging* 29(11): 1754–1762.
5. Riddle DR, Sonntag WE, and Lichtenwalner RJ (2003). Microvascular plasticity in aging. *Ageing Res Rev* 2: 149–168.
6. Frisoni GB, Galluzzi S, Pantoni L, and Filippi M (2007). The effect of white matter lesions on cognition in the elderly – small but detectable. *Nat Clin Pract Neurol* 3: 620–627.
7. Carman MB (1997). The psychology of normal aging. *Psychiatr Clin North Am* 20: 15–24.
8. Raz N and Rodrigue KM (2006). Differential aging of the brain: patterns, cognitive correlates and modifiers. *Neurosci Biobehav Rev* 30: 730–748.
9. Golden TR, Hinerfeld DA, and Melov S (2002). Oxidative stress and aging: beyond correlation. *Aging Cell* 1: 117–123.
10. Harman D (2006). Free radical theory of aging: an update: increasing the functional life span. *Ann N Y Acad Sci* 1067: 10–21.
11. Cutler RG and Mattson MP (2006). The adversities of aging. *Ageing Res Rev* 5: 221–238.
12. Heilbronn LK and Ravussin E (2003). Calorie restriction and aging: review of the literature and implications for studies in humans. *Am J Clin Nutr* 78: 361–369.
13. Hunt ND, Hyun DH, Allard JS, Minor RK, Mattson MP, Ingram DK, and de Cabo R (2006). Bioenergetics of aging and calorie restriction. *Ageing Res Rev* 5: 125–143.
14. Brand MD, Buckingham JA, Esteves TC, Green K, Lambert AJ, Miwa S, Murphy MP, Pakay JL, Talbot DA, and Echtay KS (2004). Mitochondrial superoxide and aging: uncoupling-protein activity and superoxide production. *Biochem Soc Symp* 71: 203–213.
15. Keller JN, Kindy MS, Holtsberg FW, St Clair DK, Yen HC, Germeyer A, Steiner SM, Bruce-Keller AJ, Hutchins JB, and Mattson MP (1998). Mitochondrial manganese superoxide dismutase prevents neural apoptosis and reduces ischemic brain injury: suppression of peroxynitrite production, lipid peroxidation, and mitochondrial dysfunction. *J Neurosci* 18: 687–697.
16. Sohal RS and Orr WC (1992). Relationship between antioxidants, prooxidants, and the aging process. *Ann N Y Acad Sci* 663: 74–84.
17. Mattson MP (1998). Modification of ion homeostasis by lipid peroxidation: roles in neuronal degeneration and adaptive plasticity. *Trends Neurosci* 21: 53–57.
18. Boehning D and Snyder SH (2003). Novel neural modulators. *Annu Rev Neurosci* 26: 105–131.
19. Virag L, Szabo E, Gergely P, and Szabo C (2003). Peroxynitrite-induced cytotoxicity: mechanism and opportunities for intervention. *Toxicol Lett* 140–141: 113–124.
20. Hyun DH, Hernandez JO, Mattson MP, and de Cabo R (2006). The plasma membrane redox system in aging. *Ageing Res Rev* 5: 209–220.
21. Droge W, Kinscherf R, Hildebrandt W, and Schmitt T (2006). The deficit in low molecular weight thiols as a target for antiageing therapy. *Curr Drug Targets* 7: 1505–1512.

22. Hanukoglu I (2006). Antioxidant protective mechanisms against reactive oxygen species (ROS) generated by mitochondrial P450 systems in steroidogenic cells. *Drug Metab Rev* 38: 171–196.
23. Nystrom T (2005). Role of oxidative carbonylation in protein quality control and senescence. *Embo J* 24: 1311–1317.
24. Schopfer FJ, Baker PR, and Freeman BA (2003). NO-dependent protein nitration: a cell signaling event or an oxidative inflammatory response? *Trends Biochem Sci* 28: 646–654.
25. el-Remessy AB, Bartoli M, Platt DH, Fulton D, and Caldwell RB (2005). Oxidative stress inactivates VEGF survival signaling in retinal endothelial cells via PI 3-kinase tyrosine nitration. *J Cell Sci* 118: 243–252.
26. Zhan X, Du Y, Crabb JS, Gu X, Kern TS, and Crabb JW (2008). Targets of tyrosine nitration in diabetic rat retina. *Mol Cell Proteomics* 7(5): 864–874.
27. Blanc EM, Kelly JF, Mark RJ, Waeg G, and Mattson MP (1997). 4-Hydroxynonenal, an aldehydic product of lipid peroxidation, impairs signal transduction associated with muscarinic acetylcholine and metabotropic glutamate receptors: possible action on G alpha(q/11). *J Neurochem* 69: 570–580.
28. Mark RJ, Pang Z, Geddes JW, Uchida K, and Mattson MP (1997). Amyloid beta-peptide impairs glucose transport in hippocampal and cortical neurons: involvement of membrane lipid peroxidation. *J Neurosci* 17: 1046–1054.
29. Mark RJ, Lovell MA, Markesbery WR, Uchida K, and Mattson MP (1997). A role for 4-hydroxynonenal, an aldehydic product of lipid peroxidation, in disruption of ion homeostasis and neuronal death induced by amyloid beta-peptide. *J Neurochem* 68: 255–264.
30. Mattson MP, Fu W, Waeg G, and Uchida K (1997). 4-Hydroxynonenal, a product of lipid peroxidation, inhibits dephosphorylation of the microtubule-associated protein tau. *Neuroreport* 8: 2275–2281.
31. Perez M, Hernandez F, Gomez-Ramos A, Smith M, Perry G, and Avila J (2002). Formation of aberrant phosphotau fibrillar polymers in neural cultured cells. *Eur J Biochem* 269: 1484–1489.
32. Cutler RG, Pedersen WA, Camandola S, Rothstein JD, and Mattson MP (2002). Evidence that accumulation of ceramides and cholesterol esters mediates oxidative stress-induced death of motor neurons in amyotrophic lateral sclerosis. *Ann Neurol* 52: 448–457.
33. Kruman I, Bruce-Keller AJ, Bredesen D, Waeg G, and Mattson MP (1997). Evidence that 4-hydroxynonenal mediates oxidative stress-induced neuronal apoptosis. *J Neurosci* 17: 5089–5100.
34. Lovell MA, Xie C, and Markesbery WR (2001). Acrolein is increased in Alzheimer's disease brain and is toxic to primary hippocampal cultures. *Neurobiol Aging* 22: 187–194.
35. Lodovici M, Giovannelli L, Pitozzi V, Bigagli E, Bardini G, and Rotella CM (2008). Oxidative DNA damage and plasma antioxidant capacity in type 2 diabetic patients with good and poor glycaemic control. *Mutat Res* 638: 98–102.
36. Mercer J, Mahmoudi M, and Bennett M (2007). DNA damage, p53, apoptosis and vascular disease. *Mutat Res* 621: 75–86.
37. Rass U, Ahel I, and West SC (2007). Defective DNA repair and neurodegenerative disease. *Cell* 130: 991–1004.
38. Evans MD, Dizdaroglu M, and Cooke MS (2004). Oxidative DNA damage and disease: induction, repair and significance. *Mutat Res* 567: 1–61.
39. Bohr VA (2002). Repair of oxidative DNA damage in nuclear and mitochondrial DNA, and some changes with aging in mammalian cells. *Free Radic Biol Med* 32: 804–812.
40. Bohr VA, Ottersen OP, and Tonjum T (2007). Genome instability and DNA repair in brain ageing and neurological disease. *Neuroscience* 145: 1183–1186.
41. Lovell MA and Markesbery WR (2008). Oxidatively modified RNA in mild cognitive impairment. *Neurobiol Dis* 29: 169–175.
42. Lu T, Pan Y, Kao SY, Li C, Kohane I, Chan J, and Yankner BA (2004). Gene regulation and DNA damage in the ageing human brain. *Nature* 429: 883–891.

43. Mattson MP and Magnus T (2006). Ageing and neuronal vulnerability. *Nat Rev Neurosci* 7: 278–294.
44. Keller JN and Mattson MP (1998). Roles of lipid peroxidation in modulation of cellular signaling pathways, cell dysfunction, and death in the nervous system. *Rev Neurosci* 9: 105–116.
45. Keller JN, Pang Z, Geddes JW, Begley JG, Germeyer A, Waeg G, and Mattson MP (1997). Impairment of glucose and glutamate transport and induction of mitochondrial oxidative stress and dysfunction in synaptosomes by amyloid beta-peptide: role of the lipid peroxidation product 4-hydroxynonenal. *J Neurochem* 69: 273–284.
46. Opii WO, Joshi G, Head E, Milgram NW, Muggenburg BA, Klein JB, Pierce WM, Cotman CW, and Butterfield DA (2008). Proteomic identification of brain proteins in the canine model of human aging following a long-term treatment with antioxidants and a program of behavioral enrichment: relevance to Alzheimer's disease. *Neurobiol Aging* 29: 51–70.
47. Poon HF, Calabrese V, Calvani M, and Butterfield DA (2006). Proteomics analyses of specific protein oxidation and protein expression in aged rat brain and its modulation by L-acetylcarnitine: insights into the mechanisms of action of this proposed therapeutic agent for CNS disorders associated with oxidative stress. *Antioxid Redox Signal* 8: 381–394.
48. Poon HF, Castegna A, Farr SA, Thongboonkerd V, Lynn BC, Banks WA, Morley JE, Klein JB, and Butterfield DA (2004). Quantitative proteomics analysis of specific protein expression and oxidative modification in aged senescence-accelerated-prone 8 mice brain. *Neuroscience* 126: 915–926.
49. Soreghan BA, Lu BW, Thomas SN, Duff K, Rakhmatulin EA, Nikolskaya T, Chen T, and Yang AJ (2005). Using proteomics and network analysis to elucidate the consequences of synaptic protein oxidation in a PS1 + AbetaPP mouse model of Alzheimer's disease. *J Alzheimers Dis* 8: 227–241.
50. Mattson MP (2006). Neuronal life-and-death signaling, apoptosis, and neurodegenerative disorders. *Antioxid Redox Signal* 8: 1997–2006.
51. Schindowski K, Kratzsch T, Peters J, Steiner B, Leutner S, Touchet N, Maurer K, Czech C, Pradier L, Frolich L, Muller WE, and Eckert A (2003). Impact of aging: sporadic, and genetic risk factors on vulnerability to apoptosis in Alzheimer's disease. *Neuromolecular Med* 4: 161–178.
52. Bruce-Keller AJ, Li YJ, Lovell MA, Kraemer PJ, Gary DS, Brown RR, Markesbery WR, and Mattson MP (1998). 4-Hydroxynonenal, a product of lipid peroxidation, damages cholinergic neurons and impairs visuospatial memory in rats. *J Neuropathol Exp Neurol* 57: 257–267.
53. Onodera K, Omoi NO, Fukui K, Hayasaka T, Shinkai T, Suzuki S, Abe K, and Urano S (2003). Oxidative damage of rat cerebral cortex and hippocampus, and changes in antioxidative defense systems caused by hyperoxia. *Free Radic Res* 37: 367–372.
54. Omoi NO, Arai M, Saito M, Takatsu H, Shibata A, Fukuzawa K, Sato K, Abe K, Fukui K, and Urano S (2006). Influence of oxidative stress on fusion of pre-synaptic plasma membranes of the rat brain with phosphatidyl choline liposomes, and protective effect of vitamin E. *J Nutr Sci Vitaminol (Tokyo)* 52: 248–255.
55. Rowan MJ, Klyubin I, Wang Q, and Anwyl R (2004). Mechanisms of the inhibitory effects of amyloid beta-protein on synaptic plasticity. *Exp Gerontol* 39: 1661–1667.
56. Calon F, Lim GP, Yang F, Morihara T, Teter B, Ubeda O, Rostaing P, Triller A, Salem N, Jr., Ashe KH, Frautschy SA, and Cole GM (2004). Docosahexaenoic acid protects from dendritic pathology in an Alzheimer's disease mouse model. *Neuron* 43: 633–645.
57. Eckles-Smith K, Clayton D, Bickford P, and Browning MD (2000). Caloric restriction prevents age-related deficits in LTP and in NMDA receptor expression. *Brain Res Mol Brain Res* 78: 154–162.
58. Fontan-Lozano A, Saez-Cassanelli JL, Inda MC, de los Santos-Arteaga M, Sierra-Dominguez SA, Lopez-Lluch G, Delgado-Garcia JM, and Carrion AM (2007). Caloric

restriction increases learning consolidation and facilitates synaptic plasticity through mechanisms dependent on NR2B subunits of the NMDA receptor. *J Neurosci* 27: 10185–10195.
59. Halagappa VK, Guo Z, Pearson M, Matsuoka Y, Cutler RG, Laferla FM, and Mattson MP (2007). Intermittent fasting and caloric restriction ameliorate age-related behavioral deficits in the triple-transgenic mouse model of Alzheimer's disease. *Neurobiol Dis* 26: 212–220.
60. Martin B, Pearson M, Kebejian L, Golden E, Keselman A, Bender M, Carlson O, Egan J, Ladenheim B, Cadet JL, Becker KG, Wood W, Duffy K, Vinayakumar P, Maudsley S, and Mattson MP (2007). Sex-dependent metabolic, neuroendocrine, and cognitive responses to dietary energy restriction and excess. *Endocrinology* 148: 4318–4333.
61. Swerdlow RH (2007). Treating neurodegeneration by modifying mitochondria: potential solutions to a "complex" problem. *Antioxid Redox Signal* 9: 1591–1603.
62. Toescu EC (2005). Normal brain ageing: models and mechanisms. *Philos Trans R Soc Lond B Biol Sci* 360: 2347–2354.
63. Patel MS (1977). Age-dependent changes in the oxidative metabolism in rat brain. *J Gerontol* 32: 643–646.
64. Fattoretti P, Bertoni-Freddari C, Caselli U, Paoloni R, and Meier-Ruge W (1998). Impaired succinic dehydrogenase activity of rat Purkinje cell mitochondria during aging. *Mech Ageing Dev* 101: 175–182.
65. Sanz A, Caro P, Ibanez J, Gomez J, Gredilla R, and Barja G (2005). Dietary restriction at old age lowers mitochondrial oxygen radical production and leak at complex I and oxidative DNA damage in rat brain. *J Bioenerg Biomembr* 37: 83–90.
66. Gabbita SP, Butterfield DA, Hensley K, Shaw W, and Carney JM (1997). Aging and caloric restriction affect mitochondrial respiration and lipid membrane status: an electron paramagnetic resonance investigation. *Free Radic Biol Med* 23: 191–201.
67. Curtis C, Landis GN, Folk D, Wehr NB, Hoe N, Waskar M, Abdueva D, Skvortsov D, Ford D, Luu A, Badrinath A, Levine RL, Bradley TJ, Tavare S, and Tower J (2007). Transcriptional profiling of MnSOD-mediated lifespan extension in *Drosophila* reveals a species-general network of aging and metabolic genes. *Genome Biol* 8: R262.
68. Smeyne RJ and Jackson-Lewis V (2005). The MPTP model of Parkinson's disease. *Brain Res Mol Brain Res* 134: 57–66.
69. Duan W and Mattson MP (1999). Dietary restriction and 2-deoxyglucose administration improve behavioral outcome and reduce degeneration of dopaminergic neurons in models of Parkinson's disease. *J Neurosci Res* 57: 195–206.
70. Maswood N, Young J, Tilmont E, Zhang Z, Gash DM, Gerhardt GA, Grondin R, Roth GS, Mattison J, Lane MA, Carson RE, Cohen RM, Mouton PR, Quigley C, Mattson MP, and Ingram DK (2004). Caloric restriction increases neurotrophic factor levels and attenuates neurochemical and behavioral deficits in a primate model of Parkinson's disease. *Proc Natl Acad Sci U S A* 101: 18171–18176.
71. Brouillet E, Jacquard C, Bizat N, and Blum D (2005). 3-Nitropropionic acid: a mitochondrial toxin to uncover physiopathological mechanisms underlying striatal degeneration in Huntington's disease. *J Neurochem* 95: 1521–1540.
72. Bruce-Keller AJ, Umberger G, McFall R, and Mattson MP (1999). Food restriction reduces brain damage and improves behavioral outcome following excitotoxic and metabolic insults. *Ann Neurol* 45: 8–15.
73. Melov S, Schneider JA, Day BJ, Hinerfeld D, Coskun P, Mirra SS, Crapo JD, and Wallace DC (1998). A novel neurological phenotype in mice lacking mitochondrial manganese superoxide dismutase. *Nat Genet* 18: 159–163.
74. Hyun DH, Emerson SS, Jo DG, Mattson MP, and de Cabo R (2006). Calorie restriction upregulates the plasma membrane redox system in brain cells and suppresses oxidative stress during aging. *Proc Natl Acad Sci U S A* 103: 19908–19912.
75. Hoyer S (1985). The effect of age on glucose and energy metabolism in brain cortex of rats. *Arch Gerontol Geriatr* 4: 193–203.

76. Vernace VA, Schmidt-Glenewinkel T, and Figueiredo-Pereira ME (2007). Aging and regulated protein degradation: who has the UPPer hand? *Aging Cell* 6: 599–606.
77. Fan XM, Wong BC, Wang WP, Zhou XM, Cho CH, Yuen ST, Leung SY, Lin MC, Kung HF, and Lam SK (2001). Inhibition of proteasome function induced apoptosis in gastric cancer. *Int J Cancer* 93: 481–488.
78. Vernace VA, Arnaud L, Schmidt-Glenewinkel T, and Figueiredo-Pereira ME (2007). Aging perturbs 26S proteasome assembly in *Drosophila melanogaster*. *Faseb J* 21: 2672–2682.
79. Terman A, Dalen H, and Brunk UT (1999). Ceroid/lipofuscin-loaded human fibroblasts show decreased survival time and diminished autophagocytosis during amino acid starvation. *Exp Gerontol* 34: 943–957.
80. Cuervo AM, Bergamini E, Brunk UT, Droge W, Ffrench M, and Terman A (2005). Autophagy and aging: the importance of maintaining "clean" cells. *Autophagy* 1: 131–140.
81. Melendez A, Talloczy Z, Seaman M, Eskelinen EL, Hall DH, and Levine B (2003). Autophagy genes are essential for dauer development and life-span extension in *C. elegans*. *Science* 301: 1387–1391.
82. Florez-McClure ML, Hohsfield LA, Fonte G, Bealor MT, and Link CD (2007). Decreased insulin-receptor signaling promotes the autophagic degradation of beta-amyloid peptide in *C. elegans*. *Autophagy* 3: 569–580.
83. Jia K and Levine B (2007). Autophagy is required for dietary restriction-mediated life span extension in *C. elegans*. *Autophagy* 3: 597–599.
84. Morck C and Pilon M (2006). *C. elegans* feeding defective mutants have shorter body lengths and increased autophagy. *BMC Dev Biol* 6: 39.
85. Hansen M, Taubert S, Crawford D, Libina N, Lee SJ, and Kenyon C (2007). Lifespan extension by conditions that inhibit translation in *Caenorhabditis elegans*. *Aging Cell* 6: 95–110.
86. Pan KZ, Palter JE, Rogers AN, Olsen A, Chen D, Lithgow GJ, and Kapahi P (2007). Inhibition of mRNA translation extends lifespan in *Caenorhabditis elegans*. *Aging Cell* 6: 111–119.
87. Hipkiss AR (2007). On why decreasing protein synthesis can increase lifespan. *Mech Ageing Dev* 128: 412–414.
88. Chapouton P, Jagasia R, and Bally-Cuif L (2007). Adult neurogenesis in non-mammalian vertebrates. *Bioessays* 29: 745–757.
89. Gould E (2007). How widespread is adult neurogenesis in mammals? *Nat Rev Neurosci* 8: 481–488.
90. Carlen M, Cassidy RM, Brismar H, Smith GA, Enquist LW, and Frisen J (2002). Functional integration of adult-born neurons. *Curr Biol* 12: 606–608.
91. Ge S, Goh EL, Sailor KA, Kitabatake Y, Ming GL, and Song H (2006). GABA regulates synaptic integration of newly generated neurons in the adult brain. *Nature* 439: 589–593.
92. van Praag H, Schinder AF, Christie BR, Toni N, Palmer TD, and Gage FH (2002). Functional neurogenesis in the adult hippocampus. *Nature* 415: 1030–1034.
93. Klempin F and Kempermann G (2007). Adult hippocampal neurogenesis and aging. *Eur Arch Psychiatry Clin Neurosci* 257: 271–280.
94. Lee J, Auyeung WW, and Mattson MP (2003). Interactive effects of excitotoxic injury and dietary restriction on microgliosis and neurogenesis in the hippocampus of adult mice. *Neuromolecular Med* 4: 179–196.
95. van Praag H, Shubert T, Zhao C, and Gage FH (2005). Exercise enhances learning and hippocampal neurogenesis in aged mice. *J Neurosci* 25: 8680–8685.
96. Kempermann G, Gast D, and Gage FH (2002). Neuroplasticity in old age: sustained fivefold induction of hippocampal neurogenesis by long-term environmental enrichment. *Ann Neurol* 52: 135–143.
97. Haughey NJ, Nath A, Chan SL, Borchard AC, Rao MS, and Mattson MP (2002). Disruption of neurogenesis by amyloid beta-peptide, and perturbed neural progenitor cell homeostasis, in models of Alzheimer's disease. *J Neurochem* 83: 1509–1524.

98. Pignataro G, Studer FE, Wilz A, Simon RP, and Boison D (2007). Neuroprotection in ischemic mouse brain induced by stem cell-derived brain implants. *J Cereb Blood Flow Metab* 27: 919–927.
99. Goya RL, Kuan WL, and Barker RA (2007). The future of cell therapies in the treatment of Parkinson's disease. *Expert Opin Biol Ther* 7: 1487–1498.
100. Yamasaki TR, Blurton-Jones M, Morrissette DA, Kitazawa M, Oddo S, and LaFerla FM (2007). Neural stem cells improve memory in an inducible mouse model of neuronal loss. *J Neurosci* 27: 11925–11933.
101. Mattson MP (2004). Pathways towards and away from Alzheimer's disease. *Nature* 430: 631–639.
102. Ramer MS, Priestley JV, and McMahon SB (2000). Functional regeneration of sensory axons into the adult spinal cord. *Nature* 403: 312–316.
103. Tanaka K and Webster HD (1991). Myelinated fiber regeneration after crush injury is retarded in sciatic nerves of aging mice. *J Comp Neurol* 308: 180–187.
104. Vaughan DW (1992). Effects of advancing age on peripheral nerve regeneration. *J Comp Neurol* 323: 219–237.
105. Verdu E, Ceballos D, Vilches JJ, and Navarro X (2000). Influence of aging on peripheral nerve function and regeneration. *J Peripher Nerv Syst* 5: 191–208.
106. Rawson NE (2006). Olfactory loss in aging. *Sci Aging Knowledge Environ* 2006: pe6.
107. Wallhagen MI, Pettengill E, and Whiteside M (2006). Sensory impairment in older adults: Part 1: Hearing loss. *Am J Nurs* 106: 40–48, quiz 48–49.
108. Whiteside MM, Wallhagen MI, and Pettengill E (2006). Sensory impairment in older adults: Part 2: Vision loss. *Am J Nurs* 106: 52–61, quiz 61–52.
109. Liu XZ and Yan D (2007). Ageing and hearing loss. *J Pathol* 211: 188–197.
110. Johnson KR and Zheng QY (2002). Ahl2, a second locus affecting age-related hearing loss in mice. *Genomics* 80: 461–464.
111. Mashimo T, Erven AE, Spiden SL, Guenet JL, and Steel KP (2006). Two quantitative trait loci affecting progressive hearing loss in 101/H mice. *Mamm Genome* 17: 841–850.
112. Zheng QY, Yan D, Ouyang XM, Du LL, Yu H, Chang B, Johnson KR, and Liu XZ (2005). Digenic inheritance of deafness caused by mutations in genes encoding cadherin 23 and protocadherin 15 in mice and humans. *Hum Mol Genet* 14: 103–111.
113. Street VA, McKee-Johnson JW, Fonseca RC, Tempel BL, and Noben-Trauth K (1998). Mutations in a plasma membrane Ca^{2+}-ATPase gene cause deafness in deafwaddler mice. *Nat Genet* 19: 390–394.
114. Burke SN and Barnes CA (2006). Neural plasticity in the ageing brain. *Nat Rev Neurosci* 7: 30–40.
115. Tully T, Bourtchouladze R, Scott R, and Tallman J (2003). Targeting the CREB pathway for memory enhancers. *Nat Rev Drug Discov* 2: 267–277.
116. Sekuler R, McLaughlin C, Kahana MJ, Wingfield A, and Yotsumoto Y (2006). Short-term visual recognition and temporal order memory are both well-preserved in aging. *Psychol Aging* 21: 632–637.
117. Wingfield A and Kahana MJ (2002). The dynamics of memory retrieval in older adulthood. *Can J Exp Psychol* 56: 187–199.
118. McKinney M and Jacksonville MC (2005). Brain cholinergic vulnerability: relevance to behavior and disease. *Biochem Pharmacol* 70: 1115–1124.
119. Murchison D and Griffith WH (2007). Calcium buffering systems and calcium signaling in aged rat basal forebrain neurons. *Aging Cell* 6: 297–305.
120. Tamura T, Chiang AS, Ito N, Liu HP, Horiuchi J, Tully T, and Saitoe M (2003). Aging specifically impairs amnesiac-dependent memory in *Drosophila*. *Neuron* 40: 1003–1011.
121. Yamazaki D, Horiuchi J, Nakagami Y, Nagano S, Tamura T, and Saitoe M (2007). The *Drosophila* DCO mutation suppresses age-related memory impairment without affecting lifespan. *Nat Neurosci* 10: 478–484.

122. Morgante L, Salemi G, Meneghini F, Di Rosa AE, Epifanio A, Grigoletto F, Ragonese P, Patti F, Reggio A, Di Perri R, and Savettieri G (2000). Parkinson disease survival: a population-based study. *Arch Neurol* 57: 507–512.
123. Guarente L (2005). Calorie restriction and SIR2 genes – towards a mechanism. *Mech Ageing Dev* 126: 923–928.
124. Kenyon C (2005). The plasticity of aging: insights from long-lived mutants. *Cell* 120: 449–460.
125. Partridge L, Piper MD, and Mair W (2005). Dietary restriction in *Drosophila*. *Mech Ageing Dev* 126: 938–950.
126. Sinclair DA (2005). Toward a unified theory of caloric restriction and longevity regulation. *Mech Ageing Dev* 126: 987–1002.
127. Apfeld J and Kenyon C (1998). Cell nonautonomy of *C. elegans daf-2* function in the regulation of diapause and life span. *Cell* 95: 199–210.
128. Wolkow CA, Kimura KD, Lee MS, and Ruvkun G (2000). Regulation of *C. elegans* life-span by insulinlike signaling in the nervous system. *Science* 290: 147–150.
129. Kenyon C, Chang J, Gensch E, Rudner A, and Tabtiang R (1993). A *C. elegans* mutant that lives twice as long as wild type. *Nature* 366: 461–464.
130. Kimura KD, Tissenbaum HA, Liu Y, and Ruvkun G (1997). *daf-2*, an insulin receptor-like gene that regulates longevity and diapause in *Caenorhabditis elegans*. *Science* 277: 942–946.
131. Brogiolo W, Stocker H, Ikeya T, Rintelen F, Fernandez R, and Hafen E (2001). An evolutionarily conserved function of the *Drosophila* insulin receptor and insulin-like peptides in growth control. *Curr Biol* 11: 213–221.
132. Rulifson EJ, Kim SK, and Nusse R (2002). Ablation of insulin-producing neurons in flies: growth and diabetic phenotypes. *Science* 296: 1118–1120.
133. Broughton SJ, Piper MD, Ikeya T, Bass TM, Jacobson J, Driege Y, Martinez P, Hafen E, Withers DJ, Leevers SJ, and Partridge L (2005). Longer lifespan, altered metabolism, and stress resistance in *Drosophila* from ablation of cells making insulin-like ligands. *Proc Natl Acad Sci U S A* 102: 3105–3110.
134. Bauer JH, Poon PC, Glatt-Deeley H, Abrams JM, and Helfand SL (2005). Neuronal expression of p53 dominant-negative proteins in adult *Drosophila melanogaster* extends life span. *Curr Biol* 15: 2063–2068.
135. Bauer JH, Chang C, Morris SN, Hozier S, Andersen S, Waitzman JS, and Helfand SL (2007). Expression of dominant-negative Dmp53 in the adult fly brain inhibits insulin signaling. *Proc Natl Acad Sci U S A* 104: 13355–13360.
136. Wang MC, Bohmann D, and Jasper H (2003). JNK signaling confers tolerance to oxidative stress and extends lifespan in *Drosophila*. *Dev Cell* 5: 811–816.
137. Wang MC, Bohmann D, and Jasper H (2005). JNK extends life span and limits growth by antagonizing cellular and organism-wide responses to insulin signaling. *Cell* 121: 115–125.
138. Alcedo J and Kenyon C (2004). Regulation of *C. elegans* longevity by specific gustatory and olfactory neurons. *Neuron* 41: 45–55.
139. Libert S, Zwiener J, Chu X, Vanvoorhies W, Roman G, and Pletcher SD (2007). Regulation of *Drosophila* life span by olfaction and food-derived odors. *Science* 315: 1133–1137.
140. Evason K, Huang C, Yamben I, Covey DF, and Kornfeld K (2005). Anticonvulsant medications extend worm life-span. *Science* 307: 258–262.
141. Taguchi A, Wartschow LM, and White MF (2007). Brain IRS2 signaling coordinates life span and nutrient homeostasis. *Science* 317: 369–372.
142. Zheng J, Edelman SW, Tharmarajah G, Walker DW, Pletcher SD, and Seroude L (2005). Differential patterns of apoptosis in response to aging in *Drosophila*. *Proc Natl Acad Sci U S A* 102: 12083–12088.
143. Yaari R and Corey-Bloom J (2007). Alzheimer's disease. *Semin Neurol* 27: 32–41.
144. Petersen RC (2007). Mild cognitive impairment: current research and clinical implications. *Semin Neurol* 27: 22–31.

145. Rogaeva E (2002). The solved and unsolved mysteries of the genetics of early-onset Alzheimer's disease. *Neuromolecular Med* 2: 1–10.
146. Guo Q, Fu W, Sopher BL, Miller MW, Ware CB, Martin GM, and Mattson MP (1999). Increased vulnerability of hippocampal neurons to excitotoxic necrosis in presenilin-1 mutant knock-in mice. *Nat Med* 5: 101–106.
147. Mayeux R (2003). Epidemiology of neurodegeneration. *Annu Rev Neurosci* 26: 81–104.
148. Larson EB, Wang L, Bowen JD, McCormick WC, Teri L, Crane P, and Kukull W (2006). Exercise is associated with reduced risk for incident dementia among persons 65 years of age and older. *Ann Intern Med* 144: 73–81.
149. Adlard PA, Perreau VM, Pop V, and Cotman CW (2005). Voluntary exercise decreases amyloid load in a transgenic model of Alzheimer's disease. *J Neurosci* 25: 4217–4221.
150. Scarmeas N and Stern Y (2003). Cognitive reserve and lifestyle. *J Clin Exp Neuropsychol* 25: 625–633.
151. Luchsinger JA, Tang MX, Shea S, and Mayeux R (2002). Caloric intake and the risk of Alzheimer disease. *Arch Neurol* 59: 1258–1263.
152. Patel NV, Gordon MN, Connor KE, Good RA, Engelman RW, Mason J, Morgan DG, Morgan TE, and Finch CE (2005). Caloric restriction attenuates Abeta-deposition in Alzheimer transgenic models. *Neurobiol Aging* 26: 995–1000.
153. Wang J, Ho L, Qin W, Rocher AB, Seror I, Humala N, Maniar K, Dolios G, Wang R, Hof PR, and Pasinetti GM (2005). Caloric restriction attenuates beta-amyloid neuropathology in a mouse model of Alzheimer's disease. *Faseb J* 19: 659–661.
154. Mattson MP and Cheng A (2006). Neurohormetic phytochemicals: low-dose toxins that induce adaptive neuronal stress responses. *Trends Neurosci* 29: 632–639.
155. Cole GM, Lim GP, Yang F, Teter B, Begum A, Ma Q, Harris-White ME, and Frautschy SA (2005). Prevention of Alzheimer's disease: Omega-3 fatty acid and phenolic anti-oxidant interventions. *Neurobiol Aging* 26(Suppl 1): 133–136.
156. Cooper JL (2003). Dietary lipids in the aetiology of Alzheimer's disease: implications for therapy. *Drugs Aging* 20: 399–418.
157. Friedland RP and Armon C (2007). Tales of Pacific tangles: Cycad exposure and Guamanian neurodegenerative diseases. *Neurology* 68: 1759–1761.
158. Cummings BJ, Head E, Ruehl W, Milgram NW, and Cotman CW (1996). The canine as an animal model of human aging and dementia. *Neurobiol Aging* 17: 259–268.
159. Head E, Callahan H, Muggenburg BA, Cotman CW, and Milgram NW (1998). Visual-discrimination learning ability and beta-amyloid accumulation in the dog. *Neurobiol Aging* 19: 415–425.
160. Price DL, Martin LJ, Sisodia SS, Wagster MV, Koo EH, Walker LC, Koliatsos VE, and Cork LC (1991). Aged non-human primates: an animal model of age-associated neurodegenerative disease. *Brain Pathol* 1: 287–296.
161. Pugliese M, Gangitano C, Ceccariglia S, Carrasco JL, Del Fa A, Rodriguez MJ, Michetti F, Mascort J, and Mahy N (2007). Canine cognitive dysfunction and the cerebellum: acetylcholinesterase reduction, neuronal and glial changes. *Brain Res* 1139: 85–94.
162. German DC and Eisch AJ (2004). Mouse models of Alzheimer's disease: insight into treatment. *Rev Neurosci* 15: 353–369.
163. Oddo S, Caccamo A, Shepherd JD, Murphy MP, Golde TE, Kayed R, Metherate R, Mattson MP, Akbari Y, and LaFerla FM (2003). Triple-transgenic model of Alzheimer's disease with plaques and tangles: intracellular Abeta and synaptic dysfunction. *Neuron* 39: 409–421.
164. Fonte V, Kipp DR, Yerg J, 3rd, Merin D, Forrestal M, Wagner E, Roberts CM, and Link CD (2008). Suppression of in vivo beta-amyloid peptide toxicity by overexpression of the HSP-16.2 small chaperone protein. *J Biol Chem* 283: 784–791.
165. Link CD (2005). Invertebrate models of Alzheimer's disease. *Genes Brain Behav* 4: 147–156.

166. Chade AR, Kasten M, and Tanner CM (2007). Epidemiology of Parkinson's disease. In *Parkinson's Disease: Genetics and Pathogenesis* (Dawson TM (ed)). New York: Informa Healthcare.
167. Dickson D (2007). Neuropathology and staging of Parkinson's disease. In *Parkinson's Disease: Genetics and Pathogenesis* (Dawson TM (ed)). New York: Informa Healthcare.
168. Shimura H, Schlossmacher MG, Hattori N, Frosch MP, Trockenbacher A, Schneider R, Mizuno Y, Kosik KS, and Selkoe DJ (2001). Ubiquitination of a new form of alpha-synuclein by parkin from human brain: implications for Parkinson's disease. *Science* 293: 263–269.
169. Spillantini MG, Schmidt ML, Lee VM, Trojanowski JQ, Jakes R, and Goedert M (1997). Alpha-synuclein in Lewy bodies. *Nature* 388: 839–840.
170. Wakabayashi K, Matsumoto K, Takayama K, Yoshimoto M, and Takahashi H (1997). NACP, a presynaptic protein, immunoreactivity in Lewy bodies in Parkinson's disease. *Neurosci Lett* 239: 45–48.
171. Elbaz A, Grigoletto F, Baldereschi M, Breteler MM, Manubens-Bertran JM, Lopez-Pousa S, Dartigues JF, Alperovitch A, Tzourio C, and Rocca WA (1999). Familial aggregation of Parkinson's disease: a population-based case-control study in Europe. EUROPARKINSON Study Group. *Neurology* 52: 1876–1882.
172. Chade AR, Kasten M, and Tanner CM (2006). Nongenetic causes of Parkinson's disease. *J Neural Transm Suppl* 70: 147–151.
173. Powers KM, Kay DM, Factor SA, Zabetian CP, Higgins DS, Samii A, Nutt JG, Griffith A, Leis B, Roberts JW, Martinez ED, Montimurro JS, Checkoway H, and Payami H (2008). Combined effects of smoking, coffee, and NSAIDs on Parkinson's disease risk. *Mov Disord* 23: 88–95.
174. Tan EK, Chua E, Fook-Chong SM, Teo YY, Yuen Y, Tan L, and Zhao Y (2007). Association between caffeine intake and risk of Parkinson's disease among fast and slow metabolizers. *Pharmacogenet Genomics* 17: 1001–1005.
175. Thomas B and Beal MF (2007). Parkinson's disease. *Hum Mol Genet* 16(Spec No. 2): R183–R194.
176. Polymeropoulos MH, Lavedan C, Leroy E, Ide SE, Dehejia A, Dutra A, Pike B, Root H, Rubenstein J, Boyer R, Stenroos ES, Chandrasekharappa S, Athanassiadou A, Papapetropoulos T, Johnson WG, Lazzarini AM, Duvoisin RC, Di Iorio G, Golbe LI, and Nussbaum RL (1997). Mutation in the alpha-synuclein gene identified in families with Parkinson's disease. *Science* 276: 2045–2047.
177. da Costa CA (2007). DJ-1: a new comer in Parkinson's disease pathology. *Curr Mol Med* 7: 650–657.
178. Dodson MW and Guo M (2007). Pink1, Parkin, DJ-1 and mitochondrial dysfunction in Parkinson's disease. *Curr Opin Neurobiol* 17: 331–337.
179. Hattori N and Sato S (2007). Animal models of Parkinson's disease: similarities and differences between the disease and models. *Neuropathology* 27: 479–483.
180. Schmidt E, Seifert M, and Baumeister R (2007). *Caenorhabditis elegans* as a model system for Parkinson's disease. *Neurodegener Dis* 4: 199–217.
181. Lee MK (2007). Transgenic models of alpha-synucleinopathy. In *Parkinson's Disease: Genetics and Pathogenesis* (Dawson TM (ed)). New York: Informa Healthcare.
182. Rubinsztein DC (2002). Lessons from animal models of Huntington's disease. *Trends Genet* 18: 202–209.
183. Brignull HR, Morley JF, Garcia SM, and Morimoto RI (2006). Modeling polyglutamine pathogenesis in *C. elegans*. *Methods Enzymol* 412: 256–282.
184. Walling AD (1999). Amyotrophic lateral sclerosis: Lou Gehrig's disease. *Am Fam Physician* 59: 1489–1496.
185. Mitsumoto H, Hanson MR, and Chad DA (1988). Amyotrophic lateral sclerosis. Recent advances in pathogenesis and therapeutic trials. *Arch Neurol* 45: 189–202.
186. Bruijn LI, Miller TM, and Cleveland DW (2004). Unraveling the mechanisms involved in motor neuron degeneration in ALS. *Annu Rev Neurosci* 27: 723–749.

187. Pedersen WA, Fu W, Keller JN, Markesbery WR, Appel S, Smith RG, Kasarskis E, and Mattson MP (1998). Protein modification by the lipid peroxidation product 4-hydroxynonenal in the spinal cords of amyotrophic lateral sclerosis patients. *Ann Neurol* 44: 819–824.
188. Appel SH, Beers D, Siklos L, Engelhardt JI, and Mosier DR (2001). Calcium: the Darth Vader of ALS. *Amyotroph Lateral Scler Other Motor Neuron Disord* 2(Suppl 1): S47–S54.
189. Cleveland DW and Rothstein JD (2001). From Charcot to Lou Gehrig: deciphering selective motor neuron death in ALS. *Nat Rev Neurosci* 2: 806–819.
190. Pasinelli P and Brown RH (2006). Molecular biology of amyotrophic lateral sclerosis: insights from genetics. *Nat Rev Neurosci* 7: 710–723.
191. Martin LJ (2007). Transgenic mice with human mutant genes causing Parkinson's disease and amyotrophic lateral sclerosis provide common insight into mechanisms of motor neuron selective vulnerability to degeneration. *Rev Neurosci* 18: 115–136.
192. Yim MB, Kang JH, Yim HS, Kwak HS, Chock PB, and Stadtman ER (1996). A gain-of-function of an amyotrophic lateral sclerosis-associated Cu,Zn-superoxide dismutase mutant: An enhancement of free radical formation due to a decrease in Km for hydrogen peroxide. *Proc Natl Acad Sci U S A* 93: 5709–5714.
193. Pedersen WA and Mattson MP (1999). No benefit of dietary restriction on disease onset or progression in amyotrophic lateral sclerosis Cu/Zn-superoxide dismutase mutant mice. *Brain Res* 833: 117–120.
194. Mattson MP, Cutler RG, and Camandola S (2007). Energy intake and amyotrophic lateral sclerosis. *Neuromolecular Med* 9: 17–20.
195. Plato CC, Galasko D, Garruto RM, Plato M, Gamst A, Craig UK, Torres JM, and Wiederholt W (2002). ALS and PDC of Guam: forty-year follow-up. *Neurology* 58: 765–773.
196. Cox PA and Sacks OW (2002). Cycad neurotoxins, consumption of flying foxes, and ALS-PDC disease in Guam. *Neurology* 58: 956–959.
197. Logroscino G, Marder K, Cote L, Tang MX, Shea S, and Mayeux R (1996). Dietary lipids and antioxidants in Parkinson's disease: a population-based, case-control study. *Ann Neurol* 39: 89–94.
198. Pasquier F, Boulogne A, Leys D, and Fontaine P (2006). Diabetes mellitus and dementia. *Diabetes Metab* 32: 403–414.
199. Beauquis J, Roig P, Homo-Delarche F, De Nicola A, and Saravia F (2006). Reduced hippocampal neurogenesis and number of hilar neurones in streptozotocin-induced diabetic mice: reversion by antidepressant treatment. *Eur J Neurosci* 23: 1539–1546.
200. Stranahan AM, Arumugam TV, Cutler RG, Lee K, Egan JM, and Mattson MP (2008). Diabetes impairs hippocampal function through glucocorticoid-mediated effects on new and mature neurons. *Nat Neurosci* 11(3): 309–317.
201. Duan W, Guo Z, Jiang H, Ware M, Li XJ, and Mattson MP (2003). Dietary restriction normalizes glucose metabolism and BDNF levels, slows disease progression, and increases survival in huntingtin mutant mice. *Proc Natl Acad Sci U S A* 100: 2911–2916.
202. Yu ZF and Mattson MP (1999). Dietary restriction and 2-deoxyglucose administration reduce focal ischemic brain damage and improve behavioral outcome: evidence for a preconditioning mechanism. *J Neurosci Res* 57: 830–839.
203. Mattson MP (2008). Dietary factors, hormesis and health. *Ageing Res Rev* 7: 43–48.
204. Bishop NA and Guarente L (2007). Two neurons mediate diet-restriction-induced longevity in *C. elegans*. *Nature* 447: 545–549.
205. Chen H, Zhang SM, Schwarzschild MA, Hernan MA, and Ascherio A (2005). Physical activity and the risk of Parkinson disease. *Neurology* 64: 664–669.
206. Cotman CW and Berchtold NC (2002). Exercise: a behavioral intervention to enhance brain health and plasticity. *Trends Neurosci* 25: 295–301.
207. Gareri P, De Fazio P, and De Sarro G (2002). Neuropharmacology of depression in aging and age-related diseases. *Ageing Res Rev* 1: 113–134.

208. McEwen BS (2007). Physiology and neurobiology of stress and adaptation: central role of the brain. *Physiol Rev* 87: 873–904.
209. Murakami S (2007). *Caenorhabditis elegans* as a model system to study aging of learning and memory. *Mol Neurobiol* 35: 85–94.
210. Vermeulen CJ and Loeschcke V (2007). Longevity and the stress response in *Drosophila*. *Exp Gerontol* 42: 153–159.
211. Kaplan RM and Kronick RG (2006). Marital status and longevity in the United States population. *J Epidemiol Community Health* 60: 760–765.
212. Rodriguez-Laso A, Zunzunegui MV, and Otero A (2007). The effect of social relationships on survival in elderly residents of a Southern European community: a cohort study. *BMC Geriatr* 7: 19.
213. Bennett DA, Schneider JA, Tang Y, Arnold SE, and Wilson RS (2006). The effect of social networks on the relation between Alzheimer's disease pathology and level of cognitive function in old people: a longitudinal cohort study. *Lancet Neurol* 5: 406–412.
214. Bianchi M, Fone KF, Azmi N, Heidbreder CA, Hagan JJ, and Marsden CA (2006). Isolation rearing induces recognition memory deficits accompanied by cytoskeletal alterations in rat hippocampus. *Eur J Neurosci* 24: 2894–2902.
215. Stranahan AM, Khalil D, and Gould E (2006). Social isolation delays the positive effects of running on adult neurogenesis. *Nat Neurosci* 9: 526–533.
216. Dong H, Goico B, Martin M, Csernansky CA, Bertchume A, and Csernansky JG (2004). Modulation of hippocampal cell proliferation, memory, and amyloid plaque deposition in APPsw (Tg2576) mutant mice by isolation stress. *Neuroscience* 127: 601–609.
217. Brenes JC, Rodriguez O, and Fornaguera J (2008). Differential effect of environment enrichment and social isolation on depressive-like behavior, spontaneous activity and serotonin and norepinephrine concentration in prefrontal cortex and ventral striatum. *Pharmacol Biochem Behav* 89: 85–93.
218. Scaccianoce S, Del Bianco P, Paolone G, Caprioli D, Modafferi AM, Nencini P, and Badiani A (2006). Social isolation selectively reduces hippocampal brain-derived neurotrophic factor without altering plasma corticosterone. *Behav Brain Res* 168: 323–325.
219. Amdam GV, Aase AL, Seehuus SC, Kim Fondrk M, Norberg K, and Hartfelder K (2005). Social reversal of immunosenescence in honey bee workers. *Exp Gerontol* 40: 939–947.
220. Wolkow CA (2006). Identifying factors that promote functional aging in *Caenorhabditis elegans*. *Exp Gerontol* 41: 1001–1006.
221. Link CD (2006). *C. elegans* models of age-associated neurodegenerative diseases: lessons from transgenic worm models of Alzheimer's disease. *Exp Gerontol* 41: 1007–1013.
222. Saitoe M, Horiuchi J, Tamura T, and Ito N (2005). *Drosophila* as a novel animal model for studying the genetics of age-related memory impairment. *Rev Neurosci* 16: 137–149.
223. Bilen J and Bonini NM (2005). *Drosophila* as a model for human neurodegenerative disease. *Annu Rev Genet* 39: 153–171.
224. Sang TK and Jackson GR (2005). *Drosophila* models of neurodegenerative disease. *NeuroRx* 2: 438–446.
225. Erickson CA and Barnes CA (2003). The neurobiology of memory changes in normal aging. *Exp Gerontol* 38: 61–69.
226. Kelly KM, Nadon NL, Morrison JH, Thibault O, Barnes CA, and Blalock EM (2006). The neurobiology of aging. *Epilepsy Res* 68(Suppl 1): S5–S20.
227. Melrose HL, Lincoln SJ, Tyndall GM, and Farrer MJ (2006). Parkinson's disease: a rethink of rodent models. *Exp Brain Res* 173: 196–204.
228. Boillee S, Peschanski M, and Junier MP (2003). The wobbler mouse: a neurodegeneration jigsaw puzzle. *Mol Neurobiol* 28: 65–106.
229. Hensley K, Mhatre M, Mou S, Pye QN, Stewart C, West M, and Williamson KS (2006). On the relation of oxidative stress to neuroinflammation: lessons learned from the G93A-SOD1 mouse model of amyotrophic lateral sclerosis. *Antioxid Redox Signal* 8: 2075–2087.

230. Kuhn A, Goldstein DR, Hodges A, Strand AD, Sengstag T, Kooperberg C, Becanovic K, Pouladi MA, Sathasivam K, Cha JH, Hannan AJ, Hayden MR, Leavitt BR, Dunnett SB, Ferrante RJ, Albin R, Shelbourne P, Delorenzi M, Augood SJ, Faull RL, Olson JM, Bates GP, Jones L, and Luthi-Carter R (2007). Mutant huntingtin's effects on striatal gene expression in mice recapitulate changes observed in human Huntington's disease brain and do not differ with mutant huntingtin length or wild-type huntingtin dosage. *Hum Mol Genet* 16: 1845–1861.
231. Menalled LB and Chesselet MF (2002). Mouse models of Huntington's disease. *Trends Pharmacol Sci* 23: 32–39.
232. Strand AD, Baquet ZC, Aragaki AK, Holmans P, Yang L, Cleren C, Beal MF, Jones L, Kooperberg C, Olson JM, and Jones KR (2007). Expression profiling of Huntington's disease models suggests that brain-derived neurotrophic factor depletion plays a major role in striatal degeneration. *J Neurosci* 27: 11758–11768.
233. Simola N, Morelli M, and Carta AR (2007). The 6-hydroxydopamine model of Parkinson's disease. *Neurotox Res* 11: 151–167.
234. DiBernardo AB and Cudkowicz ME (2006). Translating preclinical insights into effective human trials in ALS. *Biochim Biophys Acta* 1762: 1139–1149.
235. von Horsten S, Schmitt I, Nguyen HP, Holzmann C, Schmidt T, Walther T, Bader M, Pabst R, Kobbe P, Krotova J, Stiller D, Kask A, Vaarmann A, Rathke-Hartlieb S, Schulz JB, Grasshoff U, Bauer I, Vieira-Saecker AM, Paul M, Jones L, Lindenberg KS, Landwehrmeyer B, Bauer A, Li XJ, and Riess O (2003). Transgenic rat model of Huntington's disease. *Hum Mol Genet* 12: 617–624.
236. Chan AD, Nippak PM, Murphey H, Ikeda-Douglas CJ, Muggenburg B, Head E, Cotman CW, and Milgram NW (2002). Visuospatial impairments in aged canines (*Canis familiaris*): the role of cognitive-behavioral flexibility. *Behav Neurosci* 116: 443–454.
237. Tapp PD, Siwak CT, Estrada J, Holowachuk D, and Milgram NW (2003). Effects of age on measures of complex working memory span in the beagle dog (*Canis familiaris*) using two versions of a spatial list learning paradigm. *Learn Mem* 10: 148–160.
238. Dimakopoulos AC and Mayer RJ (2002). Aspects of neurodegeneration in the canine brain. *J Nutr* 132: 1579S–1582S.
239. Cork LC (1991). Hereditary canine spinal muscular atrophy: an animal model of motor neuron disease. *Can J Neurol Sci* 18: 432–434.
240. McFarlane D (2007). Advantages and limitations of the equine disease, pituitary pars intermedia dysfunction, as a model of spontaneous dopaminergic neurodegenerative disease. *Ageing Res Rev* 6: 54–63.
241. Peters A (2002). Structural changes that occur during normal aging of primate cerebral hemispheres. *Neurosci Biobehav Rev* 26: 733–741.
242. Emborg ME (2007). Nonhuman primate models of Parkinson's disease. *Ilar J* 48: 339–355.
243. Outeiro TF and Muchowski PJ (2004). Molecular genetics approaches in yeast to study amyloid diseases. *J Mol Neurosci* 23: 49–60.
244. Sherman MY and Muchowski PJ (2003). Making yeast tremble: yeast models as tools to study neurodegenerative disorders. *Neuromolecular Med* 4: 133–146.
245. Yeoman MS and Faragher RG (2001). Ageing and the nervous system: insights from studies on invertebrates. *Biogerontology* 2: 85–97.

Aging of the Immune System Across Different Species

Janko Nikolich-Žugich and Luka Čičin-Šain

Abstract Aging of the immune system, called immunosenescence, has been linked to reduced ability to resist infection, and increased mortality and morbidity from infectious diseases, which consistently rank in the top five causes of death in the old age, even in industrial societies. This review summarizes our knowledge about the evolution of the innate and adaptive immune systems in light of the longevity of the organism, as well as the current state of our understanding of age-related changes in each of the arms of the immune system in various model organisms. It is clear that adaptive immunity, which we propose evolved as an essential function that provides longevity, eventually erodes in the old age. The jury is still out as to whether innate immunity undergoes a similar decay, whether it stays the same, or whether, perhaps, it is able to compensate for the loss of adaptive immune function.

Keywords Immune senescence · Immunity · Innate immune system · Adaptive immunity

List of Abbreviations

Ag	Antigen
ASR	Antigen Specific Receptors
BCR	B-cell Receptor
CD	Cluster of Differentiation
CMV	Cytomegalovirus
DC	Dendritic Cell
IFNγ	Interferon γ

J. Nikolich-Žugich (✉)
Department of Immunobiology, University of Arizona College of Medicine, P.O. Pox 249221, 1501 N. Campbell Ave.Tucson, AZ, 85724, USA
e-mail: nikolich@email.arizona.edu

Supported by USPHS awards AG20719, AG21384 and AG23664 (J.N-Z.) and RR-0163 (ONPRC) from the NIA and NCRR, National Institutes of Health.

N.S. Wolf (ed.), *The Comparative Biology of Aging*,
DOI 10.1007/978-90-481-3465-6_15, © Springer Science+Business Media B.V. 2010

Ig	Immunoglobulin
IL	Interleukin
LAT	Linker of Activated T-cells
LPS	Lipopolysaccharide
LRR	Leucine Rich Region
Mf	Macrophage
MHC	Main Histocompatibility Antigens
MyD88	Myeloid Differentiation Factor 88
NF-κB	Nuclear Factor κB
NHP	Non-human Primates
NK	Natural Killer
NLR	NOD-like Receptor
NOD	Nuclear Oligomerization Domain
PAM	Pathogen Associated Molecular Pattern
PBMC	Peripheral Blood Mononuclear Cell
PGRP	Peptidoglycan Recognition Protein
PRR	Pathogen Recognition Receptors
RAG	Recombination Activating Gene
RBC	Red Blood Cell
RLH	RIG-I like helicase
RM	Rhesus Macaque
TCE	T-cell Clonal Expansion, BCE – B-cell Clonal Expansion
TCR	T-cell receptor
TEMRA	T-cell Effector Memory Cells expressing CD45RA
TIR	Toll/IL1 Receptor
TLR	Toll-like Receptor
TNFα	Tumor Necrosis Factor α
VLR	Variable Lymphocyte Receptor

Introduction

Resistance to infections is a major driver of natural selection. Only the successful host, the one surviving infections by environmental pathogens, will transfer its genes to the next generation – provided that infection occurs prior to reproduction. Microbial pathogens in turn are naturally selected for their ability to proliferate in individual hosts and spread in the host population. Therefore, it is often assumed that the co-evolution of hosts and pathogens is an arms race, where the host is evolving increasingly potent immune defenses, while pathogens follow in lock-step, by evolving novel mechanisms to evade host immunity.

At face value, survival of increasingly complex, long-living eukaryotic hosts faces exceptional microbial challenges: the extremely short generation span and quick mutation rate of bacteria and viruses allows microbial pathogens to evolve for thousands of generations over the lifetime of a single host. This should confer, at least in theory, a tremendous advantage to the pathogens, yet multicellular organisms found ways to survive against these odds and evolve in size, complexity and lifespan: while the majority of evolutionary ancient animals, like protostomes, live for few days or weeks, some recently evolved vertebrates, including the humans, enjoy lifespans measured in decades.

The r/K-selection theory by MacArthur and Wilson [1] provides a theoretical explanation to this phenomenon: the organisms that invest in fertility (r-selection), survive by numbers. In their case, large progenies offset high mortality, which allows at least some individuals to survive and reproduce. On the other hand, the organisms that invest in body maintenance (K-selection) survive longer and thrive due to their low mortality, yet this comes at the expense of a low reproduction rate. Animal species that invest a lot of time in offspring nurture, allowing the majority of their young to reach adulthood, are likely to have small litters.

Therefore, the r/K theory assumed that pluricellular hosts, investing an increasing amount of their body resources in immune defense, would be selected for, despite the increasing investments required for each individual. It has been recently proposed that strong inflammatory immune response is a K-strategy of natural selection within a species [2], improving the resistance to environmental hazards, yet lowering the fertility of individuals.

However, the r/K theory offers no insight into the cellular and molecular mechanisms of pathogen resistance in the increasingly complex and long-living, K-selected hosts.

A likely explanation of this phenomenon is offered by the discovery of two mechanisms by which hosts recognize molecular determinants of microbial pathogens – the Antigen-specific receptors (ASR), present in all craniated vertebrates, and the evolutionary more ancient Pathogen Recognition Receptors (PRR), whose members have been found in virtually all multicellular eukaryotes, including vertebrates, nematodes, arthropods and plants. We will describe here the phylogenic distribution of these receptor and their sub-classes, and offer a hypothesis on the role of ASR in the K-selection of vertebrates, both in terms of complexity and longevity, before reviewing current literature about aging of innate or adaptive immune function in the course of evolution.

Pathogen-Associated Molecular Patterns and Innate Immunity in Vertebrates and Invertebrates

PRRs are defined by their ability to recognize pathogen associated molecular patterns (PAMP) [3]. In mammalian hosts, PRR have been shown to recognize bacteria, virus and fungi derived PAMPs [4]. Bacteria derived PAMPs include lipopolysaccharides from the outer membrane of Gram-negative bacteria [5], flagellin, the major protein component of bacterial flagella [6], unmethylated CpG DNA [7], diaminopimelic acid from cell walls of Gram-negative bacteria [8] or muramyl dipeptide peptidoglycans from Gram positive and negative cell walls [9, 10].

Viral-derived PAMPs include double stranded RNA [11], single stranded uncapped RNA [12], as well as CpG DNA [13]. In invertebrates, PRR were first identified as mediators of anti-fungal immune responses in drosophila [14], yet subsequent studies identified bacterial peptidoglycans and lipopolysaccharides as PAMP motifs, stimulating immune responses in arthropods [15]. Similarly, studies of plant immune responses identified an exceptionally wide variety of bacterial and fungal PAMPs recognized by plant PRRs (for a review see [16]).

Pathogen Recognizing Receptors

PRR recognize molecular patterns that are not present in Eukaryotic hosts, but are specific for bacteria or viruses. PRR have been located on cell surface, in the cytoplasm or on intracellular membranes. In animals, they are classified by structure and presence of conserved motifs in several protein families, which include the Toll-like receptors (TLR), nucleotide oligomerization domain (NOD) like receptors (NLR), RIG-I like helicases (RLH), and peptidoglycan recognition proteins (PGRP), to name a few. Some of them, like TLRs are highly conserved, and its representatives have been found in members of both the protostome and the deuterostome branch of the subkingdom of animals with bilateral symmetry [14, 17]. NLRs are another conserved gene family, whose representatives have been identified in mammals [8] and deuterostome invertebrates, like sea urchins [18], and have closely related homologues in plants as well [19]. Most of these receptors use ancient and evolutionary conserved signaling cascades, such as that of the NF-kB pathway (TLR). Yet other gene families like the RLH, have been identified only in mammals [20]. PGRP are well known in protostomes [21, 22], but have not been found in vertebrate animals so far. The comparison of the highly conserved PRR families, like TLRs or NLRs, offers important insights into their diverging evolution in both terms of structure and function. Most of these receptors use ancient and evolutionary conserved signaling cascades, such as that of the NF-kB pathway (TLR).

Toll Like Receptors

TLR are characterized by the presence of three domains: a leucine rich region (LRR) which is involved in PAMP recognition, a transmembrane domain and a toll/IL-1 receptor (TIR) domain. TLRs have been described in a wide variety of organisms, from mammals [17, 23] to drosophila [14]. Although only one toll gene has been described in drosophila, in mammals this number is much higher, arguing for diverging evolution of TLR receptors. To date, 11 TLRs have been identified in humans and 13 in mice, varying in their localization, target PAMP motifs, and second messenger signaling pathway. In TLRs that are located on the cell surface, like TLR1, TLR2, TLR4, TLR6 and TLR9, the LRR is oriented extracellularly, allowing for PAMP recognition in the extracellular matrix and signal transduction by the TIR domain. TLRs recognizing nucleic acids, like TLR3, TLR7, TLR8, and TLR9 are located predominantly or exclusively on intracellular endosomes [23], a localization that is apparently defined by their transmembrane domain [24]. The majority of TLRs signal through the myeloid differentiation factor 88 (MyD88) to activate nuclear factor kappa-B (NF-κB), yet TLR3 and TLR4 have another signal transducer, called Trif [25], which induces STAT1 phosphorilation and strong IFNα upregulation. On the other hand, the immune function of the drosophila *toll* seems restricted to the recognition of bacterial and fungal molecular patterns, signals only through Tube, the MyD88 homologue, and stimulates Dorsal, one of the NF-κB homologues in *Drosophila*. Therefore, it appears that the TLR family has expanded in its number and scope as the vertebrate subphylum of deuterostomes evolved. It is not clear if this is a reflection of the increased need for PAMP recognition in vertebrates, or a result of the presence of the large PGRP family in insects, which may take over at least a part of the function of the TLR system.

NOD Like Receptors

NLRs are characterized by their cytosolic localization and their tripartite protein architecture consisting of (1) an N-terminal effector domain responsible for signal transduction, (2) a central nucleotide-binding oligomerization domain and (3) a C-terminal LRR region, involved in PAMP recognition. In mammals, two major NOD-like proteins have been identified in PAMP recognition and NF-κB activation: the NOD1 recognizing bacterial peptidoglycans [8], and the NOD2, recognizing the muramyl dipeptide [9, 10]. Other NLRs, like IPAF and Naip, sense bacterial flagellin in the cytosol in a TLR-6 independent manner [26] and activate caspase-1 dependent IL-1b inflammatory responses [27, 28]. Moreover, NLRs have been implied in the induction of cell death [29], arguing that the effector responses to NLR-dependent PAMP recognition can vary in biological outcomes. Nalp3, another NLR, has been shown to recognize cytosolic nucleic acids of bacterial or viral origin [30], arguing that NLRs are as broad in their spectrum of PAMP recognition as the TLRs. In invertebrates, more than 200 putative NLR genes have been identified by genome analysis of the sea urchin [18], arguing for a central role of NLRs in PAMP recognition in invertebrates. Even more interestingly, a multitude of molecules with structure similar to NLRs have been identified in plants [31]. Although the NLR-like proteins in plants are membrane bound or secreted, and not cytosolic, several similarities exist: they identify similar PAMPs as the human NLRs (e.g. flagellin), and also recognize PAMPs through their LRR-domains [19]. This argues that different kingdoms recruited similar molecular mechanisms to sense and fight microbes, arguing for converging evolution or strong evolutionary conservation. Alternatively, this overlap in gene structure and function may be a result of common ancestry of these genes. If that was indeed the case, the NLRs would emerge as the most ancient PRR known to date.

Antigen Specific Receptors

Vertebrates are an evolutionary recent subphylum of the deuterostome branch of the animal kingdom. Therefore, the anatomic and molecular specificities of vertebrates have likely evolved only recently. The immune system of virtually all known vertebrates is characterized by the existence of an adaptive immune system. This system is essentially composed of clonally diverse immune cells, called lymphocytes, whose clonal diversity is defined by the specifities of their antigen specific receptors. The molecular basis of specificity in such system is provided by somatic recombination, a process that allow the generation of a wide variety of ASR in individual lymphocytes of a single host, by random and stochastic recombination during lymphocyte maturation of modular elements encoding parts of ASR. In an animal with more than 10^8 lymphocytes, such a system offers the possibility to recognize a huge variety of molecular structures, with potential ASR diversity of up to 10^{18} different receptors [32].

The lymphocyte ASR, in terms of their structure fall in two categories: LRRs and immunoglobulins (Ig) [33]. Jawless fish, like the hagfish [34] or the lamprey [35] use LRR domains for antigen recognition. Virtually all jawed vertebrates detect antigens by means of immunoglobulin-based proteins [36].

LRR-Based ASR

The molecular architecture of the ASR in jawless fish, or agnathans, is comprised of an invariable stalk that tethers the ASR to the cell surface by means of a glycosyl-phosphatidyl-inositol anchor, and a variable LRR that is expressed on lymphocytes in monoallelic fashion [35]. Germline DNA encodes a large number of variable LRR regions located between amino- and carboxy-terminal LRRs. ASR in agnathans have been termed variable lymphocyte receptors (VLR) [35], and were shown to segregate into two distinct classes, the VLRA and VLRB. Agnathan immune response to antigen is a humoral one, and result in the secretion of VLRB in sera. Therefore VLRB are functionally similar to antibodies (Ab) in jawed vertebrates [37]. Somatic rearrangement during VLR maturation, possibly through a gene-conversion mechanism, recombines individual variable LRR regions in a seemingly stochastic manner, thus resulting in a diverse population of lymphocytes whose potential repertoire is estimated at 10^{14} clones [34]. The diversity of the anticipated VLR repertoire in agnathans is therefore limited only by the number of lymphocytes, as is the case with jawed vertebrates [33].

Immunoglobulin (Ig)-Like Antigen Specific Receptors

Concomitant with the development of the jawbone, gnathostome vertebrates have developed the immunoglobulin-based adaptive immune system [36]. Although the molecular basis of agnathans and gnathostome antigen recognition relies on radically different protein classes, the adaptive nature of both systems shows striking similarities.

The germline genome encodes a variety of variable regions from which a mature antigen receptor is recombined during lymphocyte maturation in the thymus for T-lymphocytes or in the bone marrow for B-lymphocytes [38]. Both the T-cell receptors (TCR) and the B-cell receptors (BCR) are comprised of two Ig chains, which rearrange their variable domains independently of each other, yet both contribute to antigen recognition, thus increasing the potential for diverse antigen recognition.

The molecular mechanisms of ASR maturation differ between vertebrate classes. In mammals the diversity is achieved by somatic recombination, where recombination activating genes (RAG), shown to be co-opted retrotransposon elements [39], randomly select from a multitude of variable and joining domains and excise the intervening DNA, thus giving rise to a mature rearranged chain of either the T-cell receptor (TCR) or the B-cell receptor (BCR). In birds, the clonal diversity of TCR and BCR is achieved by gene conversion during lymphocyte maturation in the thymus, or in the Bursa of Fabricius, where shorter sequences are shuffled into the mature TCR or BCR [38]. Despite this variability in the maturation process, the result is a diverse immune repertoire with the potential to recognize virtually millions of molecular structure determinants. Therefore, it is not surprising that the ASR maturation process in vertebrates must include a negative selection step, aimed to prevent self-aggression, in which the clones recognizing self-antigens are selected against and eliminated before they are allowed to exit the lymphopoietic organs [40, 41].

The BCRs from activated B-cells are secreted in the extracellular compartment in the form of antibodies. There, they recognize soluble extracellular antigens, or occasionally the antigens found directly on the cell surface, and mediate the phagocytosis of bound antigens, their neutralization, or in the case of antigens bound on host cells, they may mediate the antibody-dependent cell cytotoxicity of the infected cell. The localization of antibodies determines their relevance for the immunity against extracellular pathogens, like bacteria and fungi. Therefore, the targets of BCRs overlap with the ones of TLRs that are bound on the cell-surface and recognize extracellular PAMPs. TCRs are more similar to NLRs and intracellularly located TLRs and are specialized in the recognition of intracellular pathogens. To achieve this, TCRs recognize short oligopeptides presented on the surface of cells by a class of proteins called main histocompatibility antigens (MHC). In gnathostome cells, proteins within each cell are constantly being cleaved by the immune-proteasome complex, transferred from the cytosol into the endosomal compartment by transporter molecules and loaded onto MHC class I molecules, which are then transferred to the cell surface. Alternatively, peptides derived from the endoplasmic compartment are loaded onto MHC class II molecules and, likewise to MHC-I peptides, presented on the cell surface. The lack of MHC-like molecules and the immune proteasome is therefore believed to denote lack of the ability to target intracellular pathogens in agnathans and invertebrates for recognition by adaptive immunity [33].

Once the T-cells and B-cells leave the central hematopoietic organs, they migrate to secondary (term denoting the fact that these organs do not produce, but rather just host, lymphocytes) lymphoid organs, where they participate in scanning of tissues for possible presence of microbial antigens (Ag). Antigen encounter results in selective proliferation and long-term maintenance of lymphocytes with specificity for the encountered Ag, which results in the long-term adaptation of the host immune system to its microbiological environment, dependent on its individual history of Ag encounters.

Thus, the evolutionary youngest classes of vertebrates appear to have the most specialized and sophisticated mechanisms of pathogen recognitions, that allow them to recognize extracellular and intracellular pathogens in an anticipatory and adaptive fashion. This invites the question whether this increase in immunological complexity is a K-based evolutionary strategy, or coincidental to the increase in longevity and reduction in reproductive capacity of vertebrates compared to invertebrates.

Adaptive Immunity as a Requirement for Increased Life Span

Pathogen recognition through LRR domains of PRRs appears to be a highly conserved immune defense strategy, arguing for its efficiency and evolutionary advantage that it confers. Yet, it constitutes only one branch of the immune system of vertebrates. One inherent limitation of the PRR system is its genomic germline encoding. The PRRs can recognize a fixed and preset array of PAMPs, arguing that genetic shifts or drifts of the pathogen would allow it to avoid pattern recognition and become invisible to the host immune system. Therefore, in this system the adaptation of a host species to a novel class of pathogens can be achieved only by the evolution of novel PRRs that will recognize its specific PAMPs. Moreover, innate immune responses come at a high metabolic cost, and the genetic selection of drosophila with strong immune responses results in poor ability to acquire food [42]. Such antagonistic pleiotropy, where the same effects increase fitness through one trait, while decreasing it through another one, result in significant evolutionary trade offs [43], yet works well for short-lived species with large progenies, that can tap into the genetic diversity of a population to select for individuals with survival advantage.

This may explain the reliance of r-selected invertebrate hosts on innate immunity and the evolutional success of this immune strategy. However, this also anticipates that the K-selection of vertebrates, where each individual of a species becomes more valuable, required a more sophisticated system of immune defense.

One of the cardinal features of the ASR is its plasticity. The lymphocyte clones that encounter antigens are positively selected in the population of peripheral lymphocytes. They are allowed to proliferate and survive for long periods, even for life, in the host. The lymphocytes that do not encounter any antigen that they may recognize are selected against, and have lower chances of survival. Therefore, this system allows the immune system of a single host to evolve and adapt itself to the dynamic changes of the microbiological environment that occur during the host lifetime. It is well known that the rapidly evolving human immunodeficiency virus (HIV) mutates during chronic infection within an individual host [44]. The newly generated escape variants are selected for the loss of antigenic determinants recognized by T-cells [45]. However, the adaptive nature of the host immune system allows the evolution of T-cell responses that follow in lockstep the evolving HIV variants and mount novel immune responses to the newly generated immune escape variants [46, 47]. Therefore, the evolution of immune responses in vertebrates does not occur any more at the host population level, but within the populations of clonally diverse lymphocytes. In fact, it has been proposed that the history of antigen encounters shapes the evolution of the host CD8 T-cell repertoire [48].

Therefore, we propose that the plasticity of the adaptive immune system is a genuine K-strategy of complex hosts, which shifts selection pressure from host populations to the selection of antigen-specific populations within one host. Therefore, the adaptive nature of the immune system buys additional lifetime to the host, allowing it to adapt to long lifespan and small progenies, where each individual has a high value within his population. This also predicts that the eventual immune senescence of a host will occur when the adaptive potential of its immune system will be exhausted by the cumulative history of previous antigen encounters. This model fits well with the observations that aging-related losses of immune function are predominantly losses of adaptive immune function [49], that leave the host to depend increasingly on innate and inflammatory immune responses [2, 50]. Franceschi et al. proposed that the entire process of immune aging is akin to a host devolution [51], and coined the term "inflamm-aging" to describe the replacement of evolutionary recent adaptive immune mechanisms with ancient innate and inflammatory mechanisms [52]. However, it is not clear whether indeed there is a true increase in innate immune function in the course of mammalian aging.

In conclusion, the aging-related pathology of the immune system is a result of evolutionary K-selection that allowed us to extend our lifespan in the first place. As our society has become more protected, a large number of individuals can confidently expect to reach old age, which exposes the aging-related pathology in the general population to our attention and scrutiny [53]. Similarly, we would have never been aware of the deficiencies of the aging immune system, had it not been successful in protecting us from pathogens for decades prior to its eventual demise.

Immune Senescence

Biologists tend to overwhelmingly study certain organisms at the expense of the others, and their choices are driven by advantages offered by some, but not other, models. Aging research is no exception, as has been discussed elsewhere in this volume – research on aging is focused on a handful of favorite models, either because of their short lifespan and the ability to conduct large population-based studies, because of the availability of powerful genetic and genomic tools, or because of their relevance and similarity to humans. In that hierarchy, the dominant species studied are the yeast *S. cerevisiae*, the worm *C. elegans*, the fruitfly *Drosophilla melanogaster*, *Mus musculus* and the human. For the sake of comparative biology, one has to take one of the above models as the starting point, and to compare other models to it.

The mouse is by far the model most studied when it comes to immunosenescence, although one should bear in mind that murine senescence imperfectly recapitulates the one in humans. We shall quickly summarize essential features of immune aging in laboratory mice and humans. All comparative points thereafter will refer to these two models. Details of mouse and human immune aging have been reviewed in a number of excellent articles, and the reader is hereby encouraged to peruse these recent reviews [54–57]. Overall, there are defects and changes described in every immune function and cell type measured so far. It is unfortunate, however, that many of these changes are not consistently observed and that many studies themselves are less than satisfying. In particular, we know little about the age-related changes in the innate immune system in mice or humans. No change, or age-related declines, were reported on granulocyte and macrophage (Mf) function [58, 59]. Numbers and representation of granulocytes in fact increase in mouse spleen [60], which is consistent with the observed relative shift in stem cell commitment away from lymphoid and towards the myeloid lineage in old mice [61]. NK and NK-T cells may exhibit changes as well, but the impact and extent thereof are incompletely understood. Overall, there is an abundance of mutually contradicting descriptions of the innate immune system function in aging, and few seem to be consistently observed so far. It is likely that assay design and choice of subjects within the cohort contributed to this variability, and it will be essential to conduct standardized and rigorous studies to get to the bottom of this question. Dendritic cells (DC) are key antigen-presenting cells that function at the interface between innate and adaptive immunity. They respond to PAMPs by immediate cytokine secretion, but also capture antigens and initiate the adaptive immune response by presenting it to T and B cells. Some reports suggest that numbers and function of DC remain normal with aging in mice, but that their migratory properties degrade [60, 62]. It is unclear, however, whether this is reproducible, and whether this is due to intrinsic defects in DC (such as expression of receptors for chemokines that guide DC migration) or alterations in microenvironment (dysregulated production of chemokines and cytokines in the milieu). Moreover, DC are heterogenous and studies of DC subsets with aging, which are likely to be most informative, have been scarce.

Finally, it was proposed that a trade-off exists in between adaptive and innate immune systems so that the decline of the former is followed up by an increase in the activity of the latter [51]. While this idea is intellectually intriguing, hard experimental support is still lacking to support it.

Aging of Innate Immunity in Different model Systems

Generally, we possess insufficient knowledge of the innate immune system aging in any of the model systems, and there is a particularly glaring dearth of information on aging of innate immunity in the three favorite (*Drosophila*, *C. elegans* and yeast), or, for that matter, on any of the invertebrate models.

Young *Drosophila* shows better survival upon infection with *E. coli*, than adult or old flies [63]. The same is true for *C. elegans* [64, 65]. However, the mechanism of this age-related sensitivity to infection is not clear (For reviews see references [66, 67]). Three signaling pathways are relevant for the immune function in *C. elegans*: the insulin receptor, the p38 MAPK and the TGF-β pathway [66]. While a loss-of-function mutation in daf-2, the insulin receptor gene of *C. elegans*, results in extended life span and resistance to bacteria [68], it remains unclear whether these phenotype changes are caused by alterations in metabolic or immune function. In contrast to *C. elegans*, pathogen resistance appears to present a trade-off to longevity in *Drosophila*, because the mutations inducing strong Toll-NFκB signaling allow *Drosophila* to clear infections, yet result in chronic inflammation and shortened life-span [69]. Consistent with that, microarray analysis showed that patterns of gene expression in aging resemble those in inflammation [70]. However, much remains to be learned about the age-related losses of immune function in invertebrates. The only two studies we could find addressed insect models, and both suggested that there are age-related changes in innate immunity [71] with Kurtz demonstrating that scorpion sandfly Panorpa vulgaris exhibits a decline in phagocytic capacity of hemolymph cells, while retaining hemolymph cell numbers [71]. If confirmed, this finding will be of additional importance, because the findings were made in a free-living insect, counteracting the idea [53, 72] that free-living animals would be extremely unlikely to show signs of aging in the wild. However, with regard to comparative immunology, it is difficult to accurately assign the counterparts of hemolymph cells, so one has to stick to the phagocytic function as the functional correlate of, perhaps, neutrophils or macrophages in higher vertebrates.

By contrast, literature on innate immunity in mice, humans or other mammalian species has been replete with information. Unfortunately, the same literature is often contradictory and therefore difficult to interpret and reconcile. Thus, granulocyte numbers often increase in laboratory mice, but it is less certain whether their function is affected. Neutrophils were found to have delayed infiltration ability and were therefore less able to control bacterial loads in the cornea of 12-mo old pseudomonas-infected mice compared to 12-week old counterparts, but then tended to stay longer and mediate tissue destruction [73]. By contrast, decreased neutrophil phagocytic capacity and CD16 expression has been found in elderly humans [58], and, while hip fracture in elderly humans produces robust neutrophillia, these neutrophils exhibit decreased bactericidal activity [74]. Natural killer (NK) cells have been found to show no numerical or functional changes in aged dogs [75, 76]. By contrast, NK cells have been found to increase in old humans subjects and their total lytic capacity was preserved, although per cell basis capacity was diminished, as was their response to IL-2 and cellular activation as judged by the expression of activation marker CD69 [77]. NK-T cells were found to be increased in old mice and their depletion helped restore T-cell responses in vivo (Delayed-type hypersensitivity) [78]; Old mouse macrophages have been found to exhibit reduced production of IL-6 and TNFα in response to LPS, while increasing IL-10 production [79, 80]. Complement has been reported to play a role in macular degeneration [81] and it was suggested that it is the local production by macular neurons which leads to the disease itself; however, there is no comparative data across species to corroborate this finding. Likewise, complement defects were found in old mice and these contributed to impaired phagocytosis of bacteria from the eye in mice [82], but no comparative data exist elsewhere for comparison.

Dendritic cells have also been investigated, but no definitive consensus exists on the extent and importance of their defects. Moreover, experiments with DC have been largely limited to mice and humans, with little to no comparative studies in other species. Apparently, there is no available literature on age-related changes in innate immune function of DC populations. A recent review described studies of migration where both DC and T-cells exhibited impaired migration in the old environment [83]. Yet the same review described experiments with DC from old and adult mice which suggested that transfer of DC (adult, but to a somewhat lesser extent, old too) can ameliorate T-cell responses in vivo [83]. Of note, the (in)ability of old DC to stimulate T-cells needs to be established in a rigorous and incisive manner.

Similarities and Differences in Aging of the Adaptive Immunity in Mice and Humans

By contrast to this relative dearth of discernible and reliable information on innate immunity, there is strong consensus about the changes in adaptive immunity with aging in mice and humans. It is generally believed that overall numbers of lymphocytes are either unchanged or somewhat reduced with aging in both species. This is based upon measurements of lymphocytes in blood (humans) or blood and some secondary lymphoid organs, primarily the spleen, in mice. Pending a thorough experimental sampling of the many sites where lymphocytes reside, including the gut, bone marrow and liver, as well as all available lymph nodes, this issue remains unsettled. A highly consistent finding is the shift in population balance between antigen-experienced, or memory lymphocytes, and virgin, antigen-naïve lymphocytes, in favor of the former (rev. in [49, 84]). This reflects the operation of at least two factors: decline in production of new naïve lymphocytes by primary lymphoid organs (bone marrow for B cells [85]), and, even more markedly, the thymus for T-cells (rev. in [86]), and utilization of the existing naïve lymphocytes due to life-long exposure and immune responses to pathogens.

Other consensus findings include reduced lymphocyte proliferative responses across a diverse spectrum of stimuli, such as mitogens, phorbol ester and calcium ionophore combinations and anti-TCR and anti-BCR agonistic antibodies, with or without costimulatory signals [87, 88], rev. in [49]. Likewise, other cellular functions downstream from antigen-specific and other receptors were also blunted, delayed or impaired. Where examined, declines were found in proximal T-cell receptor signaling [89, 90], cytokine secretion (in particular that of IL-2) [91–95] and cytotoxicity [94, 96] and B-cell proliferation, somatic hypermutation and isotype switching [56, 60, 87, 97, 98], as well as in the response to homeostatic stimuli [57, 99–101].

In interpreting the above results, however, one should be mindful that many of the studies performed in mice and humans failed to control for the altered lymphocyte subset composition in the course of aging. Thus, analyzing total T – or B – or even worse (but more frequently) splenocyte or blood or peripheral blood mononuclear cell (PBMC) populations for proliferative responses will tend to read out the response of naïve lymphocytes, which are capable of the most robust proliferative burst. As these cells are invariably reduced in old animals and humans, finding that there is lower proliferation in bulk populations is not surprising. It is clearly informative to know that total proliferative capacity has been reduced, however, in the absence of controlling for the subset composition, it is impossible to tell whether this is due to population shifts or reduced responsiveness of individual populations, or both.

Fortunately, for some of the parameters, single-cell analysis studies [102] have corroborated the notion that there are cell-autonomous age-related changes that impair response to stimulation in old lymphocytes in mice. Therefore, it has been firmly established that the formation of immunological synapse is delayed and the concomitant assembly and phosphorylation of critical downstream substrates, such as the linker of activated T-cells (LAT) and the protein tyrosine kinase Zap-70,which then propagate to the MAPK and other downstream pathways [103, 104], are diminished in old naïve CD4 T-cells [102]. Other transcriptional and signaling changes were found in human CD8 cells [105, 106], particularly those that have been repeatedly stimulated by latent persistent virus CMV (cytomegalovirus) and are believed to be terminally differentiated (often called TEMRA cells, for **T**-cell **e**ffector **m**emory cells that reexpress differentiation marker CD45**RA**). Unfortunately, the huma-murine comparative data set is still incomplete, because naïve CD4 T-cells need to be analyzed in humans, memory CD4 T-cells and naïve and central memory CD8 T-cells need to be analyzed in both species, and effector memory CD8 T-cells (general population as well as those repeatedly stimulated by CMV) require analysis in mice.

As mentioned above, age-related cell population balance and homeostatic changes have been described in both mice and humans. Of note, in both species there is consensus that naïve T-cells not only are numerically reduced, but that populations of clonally identical lymphocytes expand (T- and B-cell clonal expansions, TCE and BCE, refs [97, 107]). These cells share certain characteristics in both species, and as discussed in our prior work, are likely to be of two different and distinct origins: (i) homeostatic proliferation [108, 109] in response to a likely (but not substantiated) relative surplus of homeostatic cytokines (chiefly IL-7) would give rise to "spontaneous" central memory phenotype TCE very often seen in laboratory mice older than 18–20 months [110]; and (ii) virus, mostly CMV driven accumulation of effector memory cell expansions [111, 112] seen in the vast majority of humans [113] and also in mice infected with latent persistent viruses [114, 115]. The key difference between the two species is that the second type of expansions in humans is usually marked by the loss of the costimulatory molecule CD28 [116, 117], which is rarely lost in mouse TCE unless these are very heavily and chronically stimulated. Significance of the two categories of TCE is an unsettled and incompletely understood topic, but they were found in humans to correlate inversely with residual lifespan and there is some evidence that they have the potential to impair functional immunity to unrelated infection with previously unencountered microbial pathogens [118].

Aging of Adaptive Immune Systems in Other Species

Because the analysis in many other models is not nearly as exhaustive as it is in mice and men, we will only limit ourselves to listing those results where similar processes were investigated. The reader should assume that if a given age-related change is not mentioned in a given species then means that this change was not investigated so far. The changes will be listed by species below.

Birds

Both free-living and captive-bred birds were analyzed for some of the signs of immune senescence. In the limited available published literature, Palacios et al. [119] have shown in free-ranging tree swallows (*Tachycineta bicolor*) a decline in proliferation of whole blood cells to T-cell mitogens in vitro, but no change in proliferation to B-cell mitogen LPS, and no change in Ab formation in response to sheep red blood cells (SRBC) nor in the presence of natural antibodies. In a separate study, Lavoie et al. [120] used capture-bred Japanese quails (*Coturnix coturnix japonica)* to demonstrate depressed reactivity to skin-administered PHA (a correlate of delayed-type hypersensitivity assay in birds, where components of the response were not precisely determined), which correlated well with prior results in several other bird species [121]. The same study found little to no difference in Ab response to heterologous RBC nor in Ab response to avian influenza [120], again similar to several other studies in humoral responses of other aged birds to RBC or virus challenge [122]. Most remarkably, Lavoie et al. found little if any difference in clinical response and viral shedding in the course of avian influenza infection (H9N2 virus) between the adult and aged quails [120], which is in stark contrast to human age-related susceptibility to flu and deterioration in anti-influenza immune response. On the other hand, given the known ability of certain infections to alter susceptibility of an organism to unrelated, but concurrent or subsequent infection [123], it would be of interest to determine the infection status of these birds prior to challenge.

Rats

Data obtained in rats are more detailed than in any species other than mice and humans, and they tend to parallel the observations obtained in mice. In a recent study of Fisher 344 rats; [124], there was an increase in monocyte/macrophage lineage (OX42+) cells and a decline in all subsets of T-cells (CD5+ T cells), with a significant decline of CD4/CD8 ratios which is also seen in many mouse and human studies. It is unclear why exactly CD4 cells decline significantly more than CD8 cells, but several explanations, including the inability of recent CD4 T-cell emigrants to survive after leaving the thymus [125] and the stimulation of CD8 T-cells by persistent latent viruses [113] have been put forward. Dietary interventions including caloric restriction [126] and folate supplementation [124] can reverse some of these changes, just as was reported in other species [127–129]. Rats also exhibit proliferative T-cell decline in response to ConA [124], however, it is less clear why several studies in rats did not detect a decline of IL-2 production [130, 131] and an increase in production of proinflammatory cytokines IL-6, IFNγ and TNFα. It is likely that assay design differences, including the use of different populations for assays, are at least in part responsible for differences. However, kinetics of IL-2 production and consumption is different in rats and mice (IL-2 R expression only begins 24 h post stimulation and goes up to reach maximum after 48 h and later [132], and it is possible, if not likely, that the lack of age-related defects in proliferation of rat T-cells may also in part be linked to IL-2R expression and/or signaling.

An interesting study addressed changes in mucosal immunity in rats, finding a decline in IgA-producing cell homing from Peyer's patches to lamina propria [133], a finding also made in monkeys. Other aspects of IgA production and sIgA+ B-cell biology appear not to change with age [133].

Dogs

Most of the studies in dogs have been performed on the popular breeds such as Labrador retrievers and beagles, and they came to bear many of the observations common to findings in other species [75, 76]. Thus, studies by Greeley et al. [76] in Labrador retrievers showed age-related reduction in proliferative response to mitogens, little to no change with age in white blood cell and polymorphonuclear leukocyte numbers, but reduction in lymphocyte and all lymphocyte subset numbers, with a much more pronounced loss of CD4 than CD8 cells and a consequent decrease in CD4/CD8 ratio, with a marked increase in memory T-cells over the naïve ones. All of that is consistent with a shift from lymphopoiesis to myelopoiesis and the progressive loss of naïve T-cells. In a related study by the same authors [134], caloric restriction showed the predictable improvement of proliferative function with a less pronounced effect upon T-cell subsets and a decline in B-cells, with no change in NK cells and polymorphonuclear cells. Hayek's group has shown in a heterogeneous group of dogs that proliferative T-cell responses and % of CD4 cells were depressed with aging, with an increase in serum IgA, but no changes in percentages of T and B cells or in serum IgM or G. Although the increase in post-vaccination titers appeared more robust in young dogs, that study showed no significant age-related decline in functional immunity, as evaluated by the response to vaccines against rabies, parvovirus and distemper virus.

Monkeys

It could be argued that, from the evolutionary and biological standpoint, non-human primates (NHP) provide the closest approximation of human biology of any animal model. That case was made by our group for the specific situation of immune aging [135], because, in addition to other advantages, NHP lymphocytes and, specifically, T-cells, closely parallel human counterparts with regard to marker expression and differentiation states; and because, unlike the short-lived model organisms, including mice, NHP exhibit longevity that is much more comparable to humans. With regard to aging of the immune system, as in other species, more research is needed to test different components of innate immunity. In that. regard, unpublished results from Lewinsohn et al. (Dr. Deborah Lewinsohn, Oregon Health & Science University, personal communication) suggest that there is little or no decline in functional antigen uptake, presentation or migration of skin dendritic cells to the lymph node in aged rhesus macaques (RM) as judged by the results of FITC skin painting experiment [136] The picture is far more complete with regard to T-cell immunity. There is a quantitative loss of naïve T-cells in rhesus macaques that begins with birth and is already pronounced by 3–4 years of age (RM median lifespan is 25 years, and maximal up to 40) in outdoor-living RM [137, 138], which tends to be less drastic, but significant and readily observable, in indoor-housed, and therefore presumably microbiologically less exposed, animals [129, 139]. Similarly to humans, aged CD4 T-cells mostly become central memory in RM, whereas CD8 T-cells accumulate as effector-memory phenotype cells [137, 138] due to preferential loss of CD28 by CD8 cells. CMV-specific T-cells increase with age in RM (L. Picker, personal communication) and TCE accumulate progressively with age [129, 139]. Moreover, once the naïve compartment in old macaque shrinks below a threshold level, homeostatic proliferation ensues, which further contributes to depletion of naïve T-cells from the organism [139], similar to the findings in aged human CD4 T-cells [140]. Finally, caloric restriction delays many of these changes, including, most remarkably, an increase and/or better maintenance of, naïve T-cells [129], although the time of onset of caloric restriction critically shapes its potential beneficial effects (Messaoudi et al., submitted).

Concluding Comments

Overall, immune aging in many species appears to show concordant traits, with adaptive immunity being better understood and showing more consistent changes, particularly in the realm of proliferative response decline and shifts in lymphocyte populations. There is no doubt that more studies are needed, particularly as pertaining to innate and mucosal immunity, to draw firm generalizations on cardinal traits of immune senescence in the course of evolution. Such generalizations may allow us to better understand the basic underpinnings of the aging process as well as to rationally design modes of intervention against the most important components of immune aging.

	Aging-related change in innate or adaptive immunity across species	
Species	Innate	Adaptive
C. elegans	Decreased resistance to infections [63]	*Not applicable*
Drosophila	Decreased resistance to infections [64, 65]	*Not applicable*
Birds	*No available data*	Poor T-cell response to mitogens [111]
		Depressed skin reactivity to PHA [112]
		Maintained Ab responses to Ag [114]
Rodents	Neutrophyl peak response delayed but stronger [65]	Total lymphocyte number unaltered [49]
		Decrease in CD4/CD8 ratio [116]
	Change in the cytokine profile of the macrophage responses to LPS [71]	Lymphocyte population shifts from naïve to memory subsets [49]
	Defects in DC migration? [75]	Poor T- and B-cell responses to polyclonal and Ag-specific stimulation
	Diminished complement responses [74]	Decline in IL-2 responses [49]
		Accumulation of CD28+ TCE [106, 107]
Dogs	No change in number of NK or polymorphonuclear cells [68]	Loss of T-cells, especially CD4 [67, 68]
		Poor response to mitogens [68]
		Poor proliferative CD4 response [67]
		Increased IgA responses [67]
Primates	Poor phagocytosis and bactericidal function of neutrophyls [66]	Total lymphocyte number unaltered
		Decrease in CD4/CD8 ratio
	NK-cells count increases but functional responses decrease [69]	Lymphocyte populations shift rapidly from naïve to memory subsets [129]
	Retained DC function	Poor T- and B-cell responses to polyclonal and Ag-specific stimulation.
	Diminished complement responses [73]	Decline in IL-2 responses [49]
		Accumulation of CD28– TCE [108, 109]

References

1. MacArthur RH and Wilson EO (1967). *The Theory of Island Biogeography*. Princeton, NJ: Princeton University Press.
2. Van Bodegom D, May L, Meij HJ, and Westendorp RG (2007). Regulation of human life histories: the role of the inflammatory host response. *Ann N Y Acad Sci* 1100: 84–97.
3. Medzhitov R (2007). Recognition of microorganisms and activation of the immune response. *Nature* 449(7164): 819–826.

4. Kawai T and Akira S (2006). TLR signaling. *Cell Death Differ* 13(5): 816–825.
5. Poltorak A, He X, Smirnova I, Liu MY, Van Huffel C et al. (1998). Defective LPS signaling in C3H/HeJ and C57BL/10ScCr mice: mutations in Tlr4 gene. *Science* 282(5396): 2085–2088.
6. Hayashi F, Smith KD, Ozinsky A, Hawn TR, Yi EC et al. (2001). The innate immune response to bacterial flagellin is mediated by Toll-like receptor 5. *Nature* 410(6832): 1099–1103.
7. Hacker H, Vabulas RM, Takeuchi O, Hoshino K, Akira S et al. (2000). Immune cell activation by bacterial CpG-DNA through myeloid differentiation marker 88 and tumor necrosis factor receptor-associated factor (TRAF)6. *J Exp Med* 192(4): 595–600.
8. Chamaillard M, Hashimoto M, Horie Y, Masumoto J, Qiu S et al. (2003). An essential role for NOD1 in host recognition of bacterial peptidoglycan containing diaminopimelic acid. *Nat Immunol* 4(7): 702–707.
9. Girardin SE, Boneca IG, Viala J, Chamaillard M, Labigne A et al. (2003). Nod2 is a general sensor of peptidoglycan through muramyl dipeptide (MDP) detection. *J Biol Chem* 278(11): 8869–8872.
10. Inohara N, Ogura Y, Fontalba A, Gutierrez O, Pons F et al. (2003). Host recognition of bacterial muramyl dipeptide mediated through NOD2. Implications for Crohn's disease. *J Biol Chem* 278(8): 5509–5512.
11. Alexopoulou L, Holt AC, Medzhitov R, and Flavell RA (2001). Recognition of double-stranded RNA and activation of NF-kappaB by Toll-like receptor 3. *Nature* 413(6857): 732–738.
12. Heil F, Hemmi H, Hochrein H, Ampenberger F, Kirschning C et al. (2004). Species-specific recognition of single-stranded RNA via toll-like receptor 7 and 8. *Science* 303(5663): 1526–1529.
13. Krug A, French AR, Barchet W, Fischer JA, Dzionek A et al. (2004). TLR9-dependent recognition of MCMV by IPC and DC generates coordinated cytokine responses that activate antiviral NK cell function. *Immunity* 21(1): 107–119.
14. Lemaitre B, Nicolas E, Michaut L, Reichhart JM, and Hoffmann JA (1996). The dorsoventral regulatory gene cassette spatzle/Toll/cactus controls the potent antifungal response in *Drosophila* adults. *Cell* 86(6): 973–983.
15. Kurata S, Ariki S, and Kawabata S (2006). Recognition of pathogens and activation of immune responses in *Drosophila* and horseshoe crab innate immunity. *Immunobiology* 211(4): 237–249.
16. Nurnberger T, Brunner F, Kemmerling B, and Piater L (2004). Innate immunity in plants and animals: striking similarities and obvious differences. *Immunol Rev* 198: 249–266.
17. Medzhitov R and Janeway CA, Jr. (1997). Innate immunity: the virtues of a nonclonal system of recognition. *Cell* 91(3): 295–298.
18. Rast JP, Smith LC, Loza-Coll M, Hibino T, and Litman GW (2006). Genomic insights into the immune system of the sea urchin. *Science* 314(5801): 952–956.
19. Zipfel C (2008). Pattern-recognition receptors in plant innate immunity. *Curr Opin Immunol* 20(1): 10–16.
20. Takeuchi O and Akira S (2008). MDA5/RIG-I and virus recognition. *Curr Opin Immunol* 20(1): 17–22.
21. Zhang SM, Zeng Y, and Loker ES (2007). Characterization of immune genes from the schistosome host snail Biomphalaria glabrata that encode peptidoglycan recognition proteins and gram-negative bacteria binding protein. *Immunogenetics* 59(11): 883–898.
22. Aggrawal K and Silverman N (2007). Peptidoglycan recognition in *Drosophila*. *Biochem Soc Trans* 35(Pt 6): 1496–1500.
23. Kawai T and Akira S (2007). Antiviral signaling through pattern recognition receptors. *J Biochem* 141(2): 137–145.
24. Nishiya T, Kajita E, Miwa S, and Defranco AL (2005). TLR3 and TLR7 are targeted to the same intracellular compartments by distinct regulatory elements. *J Biol Chem* 280(44): 37107–37117.

25. Hoebe K, Du X, Georgel P, Janssen E, Tabeta K et al. (2003). Identification of Lps2 as a key transducer of MyD88-independent TIR signalling. *Nature* 424(6950): 743–748.
26. Molofsky AB, Byrne BG, Whitfield NN, Madigan CA, Fuse ET et al. (2006). Cytosolic recognition of flagellin by mouse macrophages restricts Legionella pneumophila infection. *J Exp Med* 203(4): 1093–1104.
27. Amer A, Franchi L, Kanneganti TD, Body-Malapel M, Ozoren N et al. (2006). Regulation of Legionella phagosome maturation and infection through flagellin and host Ipaf. *J Biol Chem* 281(46): 35217–35223.
28. Franchi L, Amer A, Body-Malapel M, Kanneganti TD, Ozoren N et al. (2006). Cytosolic flagellin requires Ipaf for activation of caspase-1 and interleukin 1beta in salmonella-infected macrophages. *Nat Immunol* 7(6): 576–582.
29. Ting JP, Willingham SB, and Bergstralh DT (2008). NLRs at the intersection of cell death and immunity. *Nat Rev Immunol* 8(5): 372–379.
30. Kanneganti TD, Ozoren N, Body-Malapel M, Amer A, Park JH et al. (2006). Bacterial RNA and small antiviral compounds activate caspase-1 through cryopyrin/Nalp3. *Nature* 440(7081): 233–236.
31. Belkhadir Y, Subramaniam R, and Dangl JL (2004). Plant disease resistance protein signaling: NBS-LRR proteins and their partners. *Curr Opin Plant Biol* 7(4): 391–399.
32. Davis MM and Bjorkman PJ (1988). T-cell antigen receptor genes and T-cell recognition. *Nature* 334: 395.
33. Cooper MD and Alder MN (2006). The evolution of adaptive immune systems. *Cell* 124(4): 815–822.
34. Alder MN, Rogozin IB, Iyer LM, Glazko GV, Cooper MD et al. (2005). Diversity and function of adaptive immune receptors in a jawless vertebrate. *Science* 310(5756): 1970–1973.
35. Pancer Z, Amemiya CT, Ehrhardt GR, Ceitlin J, Gartland GL et al. (2004). Somatic diversification of variable lymphocyte receptors in the agnathan sea lamprey. *Nature* 430(6996): 174–180.
36. Du Pasquier L (2005). Meeting the demand for innate and adaptive immunities during evolution. *Scand J Immunol* 62(Suppl 1): 39–48.
37. Alder MN, Herrin BR, Sadlonova A, Stockard CR, Grizzle WE et al. (2008). Antibody responses of variable lymphocyte receptors in the lamprey. *Nat Immunol* 9(3): 319–327.
38. Janeway CA, Travers P, Wahlport M, and Capra JD (2000). *Immunobiology: The Immune System in Health and Disease* (4th ed.). Florence, KY: Garland.
39. Agrawal A, Eastman QM, and Schatz DG (1998). Transposition mediated by RAG1 and RAG2 and its implications for the evolution of the immune system. *Nature* 394(6695): 744–751.
40. Gallegos AM and Bevan MJ (2006). Central tolerance: good but imperfect. *Immunol Rev* 209: 290–296.
41. Monroe JG, Bannish G, Fuentes-Panana EM, King LB, Sandel PC et al. (2003). Positive and negative selection during B lymphocyte development. *Immunol Res* 27(2–3): 427–442.
42. Kraaijeveld AR and Godfray HC (1997). Trade-off between parasitoid resistance and larval competitive ability in *Drosophila melanogaster*. *Nature* 389(6648): 278–280.
43. Rolff J and Siva-Jothy MT (2003). Invertebrate ecological immunology. *Science* 301(5632): 472–475.
44. Wolinsky SM, Korber BT, Neumann AU, Daniels M, Kunstman KJ et al. (1996). Adaptive evolution of human immunodeficiency virus-type 1 during the natural course of infection. *Science* 272(5261): 537–542.
45. Koenig S, Conley AJ, Brewah YA, Jones GM, Leath S et al. (1995). Transfer of HIV-1-specific cytotoxic T lymphocytes to an AIDS patient leads to selection for mutant HIV variants and subsequent disease progression. *Nat Med* 1(4): 330–336.
46. Allen TM, Yu XG, Kalife ET, Reyor LL, Lichterfeld M et al. (2005). De novo generation of escape variant-specific CD8+ T-cell responses following cytotoxic T-lymphocyte escape in chronic human immunodeficiency virus type 1 infection. *J Virol* 79(20): 12952–12960.

47. Feeney ME, Tang Y, Pfafferott K, Roosevelt KA, Draenert R et al. (2005). HIV-1 viral escape in infancy followed by emergence of a variant-specific CTL response. *J Immunol* 174(12): 7524–7530.
48. Lin MY, Selin LK, and Welsh RM (2000). Evolution of the CD8 T-cell repertoire during infections. *Microbes Infect* 2(9): 1025–1039.
49. Miller RA (1996). The aging immune system: primer and prospectus. *Science* 273(5271): 70–74.
50. Salminen A, Huuskonen J, Ojala J, Kauppinen A, Kaarniranta K et al. (2008). Activation of innate immunity system during aging: NF-kB signaling is the molecular culprit of inflamm-aging. *Ageing Res Rev* 7(2): 83–105.
51. Franceschi C, Bonafe M, and Valensin S (2000). Human immunosenescence: the prevailing of innate immunity, the failing of clonotypic immunity, and the filling of immunological space. *Vaccine* 18(16): 1717–1720.
52. Franceschi C, Valensin S, Bonafe M, Paolisso G, Yashin AI et al. (2000). The network and the remodeling theories of aging: historical background and new perspectives. *Exp Gerontol* 35(6–7): 879–896.
53. Kirkwood TB and Austad SN (2000). Why do we age? *Nature* 408(6809): 233–238.
54. Gruver AL, Hudson LL, and Sempowski GD (2007). Immunosenescence of ageing. *J Pathol* 211(2): 144–156.
55. Wagner WM, Ouyang Q, Sekeri-Pataryas K, Sourlingas TG, and Pawelec G (2004). Basic biology and clinical impact of immunosenescence. *Biogerontology* 5(1): 63–66.
56. Cambier J (2005). Immunosenescence: a problem of lymphopoiesis, homeostasis, microenvironment, and signaling. *Immunol Rev* 205: 5–6.
57. Nikolich-Žugich J (2008). Ageing and life-long maintenance of T-cell subsets in the face of latent persistent infections. *Nat Rev Immunol* 8(7): 512–522.
58. Butcher SK, Chahal H, Nayak L, Sinclair A, Henriquez NV et al. (2001). Senescence in innate immune responses: reduced neutrophil phagocytic capacity and CD16 expression in elderly humans. *J Leukoc Biol* 70(6): 881–886.
59. Solana R, Pawelec G, and Tarazona R (2006). Aging and innate immunity. *Immunity* 24(5): 491–494.
60. Linton PJ and Dorshkind K (2004). Age-related changes in lymphocyte development and function. *Nat Immunol* 5(2): 133–139.
61. Min H, Montecino-Rodriguez E, and Dorshkind K (2004). Reduction in the developmental potential of intrathymic T cell progenitors with age. *J Immunol* 173(1): 245–250.
62. Agrawal A, Agrawal S, Tay J, and Gupta S (2008). Biology of dendritic cells in aging. *J Clin Immunol* 28(1): 14–20.
63. Ramsden S, Cheung YY, and Seroude L (2008). Functional analysis of the *Drosophila* immune response during aging. *Aging Cell* 7(2): 225–236.
64. Laws TR, Harding SV, Smith MP, Atkins TP, and Titball RW (2004). Age influences resistance of Caenorhabditis elegans to killing by pathogenic bacteria. *FEMS Microbiol Lett* 234(2): 281–287.
65. Kurz CL and Ewbank JJ (2003). Caenorhabditis elegans: an emerging genetic model for the study of innate immunity. *Nat Rev Genet* 4(5): 380–390.
66. Kurz CL and Tan MW (2004). Regulation of aging and innate immunity in C. elegans. *Aging Cell* 3(4): 185–193.
67. Zerofsky M, Harel E, Silverman N, and Tatar M (2005). Aging of the innate immune response in *Drosophila melanogaster*. *Aging Cell* 4(2): 103–118.
68. Garsin DA, Villanueva JM, Begun J, Kim DH, Sifri CD et al. (2003). Long-lived C. elegans daf-2 mutants are resistant to bacterial pathogens. *Science* 300(5627): 1921.
69. Libert S, Chao Y, Chu X, and Pletcher SD (2006). Trade-offs between longevity and pathogen resistance in *Drosophila melanogaster* are mediated by NFkappaB signaling. *Aging Cell* 5(6): 533–543.

70. Landis GN, Abdueva D, Skvortsov D, Yang J, Rabin BE et al. (2004). Similar gene expression patterns characterize aging and oxidative stress in *Drosophila melanogaster*. *Proc Natl Acad Sci U S A* 101(20): 7663–7668.
71. Kurtz J (2002). Phagocytosis by invertebrate hemocytes: causes of individual variation in Panorpa vulgaris scorpionflies. *Microsc Res Tech* 57(6): 456–468.
72. Holmes DJ and Austad SN (1995). Birds as animal models for the comparative biology of aging: a prospectus. *J Gerontol A Biol Sci Med Sci* 50(2): B59–B66.
73. Kernacki KA, Barrett RP, McClellan SA, and Hazlett LD (2000). Aging and PMN response to P. aeruginosa infection. *Invest Ophthalmol Vis Sci* 41(10): 3019–3025.
74. Butcher SK, Killampalli V, Lascelles D, Wang K, Alpar EK et al. (2005). Raised cortisol:DHEAS ratios in the elderly after injury: potential impact upon neutrophil function and immunity. *Aging Cell* 4(6): 319–324.
75. HogenEsch H, Thompson S, Dunham A, Ceddia M, and Hayek M (2004). Effect of age on immune parameters and the immune response of dogs to vaccines: a cross-sectional study. *Vet Immunol Immunopathol* 97(1–2): 77–85.
76. Greeley EH, Ballam JM, Harrison JM, Kealy RD, Lawler DF et al. (2001). The influence of age and gender on the immune system: a longitudinal study in labrador retriever dogs. *Vet Immunol Immunopathol* 82(1–2): 57–71.
77. Solana R, Alonso MC, and Pena J (1999). Natural killer cells in healthy aging. *Exp Gerontol* 34(3): 435–443.
78. Faunce DE, Palmer JL, Paskowicz KK, Witte PL, and Kovacs EJ (2005). CD1d-restricted NKT cells contribute to the age-associated decline of T cell immunity. *J Immunol* 175(5): 3102–3109.
79. Boehmer ED, Goral J, Faunce DE, and Kovacs EJ (2004). Age-dependent decrease in Toll-like receptor 4-mediated proinflammatory cytokine production and mitogen-activated protein kinase expression. *J Leukoc Biol* 75(2): 342–349.
80. Boehmer ED, Meehan MJ, Cutro BT, and Kovacs EJ (2005). Aging negatively skews macrophage TLR2- and TLR4-mediated pro-inflammatory responses without affecting the IL-2-stimulated pathway. *Mech Ageing Dev* 126(12): 1305–1313.
81. Richards A, Kavanagh D, and Atkinson JP (2007). Inherited complement regulatory protein deficiency predisposes to human disease in acute injury and chronic inflammatory states the examples of vascular damage in atypical hemolytic uremic syndrome and debris accumulation in age-related macular degeneration. *Adv Immunol* 96: 141–177.
82. Hazlett LD, Masinick-McClellan SA, and Barrett RP (1999). Complement defects in aged mice compromise phagocytosis of Pseudomonas aeruginosa. *Curr Eye Res* 19(1): 26–32.
83. Linton PJ, Li SP, Zhang Y, Bautista B, Huynh Q et al. (2005). Intrinsic versus environmental influences on T-cell responses in aging. *Immunol Rev* 205: 207–219.
84. Goronzy JJ and Weyand CM (2003). Aging, autoimmunity and arthritis: T-cell senescence and contraction of T-cell repertoire diversity – catalysts of autoimmunity and chronic inflammation. *Arthritis Res Ther* 5(5): 225–234.
85. Miller JP and Allman D (2003). The decline in B lymphopoiesis in aged mice reflects loss of very early B-lineage precursors. *J Immunol* 171(5): 2326–2330.
86. Chidgey AP and Boyd RL (2006). Stemming the tide of thymic aging. *Nat Immunol* 7(10): 1013–1016.
87. Mathies M, Lipps L, Smith GS, and Walford RL (1973). Age-related decline in response to phytohemagglutinin and pokeweed mitogen by spleen cells from hamsters and a long-lived mouse strain. *J Gerontol* 28(4): 425–430.
88. Perillo NL, Naeim F, Walford RL, and Effros RB (1993). The in vitro senescence of human T lymphocytes: failure to divide is not associated with a loss of cytolytic activity or memory T cell phenotype. *Mech Ageing Dev* 67(1–2): 173–185.
89. Miller RA (2000). Effect of aging on T lymphocyte activation. *Vaccine* 18(16): 1654–1660.
90. Trebilcock GU and Ponnappan U (1998). Nuclear factor-kappaB induction in $CD45RO^+$ and $CD45RA^+$ T cell subsets during aging. *Mech Ageing Dev* 102(2–3): 149–163.

91. Miller RA and Stutman O (1981). Decline, in aging mice, of the anti-2,4,6-trinitrophenyl (TNP) cytotoxic T cell response attributable to loss of Lyt-2-, interleukin 2-producing helper cell function. *Eur J Immunol* 11(10): 751–756.
92. Thoman ML and Weigle WO (1981). Lymphokines and aging: interleukin-2 production and activity in aged animals. *J Immunol* 127(5): 2102–2106.
93. Gillis S, Kozak R, Durante M, and Weksler ME (1981). Immunological studies of aging. Decreased production of and response to T cell growth factor by lymphocytes from aged humans. *J Clin Invest* 67(4): 937–942.
94. Effros RB and Walford RL (1983). The immune response of aged mice to influenza: diminished T-cell proliferation, interleukin 2 production and cytotoxicity. *Cell Immunol* 81(2): 298–305.
95. Warren RP, Yonk LJ, Burger RA, and Singh VK (1990). Age-related changes in CD45R and CDw29 helper T cells in human subjects. *Aging Immunol Infect Dis* 2: 91–94.
96. Han SN and Meydani SN (2000). Antioxidants, cytokines, and influenza infection in aged mice and elderly humans. *J Infect Dis* 182(Suppl 1): S74–S80.
97. LeMaoult J, Szabo P, and Weksler ME (1997). Effect of age on humoral immunity, selection of the B-cell repertoire and B-cell development. *Immunol Rev* 160: 115–126.
98. Song H, Price PW, and Cerny J (1997). Age-related changes in antibody repertoire: contribution from T cells. *Immunol Rev* 160: 55.
99. Boyman O, Purton JF, Surh CD, and Sprent J (2007). Cytokines and T-cell homeostasis. *Curr Opin Immunol* 19(3): 320–326.
100. Kim HR, Hong MS, Dan JM, and Kang I (2005). Altered IL-7R{alpha} expression with aging and the potential implications of IL-7 therapy on CD8+ T cell immune responses. *Blood* 107: 2855–2862.
101. Kim HR, Hwang KA, Kim KC, and Kang I (2007). Down-regulation of IL-7R{alpha} expression in human T cells via DNA methylation. *J Immunol* 178(9): 5473–5479.
102. Garcia GG and Miller RA (2001). Single-cell analyses reveal two defects in peptide-specific activation of naive T cells from aged mice. *J Immunol* 166(5): 3151–3157.
103. Kirk CJ and Miller RA (1998). Analysis of Raf-1 activation in response to TCR activation and costimulation in murine T-lymphocytes: effect of age. *Cell Immunol* 190(1): 33–42.
104. Gorgas G, Butch ER, Guan KL, and Miller RA (1997). Diminished activation of the MAP kinase pathway in CD3-stimulated T lymphocytes from old mice. *Mech Ageing Dev* 94((1–3)): 71–83.
105. Dunne PJ, Belaramani L, Fletcher JM, Fernandez de Mattos S, Lawrenz M et al. (2005). Quiescence and functional reprogramming of Epstein-Barr virus (EBV)-specific CD8+ T cells during persistent infection. *Blood* 106(2): 558–565.
106. Plunkett FJ, Franzese O, Finney HM, Fletcher JM, Belaramani LL et al. (2007). The loss of telomerase activity in highly differentiated CD8+CD28-CD27- T cells is associated with decreased Akt (Ser473) phosphorylation. *J Immunol* 178(12): 7710–7719.
107. Clambey ET, van Dyk LF, Kappler JW, and Marrack P (2005). Non-malignant clonal expansions of CD8+ memory T cells in aged individuals. *Immunol Rev* 205: 170–189.
108. Messaoudi I, Warner J, Nikolich-Žugich D, Fischer M, and Nikolich-Žugich J (2006). Molecular, cellular, and antigen requirements for development of age-associated T cell clonal expansions in vivo. *J Immunol* 176(1): 301–308.
109. Messaoudi I, Warner J, and Nikolich-Žugich J (2006). Age-related CD8+ T cell clonal expansions express elevated levels of CD122 and CD127 and display defects in perceiving homeostatic signals. *J Immunol* 177(5): 2784–2792.
110. Callahan JE, Kappler JW, and Marrack P (1993). Unexpected expansions of CD8-bearing cells in old mice. *J Immunol* 151: 6657–6669.
111. Hingorani R, Choi I-H, Akolka P, Gulwani-Akolkar B, Pergolizzi R et al. (1993). Clonal predominance of T cell receptors within the CD8+ CD45RO+ subset in normal human subjects. *J Immunol* 151: 5762–5769.

112. Posnett DN, Sinha R, Kabak S, and Russo C (1994). Clonal populations of T cells in normal elderly humans: the T cell equivalent to "benign monoclonal gammapathy". *J Exp Med* 179(2): 609–618.
113. Pawelec G, Akbar A, Caruso C, Effros R, Grubeck-Loebenstein B et al. (2004). Is immunosenescence infectious? *Trends Immunol* 25(8): 406–410.
114. Karrer U, Sierro S, Wagner M, Oxenius A, Hengel H et al. (2003). Memory inflation: continuous accumulation of antiviral CD8+ T cells over time. *J Immunol* 170(4): 2022–2029.
115. Lang A, Brien JD, Messaoudi I, and Nikolich-Žugich J (2008). Age-related dysregulation of CD8+ T cell memory specific for a persistent virus is independent of viral replication. *J Immunol* 180(7): 4848–4857.
116. Effros RB (1997). Loss of CD28 expression on T lymphocytes: a marker of replicative senescence. *Dev Comp Immunol* 21(6): 471–478.
117. Vallejo AN, Nestel AR, Schirmer M, Weyand CM, and Goronzy JJ (1998). Aging-related deficiency of CD28 expression in CD4+ T cells is associated with the loss of gene-specific nuclear factor binding activity. *J Biol Chem* 273(14): 8119–8129.
118. Messaoudi I, Lemaoult J, Guevara-Patino JA, Metzner BM, and Nikolich-Žugich J (2004). Age-related CD8 T cell clonal expansions constrict CD8 T cell repertoire and have the potential to impair immune defense. *J Exp Med* 200(10): 1347–1358.
119. Palacios MG, Cunnick JE, Winkler DW, and Vleck CM (2007). Immunosenescence in some but not all immune components in a free-living vertebrate, the tree swallow. *Proc Biol Sci* 274(1612): 951–957.
120. Lavoie ET, Sorrell EM, Perez DR, and Ottinger MA (2007). Immunosenescence and age-related susceptibility to influenza virus in Japanese quail. *Dev Comp Immunol* 31(4): 407–414.
121. Haussmann MF, Winkler DW, Huntington CE, Vleck D, Sanneman CE et al. (2005). Cell-mediated immunosenescence in birds. *Oecologia* 145(2): 270–275.
122. Cichon M, Sendecka J, and Gustafsson L (2003). Age-related decline in humoral immune function in Collared Flycatchers. *J Evol Biol* 16(6): 1205–1210.
123. Barton ES, White DW, Cathelyn JS, Brett-McClellan KA, Engle M et al. (2007). Herpesvirus latency confers symbiotic protection from bacterial infection. *Nature* 447(7142): 326–329.
124. Field CJ, Van Aerde A, Drager KL, Goruk S, and Basu T (2006). Dietary folate improves age-related decreases in lymphocyte function. *J Nutr Biochem* 17(1): 37–44.
125. Hale JS, Boursalian TE, Turk GL, and Fink PJ (2006). Thymic output in aged mice. *Proc Natl Acad Sci U S A* 103(22): 8447–8452.
126. Higami Y, Yu BP, Shimokawa I, Masoro EJ, and Ikeda T (1994). Duration of dietary restriction: an important determinant for the incidence and age of onset of leukemia in male F344 rats. *J Gerontol* 49(5): B239–B244.
127. Masoro EJ (2005). Overview of caloric restriction and ageing. *Mech Ageing Dev* 126(9): 913–922.
128. Chen J, Astle CM, and Harrison DE (1998). Delayed immune aging in diet-restricted B6CBAT6 F1 mice is associated with preservation of naive T cells. *J Gerontol A Biol Sci Med Sci* 53A: B330–B337.
129. Messaoudi I, Warner J, Fischer M, Park B, Hill B et al. (2006). Delay of T cell senescence by caloric restriction in aged long-lived nonhuman primates. *Proc Natl Acad Sci USA* 103(51): 19448–19453.
130. Imani F, Horii Y, Suthanthiran M, Skolnik EY, Makita Z et al. (1993). Advanced glycosylation endproduct-specific receptors on human and rat T-lymphocytes mediate synthesis of interferon gamma: role in tissue remodeling. *J Exp Med* 178(6): 2165–2172.
131. Holbrook NJ, Chopra RK, McCoy MT, Nagel JE, Powers DC et al. (1989). Expression of interleukin 2 and the interleukin 2 receptor in aging rats. *Cell Immunol* 120(1): 1–9.
132. Smith KA, Baker PE, Gillis S, and Ruscetti FW (1980). Functional and molecular characteristics of T-cell growth factor. *Mol Immunol* 17(5): 579–589.
133. Schmucker DL (2002). Intestinal mucosal immunosenescence in rats. *Exp Gerontol* 37(2–3): 197–203.

134. Kealy RD, Lawler DF, Ballam JM, Mantz SL, Biery DN et al. (2002). Effects of diet restriction on life span and age-related changes in dogs. *J Am Vet Med Assoc* 220(9): 1315–1320.
135. Nikolich-Žugich J (2007). Non-human primate models of T-cell reconstitution. *Semin Immunol* 19(5): 310–317.
136. Harshyne LA, Watkins SC, Gambotto A, and Barratt-Boyes M (2001). Dendritic cells acquire antigens from live cells for cross-presentation to CTL. *J Immunol* 166: 3717–3723.
137. Pitcher CJ, Hagen SI, Walker JM, Lum R, Mitchell BL et al. (2002). Development and homeostasis of T cell memory in rhesus macaque. *J Immunol* 168(1): 29–43.
138. Jankovic V, Messaoudi I, and Nikolich-Žugich J (2003). Phenotypic and functional T-cell aging in rhesus macaques (Macaca mulatta): differential behavior of CD4 and CD8 subsets. *Blood* 102(9): 3244–3251.
139. Čičin-Šain L, Messaoudi I, Park B, Currier N, Planer S et al. (2007). Dramatic increase in naive T cell turnover is linked to loss of naive T cells from old primates. *Proc Natl Acad Sci USA* 104: 19960–19965.
140. Naylor K, Li G, Vallejo AN, Lee WW, Koetz K et al. (2005). The influence of age on T cell generation and TCR diversity. *J Immunol* 174(11): 7446–7452.

Index

N.S. Wolf (ed.), *The Comparative Biology of Aging*,
DOI 10.1007/978-90-481-3465-6, © Springer Science+Business Media B.V. 2010

MIX
Papier aus verantwortungsvollen Quellen
Paper from responsible sources
FSC® C105338

If you have any concerns about our products,
you can contact us on
ProductSafety@springernature.com

In case Publisher is established outside the EU,
the EU authorized representative is:
Springer Nature Customer Service Center GmbH
Europaplatz 3, 69115 Heidelberg, Germany

Printed by Libri Plureos GmbH
in Hamburg, Germany